W9-DAY-529

Applied Math
FOR WASTEWATER PLANT OPERATORS

JOANNE KIRKPATRICK PRICE
Training Consultant

CRC PRESS
Boca Raton London New York Washington, D.C.

Published in 1991 by
CRC Press
Taylor & Francis Group
6000 Broken Sound Parkway NW, Suite 300
Boca Raton, FL 33487-2742

© 1991 by Taylor & Francis Group, LLC
CRC Press is an imprint of Taylor & Francis Group

No claim to original U.S. Government works
Printed in the United States of America on acid-free paper
15 14 13 12 11 10 9 8

International Standard Book Number-10: 0-8776-2809-2 (Hardcover)
International Standard Book Number-13: 978-0-8776-2809-5(Hardcover)
Library of Congress Card Number 90-71881
Cover art adapted from photograph of Fallbrook Sanitary District Reclaimed Water System

This book contains information obtained from authentic and highly regarded sources. Reprinted material is quoted with permission, and sources are indicated. A wide variety of references are listed. Reasonable efforts have been made to publish reliable data and information, but the author and the publisher cannot assume responsibility for the validity of all materials or for the consequences of their use.

No part of this book may be reprinted, reproduced, transmitted, or utilized in any form by any electronic, mechanical, or other means, now known or hereafter invented, including photocopying, microfilming, and recording, or in any information storage or retrieval system, without written permission from the publishers.

For permission to photocopy or use material electronically from this work, please access www.copyright.com (http://www.copyright.com/) or contact the Copyright Clearance Center, Inc. (CCC) 222 Rosewood Drive, Danvers, MA 01923, 978-750-8400. CCC is a not-for-profit organization that provides licenses and registration for a variety of users. For organizations that have been granted a photocopy license by the CCC, a separate system of payment has been arranged.

Trademark Notice: Product or corporate names may be trademarks or registered trademarks, and are used only for identification and explanation without intent to infringe.

Library of Congress Cataloging-in-Publication Data

Visit the Taylor & Francis Web site at
http://www.taylorandfrancis.com

and the CRC Press Web site at
http://www.crcpress.com

Dedication

This book is dedicated to my family:

To my husband Benton C. Price who was patient and supportive during the two years it took to write these texts, and who not only had to carry extra responsibilities at home during this time, but also, as a sanitary engineer, provided frequent technical critique and suggestions.

To our children Lisa, Derek, Kimberly, and Corinne, who so many times had to pitch in while I was busy writing, and who frequently had to wait for my attention.

To my mother who has always been so encouraging and who helped in so many ways throughout the writing process.

To my father, who passed away since the writing of the first edition, but who, I know, would have had just as instrumental a role in these books.

To the other members of my family, who have had to put up with this and many other projects, but who maintain a sense of humor about it.

Thank you for your love in allowing me to do something that was important to me.

J.K.P.

Contents

Contents—Cont'd

Contents—Cont'd

Contents—Cont'd

Preface to the Second Edition

The first edition of these texts was written at the conclusion of three and a half years of instruction at Orange Coast College, Costa Mesa, California, for two different water and wastewater technology courses. The fundamental philosophy that governed the writing of these texts was that those who have difficulty in math often do not lack the ability for mathematical calculation, they merely have not learned, or have not been taught, the "language of math." The books, therefore, represent an attempt to bridge the gap between the reasoning processes and the language of math that exists for students who have difficulty in mathematics.

In the years since the first edition, I have continued to consider ways in which the texts could be improved. In this regard, I researched several topics including how people learn (learning styles, etc.), how the brain functions in storing and retrieving information, and the fundamentals of memory systems. Many of the changes incorporated in this second edition are a result of this research.

Two features of this second edition are of particular importance:

- the **skills check section** provided at the beginning of every basic math chapter

- a **grouping of similar types of calculations** in the applied math texts

The skills check feature of the basic math text enables the student to pinpoint the areas of math weakness, and thereby customizes the instruction to the needs of the individual student.

The first six chapters of each applied math text include calculations grouped by type of problem. These chapters have been included so that students could see the common thread in a variety of seemingly different calculations.

The changes incorporated in this second edition were field-tested during a three-year period in which I taught a water and wastewater mathematics course for Palomar Community College, San Marcos, California.

Written comments or suggestions regarding the improvement of any section of these texts or workbooks will be greatly appreciated by the author.

Joanne Kirkpatrick Price

Acknowledgments

"From the original planning of a book to its completion, the continued encouragement and support that the author receives is instrumental to the success of the book." This quote from the acknowledgments page of the first edition of these texts is even more true of the second edition.

First Edition

Those who assisted during the development of the first edition are: Walter S. Johnson and Benton C. Price, who reviewed both texts for content and made valuable suggestions for improvements; Silas Bruce, with whom the author team-taught for two and a half years, and who has a down-to-earth way of presenting wastewater concepts; Mariann Pape, Samuel R. Peterson and Robert B. Moore of Orange Coast College, Costa Mesa, California, and Jim Catania and Wayne Rodgers of the California State Water Resources Control Board, all of whom provided much needed support during the writing of the first edition.

The first edition was typed by Margaret Dionis, who completed the typing task with grace and style. Adele B. Reese, my mother, proofed both books from cover to cover and Robert V. Reese, my father, drew all diagrams (by hand) shown in both books.

Second Edition

The second edition was an even greater undertaking due to many additional calculations and because of the complex layout required. I would first like to acknowledge and thank Laurie Pilz, who did the computer work for all three texts and the two workbooks. Her skill, patience, and most of all perseverance has been instrumental in providing this new format for the texts. Her husband, Herb Pilz, helped in the original format design and he assisted frequently regarding questions of graphics design and computer software.

Those who provided technical review of various portions of the texts include Benton C. Price, Kenneth D. Kerri, Lynn Marshall, Wyatt Troxel and Mike Hoover. Their comments and suggestions are appreciated and have improved the current edition.

Many thanks also to the staff of the Fallbrook Sanitary District, Fallbrook, California, especially Virginia Grossman, Nancy Hector, Joyce Shand, Mike Page, and Weldon Platt for the numerous times questions were directed their way during the writing of these texts.

The staff of Technomic Publishing Company, Inc., also provided much advice and support during the writing of these texts. First, Melvyn Kohudic, President of Technomic Publishing Company, contacted me several times over the last few years, suggesting that the texts be revised. It was his gentle nudging that finally got the revision underway. Joseph Eckenrode helped work out some of the details in the initial stages and was a constant source of encouragement. Jeff Perini was copy editor for the texts. His keen attention to detail has been of great benefit to the final product. Leo Motter had the arduous task of final proof reading.

I wish to thank all my friends, but especially those in our Bible study group (Gene and Judy Rau, Floyd and Juanita Miller, Dick and Althea Birchall, and Mark and Penny Gray) and our neighbors, Herb and Laurie Pilz, who have all had to live with this project as it progressed slowly chapter by chapter, but who remained a source of strength and support when the project sometimes seemed overwhelming.

Lastly, the many students who have been in my classes or seminars over the years have had no small part in the final form these books have taken. The format and content of these texts is in response to their questions, problems, and successes over the years.

To all of these I extend my heartfelt thanks.

How To Use These Books

The *Mathematics for Water and Wastewater Treatment Plant Operators* series includes three texts and two workbooks:

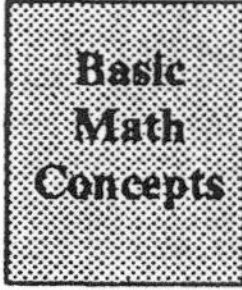

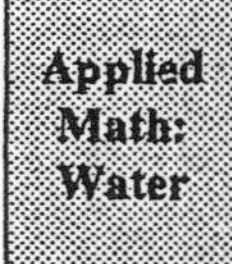

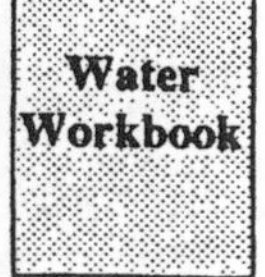

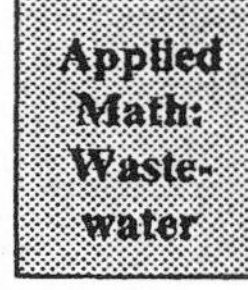

- Basic Math Concepts for Water and Wastewater Plant Operators
- Applied Math for Water Plant Operators
- Workbook—Applied Math for Water Plant Operators
- Applied Math for Wastewater Plant Operators
- Workbook—Applied Math for Wastewater Plant Operators

Basic Math Concepts

All the basic math you will need to become adept in water and wastewater calculations has been included in the Basic Math Concepts text. This section has been expanded considerably from the basic math included in the first edition. For this reason, students are provided with more methods by which they may solve the problems.

Many people have weak areas in their math skills. It is therefore advisable to take the skills test at the beginning of each chapter in the basic math book to pinpoint areas that require review or study. If possible, it is best to resolve these weak areas before beginning either of the applied math texts. However, when this is not possible, the Basic Math Concepts text can be used as a reference resource for the applied math texts. For example, when making a calculation that includes tank volume, you may wish to refer to the basic math section on volumes.

Applied Math Texts and Workbooks

The applied math texts and workbooks are companion volumes. There is one set for water treatment plant operators and another for wastewater treatment plant operators. Each applied math text has two sections:

- Chapters 1 through 6 present various calculations **grouped by type of math problem.** Perhaps 70 percent of all water and wastewater calculations are represented by these six types. Chapter 7 groups various types of pumping problems into a single chapter. The calculations presented in these seven chapters are common to the water and wastewater fields and have therefore been included in both applied math texts.

  Since the calculations described in Chapters 1 through 6 represent the heart of water and wastewater treatment math, if possible, it is advisable that you master these general types of calculations before continuing with other calculations. Once completed, a review of these calculations in subsequent chapters will further strengthen your math skills.

- The remaining chapters in each applied math text include calculations **grouped by unit processes**. The calculations are presented in the order of the flow through a plant. Some of the calculations included in these chapters are not incorporated in Chapters 1 through 7, since they do not fall into any general problem-type grouping. These chapters are particularly suited for use in a classroom or seminar setting, where the math instruction must parallel unit process instruction.

The workbooks support the applied math texts section by section. They have also been vastly expanded in this edition so that the student can build strength in each type of calculation. A detailed answer key has been provided for all problems. The workbook pages have been perforated so that they may be used in a classroom setting as hand-in assignments. The pages have also been hole-punched so that the student may retain the pages in a notebook when they are returned.

The workbooks may be useful in preparing for a certification exam. However, because theses texts include both fundamental and advanced calculations, and because the requirements for each certification level vary somewhat from state to state, it is advisable that you first determine the types of problems to be covered in your exam, then focus on those types of calculations in these texts.

1 Applied Volume Calculations

SUMMARY

The **general equation** for most volume calculations is:

$$\text{Volume} = \left[\begin{array}{l}\text{Representative}\\ \text{Surface Area}\end{array}\right]\left[\begin{array}{l}\text{Depth or}\\ \text{Height}\end{array}\right]$$

1. **Tank volume calculations.**
   Most tank volume calculations are for tanks that are either rectangular or cylindrical in shape.

Rectangular Tank

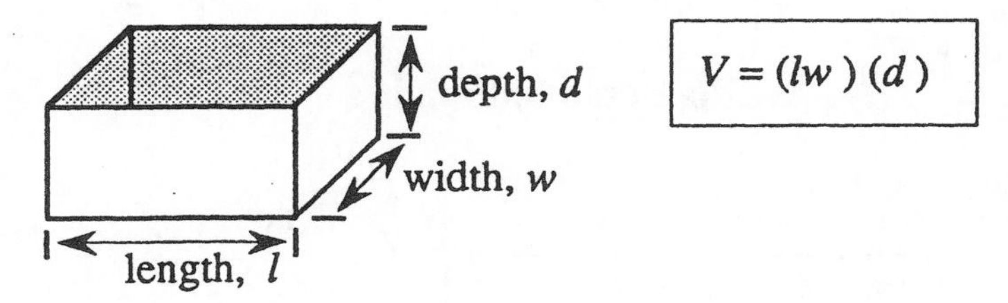

$V = (lw)(d)$

Cylindrical Tank

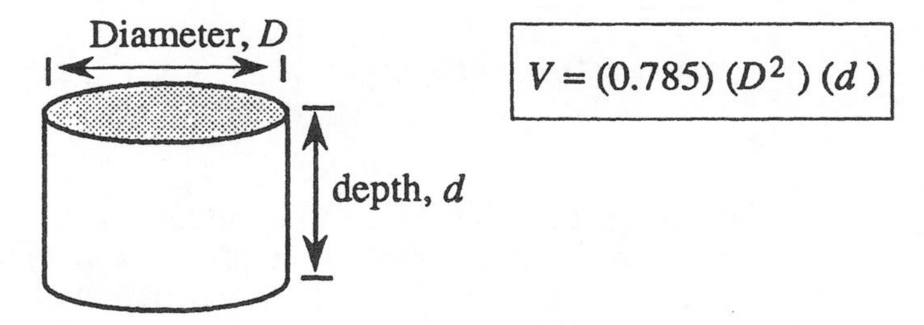

$V = (0.785)(D^2)(d)$

2. **Channel or pipeline volume calculations** are very similar to tank volume calculations. The principal shapes are shown below.

Portion of a Rectangular Channel

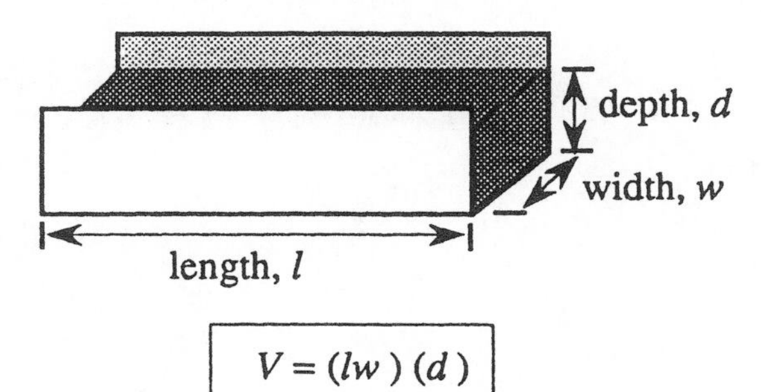

$V = (lw)(d)$

Three general types of water and wastewater volume calculations are:

- Tank Volume
- Channel or Pipeline Volume
- Pit, Trench, or Pond Volume

Each of these calculations is simply a specific application of volume calculations. For a more detailed discussion of volume calculations, refer to Chapter 11 in *Basic Math Concepts*.

For many calculations, the volumes must be expressed in terms of **gallons**. To convert from cubic feet to gallons volume, a factor of 7.48 gal/cu ft is used. Refer to Chapter 8 of *Basic Math Concepts* for a detailed discussion of cubic feet to gallons conversions.

SUMMARY—Cont'd

Portion of a Trapezoidal Channel

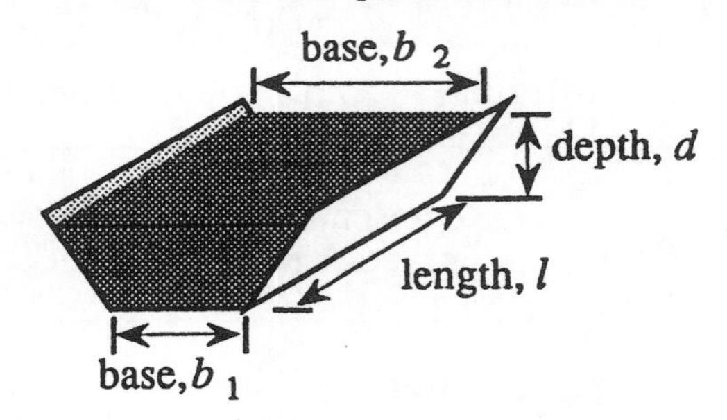

$$V = \frac{(b_1 + b_2)(d)(l)}{2}$$

Portion of a Pipeline

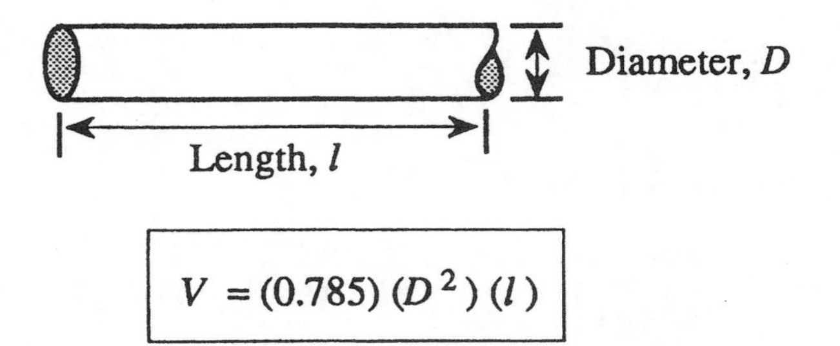

$$V = (0.785)(D^2)(l)$$

3. **Other volume calculations** involving ditches or ponds depend on the shape of the ditch or pond. A pit or trench is often rectangular in shape. A pond or oxidation ditch may have a trapezoidal cross section.

NOTES:

1.1 TANK VOLUME CALCULATIONS

The two common tank shapes in water and wastewater treatment are rectangular and cylindrical tanks.

Rectangular Tank:

$$V = (lw)(d)$$

Where:

V = Volume, cu ft
l = Length, ft
w = Width, ft
d = Depth, ft

Cylindrical Tank:

$$V = (0.785)(D^2)(d)$$

Where:

V = Volume, cu ft
D = Diameter, ft
d = Depth, ft

The volume of these tanks can be expressed in cubic feet or gallons. The equations shown above are for cubic feet volume. Since each cubic foot of water contains 7.48 gallons, to convert cubic feet volume to gallons volume, multiply by 7.48 gal/cu ft, as illustrated in Example 2. As an alternative, you may wish to include the 7.48 gal/cu ft factor in the volume equation, as shown in Example 3.

Example 1: (Tank Volume)

❑ The dimensions of a tank are given below. Calculate the volume of the tank in cubic feet.

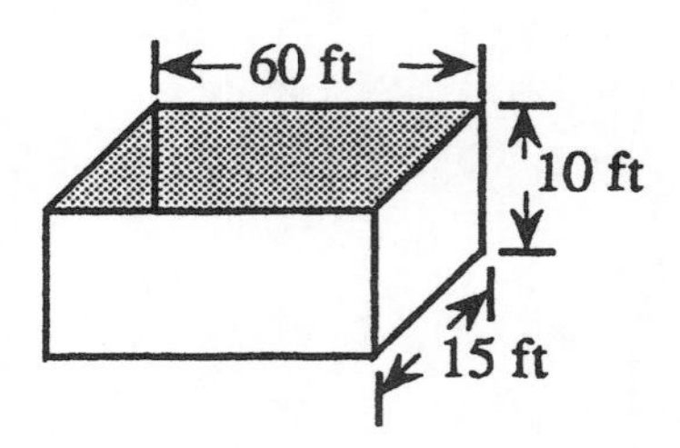

Vol., cu ft = $(lw)(d)$

= (60 ft) (15 ft) (10 ft)

= **9000 cu ft**

Example 2: (Tank Volume)

❑ A tank is 25 ft wide, 75 ft long, and can hold water to a depth of 10 ft. What is the volume of the tank, in gallons?

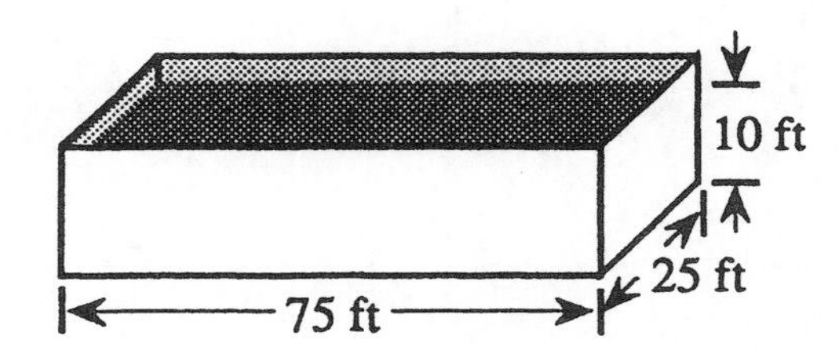

Vol., cu ft = $(lw)(d)$

= (75 ft) (25 ft) (10 ft)

= 18,750 cu ft

Now convert cu ft volume to gal:

(18,750 cu ft) (7.48 gal/cu ft) = **140,250 gal**

Example 3: (Tank Volume)

❑ The diameter of a tank is 60 ft. When the water depth is 25 ft, what is the volume of water in the tank, in gallons?

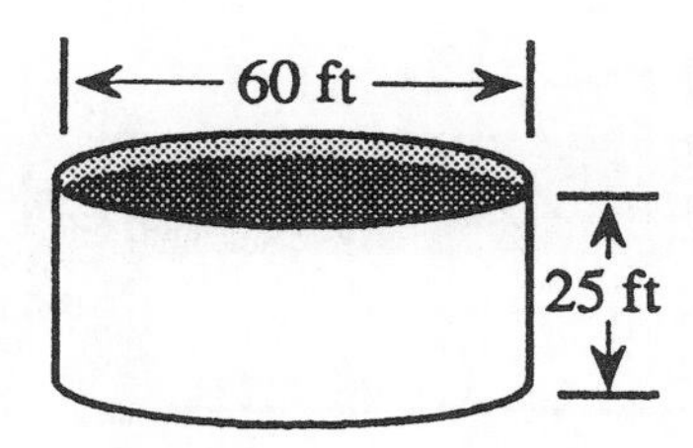

Vol., gal = (0.785) (D^2) (d) (7.48 gal/cu ft)

= (0.785) (60 ft) (60 ft) (25 ft) (7.48 gal/cu ft)

= **528,462 gal**

Example 4: (Tank Volume)

❑ A tank is 12 ft wide and 20 ft long. If the depth of water is 11 ft, what is the volume of water in the tank?

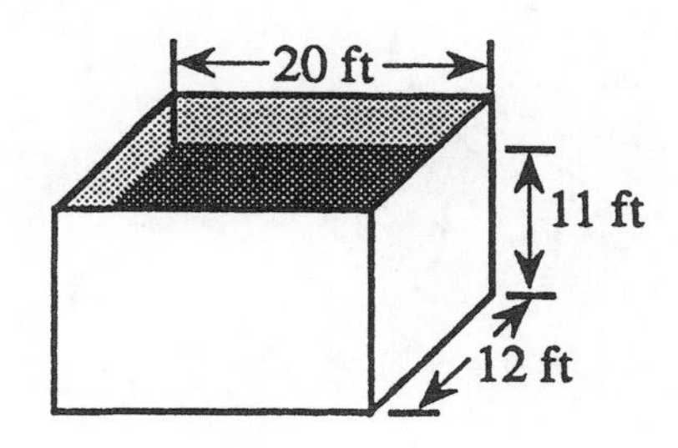

Vol., gal = (lw) (d) (7.48 gal/cu ft)

= (20 ft) (12 ft) (11 ft) (7.48 gal/cu ft)

= **19,747 gal**

1.2 CHANNEL OR PIPELINE VOLUME CALCULATIONS

Channel or pipeline volume calculations are similar to tank volume calculations.* The equations to be used in calculating these volumes are given below.

Channel or pipeline volumes may be expressed as cubic feet, as shown in Example 1, or as gallons, shown in Examples 2-4.

Channel with Rectangular Cross Section:

$$\text{Volume, cu ft} = (lw)(d)$$

Channel with Trapezoidal Cross Section:

$$\text{Volume, cu ft} = \frac{(b_1 + b_2)}{2}(d)(l)$$

Pipeline with Circular Cross Section:

$$\text{Volume, cu ft} = (0.785)(D^2)(l)$$

Example 1: (Channel or Pipe Volume)

❑ Calculate the volume of water (in cu ft) in the section of rectangular channel shown below when the water is 4 ft deep.

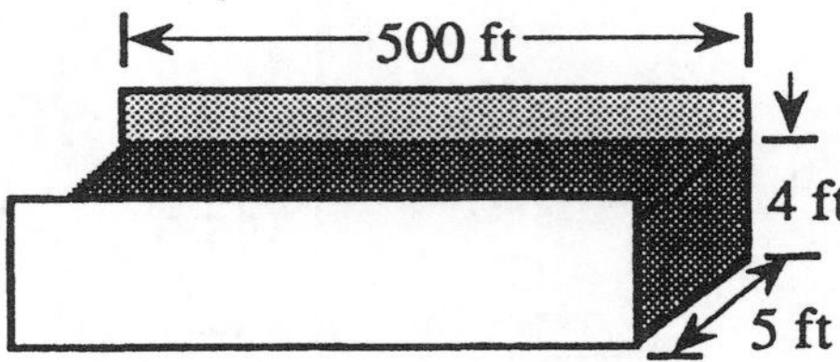

$$\text{Vol., cu ft} = (lw)(d)$$

$$= (500 \text{ ft})(5 \text{ ft})(4 \text{ ft})$$

$$= \boxed{10{,}000 \text{ cu ft}}$$

Example 2: (Channel or Pipe Volume)

❑ Calculate the volume of water (in gallons) in the section of trapezoidal channel shown below when the water depth is 4 ft.

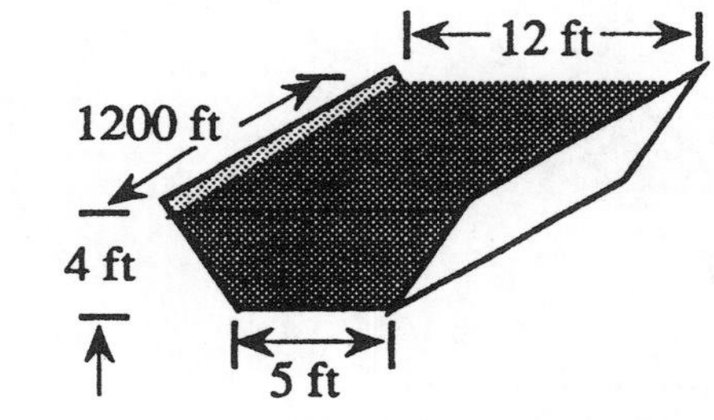

$$\text{Vol., gal} = \frac{(b_1 + b_2)}{2}(d)(l)(7.48 \text{ gal/cu ft})$$

$$= \frac{(5 \text{ ft} + 12 \text{ ft})}{2}(4 \text{ ft})(1200 \text{ ft})(7.48 \text{ gal/cu ft})$$

$$= (8.5)(4)(1200)(7.48)$$

$$= \boxed{305{,}184 \text{ gal}}$$

* For a detailed review of volume calculations, refer to Chapter 11 in *Basic Math Concepts*.

Example 3: (Channel or Pipe Volume)

❑ A new section of 12-inch diameter pipe is to be disinfected before it is put into service. If the length of pipeline is 2000 ft, how many gallons of water will be needed to fill the pipeline?

2000 ft

12 in
= 1 ft

$$\text{Vol., gal} = (0.785)\,(D^2)\,(l)\,(7.48\ \text{gal/cu ft})$$

$$= (0.785)\,(1\ \text{ft})\,(1\ \text{ft})\,(2000\ \text{ft})\,(7.48\ \text{gal/cu ft})$$

$$= \boxed{11{,}744\ \text{gal}}$$

Example 4: (Channel and Pipe Volume)

❑ A section of 6-inch diameter pipeline is to be filled with chlorinated water for disinfection. If 1320 ft of pipeline is to be disinfected, how many gallons of water will be required?

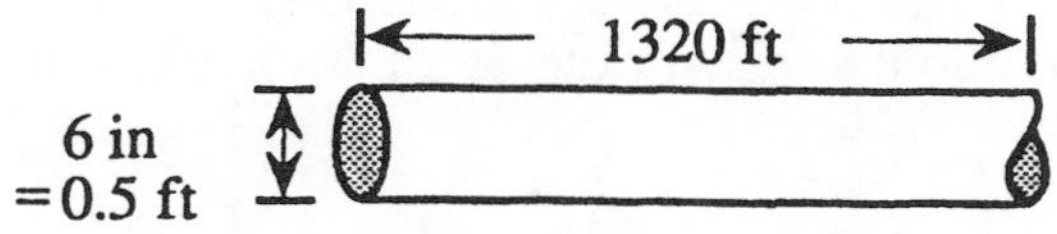

1320 ft

6 in
= 0.5 ft

$$\text{Vol., gal} = (0.785)\,(D^2)\,(l)\,(7.48\ \text{gal/cu ft})$$

$$= (0.785)\,(0.5\ \text{ft})\,(0.5\ \text{ft})\,(1320\ \text{ft})\,(7.48\ \text{gal/cu ft})$$

$$= \boxed{1{,}938\ \text{gal}}$$

1.3 OTHER VOLUME CALCULATIONS

PIT OR TRENCH VOLUMES

These volume calculations are similar to tank and channel volume calculations, with one exception—the volume is often expressed as cubic yards rather than cubic feet or gallons.

There are two approaches to calculating cubic yard volume:

- Calculate the cubic feet volume, then convert to cubic yards volume. (See Example 1.)

$$\frac{\text{(cu ft)}}{\text{27 cu ft/cu yd}} = \text{cu yds}$$

- Express all dimensions in yards so that the resulting volume calculated will be cubic yards. (See Example 2.)

$$\text{(yds) (yds) (yds)} = \text{cu yds}$$

Example 1: (Other Volume Calculations)

❑ A trench is to be excavated 2.5 ft wide, 4 ft deep and 900 ft long. What is the cubic yards volume of the trench?

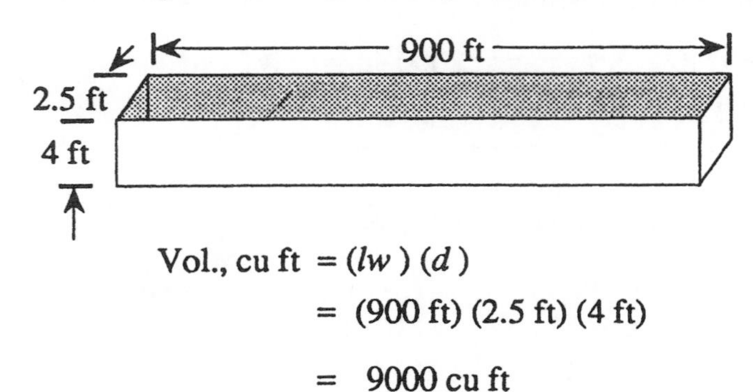

$$\text{Vol., cu ft} = (lw)(d)$$

$$= (900 \text{ ft})(2.5 \text{ ft})(4 \text{ ft})$$

$$= 9000 \text{ cu ft}$$

Now convert cu ft volume to cu yds:

$$\frac{\text{9000 cu ft}}{\text{27 cu ft/cu yds}} = \boxed{\text{333 cu yds}}$$

Example 2: (Other Volume Calculations)

❑ What is the cubic yard volume of a trench 500 ft long, 2.25 ft wide and 4 ft deep?

Convert dimensions in ft to yds before beginning the volume calculation:

$$\text{Length: } \frac{\text{500 ft}}{\text{3 ft/yd}} = \text{166.7 yds}$$

$$\text{Width: } \frac{\text{2.25 ft}}{\text{3 ft/yd}} = \text{0.75 yds}$$

$$\text{Depth: } \frac{\text{4 ft}}{\text{3 ft/yd}} = \text{1.33 yds}$$

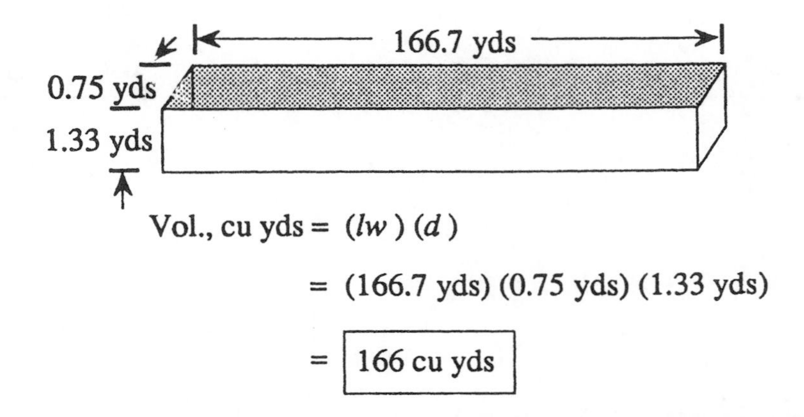

$$\text{Vol., cu yds} = (lw)(d)$$

$$= (166.7 \text{ yds})(0.75 \text{ yds})(1.33 \text{ yds})$$

$$= \boxed{\text{166 cu yds}}$$

Example 3: (Other Volume Calculations)

❑ Calculate the volume of the oxidation ditch shown below, in cu ft. The cross section of the ditch is trapezoidal.

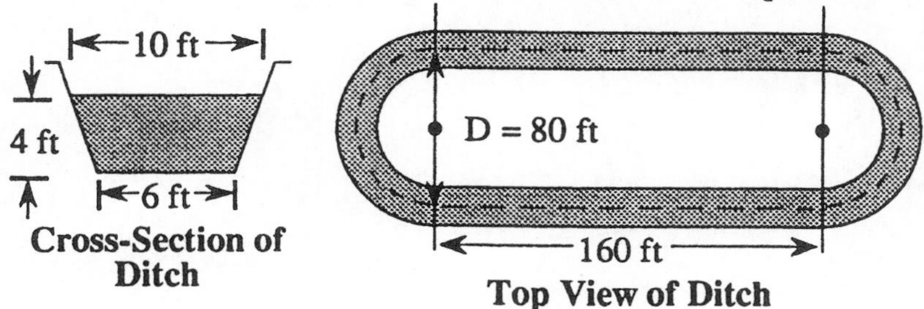

$$\text{Total Volume} = (\text{Trapezoidal Area})\,(\text{Total Length})$$

$$= \left[\frac{(b_1+b_2)}{2}(h)\right]\left[(\text{Length of 2 Sides}) + (\text{Length Around 2 Half Circles})\right]$$

$$= \left[\frac{(6\text{ ft} + 10\text{ ft})}{2}(4\text{ ft})\right]\left[320\text{ ft} + (3.14)(80\text{ ft})\right]$$

$$= \left[(8\text{ ft})(4\text{ ft})\right]\left[571.2\text{ ft}\right]$$

$$= \boxed{18{,}278\text{ cu ft}}$$

Example 4: (Other Volume Calculations)

❑ A pond is 5 ft deep with side slopes of 2:1 (2 ft horizontal: 1 ft vertical). Given the following data, calculate the volume of the pond in cubic feet.

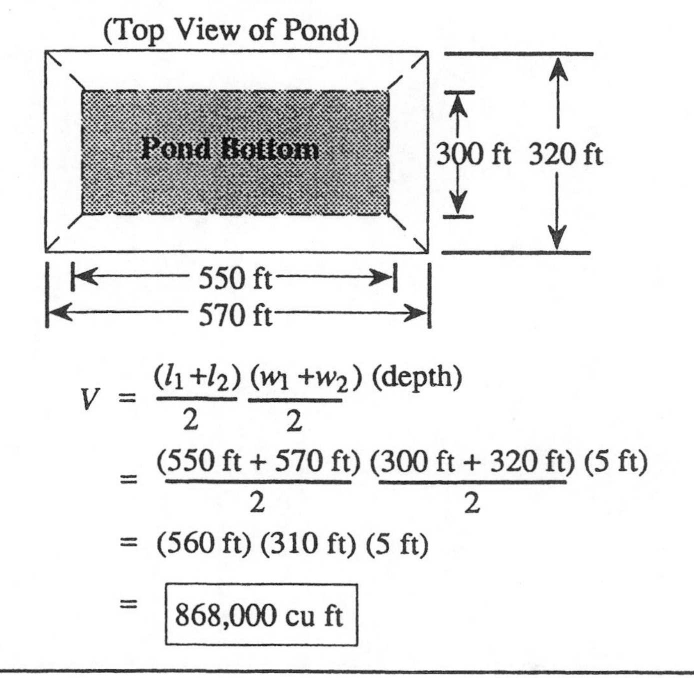

$$V = \frac{(l_1+l_2)}{2}\,\frac{(w_1+w_2)}{2}\,(\text{depth})$$

$$= \frac{(550\text{ ft} + 570\text{ ft})}{2}\,\frac{(300\text{ ft} + 320\text{ ft})}{2}\,(5\text{ ft})$$

$$= (560\text{ ft})(310\text{ ft})(5\text{ ft})$$

$$= \boxed{868{,}000\text{ cu ft}}$$

OXIDATION DITCH OR POND VOLUMES

Many times oxidation ditches and ponds are trapezoidal in configuration. Examples 3 and 4 illustrate these calculations.

In Example 3, the oxidation ditch has sloping sides (trapezoidal cross section). The total volume of the oxidation ditch is the trapezoidal area times the total length:

$$\text{Total Vol.} = \frac{(b_1+b_2)}{2}\,(d)\,(\text{Total Length})$$

(The total length is measured at the center of the ditch; it is equal to the length of the two straight lengths plus two half-circle lengths. Note that the length around the two half circles is equal to the circumference of <u>one</u> full circle.)

WHEN ALL SIDES SLOPE

In many calculations of trapezoidal volume, such as for a trapezoidal channel, only two of the sides slope and the ends are vertical. To calculate the volume for such a shape, the following equation is normally used:

$$V = \frac{(b_1+b_2)}{2}\,(d)\,(l)$$

Another way of thinking of this calculation is average width times the depth of water times the length :

$$V = (\text{aver. width})\,(d)\,(\text{length})$$

In Example 4, however, since both length and width sides are trapezoidal, **the equation must include average length and average width dimensions:**

$$\boxed{V = \frac{(l_1+l_2)}{2}\,\frac{(w_1+w_2)}{2}\,(\text{depth})}$$

NOTES:

2 Flow and Velocity Calculations

SUMMARY

1. **Instantaneous Flow Rates**
   The Q=AV equation can be used to estimate flow rates through channels, tanks and pipelines. In all Q=AV calculations, the units on the left side of the equation (Q) must match the combined units on the right side of the equation (A and V) with respect to volume (cubic feet or gallons) and time (sec, min, hrs, or days). The $Q = AV$ equations in this summary will be expressed in terms of cubic feet per minute.

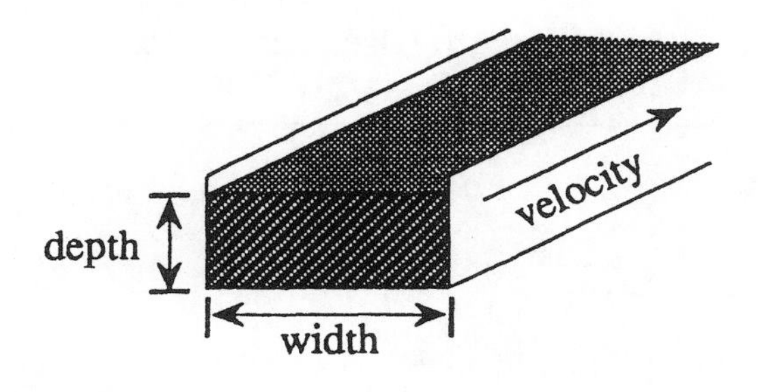

$$Q_{cfm} = AV_{fpm}$$

$$\underset{\text{cfm}}{Q} = \underset{\text{ft}}{(\text{width})}\ \underset{\text{ft}}{(\text{depth})}\ \underset{\text{fpm}}{(\text{velocity})}$$

Flow Through A Trapezoidal Channel

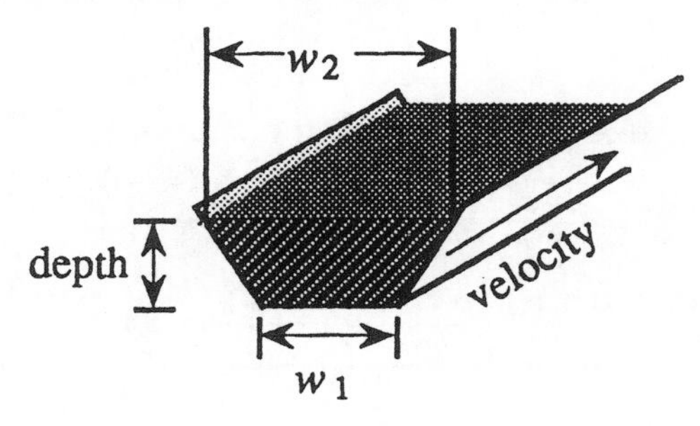

$$Q_{cfm} = AV_{fpm}$$

$$\underset{\text{cfm}}{Q} = \frac{(w_1 + w_2)}{2}\ \underset{\text{ft}}{(\text{depth})}\ \underset{\text{fpm}}{(\text{velocity})}$$

There are several ways to determine flow and velocity. Various flow metering devices may be used to measure water or wastewater flows at a particular moment (instantaneous flow) or over a specified time period (total flow). Instantaneous flow can also be determined using the Q=AV equation.

This chapter includes discussions of Q=AV, velocity, average flow rates, and flow conversions.

SUMMARY—Cont'd

Flow Into Or Out Of A Tank

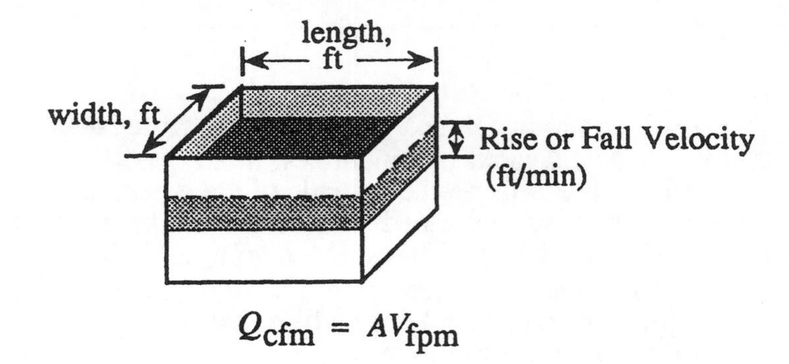

$$Q_{\text{cfm}} = AV_{\text{fpm}}$$

$$\underset{\text{cfm}}{Q} = \underset{\text{ft}}{(\text{length})}\ \underset{\text{ft}}{(\text{width})}\ \underset{\text{fpm}}{(\text{rise or fall velocity})}$$

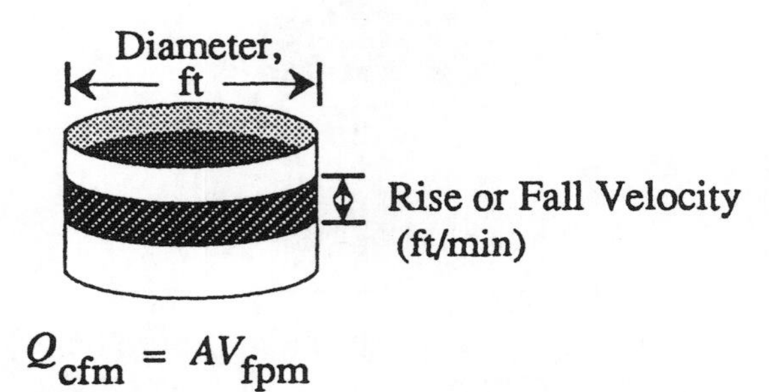

$$Q_{\text{cfm}} = AV_{\text{fpm}}$$

$$\underset{\text{cfm}}{Q} = (0.785)\ (D^2)\ \underset{\text{fpm}}{(\text{rise or fall velocity})}$$

Flow Through A Pipeline—When Flowing Full

Diameter, ft

velocity, fpm

$$Q_{\text{cfm}} = AV_{\text{fpm}}$$

$$\underset{\text{cfm}}{Q} = (0.785)\ (D^2)\ \underset{\text{fpm}}{(\text{velocity})}$$

SUMMARY—Cont'd

Flow Through A Pipeline—When Flowing When Less Than Full

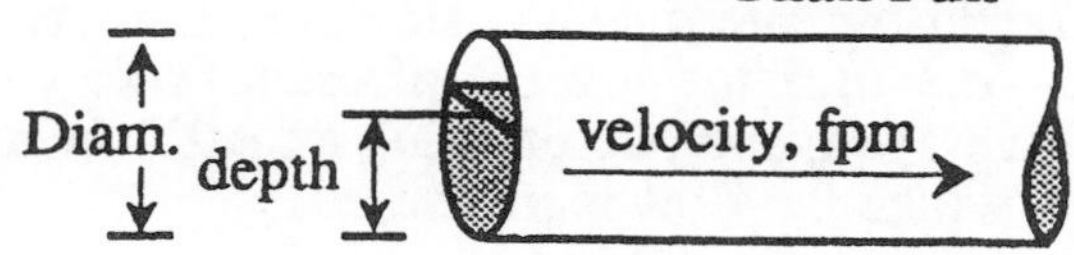

$$\underset{\text{cfm}}{Q} = \underset{\text{factor}}{(\text{new})} (D^2) \underset{\text{fpm}}{(\text{velocity})}$$

*Based on *d/D* Table

2. **Velocity Calculations**
   The $Q=AV$ equation can be used to determine the velocity of water at a particular moment. Use the same $Q=AV$ equation, fill in the known information, then solve for velocity.

$$Q = AV$$

Another method for determining velocity is to time the movement of a float or dye through the water.

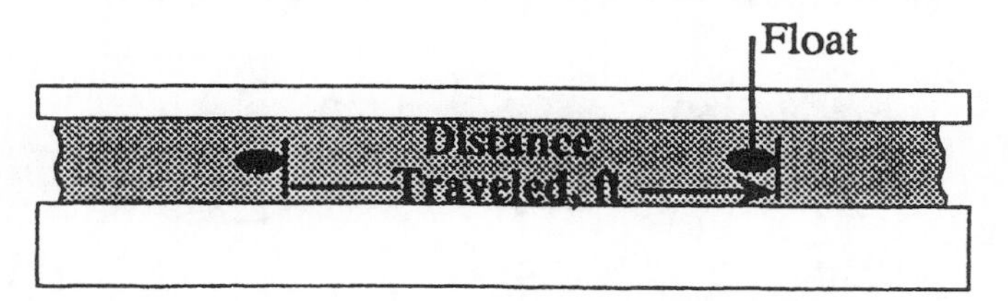

$$\text{Velocity} = \frac{\text{Distance Traveled, ft}}{\text{Duration of Test, min}}$$

$$\text{Velocity} = \frac{\text{ft}}{\text{min}}$$

The $Q=AV$ equation can also be used to determine velocity changes due to differences in pipe diameters. The AV in one pipe is equal to the AV in the other pipe.

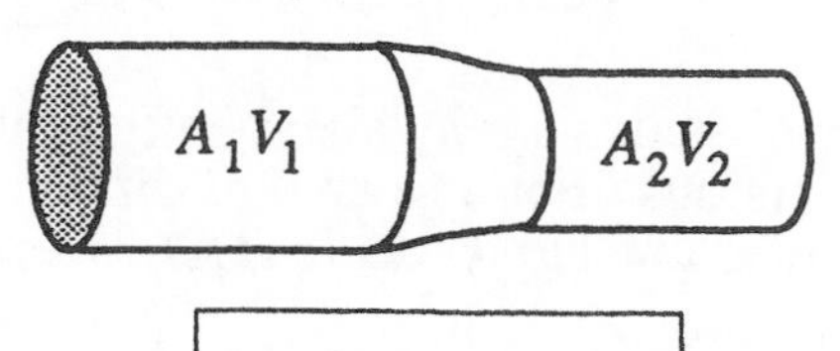

$$A_1V_1 = A_2V_2$$

SUMMARY—Cont'd

3. Average Flow Rates

The average flow rate may be calculated using two methods—one utilizing several different flows which are then averaged; and one utilizing a total flow and the time over which the flow is measured.

$$\text{Average Flow} = \frac{\text{Total of all Sample Flows}}{\text{Number of Samples}}$$

Or

$$\text{Average Flow} = \frac{\text{Total Flow}}{\text{Time Flow Measured}}$$

4. Flow Conversions

The box method may be used to convert from one flow expression to another. When using the box method of conversions, **multiply** when moving from a smaller box to a larger box, and **divide** when moving from a larger box to a smaller box. The common conversions are shown below:*

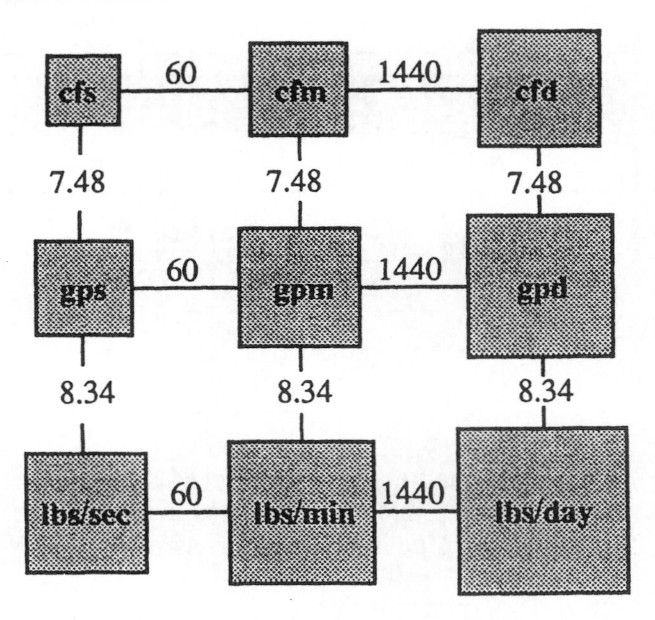

Dimensional analysis may also be used in making flow conversions. For a review of dimensional analysis, refer to Chapter 15 in *Basic Math Concepts*.

* The factors shown in the diagram have the following units associated with them: 60 min/sec, 1440 min/day, 7.48 gal/cu ft, and 8.34 lbs/gal.

NOTES:

2.1 INSTANTANEOUS FLOW RATES CALCULATIONS

The flow rate through channels and pipelines is normally measured by some type of flow metering device. However the flow rate for any particular moment can also be determined by using the $Q=AV$ equation.

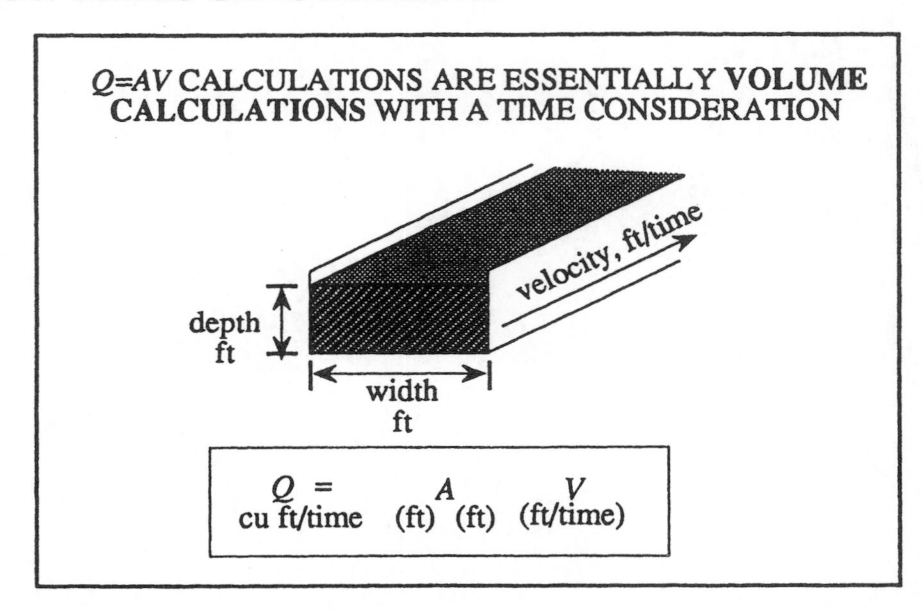

The flow rate (Q) is equal to the cross-sectional area (A) of the channel or pipeline multiplied by the velocity through the channel or pipeline. There are two important considerations in these calculations:

1. Remember that volume is calculated by multiplying the representative area by a third dimension, often depth or height.* The $Q=AV$ calculation is essentially a volume calculation. The length dimension is a velocity length (length/time):

Vol = (Cross Sectional Area) (3rd Dim.)

Q = A V

2. The units used for volume and time must be the same on both sides of the equation, as shown in the diagram to the right.

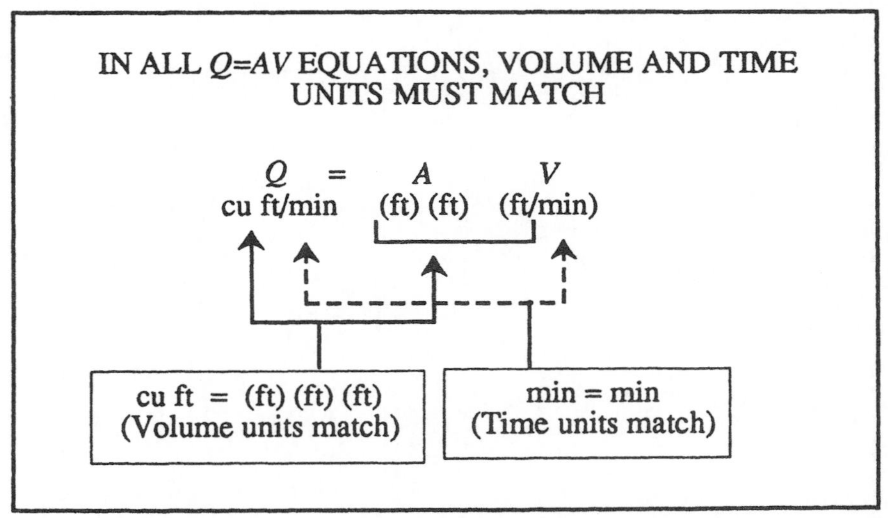

FLOW THROUGH A RECTANGULAR CHANNEL

The principal difference among various $Q=AV$ calculations is the shape of the cross-sectional area. Channels normally have rectangular or trapezoidal cross sections, whereas pipelines have circular cross sections. Examples 1-3 illustrate the $Q=AV$ calculation when the channel is rectangular.

Example 1: (Instantaneous Flow)

❑ A channel 3 ft wide has water flowing to a depth of 2.5 ft. If the velocity through the channel is 2 fps, what is the cfs flow rate through the channel?

$$Q_{cfs} = AV_{fps}$$

$$= (3 \text{ ft})\ (2.5 \text{ ft})\ (2 \text{ fps})$$

$$= \boxed{15 \text{ cfs}}$$

* For a review of volume calculations, refer to Chapter 11 in *Basic Math Concepts*.

Example 2: (Instantaneous Flow)
❑ A channel 40 inches wide has water flowing to a depth of 1.5 ft. If the velocity of the water is 2.3 fps, what is the cfs flow in the channel?

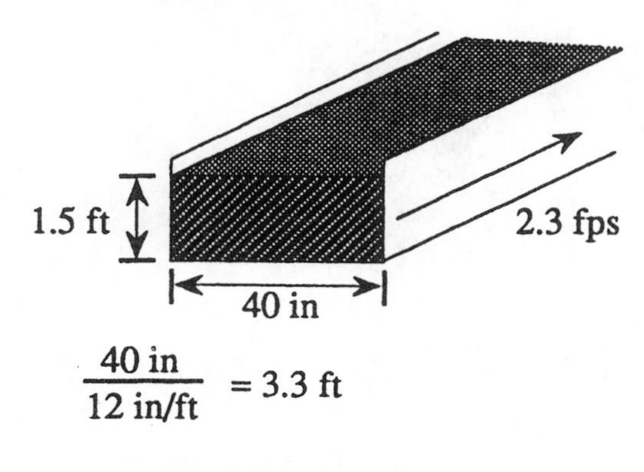

$$\frac{40 \text{ in}}{12 \text{ in/ft}} = 3.3 \text{ ft}$$

$$Q_{cfs} = AV_{fps}$$

$$= (3.3 \text{ ft})\,(1.5 \text{ ft})\,(2.3 \text{ fps})$$

$$= \boxed{11.4 \text{ cfs}}$$

Example 3: (Instantaneous Flow)
❑ A channel 3 ft wide has water flowing at a velocity of 1.5 fps. If the flow through the channel is 8.1 cfs, what is the depth of the water?

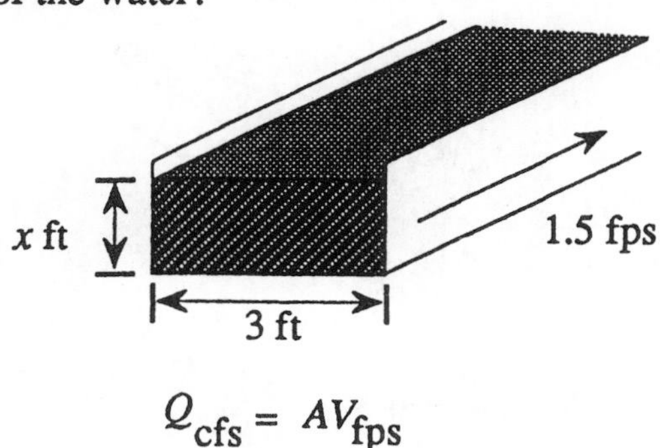

$$Q_{cfs} = AV_{fps}$$

$$8.1 \text{ cfs} = (3 \text{ ft})\,(x \text{ ft})\,1.5 \text{ fps})$$

$$\frac{8.1}{(3)\,(1.5)} = x \text{ ft}$$

$$\boxed{1.8 \text{ ft}} = x$$

DIMENSIONS SHOULD BE EXPRESSED AS FEET

The dimensions in a $Q=AV$ calculation should always be expressed in feet because (ft) (ft) (ft) = cu ft. Therefore, when dimensions are given as inches, first convert all dimensions to feet before beginning the $Q=AV$ calculation.

Note that velocity may be written in either of two forms:

- fps or fpm
- ft/sec or ft/min

The first form is shorter. The second form is useful when dimensional analysis* is desired.

CALCULATING OTHER UNKNOWN VARIABLES

There are four variables in $Q=AV$ calculations for rectangular channels: flow rate, width, depth, and velocity. In Examples 1 and 2, the unknown variable was flow rate, Q. However, any of the other variables can also be unknown.**Example 3 illustrates a calculation when depth is the unknown factor. Section 2.2 of this chapter illustrates calculations when velocity is the unknown factor.

* The concept of dimensional analysis is discussed in Chapter 15 in *Basic Math Concepts*.

** For a review of solving for the unknown variable, refer to Chapter 2 in *Basic Math Concepts*. These type problems are primarily theoretical, since channel size is normally a given and water depth is measured.

FLOW THROUGH A TRAPEZOIDAL CHANNEL

Calculating the flow rate for a trapezoidal channel is similar to calculating the flow rate for a rectangular channel **except that the cross-sectional area, *A*, is a trapezoid.***

$$A = \underset{\text{width}}{(\text{average})}\ \underset{\text{depth}}{(\text{water})}$$

$$A = \frac{(w_1 + w_2)}{2}\ (\text{depth})$$

$Q = AV$ FOR A TRAPEZOIDAL CHANNEL

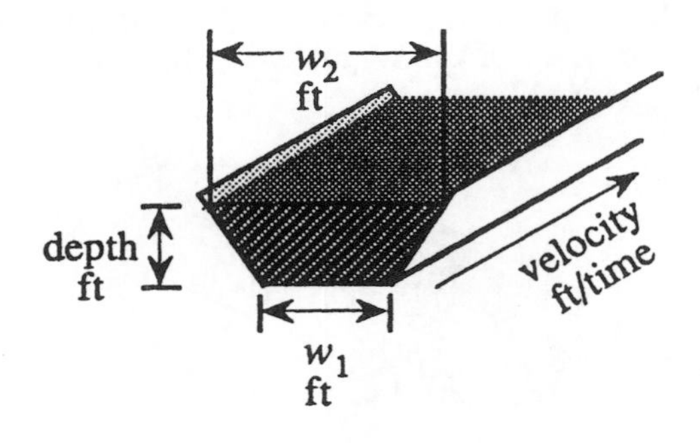

$$Q = \overbrace{\quad A \quad}\ \overbrace{\quad V \quad}$$

$$\underset{\text{cfm}}{Q} = \frac{(w_1 + w_2)}{2}\ \underset{\text{ft}}{(\text{depth})}\ \underset{\text{fpm}}{(\text{velocity})}$$

Example 4: (Instantaneous Flow)

❑ A trapezoidal channel has water flowing to a depth of 2 ft. The width of the channel at the water surface is 6 ft and the width of the channel at the bottom is 4 ft. What is the cfm flow rate in the channel if the velocity is 132 fpm?

6 ft

2 ft

132 fpm

4 ft

$$\underset{\text{cfm}}{Q} = \frac{(w_1 + w_2)}{2}\ \underset{\text{ft}}{(\text{depth})}\ \underset{\text{fpm}}{(\text{velocity})}$$

$$= \frac{(4\ \text{ft} + 6\ \text{ft})}{2}\ (2\ \text{ft})\ (132\ \text{fpm})$$

$$= (5\ \text{ft})\ (2\ \text{ft})\ (132\ \text{fpm})$$

$$= \boxed{1320\ \text{cfm}}$$

* For a review of trapezoid area calculations, refer to Chapter 10 in *Basic Math Concepts*.

WHEN DATA IS NOT GIVEN IN DESIRED TERMS

Many times the data to be used in a *Q=AV* equation is not in the form desired. For example, dimensions might be given in inches rather than in feet, as desired. Or perhaps the velocity is expressed as fps yet the flow rate is desired in cfm. (The time element does not match—seconds vs. minutes.) These type calculations are illustrated in Examples 5 and 6.

Example 5: (Instantaneous Flow)

❑ A trapezoidal channel is 3 ft wide at the bottom and 5.5 ft wide at the water surface. The water depth is 30 inches. If the flow velocity through the channel is 168 ft/min, what is the cfm flow rate through the channel?

5.5 ft

$$\frac{30 \text{ in.}}{12 \text{ in./ft}} = 2.5 \text{ ft}$$

168 ft/min

3 ft

$$\underset{\text{cfm}}{Q} = \frac{(w_1 + w_2)}{2} \underset{\text{ft}}{(\text{depth})} \underset{\text{fpm}}{(\text{velocity})}$$

$$= \frac{(3 \text{ ft} + 5.5 \text{ ft})}{2} (2.5 \text{ ft}) (168 \text{ fpm})$$

$$= \boxed{1785 \text{ cfm}}$$

When the velocity and flow rate time frames do not match, you must convert one of the terms to match the other.

Since flow rate conversions are quite common, you may find it easiest to leave the velocity expression as is and then convert the flow rate to match the velocity time frame. Example 6 illustrates such a process.

Example 6: (Instantaneous Flow)

❑ A trapezoidal channel has water flowing to a depth of 16 inches. The width of the channel at the bottom is 3 ft and the width of the channel at the water surface is 4.5 ft. If the velocity of flow through the channel is 2.5 ft/sec, what is the cfm flow through the channel?

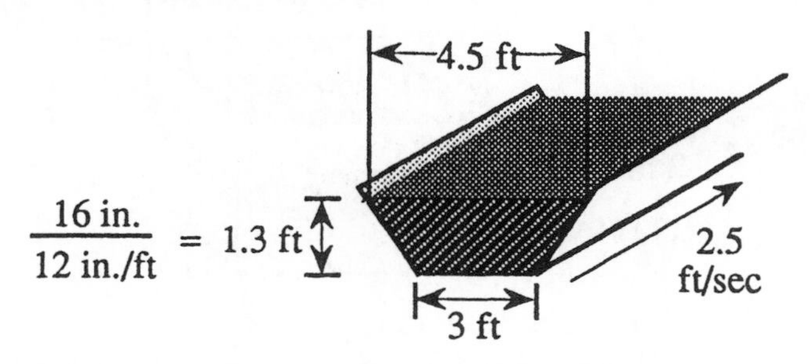

First calculate the flow rate that matches the velocity time frame:

$$\underset{\text{cfs}}{Q} = \frac{(3 \text{ ft} + 4.5 \text{ ft})}{2} (1.3 \text{ ft}) (2.5 \text{ fps})$$

$$= 12.2 \text{ cfs}$$

Now convert cfs flow rate to cfm flow rate:

$$(12.2 \text{ cfs}) \frac{(60 \text{ sec})}{\text{min}} = \boxed{732 \text{ cfm}}$$

FLOW INTO A TANK

Flow through a tank can be considered a type of $Q=AV$ calculation.* If the discharge valve to a tank were closed, the water level would begin to rise. Timing how fast the water rises would give you an indication of the **velocity of flow into the tank**. The $Q=AV$ equation could then be used to determine the flow rate into the tank, as illustrated in Example 7.

If the influent valve to the tank were closed, rather than the discharge valve, and a pump continued discharging water from the tank, the water level in the tank would begin to drop. The rate of this drop in water level could be timed so that **the velocity of flow from the tank** could be calculated. Then the $Q=AV$ equation could be used to determine the flow rate out of the tank, as illustrated in Example 8.

Example 7: (Instantaneous Flow)

❑ A tank is 12 ft by 12 ft. With the discharge valve closed, the influent to the tank causes the water level to rise 1.25 feet in one minute. What is the gpm flow into the tank?

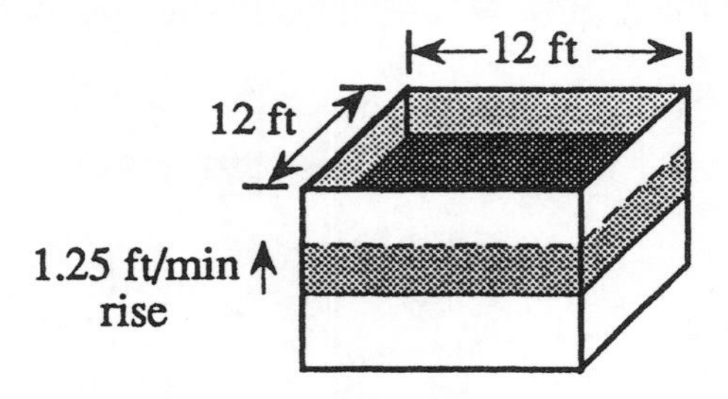

First, calculate the cfm flow rate:

$$Q_{cfm} = AV_{fpm}$$

$$= (12 \text{ ft})\ (12 \text{ ft})\ (1.25 \text{ fpm})$$

$$= 180 \text{ cfm}$$

Then convert cfm flow rate to gpm flow rate:

$$(180 \text{ cfm})\ (7.48 \text{ gal/cu ft}) = \boxed{1346 \text{ gpm}}$$

Example 8: (Instantaneous Flow)

❑ A tank is 8 ft wide and 10 ft long. The influent valve to the tank is closed and the water level drops 2.8 ft in 2 minutes. What is the gpm flow from the tank?

10 ft

8 ft

Drop: $= \dfrac{2.8 \text{ ft}}{2 \text{ min}}$

$= 1.4$ ft/min

First, calculate the cfm flow rate:

$$Q_{cfm} = AV_{fpm}$$

$$= (10 \text{ ft})\ (8 \text{ ft})\ (1.4 \text{ fpm})$$

$$= 112 \text{ cfm}$$

Then convert cfm flow rate to gpm flow rate:

$$(112 \text{ cfm})\ (7.48 \text{ gal/cu ft}) = \boxed{838 \text{ gpm}}$$

---

* This is the same type of calculation described in Chapter 7 as pump capacity calculations.

The same basic method is used to determine the flow rate when the tank is cylindrical in shape. Examples 9 and 10 illustrate the calculation for cylindrical tanks.

Example 9: (Instantaneous Flow)

❑ The discharge valve to a 30-ft diameter tank is closed. If the water rises at a rate of 10 inches in 5 minutes. What is the gpm flow into the tank?

|←30 ft→|

Rise:*
(10 in. = 0.83 ft)

$$= \frac{0.83 \text{ ft}}{5 \text{ min}}$$

$$= 0.17 \text{ ft/min}$$

First calculate the cfm flow into the tank:

$$Q_{\text{cfm}} = AV_{\text{fpm}}$$

$$= (0.785)\,(30 \text{ ft})\,(30 \text{ ft})\,(0.17 \text{ ft/min})$$

$$= 120 \text{ cfm}$$

Then convert cfm flow rate to gpm flow rate:

$$(120 \text{ cfm})\,(7.48 \text{ gal/cu ft}) = \boxed{898 \text{ gpm}}$$

Example 10: (Instantaneous Flow)

❑ A pump discharges into a 2-ft diameter barrel. If the water level in the barrel rises 2 ft in 30 seconds, what is the gpm flow into the barrel?

|←2 ft→|

Rise: $= \frac{2 \text{ ft}}{30 \text{ sec}}$

$= 4 \text{ ft/min}$

First calculate the cfm flow into the tank:

$$Q_{\text{cfm}} = AV_{\text{fpm}}$$

$$= (0.785)\,(2 \text{ ft})\,(2 \text{ ft})\,(4 \text{ fpm})$$

$$= 12.6 \text{ cfm}$$

Then convert cfm flow rate to gpm flow rate:

$$(12.6 \text{ cfm})\,(7.48 \text{ gal/cu ft}) = \boxed{94 \text{ gpm}}$$

* Refer to Chapter 8, "Linear Measurement Conversions", in *Basic Math Concepts.*

FLOW THROUGH A PIPELINE—WHEN FLOWING FULL

The flow rate through a pipeline can be calculated using the $Q{=}AV$ equation. The cross-sectional area is a circle, so the area, A, is represented by (0.785) (D^2). **Pipe diameters should generally be expressed as feet to avoid errors in terms.**

$Q{=}AV$ CALCULATIONS FOR A PIPELINE FLOWING FULL

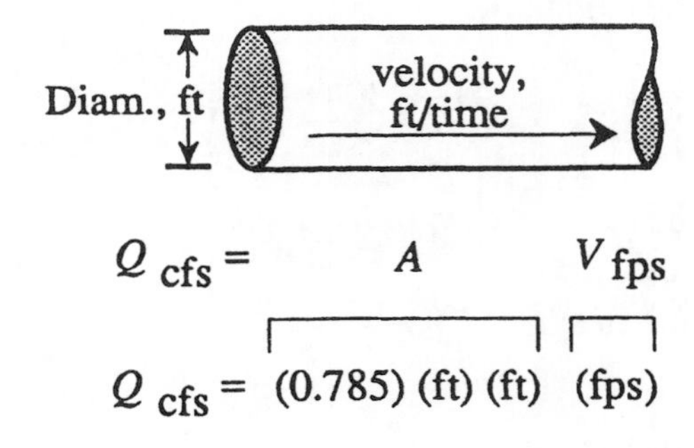

Example 11: (Instantaneous Flow)

❑ The flow through a 6-inch diameter pipeline is moving at a velocity of 3 ft/sec. What is the cfs flow rate through the pipeline? (Assume the pipe is flowing full.)

6 in = 0.5 ft

3 fps

$$Q_{cfs} = AV_{fps}$$

$$= (0.785)\ (0.5\ \text{ft})\ (0.5\ \text{ft})\ (3\ \text{fps})$$

$$= \boxed{0.59\ \text{cfs}}$$

Example 12: (Instantaneous Flow)

❑ An 8-inch diameter pipeline has water flowing at a velocity of 3.4 fps. What is the gpm flow rate through the pipeline?

8 in = 0.67 ft

3.4 fps

First, calculate the cfs flow rate:

$$Q_{cfs} = AV_{fps}$$

$$= (0.785)\ (0.67\ \text{ft})\ (0.67\ \text{ft})\ (3.4\ \text{fps})$$

$$= 1.2\ \text{cfs}$$

Then convert cfs flow rate to gpm flow rate:

$$\frac{(1.2\ \text{cu ft})}{\text{sec}}\ \frac{(60\ \text{sec})}{\text{min}}\ \frac{(7.48\ \text{gal})}{\text{cu ft}} = \boxed{539\ \text{gpm}}$$

Example 13: (Instantaneous Flow)

❑ The flow through a pipeline is 0.7 cfs. If the velocity of flow is 3.6 ft/sec and the pipe is flowing full, what is the diameter (inches) of the pipeline?

x ft

3.6 fps

Flow Rate = 0.7 cfs

First calculate the diameter in feet:

$$Q_{cfs} = AV_{fps}$$

$$0.7\ \text{cfs} = (0.785)\ (x^2\ \text{sq ft})\ (3.6\ \text{fps})$$

$$\frac{0.7}{(0.785)\ (3.6)} = x^2$$

$$0.25\ \text{ft}^2 = x^2$$

$$0.5\ \text{ft} = x$$

Now convert feet to inches:

$$(0.5\ \text{ft})\ \frac{(12\ \text{in.})}{\text{ft}} = \boxed{6\ \text{inches}}$$

SOLVING FOR OTHER UNKNOWN VARIABLES

There are three variables in *Q=AV* calculations for pipelines: flow rate (*Q*), diameter (*D*), and velocity (*V*). In Examples 11 and 12, the unknown factor is flow rate. In Example 13, the unknown factor is pipeline diameter. (Section 2.2 illustrates calculations when velocity is the unknown variable.)

When the diameter is the unknown variable, first solve for x^2. Then, by taking the square root* of both sides of the equation, x may be determined.

* For a review of square roots, refer to Chapter 13 in *Basic Math Concepts*.

FLOW THROUGH A PIPELINE—WHEN FLOWING LESS THAN FULL

Calculating the flow rate through a pipeline flowing less than full is similar to calculating the flow rate for a pipeline flowing full with one exception—**instead of using 0.785 as a factor in the area calculation, a different factor is used.** This factor is based on the ratio of water depth (*d*) to the pipe diameter (*D*). Calculate the *d*/*D* value, then use the table to determine the factor to be used instead of 0.785.

WHEN MAKING *Q*=*AV* CALCULATIONS FOR A PIPELINE FLOWING LESS THAN FULL —A DIFFERENT FACTOR THAN 0.785 IS USED

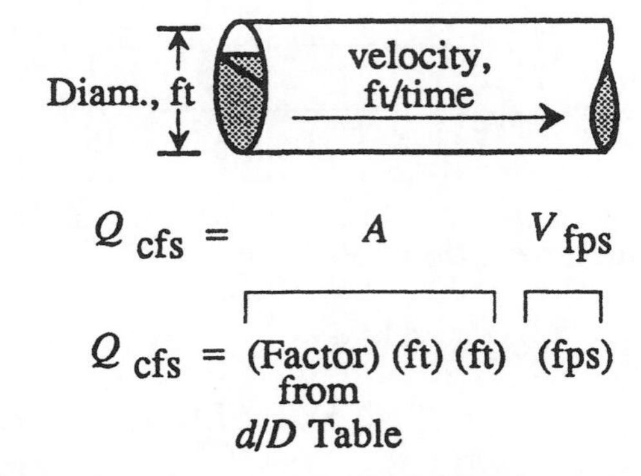

depth/Diameter Table

d/*D*	Factor	*d*/*D*	Factor	*d*/*D*	Factor	*d*/*D*	Factor
0.01	0.0013	0.26	0.1623	0.51	0.4027	0.76	0.6404
0.02	0.0037	0.27	0.1711	0.52	0.4127	0.77	0.6489
0.03	0.0069	0.28	0.1800	0.53	0.4227	0.78	0.6573
0.04	0.0105	0.29	0.1890	0.54	0.4327	0.79	0.6655
0.05	0.0147	0.30	0.1982	0.55	0.4426	0.80	0.6736
0.06	0.0192	0.31	0.2074	0.56	0.4526	0.81	0.6815
0.07	0.0242	0.32	0.2167	0.57	0.4625	0.82	0.6893
0.08	0.0294	0.33	0.2260	0.58	0.4724	0.83	0.6969
0.09	0.0350	0.34	0.2355	0.59	0.4822	0.84	0.7043
0.10	0.0409	0.35	0.2450	0.60	0.4920	0.85	0.7115
0.11	0.0470	0.36	0.2546	0.61	0.5018	0.86	0.7186
0.12	0.0534	0.37	0.2642	0.62	0.5115	0.87	0.7254
0.13	0.0600	0.38	0.2739	0.63	0.5212	0.88	0.7320
0.14	0.0668	0.39	0.2836	0.64	0.5308	0.89	0.7384
0.15	0.0739	0.40	0.2934	0.65	0.5404	0.90	0.7445
0.16	0.0811	0.41	0.3032	0.66	0.5499	0.91	0.7504
0.17	0.0885	0.42	0.3130	0.67	0.5594	0.92	0.7560
0.18	0.0961	0.43	0.3229	0.68	0.5687	0.93	0.7612
0.19	0.1039	0.44	0.3328	0.69	0.5780	0.94	0.7662
0.20	0.1118	0.45	0.3428	0.70	0.5872	0.95	0.7707
0.21	0.1199	0.46	0.3527	0.71	0.5964	0.96	0.7749
0.22	0.1281	0.47	0.3627	0.72	0.6054	0.97	0.7785
0.23	0.1365	0.48	0.3727	0.73	0.6143	0.98	0.7816
0.24	0.1449	0.49	0.3827	0.74	0.6231	0.99	0.7841
0.25	0.1535	0.50	0.3927	0.75	0.6318	1.00	0.7854

Examples 14 and 15 illustrate use of the $Q = AV$ equation when the pipeline is flowing less than full.

Example 14: (Instantaneous Flow)

❑ The flow through a 6-inch diameter pipeline is moving at a velocity of 3 ft/sec. If the water is flowing at a depth of 4 inches, what is the cfs flow rate through the pipeline?

4 in. — 3 fps — 6 in. = 0.5 ft

First use the d/D ratio to determine the factor to be used instead of 0.785 in the $Q=AV$ calculation:

$$\frac{d}{D} = \frac{4}{6} = 0.67$$

The factor shown in the table corresponding to a d/D of 0.67 is 0.5594. Now calculate the flow rate using $Q=AV$:

$$Q_{cfs} = AV_{fps}$$

$$= (0.5594)\ (0.5\text{ ft})\ (0.5\text{ ft})\ (3\text{ fps})$$

$$= \boxed{0.42\text{ cfs}}$$

Example 15: (Instantaneous Flow)

❑ An 8-inch diameter pipeline has water flowing at a velocity of 3.4 fps. What is the gpm flow rate through the pipeline if the water is flowing at a depth of 5 inches?

5 in — 3.4 fps — 8 in = 0.67 ft

First, determine the factor to be used instead of 0.785. Since d/D=0.63, the factor listed in the table is 0.5212. Now calculate the flow rate using $Q=AV$. Although gpm flow rate is desired, first calculate cfs flow rate, then convert cfs to gpm flow rate:

$$Q_{cfs} = AV_{fps}$$

$$= (0.5212)\ (0.67\text{ ft})\ (0.67\text{ ft})\ (3.4\text{ fps})$$

$$= 0.8\text{ cfs}$$

Then convert cfs flow rate to gpm flow rate:

$$\left(\frac{0.8\text{ cu ft}}{\text{sec}}\right)\left(\frac{60\text{ sec}}{\text{min}}\right)\left(\frac{7.48\text{ gal}}{\text{cu ft}}\right) = \boxed{359\text{ gpm}}$$

NWTC Library
2740 W. Mason St.
Green Bay, WI 54307

2.2 VELOCITY CALCULATIONS

VELOCITY USING $Q=AV$

The $Q=AV$ equation may be used to estimate the velocity of flow in a channel or pipeline. Write the equation as usual, filling in the known data, then solve for the unknown factor (velocity in this case).

Be sure that the volume and time expressions match on both sides of the equation. For instance, if the velocity is desired in ft/sec, then the flow rate should be converted to cfs before beginning the $Q=AV$ calculation. Examples 1 and 2 illustrate a velocity estimate for a pipeline.

Example 1: (Velocity Calculations)

❑ A channel has a rectangular cross section. The channel is 4 ft wide with water flowing to a depth of 1.8 ft. If the flow rate through the channel is 9050 gpm, what is the velocity of the water in the channel (ft/sec)?

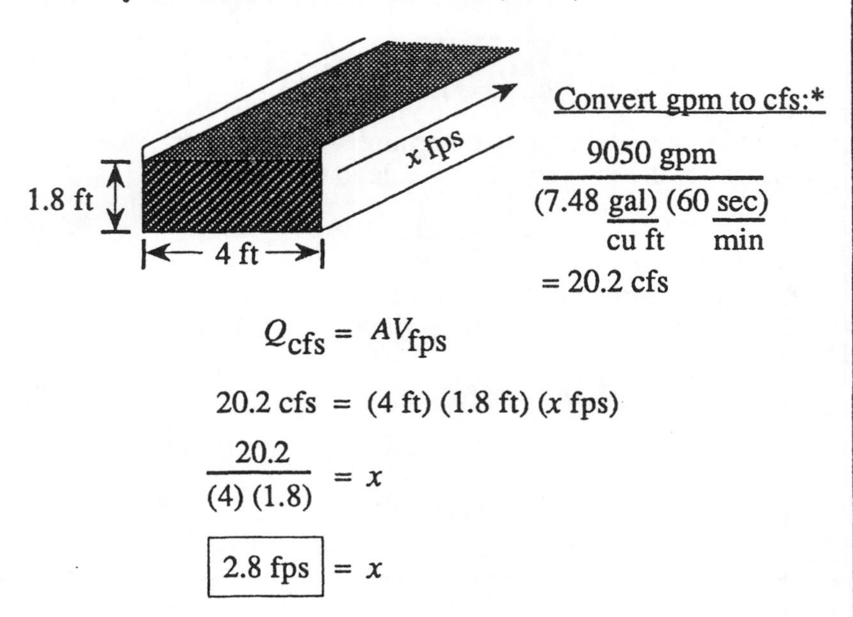

Convert gpm to cfs:*

$$\frac{9050 \text{ gpm}}{(7.48 \frac{\text{gal}}{\text{cu ft}})(60 \frac{\text{sec}}{\text{min}})}$$

$= 20.2$ cfs

$$Q_{cfs} = AV_{fps}$$

$$20.2 \text{ cfs} = (4 \text{ ft})(1.8 \text{ ft})(x \text{ fps})$$

$$\frac{20.2}{(4)(1.8)} = x$$

$$\boxed{2.8 \text{ fps}} = x$$

Example 2: (Velocity Calculations)

❑ A 6-inch diameter pipe flowing full delivers 280 gpm. What is the velocity of flow in the pipeline (ft/sec)?

6 in = 0.5 ft

x fps

Convert gpm to cfs flow:

$$\frac{280 \text{ gpm}}{(7.48 \frac{\text{gal}}{\text{cu ft}})(60 \frac{\text{sec}}{\text{min}})}$$

$= 0.62$ cfs

$$Q_{cfs} = AV_{fps}$$

$$0.62 \text{ cfs} = (0.785)(0.5 \text{ ft})(0.5 \text{ ft})(x \text{ fps})$$

$$\frac{0.62}{(0.785)(0.5)(0.5)} = x$$

$$\boxed{3.2 \text{ fps}} = x$$

* For a review of flow conversions, refer to Chapter 8 in *Basic Math Concepts*.

Example 3: (Velocity Calculations)

❑ A float travels 300 ft in a channel in 2 min 14 sec. What is the estimated velocity in the channel (ft/sec)?

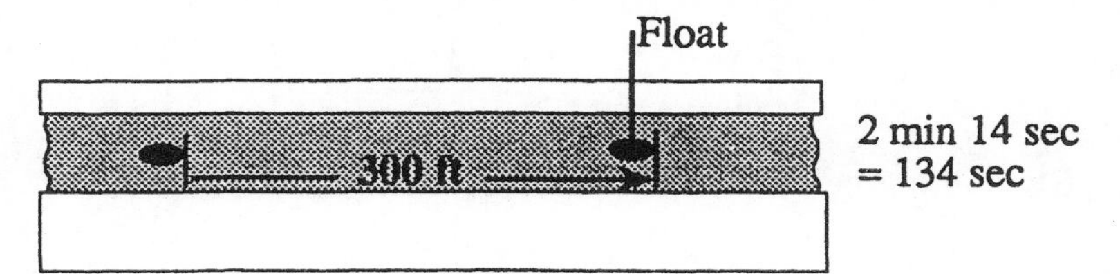

$$\text{Velocity, ft/sec} = \frac{\text{Distance, ft}}{\text{Time, sec}}$$

$$= \frac{300 \text{ ft}}{134 \text{ sec}}$$

$$= \boxed{2.24 \text{ ft/sec}}$$

Example 4: (Velocity Calculations)

❑ A fluorescent dye is used to estimate the velocity of flow in a sewer. The dye is injected in the water at one manhole and the travel time to the next manhole 400 ft away is noted. The dye first appears at the downstream manhole in 128 seconds. The dye continues to be visible until a total elapsed time of 148 seconds. What is the ft/sec velocity of flow through the pipeline?

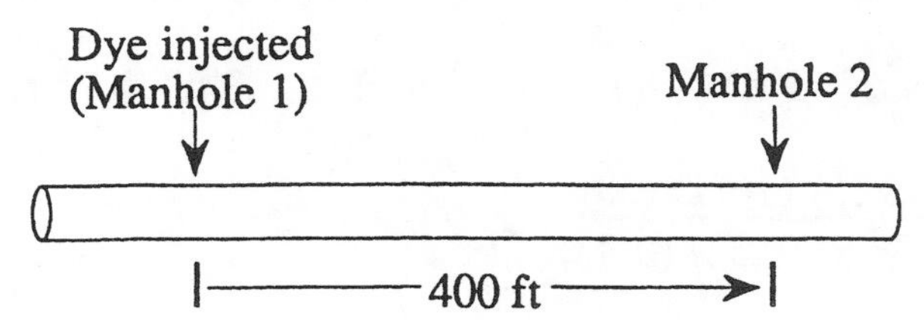

First calculate the average travel time of the dye:

$$\frac{128 \text{ sec} + 148 \text{ sec}}{2} = 138 \text{ sec}$$

Then calculate the ft/sec velocity:

$$\text{Velocity, ft/sec} = \frac{400 \text{ ft}}{138 \text{ sec}}$$

$$= \boxed{2.9 \text{ ft/sec}}$$

VELOCITY USING THE FLOAT OR DYE METHOD

The $Q = AV$ calculation estimates the theoretical velocity of flow in a channel or pipeline. Actual velocities in the pipeline can be measured by metering devices. Velocities can also be estimated by the use of a float or dye placed in the water. Then, by timing the distance traveled using a float or dye, the velocity of flow can be determined:

$$\text{Velocity, ft/sec} = \frac{\text{Distance, ft}}{\text{Time, sec}}$$

A float is perhaps less accurate in estimating velocities in a pipeline than use of fluorescent tracer dyes. In channels, floats move along with the faster surface waters and can be as much as 10 or 15 percent faster than the actual average flow rate. Some floats designed for use in channels include segments that extend into the water, thus responding to a more average velocity through the channel.

In pipelines, floats can become entangled or slowed down by obstructions.

Tracer dyes tend to give a better estimate of velocity. Since some of the dye will travel faster and some slower, you will need to determine the **average time** required to travel from one point to the next. To calculate the average time for travel:

$$\text{Average Time, sec} = \frac{\text{Total elapsed time 'til dye first appears} + \text{Total elapsed time 'til dye no longer seen}}{2}$$

USING $Q=AV$ TO ESTIMATE CHANGES IN VELOCITY

In addition to estimating flow in a channel or pipeline, the $Q=AV$ equation can be used to estimate the change in velocity as the water flows from one diameter pipeline to another.

FLOW RATE (Q) IN PIPES REMAIN CONSTANT

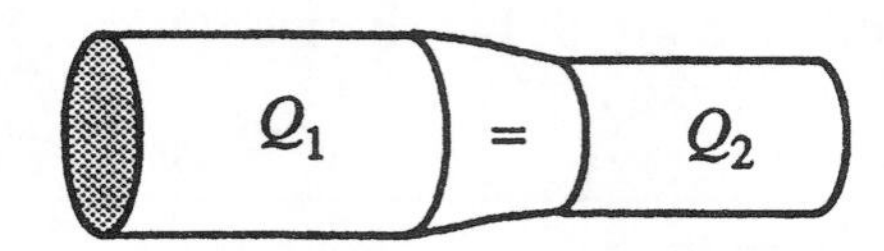

Since the total flow in the pipeline must remain constant:

$$Q_1 = Q_2$$

$$A_1V_1 = A_2V_2$$

A_1V_1 = A_2V_2

When water flows from a larger diameter pipe to a smaller diameter pipe, the velocity increases. (The water must move faster since the same amount of water is flowing through a smaller space.) Example 5 illustrates this calculation.

Example 5: (Velocity Calculations)

❑ The velocity in a 12-inch diameter pipeline is 3.8 ft/sec. If the 12-inch pipeline flows into a 10-inch diameter pipeline, what is the velocity in the 10-inch pipeline?

12 in = 1 ft $\quad A_1V_1 = A_2V_2 \quad \frac{10 \text{ in}}{12 \text{ in/ft}} = 0.83 \text{ ft}$

$$A_1V_1 = A_2V_2$$

$$(0.785)(1 \text{ ft})(1 \text{ ft})(3.8 \text{ fps}) = (0.785)(0.83 \text{ ft})(0.83 \text{ ft})(x \text{ fps})$$

$$\frac{(\cancel{0.785})(1)(1)(3.8)}{(\cancel{0.785})(0.83)(0.83)} = x \text{ fps}$$

$$\boxed{5.5 \text{ fps}} = x$$

Example 6: (Velocity Calculations)
❑ The velocity in a 6-inch diameter pipe is 4.8 ft/sec. If the flow travels from a 6-inch pipeline to an 8-inch pipeline, what is the velocity in the 8-inch pipeline?

$$\frac{6 \text{ in}}{12 \text{ in/ft}} = 0.5 \text{ ft} \quad A_1V_1 = A_2V_2 \quad \frac{8 \text{ in}}{12 \text{ in/ft}} = 0.67 \text{ ft}$$

$$A_1V_1 = A_2V_2$$

$$(0.785)(0.5 \text{ ft})(0.5 \text{ ft})(4.8 \text{ fps}) = (0.785)(0.67 \text{ ft})(0.67 \text{ ft})(x \text{ fps})$$

$$\frac{(0.785)(0.5)(0.5)(4.8)}{(0.785)(0.67)(0.67)} = x \text{ fps}$$

$$\boxed{2.7 \text{ fps}} = x$$

When water flows from a smaller diameter pipe to a larger diameter pipe, the velocity decreases. Example 6 illustrates this principle.

WHEN *Q* DATA IS GIVEN FOR ONE PIPE AND *AV* DATA GIVEN FOR THE OTHER

In Examples 5 and 6, the *AV* of one pipeline was set equal to the *AV* of the second pipeline:

$$\boxed{A_1V_1 = A_2V_2}$$

This equation is possible since the same flow travels through both pipes:

$$\boxed{Q_1 = Q_2}$$

Since $Q_1 = A_1V_1$, either of these terms (Q_1 **or** A_1V_1) can be used on the left side of the equation. Similarly since $Q_2 = A_2V_2$, either of these terms (Q_2 **or** A_2V_2) can be used interchangeably on the right side of the equation:

$$\boxed{Q_1 = A_2V_2}$$

$$\boxed{A_1V_1 = Q_2}$$

Example 7 illustrates a calculation where the *Q* of one pipeline is set equal to the *AV* of the second pipeline.

Example 7: (Velocity Calculations)
❑ The flow through a 6-inch diameter pipeline is 220 gpm. What is the estimated velocity of flow (fps) through the 4-inch diameter pipeline shown below?

$$A_1V_1 = A_2V_2 \quad \frac{4 \text{ in}}{12 \text{ in/ft}} = 0.33 \text{ ft}$$

Since flow data is given for the first pipeline and velocity (part of the *AV*) is unknown for the second pipeline, the equation to be used is:

$$Q_1 = A_2V_2$$

Remember that *Q* must be expressed in cfs since the right side of the equation is (sq ft) (ft/sec) or cfs:

$$\frac{220 \text{ gpm}}{(7.48 \text{ gal/cu ft})(60 \text{ sec/min})} = 0.49 \text{ cfs}$$

Now complete the *Q*=*AV* equation:

$$Q_1 = A_2V_2$$

$$0.49 \text{ cfs} = (0.785)(0.33 \text{ ft})(0.33 \text{ ft})(x \text{ fps})$$

$$\frac{0.49}{(0.785)(0.33)(0.33)} = x$$

$$\boxed{5.7 \text{ fps}} = x$$

2.3 AVERAGE FLOW RATES CALCULATIONS

Flow rates in a treatment system may vary considerably during the course of a day. Calculating an **average flow rate** is a way to determine the **typical flow rate** for a given time frame such as: average daily flow, average weekly flow, average monthly flow, or even average yearly flow.

There are two ways to calculate an average flow rate. In the first method, several flow rate values are used to determine an average value, as illustrated in Examples 1 and 2.

$$\text{Average Flow} = \frac{\text{Tot. of all Sample Flows}}{\text{No. of Samples}}$$

In the second method, a total flow is used (from a totalizer reading) to determine an average flow rate. Examples 3 and 4 illustrate this type of calculation.

$$\text{Average Flow} = \frac{\text{Tot. Flow from Totalizer}}{\text{Time Over Which Flow Measured}}$$

Example 1: (Average Flow Rates)

❑ The following flows were recorded for the week: Monday—8.6 MGD; Tuesday—7.6 MGD; Wednesday—7.2 MGD; Thursday—7.8 MGD; Friday—8.4 MGD; Saturday—8.6 MGD; Sunday—7.5 MGD. What was the average daily flow for the week?

$$\text{Average Daily Flow} = \frac{\text{Total of all Sample Flows}}{\text{Number of Days}}$$

$$= \frac{55.7 \text{ MGD}}{7 \text{ Days}}$$

$$= \boxed{8.0 \text{ MGD}}$$

Example 2: (Average Flow Rates)

❑ The following flows were recorded for the months of September, October and November: September—120.8 MG; October—136.4 MG; November—156.1 MG. What was the average daily flow for this three-month period?

Since average **daily** flow is desired, you must divide by the **number of days** represented by the three-month period, rather than by the number of months represented. (If average monthly flow had been desired, then you would divide by the number of months represented.)

$$\text{Average Daily Flow} = \frac{\text{Total of all Sample Flows}}{\text{Number of Days}}$$

$$= \frac{413.3 \text{ MG}}{91 \text{ days}}$$

$$= \boxed{4.5 \text{ MGD}}$$

Example 3: (Average Flow Rates)

❑ The totalizer reading for the month of November was 142.8 MG. What was the average daily flow (ADF) for the month of November?

Since average daily flow is desired, the denominator must reflect the number of days represented by the totalizer flow:

$$\text{Average Daily Flow} = \frac{\text{Tot. Flow from Totalizer}}{\text{Number of Days}}$$

$$= \frac{142.8 \text{ MG}}{30 \text{ days}}$$

$$= \boxed{4.8 \text{ MGD}}$$

Example 4: (Average Flow Rates)

❑ The total flow for one day at a plant was 2,600,000 gallons. What was the average gpm flow for that day?

The average flow per **minute** is desired. Therefore, the denominator must reflect the **number of minutes** represented by the totalizer flow:

$$\text{Average Flow, gpm} = \frac{\text{Tot. Flow from Totalizer}}{\text{Number of Minutes}}$$

$$= \frac{2{,}600{,}000 \text{ gallons}}{1440 \text{ minutes}}$$

$$= \boxed{1805.6 \text{ gpm}}$$

CALCULATING AVERAGE FLOWS USING TOTALIZER INFORMATION

When totalizer data is used to calculate average flows, the number in the denominator depends on the time frame desired. If you desire to know the average **flow per minute**, divide by the **number of minutes** represented by the totalizer flow. If you wish to determine average **daily** flow, divide by the **number of days** represented by the totalizer flow. To calculate the average **weekly** flow, divide by the **number of weeks** represented by the totalizer flow.

2.4 FLOW CONVERSIONS CALCULATIONS

Flow rates may be expressed in several different ways—cubic feet per second (cfs), cubic feet per minute (cfm), gallons per minute (gpm), gallons per day (gpd), etc. as shown in the diagram to the right.

To excel in water and wastewater math, it is essential that you be able to convert one expression of flow to any other.

The box method was designed to make these conversions easy to visualize. **Moving from a smaller box to a larger box requires multiplication by the factor indicated. Moving from a larger box to a smaller box requires division by the factor indicated.**

FLOW CONVERSIONS USING THE BOX METHOD*

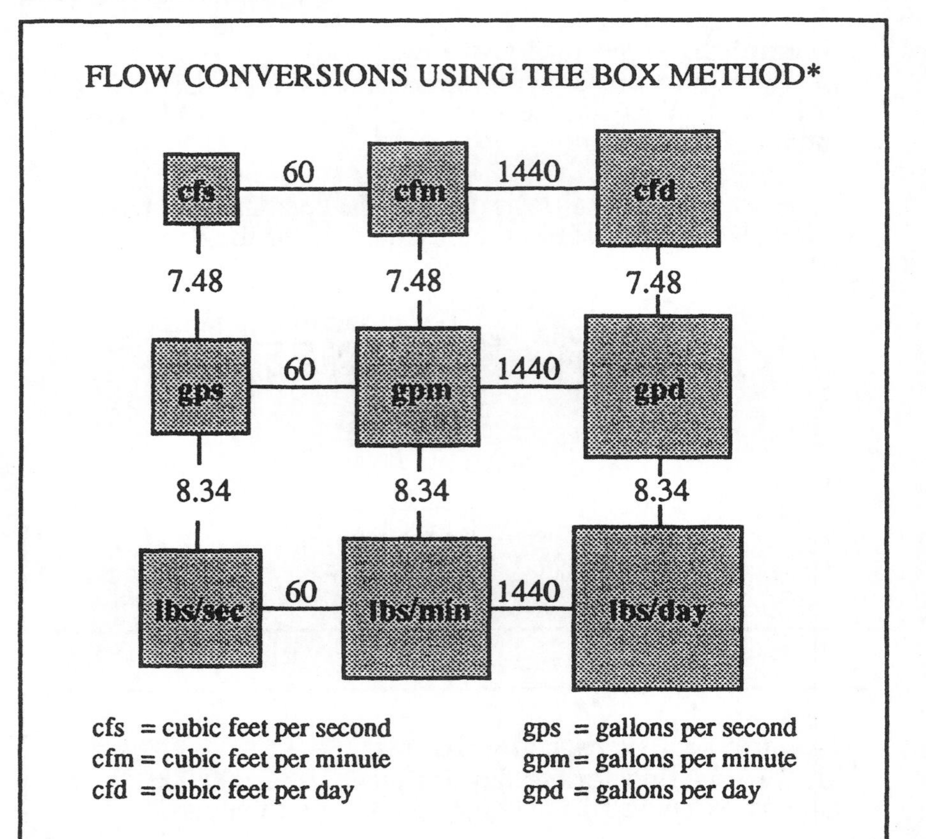

cfs = cubic feet per second
cfm = cubic feet per minute
cfd = cubic feet per day

gps = gallons per second
gpm = gallons per minute
gpd = gallons per day

Example 1: (Flow Conversions)

❑ Convert a flow of 3 cfs to gpm.

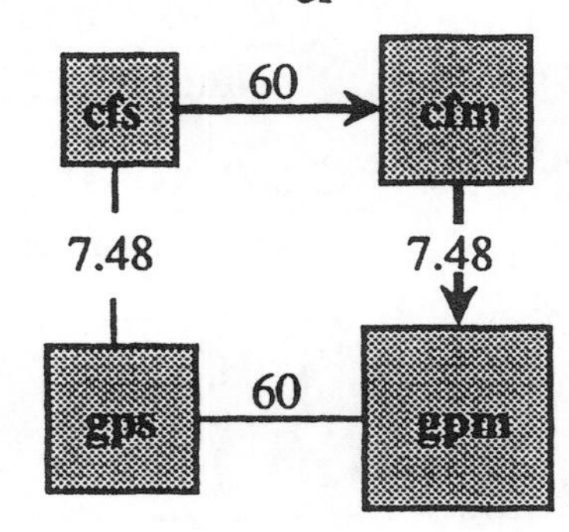

First write the flow rate to be converted (3 cfs). Then any factors to be multiplied must be placed in the numerator with 3 cfs. Any division factors are placed in the denominator. There are two different paths to gpm. **Either path will result in the same answer:**

$$(3\ \text{cfs})\left(60\ \frac{\text{sec}}{\text{min}}\right)\left(7.48\ \frac{\text{gal}}{\text{cu ft}}\right) = \boxed{1346\ \text{gpm}}$$

* The factors shown in the diagram have the following units associated with them: 60 sec/min, 1440 min/day, 7.48 gal/cu ft, and 8.34 lbs/gal.

Example 2: (Flow Conversions)

❑ Convert a flow of 45 gps to gpd. Use dimensional analysis to check the set up of the problem.

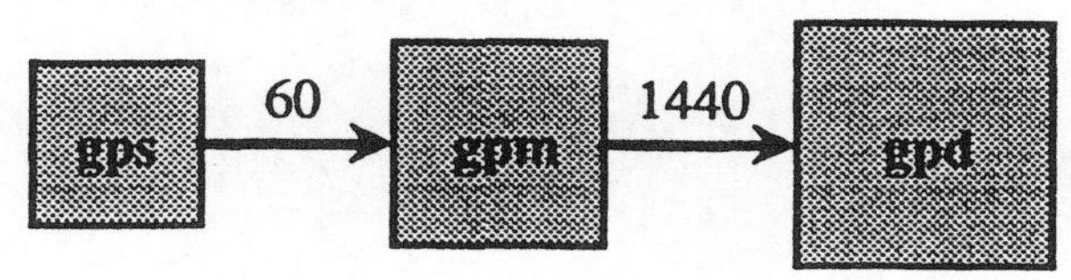

Write the flow rate to be converted, then place multiplication or division factors in the numerator or denominator, as required:

$$(45 \text{ gps}) \left(60 \frac{\text{sec}}{\text{min}}\right) \left(1440 \frac{\text{min}}{\text{day}}\right) = \boxed{3{,}888{,}000 \text{ gpd}}$$

Now use dimensional analysis to check the math set up of this problem:

$$\frac{\text{gal}}{\cancel{\text{sec}}} \cdot \frac{\cancel{\text{sec}}}{\cancel{\text{min}}} \cdot \frac{\cancel{\text{min}}}{\text{day}} = \frac{\text{gal}}{\text{day}}$$

Example 3: (Flow Conversions)

❑ Convert a flow of 3,200,000 gpd to cfm. Use dimensional analysis to check the set up of the problem.

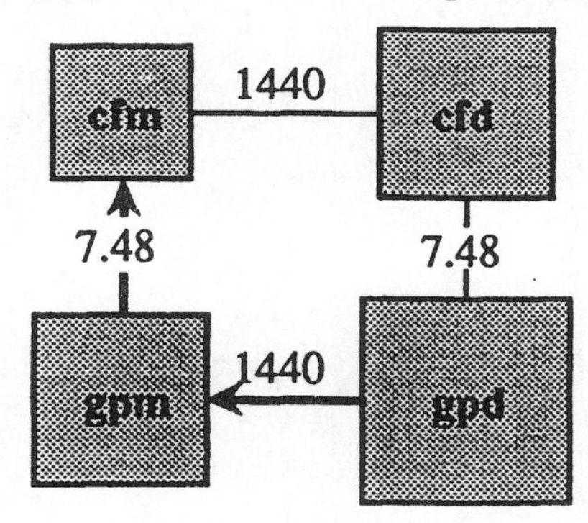

Two different paths may be used from gpd to cfm. Either path will result in the same answer:

$$\frac{3{,}200{,}000 \text{ gpd}}{\left(1440 \frac{\text{min}}{\text{day}}\right)\left(7.48 \frac{\text{gal}}{\text{cu ft}}\right)} = \boxed{297 \text{ cfm}}$$

Now use dimensional analysis to check the math set up of this problem:*

$$\frac{\frac{\text{gal}}{\text{day}}}{\frac{\text{min}}{\text{day}} \cdot \frac{\text{gal}}{\text{cu ft}}} = \frac{\cancel{\text{gal}}}{\cancel{\text{day}}} \cdot \frac{\cancel{\text{day}}}{\text{min}} \cdot \frac{\text{cu ft}}{\cancel{\text{gal}}} = \frac{\text{cu ft}}{\text{min}}$$

* For a review of complex fractions, refer to Chapter 3 in *Basic Math Concepts*.

USING DIMENSIONAL ANALYSIS TO CHECK THE MATH SET UP

Dimensional analysis is often used to check the mathematical set up of conversions. Examples 2 and 3 illustrate how to use dimensional analysis in checking the problem set up. Refer to Chapter 15 in *Basic Math Concepts* for a further discussion of dimensional analysis.

QUICK CONVERSIONS

There are two conversion equations used quite frequently in water and wastewater treatment calculations. You would be well advised to memorize these equations for use in quick conversions:

1 MGD = 1.55 cfs
1 MGD = 694 gpm

Should you forget these numbers, you can always derive them yourself using the box method of conversions.

NOTES:

3 Milligrams per Liter to Pounds per Day Calculations

SUMMARY

1. The five general types of mg/*L* to lbs/day or lbs calculations are:

   - Chemical Dosage
   - BOD, COD, or SS Loading
   - BOD, COD, or SS Removal
   - Pounds of Solids Under Aeration
   - WAS Pumping Rate

2. All of the calculations listed above use one of two equations:

   (mg/*L*) (MGD flow) (8.34 lbs/gal) = lbs/day

   or

   (mg/*L*) (MG volume) (8.34 lbs/gal) = lbs

3. To determine which of the two equations to use, you must first determine whether the mg/*L* concentration pertains to a **flow** or a **tank or pipeline volume.** If the mg/*L* concentration represents a concentration in a flow, then million gallons per day (MGD) flow is used as the second factor. If the concentration pertains to a tank or pipeline volume, then million gallons (MG) volume is used as the second factor.

One of the most frequently used calculations in water and water mathematics is the conversion of milligrams per liter (mg/*L*) concentration to pounds per day (lbs/day) or pounds (lbs) dosage or loading. This calculation is the basis of five general types of calculations, as noted in the summary to the left.

3.1 CHEMICAL DOSAGE CALCULATIONS

CHLORINE DOSAGE

In chemical dosing, a measured amount of chemical is added to the water or wastewater. The amount of chemical required depends on such factors as the type of chemical used, the reason for dosing, and the flow rate being treated.

Two ways to describe the amount of chemical added or required are:

- milligrams per liter (mg/*L*)
- pounds per day (lbs/day)

To convert from mg/*L* (or ppm) concentration to lbs/day, use the following equation:

(mg/*L*) (MGD) (8.34) = lbs/day
Conc. flow lbs/gal

In previous years, parts per million (ppm) was also used as an expression of concentration. In fact, it was used interchangeably with mg/*L* concentration, since 1 mg/*L* = 1 ppm.* However, because *Standard Methods* no longer uses ppm, mg/*L* is the preferred expression of concentration.

MILLIGRAMS PER LITER IS A MEASURE OF CONCENTRATION

Assume each liter below is divided into 1 million parts. Then:

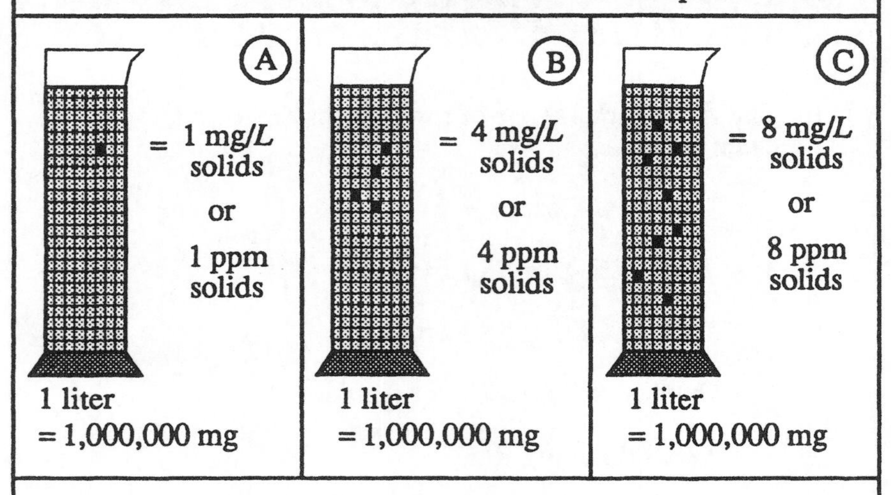

Assuming the liter in these three examples has been divided into 1 million parts (each part representing 1 milligram, mg), the **concentration of solids** in each liter could be expressed as:

- The number of mg solids per liter (mg/*L*) or
- The number of mg solids per 1,000,000 mg (ppm).

The concentration of solids shown in diagram A is 1 milligram per liter (1 mg/*L*). The solids concentration shown in diagrams B and C are 4 mg/*L* and 8 mg/*L*, respectively.

Example 1: (Chemical Dosage)

❑ Determine the chlorinator setting (lbs/day) needed to treat a flow of 3 MGD with a chlorine dose of 4 mg/*L*.

First write the equation. Then fill in the information given:

(mg/*L*) (MGD) (8.34) = lbs/day
Conc. flow lbs/gal

(4 mg/*L*) (3 MGD) (8.34 lbs/gal) = lbs/day

= **100 lbs/day**

$$* \ \frac{1\text{mg}}{L} = \frac{1\text{ mg}}{1{,}000{,}000\text{ mg}} = \frac{1\text{ lb}}{1{,}000{,}000\text{ lbs}} = \frac{1\text{ part}}{1{,}000{,}000\text{ parts}} = 1\text{ ppm}$$

Example 2: (Chemical Dosage)

❑ Determine the chlorinator setting (lbs/day) if a flow of 3.8 MGD is to be treated with a chlorine dose of 2.7 mg/*L*.

Write the equation then fill in the information given:

$$\underset{\text{Conc.}}{(\text{mg}/L)}\ \underset{\text{flow}}{(\text{MGD})}\ \underset{\text{lbs/gal}}{(8.34)} = \text{lbs/day}$$

$$(2.7\ \text{mg}/L)\ (3.8\ \text{MGD})\ (8.34\ \text{lbs/gal}) = \text{lbs/day}$$

$$= \boxed{85.6\ \text{lbs/day}}$$

Example 3: (Chemical Dosage)

❑ What should the chlorinator setting be (lbs/day) to treat a flow of 2 MGD if the chlorine demand is 10 mg/*L* and a chlorine residual of 2 mg/*L* is desired?

First, write the mg/*L* to lbs/day equation:

$$\underset{\text{Conc.}}{(\text{mg}/L)}\ \underset{\text{flow}}{(\text{MGD})}\ \underset{\text{lbs/gal}}{(8.34)} = \text{lbs/day}$$

In this problem the unknown value is lbs/day. Information is given for each of the other two variables: mg/*L* and flow. Notice that information for the mg/*L* dose is given only indirectly, as chlorine demand and residual and can be found using the equation:

$$\underset{\text{mg}/L}{Cl_2\ \text{Dose}} = \underset{\text{mg}/L}{Cl_2\ \text{Demand}} + \underset{\text{mg}/L}{Cl_2\ \text{Residual}}$$

$$= 10\ \text{mg}/L + 2\ \text{mg}/L$$

$$= 12\ \text{mg}/L$$

The mg/*L* to lbs/day calculation may now be completed:

$$(12\ \text{mg}/L)\ (2\ \text{MGD})\ (8.34\ \text{lbs/gal}) = \boxed{200\ \text{lbs/day}}$$

CHLORINE DOSAGE, DEMAND, AND RESIDUAL

In some chlorination calculations, the mg/*L* chlorine dose is not given directly but indirectly as chlorine demand and residual information.

Chlorine dose depends on two considerations—the chlorine demand and the desired chlorine residual such that:

$$\boxed{\underset{\text{mg}/L}{\text{Dose}} = \underset{\text{mg}/L}{\text{Demand}} + \underset{\text{mg}/L}{\text{Resid.}}}$$

The **chlorine demand** is the amount of chlorine used in reacting with various components of the water such as harmful organisms and other organic and inorganic substances. When the chlorine demand has been satisfied, these reactions stop.

In some cases, such as perhaps during pretreatment, chlorinating just enough to meet some or all of the chlorine demand is sufficient. However, in other cases, it is desirable to have an additional amount of chlorine available for disinfection.

Using the equation shown above, if you are given information about any two of the variables, you can determine the value of the third variable. For example, if you know that the chlorine dose is 3 mg/L and the chlorine residual is 0.5 mg/*L*, the chlorine demand must therefore be 2.5 mg/*L*:

$$3\ \text{mg}/L = 2.5\ \text{mg}/L + 0.5\ \text{mg}/L$$

If chlorine demand and residual are known, then chlorine dose (mg/*L*) can be determined, as illustrated in Example 3.

CHEMICAL DOSAGE FOR OTHER CHEMICALS

Examples 1-3 illustrated chemical dosage calculations for chlorine. The same method is used in calculating dosages for other chemicals, as shown in Examples 4 and 5.

Example 4: (Chemical Dosage)

❑ A jar test indicates that the best dry alum dose is 12 mg/*L*. If the flow is 3.5 MGD, what is the desired alum feed rate? (lbs/day)

$$\underset{\text{Conc.}}{(\text{mg}/L)}\ \underset{\text{flow}}{(\text{MGD})}\ \underset{\text{lbs/gal}}{(8.34)} = \text{lbs/day}$$

$$(12\ \text{mg}/L)\ (3.5\ \text{MGD})\ (8.34\ \text{lbs/gal}) = \text{lbs/day}$$

$$= \boxed{350\ \text{lbs/day}}$$

Example 5: (Chemical Dosage)

❑ To dechlorinate a wastewater, sulfur dioxide is to be applied at a level 3 mg/*L* more than the chlorine residual. What should the sulfonator feed rate be (lbs/day) for a flow of 4 MGD with a chlorine residual of 4.2 mg/*L*?

Since the chlorine residual is 4.2 mg/*L*, the sulfur dioxide dosage should be 4.2 + 3 = 7.2 mg/*L*:

$$\underset{\text{Conc.}}{(\text{mg}/L)}\ \underset{\text{flow}}{(\text{MGD})}\ \underset{\text{lbs/gal}}{(8.34)} = \text{lbs/day}$$

$$(7.2\ \text{mg}/L)\ (4\ \text{MGD})\ (8.34\ \text{lbs/gal}) = \text{lbs/day}$$

$$= \boxed{240\ \text{lbs/day}}$$

CALCULATING mg/*L* GIVEN lbs /day

In some chemical dosage calculations, you will know the dosage in lbs/day and the flow rate, but the mg/*L* dosage will be unknown. Approach these problems as any other mg/*L* to lbs/day problem:

- Write the equation,
- Fill in the known information,
- Solve for the unknown value.

Example 6: (Chemical Dosage)

❑ The chlorine feed rate at a plant is 175 lbs/day. If the flow is 2,450,000 gpd, what is this dosage in mg/*L*?

$$\underset{\text{Conc.}}{(\text{mg}/L)}\ \underset{\text{flow}}{(\text{MGD})}\ \underset{\text{lbs/gal}}{(8.34)} = \text{lbs/day}$$

$$(x\ \text{mg}/L)\ (2.45\ \text{MGD})\ (8.34\ \text{lbs/gal}) = 175\ \text{lbs/day}$$

$$x = \frac{175\ \text{lbs/day}}{(2.45\ \text{MGD})\ (8.34\ \text{lbs/gal})}$$

$$x = \boxed{8.6\ \text{mg}/L}$$

Example 7: (Chemical Dosage)

❑ A storage tank is to be disinfected with a 50 mg/*L* chlorine solution. If the tank holds 70,000 gallons, how many pounds of chlorine (gas) will be needed?

$$\underset{\text{Conc.}}{(\text{mg}/L)}\ \underset{\text{Vol}}{(\text{MG})}\ \underset{\text{lbs/gal}}{(8.34)} = \text{lbs}$$

$$(50\ \text{mg}/L)\ (0.07\ \text{MG})\ (8.34\ \text{lbs/gal}) = \text{lbs}$$

$$= \boxed{29.2\ \text{lbs}}$$

Example 8: (Chemical Dosage)

❑ To neutralize a sour digester, one pound of lime is to be added for every pound of volatile acids in the digester liquor. If the digester contains 250,000 gal of sludge with a volatile acid (VA) level of 2,300 mg/*L*, how many pounds of lime should be added?

Since the VA concentration is 2300 mg/*L*, the lime concentration should also be 2300 mg/*L*:

$$\underset{\text{Conc.}}{(\text{mg}/L)}\ \underset{\text{Vol}}{(\text{MG})}\ \underset{\text{lbs/gal}}{(8.34)} = \text{lbs}$$

$$(2300\ \text{mg}/L)\ (0.25\ \text{MG})\ (8.34\ \text{lbs/gal}) = \text{lbs}$$

$$= \boxed{4{,}796\ \text{lbs}}$$

CHEMICAL DOSAGE IN WELLS, TANKS, RESERVOIRS, OR PIPELINES

Wells are disinfected (chlorinated) during and after construction and also after any well or pump repairs. Tanks and reservoirs are chlorinated after initial inspection and after any time they have been drained for cleaning, repair or maintenance. Similarly, a pipeline is chlorinated after initial installation and after any repair.

Digesters may also require chemical dosing, although the chemical used is not chlorine but lime or some other chemical.

For calculations such as these, use the mg/*L* to lbs equation:

$$\boxed{\underset{\text{Conc.}}{(\text{mg}/L)}\ \underset{\text{Vol}}{(\text{MG})}\ \underset{\text{lbs/gal}}{(8.34)} = \text{lbs}}$$

Notice that this equation is very similar to that used in Examples 1-6. The only difference is that MG volume is used rather than MGD flow; therefore, the result is lbs rather than lbs/day. (When dosing a volume, there is no time factor consideration.) Examples 7-8 illustrate these calculations.

HYPOCHLORITE COMPOUNDS

When chlorinating water or wastewater with chlorine gas, you are chlorinating with 100% available chlorine. Therefore, if the chlorine demand and residual requires 50 lbs/day chlorine, the chlorinator setting would be just that—50 lbs/24 hrs.

Many times, however, a chlorine compound called hypochlorite is used to chlorinate water or wastewater. Hypochlorite compounds contain chlorine and are similar to a strong bleach. They are available in liquid form or as powder or granules. Calcium hypochlorite, sometimes referred to as HTH is the most commonly used dry hypochlorite. It contains about 65% available chlorine. Sodium hypochlorite, or liquid bleach, contains about 12-15% available chlorine as commercial bleach or 3-5.25% as household bleach.

Because hypochlorite is not 100% pure chlorine, **more lbs/day must be fed into the system to obtain the same amount of chlorine for disinfection.**

To calculate the lbs/day hypochlorite required:

1. First calculate the lbs/day chlorine required.

$$\underset{\text{Conc.}}{(\text{mg}/L)}\ \underset{\text{flow}}{(\text{MGD})}\ \underset{\text{lbs/gal}}{(8.34)} = \text{lbs/day}$$

2. Then calculate the lbs/day hypochlorite needed by dividing the lbs/day chlorine by the percent available chlorine.

$$\frac{\text{Chlorine, lbs/day}}{\frac{\%\ \text{Available}}{100}} = \text{Hypochlorite lbs/day}$$

Example 9: (Chemical Dosage)

❑ A total chlorine dosage of 12 mg/L is required to treat a particular water. If the flow is 1.2 MGD and the hypochlorite has 65% available chlorine how many lbs/day of hypochlorite will be required?

First, calculate the lbs/day chlorine required using the mg/L to lbs/day equation:

$$\underset{\text{Conc.}}{(\text{mg}/L)}\ \underset{\text{flow}}{(\text{MGD})}\ \underset{\text{lbs/gal}}{(8.34)} = \text{lbs/day}$$

$$(12\ \text{mg}/L)\ (1.2\ \text{MGD})\ (8.34\ \text{lbs/gal}) = \text{lbs/day}$$

$$= \boxed{120\ \text{lbs/day}}$$

Now calculate the lbs/day hypochlorite required. Since only 65% of the hypochlorite is chlorine, more than 120 lbs/day will be required:

$$\frac{120\ \text{lbs/day}\ Cl_2}{\frac{65}{100}\ \text{Avail.}\ Cl_2} = \boxed{\begin{array}{l}185\ \text{lbs/day}\\ \text{Hypochlorite}\end{array}}$$

Example 10: (Chemical Dosage)

❑ A wastewater flow of 850,000 gpd requires a chlorine dose of 25 mg/L. If sodium hypochlorite (15% available chlorine) is to be used, how many lbs/day of sodium hypochlorite are required? How many gal/day of sodium hypochlorite is this?

First, calculate the lbs/day chlorine required:

$$\underset{\text{Conc.}}{(\text{mg}/L)}\ \underset{\text{flow}}{(\text{MGD})}\ \underset{\text{lbs/gal}}{(8.34)} = \text{lbs/day}$$

$$(25\ \text{mg}/L)\ (0.85\ \text{MGD})\ (8.34\ \text{lbs/gal}) = \boxed{\begin{array}{l}177\ \text{lbs/day}\\ \text{Chlorine}\end{array}}$$

Then calculate the lbs/day sodium hypochlorite:

$$\frac{177\ \text{lbs/day}\ Cl_2}{\frac{15}{100}\ \text{Avail.}\ Cl_2} = \boxed{\begin{array}{l}1180\ \text{lbs/day}\\ \text{Hypochlorite}\end{array}}$$

Then calculate the gal/day sodium hypochlorite:

$$\frac{1180\ \text{lbs/day}}{8.34\ \text{lbs/gal}} = \boxed{\begin{array}{l}141\ \text{gal/day}\\ \text{Sodium Hypochlorite}\end{array}}$$

Example 11: (Chemical Dosage)

❑ A flow of 800,000 gpd requires a chlorine dose of 9 mg/*L*. If chlorinated lime (34% available chlorine) is to be used, how many lbs/day of chlorinated lime will be required?

$$\underset{\text{Conc.}}{(\text{mg}/L)}\ \underset{\text{flow}}{(\text{MGD})}\ \underset{\text{lbs/gal}}{(8.34)} = \text{lbs/day}$$

$$(9\ \text{mg}/L)\ (0.8\ \text{MGD})\ (8.34\ \text{lbs/gal}) = \text{lbs/day}$$

$$= \boxed{\begin{array}{l}60\ \text{lbs/day}\\ \text{Chlorine}\end{array}}$$

Then calculate the lbs/day chlorinated lime needed:

$$\frac{60\ \text{lbs/day}\ Cl_2}{\frac{34}{100}\ \text{Avail.}\ Cl_2} = \boxed{\begin{array}{l}176\ \text{lbs/day}\\ \text{Chlorinated Lime}\end{array}}$$

Example 12: (Chemical Dosage)

❑ A small reservoir holds 70 acre-feet of water. To treat the reservoir for algae control, 0.5 mg/*L* of copper is required. How many pounds of copper sulfate will be required if the copper sulfate to be used contains 25% copper?

Before the mg/*L* to lbs equation can be used, the reservoir volume must be converted from ac-ft to cu ft to gal:

$$(70\ \text{ac-ft})\ (43{,}560\ \frac{\text{cu ft}}{\text{ac-ft}}) = 3{,}049{,}200\ \text{cu ft}$$

$$(3{,}049{,}200\ \text{cu ft})\ (7.48\ \frac{\text{gal}}{\text{cu ft}}) = 22{,}808{,}016\ \text{gal}$$

Now calculate the lbs/day copper required:

$$\underset{\text{copper}}{(0.5\ \text{mg}/L)}\ (22.8\ \text{MG})\ (8.34\ \text{lbs/gal}) = \boxed{\begin{array}{l}95\ \text{lbs}\\ \text{Copper}\end{array}}$$

And then the lbs/day copper sulfate required:

$$\frac{95\ \text{lbs Copper}}{\frac{25}{100}\ \text{Avail. Copper}} = \boxed{\begin{array}{l}380\ \text{lbs}\\ \text{Copper Sulfate}\end{array}}$$

OTHER CHEMICAL COMPOUNDS

Other chemical compounds used in water and wastewater treatment are like hypochlorite compounds. For example, chlorinated lime contains only about 34% available chlorine. And copper sulfate pentahydrate contains about 25% copper (the chemical of interest for algae control).

Calculating the lbs or lbs/day of chlorinated lime, copper sulfate, or other similar compound, you follow the same procedure as with the hypochlorite problems:

1. First calculate the lbs/day of chemical desired (such as chlorine or copper). Using the usual mg/*L* to lbs/day or lbs equations:

$$\boxed{\underset{\text{Conc}}{(\text{mg}/L)}\ \underset{\text{flow}}{(\text{MGD})}\ \underset{\text{lbs/gal}}{(8.34)} = \text{lbs/day}}$$

or

$$\boxed{\underset{\text{Conc.}}{(\text{mg}/L)}\ \underset{\text{Vol}}{(\text{MG})}\ \underset{\text{lbs/gal}}{(8.34)} = \text{lbs}}$$

2. Then calculate the lbs/day or lbs compound required:

$$\boxed{\frac{\text{Chemical, lbs/day}}{\frac{\%\ \text{Available}}{100}} = \begin{array}{c}\text{Compound}\\ \text{lbs/day}\end{array}}$$

Examples 11 and 12 illustrate these calculations. Note that in Example 11 a flow is being dosed, and in Example 12 a reservoir is being dosed.

3.2 LOADING CALCULATIONS—BOD, COD AND SS

When calculating BOD (Biochemical Oxygen Demand), COD (Chemical Oxygen Demand), or SS (Suspended Solids) loading on a treatment system, the following equation is used:

(mg/*L*) (MGD) (8.34) = lbs/day
Conc. flow lbs/gal

Loading on a system is usually calculated as lbs/day. Given the BOD, COD, or SS concentration and flow information, the lbs/day loading may be calculated as demonstrated in Examples 1-3.

Example 1: (Loading Calculations)

❑ Calculate the BOD loading (lbs/day) on a stream if the secondary effluent flow is 2.5 MGD and the BOD of the secondary effluent is 20 mg/*L*.

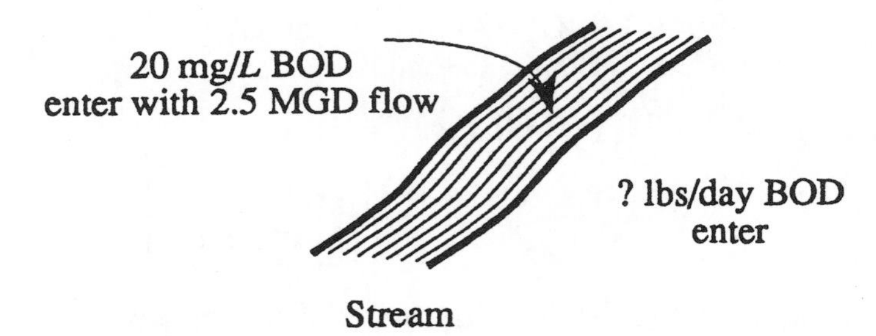

First, select the appropriate equation:

(mg/*L*) (MGD flow) (8.34 lbs/gal) = lbs/day

Then fill in the information given in the problem:

(20 mg/*L*) (2.5 MGD) (8.34 lbs/gal) = **417 lbs/day BOD**

Example 2: (Loading Calculations)

❑ The suspended solids concentration of the wastewater entering the primary system is 480 mg/*L*. If the plant flow is 3,600,000 gpd, how many lbs/day suspended solids enter the primary system?

480 mg/*L* SS
enter with 3.6 MGD flow

Primary System

? lbs/day SS
enter

First write the equation:

(mg/L) (MGD flow) (8.34 lbs/gal) = lbs/day

Then fill in the data given in the problem:

(480 mg/L) (3.6 MGD) (8.34 lbs/gal) = **14,412 lbs/day SS**

Example 3: (Loading Calculations)

❑ The flow to an aeration tank is 7 MGD. If the COD concentration of the water is 110 mg/*L*, how many pounds of COD are applied to the aeration tank daily?

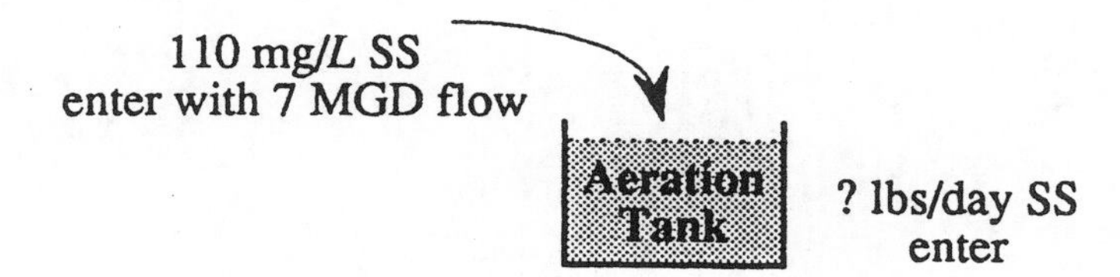

Use the mg/*L* to lbs/day equation to solve the problem:

(mg/*L*) (MGD flow) (8.34 lbs/gal) = lbs/day

(110 mg/*L*) (7 MGD) (8.34 lbs/gal) = **6422 lbs/day COD**
COD

Example 4: (Loading Calculations)

❑ The daily flow to a trickling filter is 4,500,000 gpd. If the BOD concentration of the trickling filter influent is 213 mg/*L*, how many lbs BOD enter the trickling filter daily?

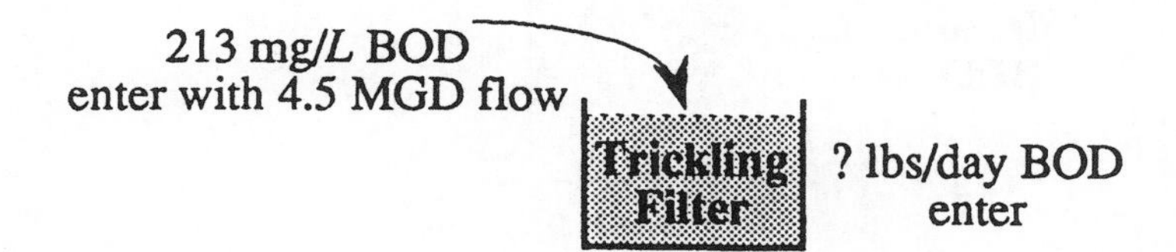

Write the equation, fill in the given information, then solve for the unknown value:

(mg/*L*) (MGD flow) (8.34 lbs/gal) = lbs/day

(213 mg/*L*) (4.5 MGD) (8.34 lbs/gal) = **7994 lbs/day BOD**

3.3 BOD AND SS REMOVAL CALCULATIONS, lbs/day

To calculate the pounds of BOD or suspended solids removed each day, you will need to know the mg/*L* BOD or SS removed and the plant flow. Then you can use the mg/*L* to lbs/day equation:

(mg/*L*) Removed	(MGD) flow	(8.34) lbs/gal	= lbs/day

For most calculations of BOD or SS removal, you will not be given information stating how many mg/*L* BOD or SS have been removed. This is something you will calculate based on the mg/*L* concentrations entering (influent) and leaving (effluent) the system.

The influent BOD or SS concentration indicates how much BOD or SS is entering the system. The effluent concentration indicates how much is still in the wastewater (the part not removed). The mg/*L* SS or BOD removed would therefore be:

$$\text{Influent SS mg/}L - \text{Effluent SS mg/}L = \text{Removed SS mg/}L$$

Once you have determined the mg/*L* BOD or SS removed, you can then continue with the usual mg/*L* to lbs/day equation to calculate lbs/day BOD or SS removed. Examples 2-4 illustrate this calculation.

Example 1: (BOD and SS Removal)

❑ If 130 mg/*L* suspended solids are removed by a primary clarifier, how many lbs/day suspended solids are removed when the flow is 7.4 MGD?

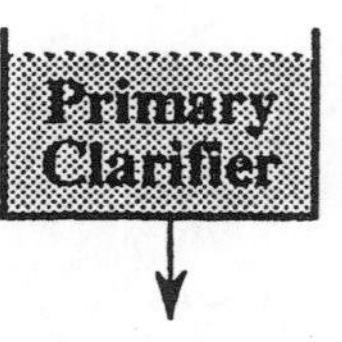

130 mg/*L*
SS Removed

(mg/*L*) (MGD flow) (8.34 lbs/gal) = lbs/day

(130 mg/*L*) (7.4 MGD) (8.34 lbs/gal) = **8023 lbs/day SS Removed**

Example 2: (BOD and SS Removal)

❑ The flow to a trickling filter is 3.7 MGD. If the primary effluent has a BOD concentration of 180 mg/*L* and the trickling filter effluent has a BOD concentration of 28 mg/*L*, how many pounds of BOD are removed daily?

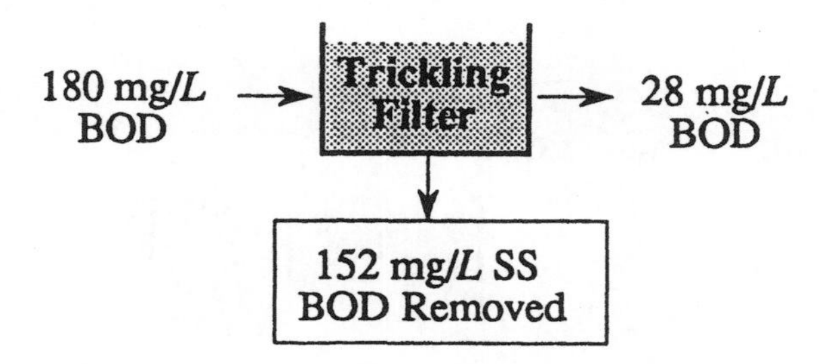

After calculating mg/*L* BOD removed, you can now calculate lbs/day BOD removed:

(mg/*L* Removed) (MGD flow) (8.34 lbs/gal) = lbs/day Removed

(152 mg/*L*) (3.7 MGD) (8.34 lbs/gal) = **4690 lbs/day BOD Removed**

Example 3: (BOD and SS Removal)

❑ The flow to a primary clarifier is 2.7 MGD. If the influent to the clarifier has a suspended solids concentration of 230 mg/*L* and the primary effluent has 110 mg/*L* SS, how many lbs/day suspended solids are removed by the clarifier?

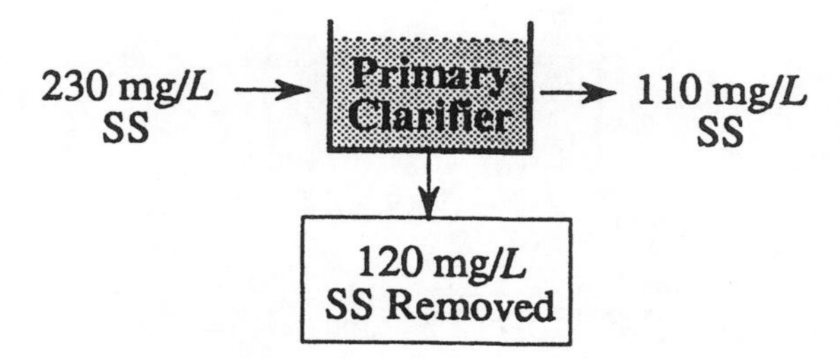

Now calculate lbs/day SS removed:

(mg/*L*) (MGD flow) (8.34 lbs/gal) = lbs/day

(120 mg/*L*) (2.7 MGD) (8.34 lbs/gal) = **2702 lbs/day SS Removed**

Example 4: (BOD and SS Removal)

❑ The flow to a trickling filter is 4,600,000 gpd, with a BOD concentration of 195 mg/*L*. If theBOD of the trickling filter effluent is 98 mg/*L*, how many lbs/day BOD are removed by the trickling filter ?

195 mg/*L* BOD → Trickling Filter → 98 mg/*L* BOD

97 mg/*L* SS BOD Removed

Now calculate lbs/day BOD removed:

(mg/*L*) (MGD flow) (8.34 lbs/gal) = lbs/day

(97 mg/*L*) (4.6 MGD) (8.34 lbs/gal) = **3721 lbs/day BOD Removed**

3.4 POUNDS OF SOLIDS UNDER AERATION CALCULATIONS

In any activated sludge system it is important to control the amount of solids under aeration (solids inventory). The suspended solids in an aerator are called Mixed Liquor Suspended Solids (MLSS). To calculate the pounds of suspended solids in the aeration tank, you will need to know the mg/*L* concentration of the MLSS. Then mg/*L* MLSS can be expressed as lbs MLSS, using the mg/*L* to lbs equation:

(mg/*L*) MLSS	(MG) Vol	(8.34) lbs/gal	= lbs

Notice that the mixed liquor suspended solids concentration is **concentration within a tank.** Therefore, the equation using **MG volume** is used.

Another important measure of solids in the aeration tank is the amount of volatile suspended solids.* The volatile solids content of the aeration tank is used as an estimate of the microorganism population of the aeration tank. The Mixed Liquor Volatile Suspended Solids (MLVSS) usually comprises about 70% of the MLSS. The other 30% of the MLSS are fixed (inorganic) solids. To calculate the lbs MLVSS, use the mg/*L* to lbs equation:

(mg/*L*) MLVSS	(MG) Vol	(8.34) lbs/gal	= lbs

Example 1: (lbs Solids Under Aeration)

❑ An aeration tank has a volume of 450,000 gallons. If the mixed liquor suspended solids are 1820 mg/*L*, how many pounds of suspended solids are in the aerator?

Aeration Tank
1820 mg/*L*
MLSS

Vol = 0.45 MG

(mg/*L*) (MG vol) (8.34 lbs/gal) = lbs

(1820 mg/*L*) (0.45 MG) (8.34 lbs/gal) = **6830 lbs MLSS**

Example 2: (lbs Solids Under Aeration)

❑ An oxidation ditch contains 23,040 cubic feet of wastewater. If the MLVSS concentration is 3800 mg/*L*, how many pounds of volatile suspended solids are under aeration?

Before the mg/*L* to lbs equation can be used cubic feet oxidation ditch volume must be converted to million gallons:

Oxidation Ditch
3800 mg/*L*

$$\text{Vol} = (23{,}040 \text{ cu ft}) \left(7.48 \frac{\text{gal}}{\text{cu ft}}\right)$$

$$= 172{,}339 \text{ gal}$$

$$\text{or} = 0.17 \text{ MG}$$

The lbs volatile solids calculation can now be completed:

(3800 mg/*L*) (0.17 MG vol) (8.34 lbs/gal) = **5388 lbs MLVSS**

---

* For a discussion of volatile suspended solids calculations, refer to Chapter 6, Efficiency and Other Percent Calculations.

Example 3: (lbs Solids Under Aeration)

❑ The aeration tank of a conventional activated sludge plant has a mixed liquor volatile suspended solids concentration of 2300 mg/L. If the aeration basin is 110 ft long, 35 ft wide and has wastewater to a depth of 13 ft, how many pounds of MLVSS are under aeration?

Aeration Tank
2300 mg/*L*

$$\text{Vol} = (110 \text{ ft}) (35 \text{ ft}) (13 \text{ ft}) \left(7.48 \frac{\text{gal}}{\text{cu ft}}\right)$$

$$= 374,374 \text{ gal}$$

$$\text{or} = 0.37 \text{ MG}$$

Now calculate lbs MLVSS using the usual equation and fill in the given information:

$$(2300 \text{ mg/}L) (0.37 \text{ MG}) (8.34 \text{ lbs/gal}) = \boxed{7097 \text{ lbs MLVSS}}$$

Example 4: (lbs Solids Under Aeration)

❑ An aeration tank is 90 ft long and 40 ft wide. The depth of wastewater in the tank is 16 ft. If the concentration of MLSS is 1980 mg*L*, how many pounds of MLSS are under aeration?

Aeration Tank
1980 mg/*L*

$$\text{Vol} = (90 \text{ ft}) (40 \text{ ft}) (16 \text{ ft}) \left(7.48 \frac{\text{gal}}{\text{cu ft}}\right)$$

$$= 430,848 \text{ gal}$$

$$\text{or} = 0.43 \text{ MG}$$

Now fill in the mg/*L* to lbs equation with known information:

$$(1980 \text{ mg/}L) (0.43 \text{ MG}) (8.34 \text{ lbs/gal}) = \boxed{7101 \text{ lbs MLSS}}$$

3.5 WAS PUMPING RATE CALCULATIONS

Waste Activated Sludge (WAS) pumping rate calculations are calculations that involve mg/*L* and flow. Therefore the equation used in these calculations is:

(mg/*L*) (MGD) (8.34) = lbs/day flow lbs/gal

In WAS pumping rate calculations, the "WAS SS" refers to the suspended solids content of the Waste Activated Sludge being pumped away, and "MGD flow" refers to the WAS pumping rate of the sludge being wasted.

Sometimes waste activated sludge SS is not known but return activated sludge SS is known. Remember that **RAS SS and WAS SS are the same measurement**. It is a measurement taken of secondary clarifier sludge. This sludge is either pumped back to the aerator (RAS) or wasted (WAS).

WAS PUMPING RATE CALCULATIONS ARE mg/*L* TO lbs/day PROBLEMS:

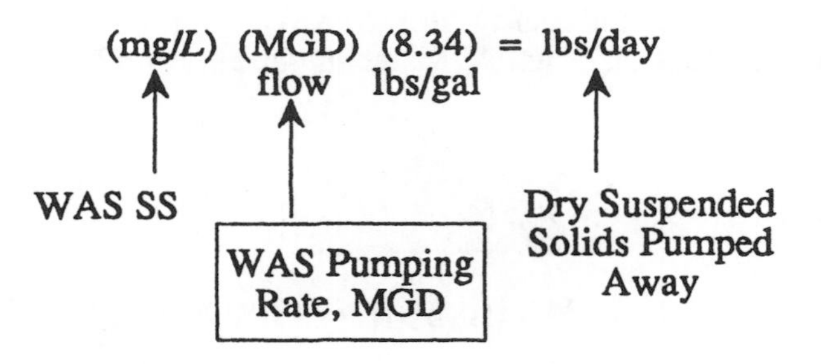

Example 1: (WAS Pumping Rate Calculations)

❑ The WAS suspended solids concentration is 5860 mg/*L*. If 3800 lbs/day dry solids are to be wasted, (a) What must the WAS pumping rate be, in MGD? (b) What is this rate expressed in gpm?

(a) First calculate the MGD pumping rate required, using the mg/*L* to lbs/day equation:

$$(\text{mg}/L)\ (\text{MGD flow})\ (8.34\ \text{lbs/gal}) = \text{lbs/day}$$

$$\underset{\text{mg}/L}{(5860)}\ \underset{\text{flow}}{(x\ \text{MGD})}\ \underset{\text{lbs/day}}{(8.34)} = 3800\ \text{lbs/day}$$

$$x = \frac{3800\ \text{lbs/day}}{\underset{\text{mg}/L}{(5600)}\ \underset{\text{lbs/gal}}{(8.34)}}$$

$$x = 0.0814\ \text{MGD}$$

(b) Then convert the MGD flow to gpm flow: *

$$0.0814\ \text{MGD} = 81{,}400\ \text{gpd}$$

$$= \frac{81{,}400\ \text{gpd}}{1440\ \text{min/day}}$$

$$= \boxed{57\ \text{gpm}}$$

* Refer to Chapter 8 in *Basic Math Concepts* for a review of flow conversions.

Example 2: (WAS Pumping Rate Calculations)

❑ It has been determined that 4700 lbs/day of dry solids must be removed from the secondary system. If the WAS SS concentration is 7340 mg/*L*, what must be the WAS pumping rate, in gpm?

First calculate the MGD pumping rate required:

$$\underset{mg/L}{(7340)}\ \underset{flow}{(x\ MGD)}\ \underset{lbs/day}{(8.34)} = 4700\ lbs/day$$

$$x = \frac{4700\ lbs/day}{\underset{mg/L}{(7340)}\ \underset{lbs/gal}{(8.34)}}$$

$$x = 0.0768\ MGD$$

Then convert MGD pumping rate to gpm pumping rate:

$$0.0768\ MGD = 76{,}800\ gpd$$

$$= \frac{76{,}800\ gpd}{1440\ min/day}$$

$$= \boxed{53\ gpm}$$

Example 3: (WAS Pumping Rate Calculations)

❑ The WAS suspended solids concentration is 6980 mg/*L*. If 5300 lbs/day dry sludge solids are to be wasted, what must be the WAS pumping rate, in gpm?

First calculate the MGD pumping rate required:

$$(mg/L)\ (MGD\ flow)\ (8.34\ lbs/gal) = lbs/day$$

$$\underset{mg/L}{(6980)}\ \underset{flow}{(x\ MGD)}\ \underset{lbs/day}{(8.34)} = 5300\ lbs/day$$

$$x = \frac{5300\ lbs/day}{\underset{mg/L}{(6980)}\ \underset{lbs/gal}{(8.34)}}$$

$$x = 0.091\ MGD$$

Then convert the MGD flow to gpm flow:

$$0.091\ MGD = 91{,}000\ gpd$$

$$= \frac{91{,}000\ gpd}{1440\ min/day}$$

$$= \boxed{63\ gpm}$$

NOTES:

4 Loading Rate Calculations

SUMMARY

1. The **hydraulic loading rate** of a treatment system is a measure of the flow treated per square foot of surface area. The most common expression of hydraulic loading rate is gpd/sq ft. Recirculated flows are included as part of the gpd flow to the system.

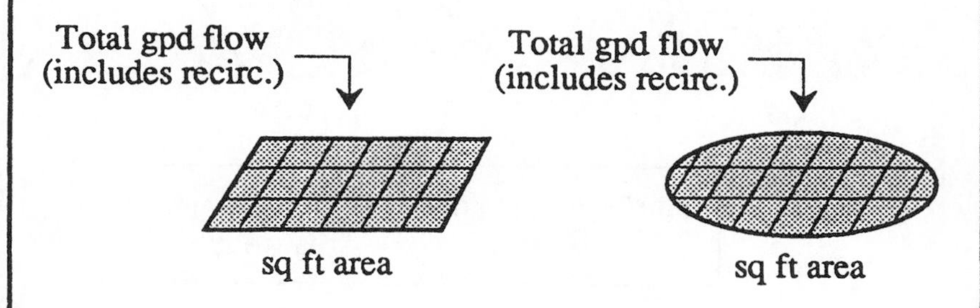

$$\text{Hydraulic Loading Rate} = \frac{\text{Flow, gpd}}{\text{Area, sq ft}}$$

Hydraulic loading rate for ponds is generally expressed as inches/day:

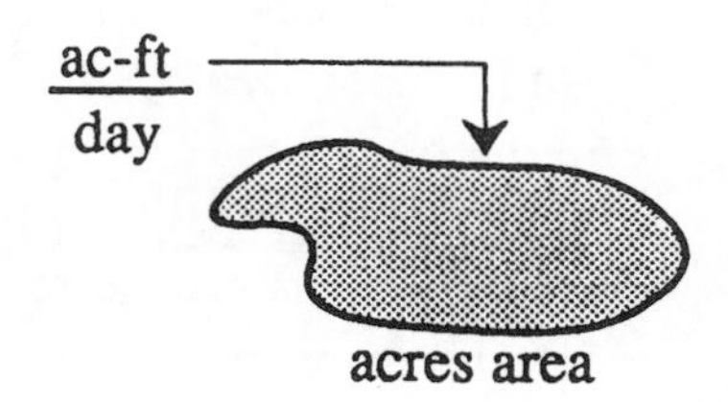

$$\text{Hydraulic Loading Rate} = \frac{\text{Flow, ac-ft/day}}{\text{Area, ac}}$$

$$= \text{ft/day}$$

Then, multiplying by 12 in./ft, the hydraulic loading rate can be expressed in in./day.

$$\text{Hydraulic Loading Rate} = \frac{\text{in.}}{\text{day}}$$

There are several calculations that measure the water and solids loading on the treatment system. When the water and solids loading consistently exceed design values, the efficiency of the treatment system begins to deteriorate.

Calculations that reflect various types of **water loading** on the system include:

- Hydraulic Loading Rate (gpd/sq ft)
- Surface Loading Rate (gpd/sq ft, gpm/sq ft or in./day)
- Filtration Rate (gpm/sq ft or in./min)
- Backwash Rate (gpm/sq ft or in./min)
- Unit Filter Run Volume, UFRV, (gal/sq ft)
- Weir Overflow Rate (gpd/ft)

Calculations that reflect various types of **solids loading** are:

- Organic Loading Rate (lbs BOD/day/1000 cu ft)
- Food/Microorganism Ratio (lbs BOD/day/lb MLVSS)
- Solids Loading Rate (lbs SS/day/sq ft)
- Digester Loading Rate (lbsVS/day/cu ft)
- Digester Volatile Solids Loading Ratio (lbs VS/day Added/lb VS in Digester)
- Population Loading and Population Equivalent

SUMMARY—Cont'd

2. **Surface overflow rate** is a calculation similar to hydraulic loading rate—flow rate per unit area. The difference between these calculations pertains to recirculation rates. Recirculation is not included in the surface overflow rate calculation.

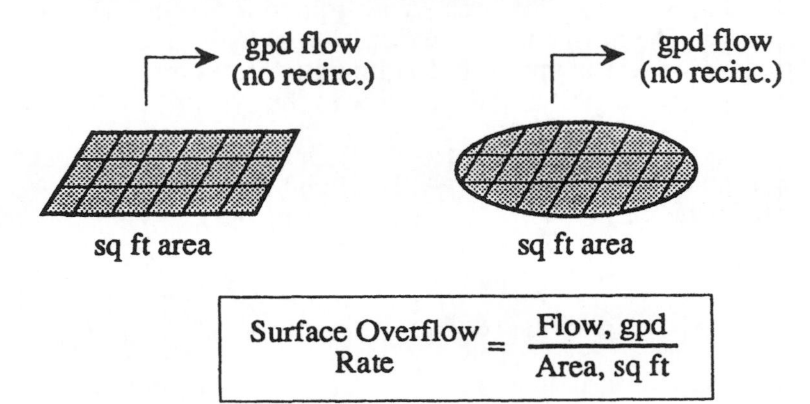

3. **Filtration rate** is a measure of the gallons per minute flow filtered by each square foot of filter.

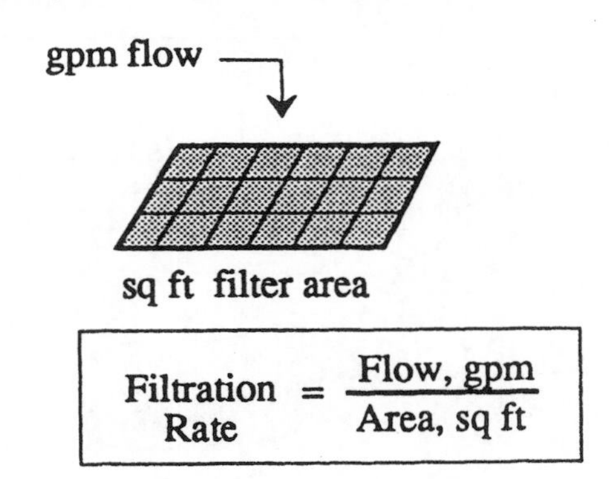

4. **Backwash rate** is a very similar calculation to filtration rate. It is the gallons per minute of backwash water flowing through each square foot of filter area.

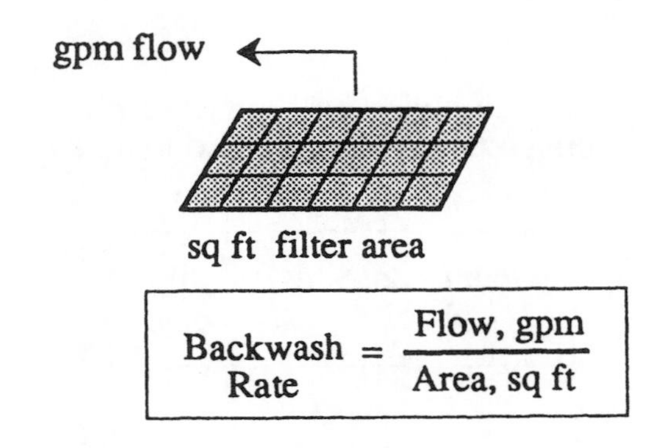

SUMMARY—Cont'd

5. **Unit filter run volume** (UFRV) is a measure of the total gallons of water filtered by each square foot of filter surface area during a filter run.

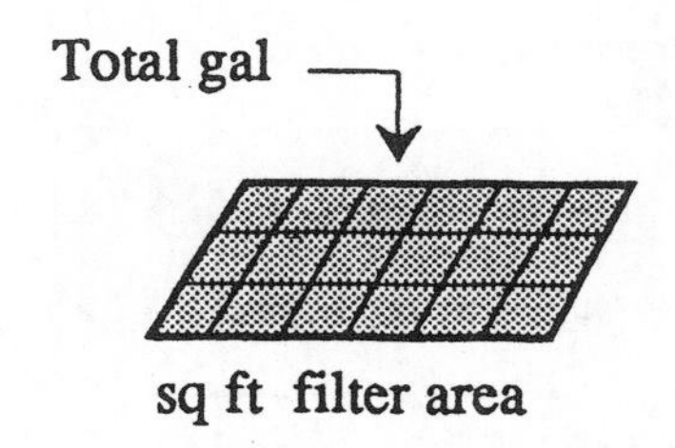

$$\text{UFRV} = \frac{\text{Total gal}}{\text{Area, sq ft}}$$

6. The **weir overflow rate** is a measure of the gallons per day flowing over each foot of weir.

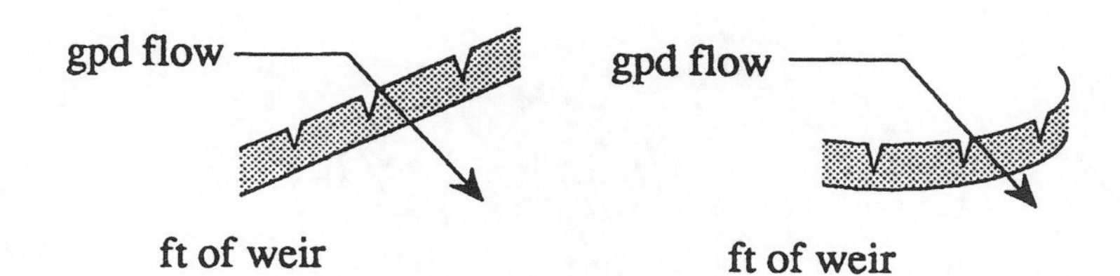

$$\text{Weir Overflow Rate} = \frac{\text{Flow, gpd}}{\text{ft of weir}}$$

7. The **organic loading rate** on a system is the pounds per day of BOD applied to each 1000 cu ft volume.

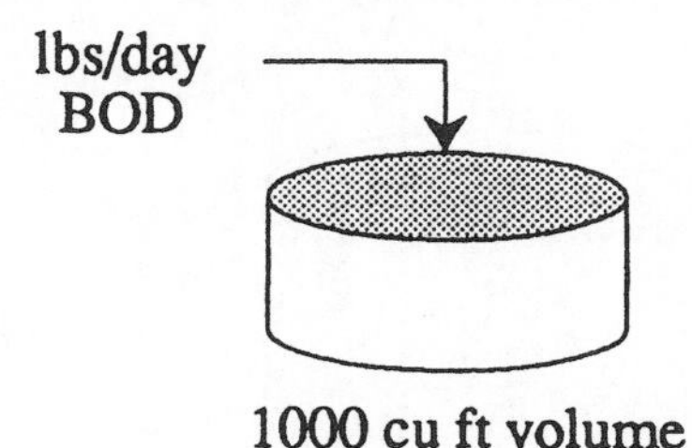

$$\text{Organic Loading} = \frac{\text{BOD, lbs/day}}{\text{Volume, 1000 cu ft}}$$

SUMMARY—Cont'd

8. The **food/microorganism ratio** indicates the relative balance between the food entering the secondary system and the microorganisms present.

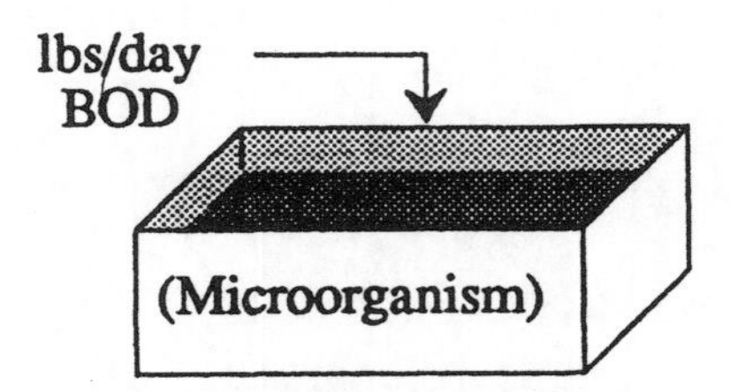

$$F/M = \frac{\text{BOD entering, lbs/day}}{\text{MLVSS, lbs}}$$

9. **Solids loading rate** indicates the pounds of solids that are loaded daily per square foot of secondary clarifier surface area.

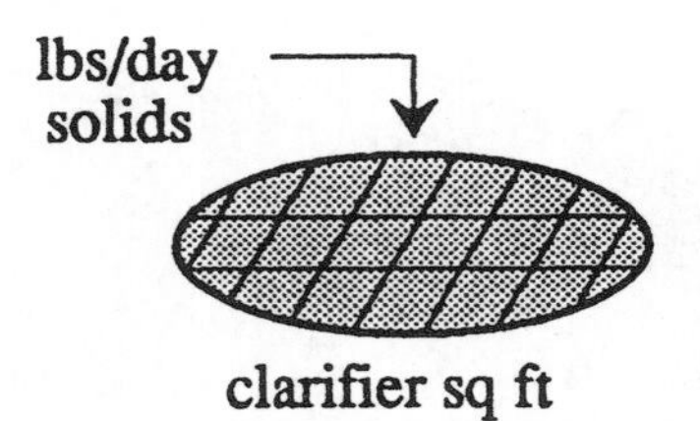

$$\text{Solids Loading Rate} = \frac{\text{Solids, lbs/day}}{\text{Area, sq ft}}$$

10. The **digester loading rate** measures the lbs/day volatile solids entering the digester per cubic foot of digester volume.

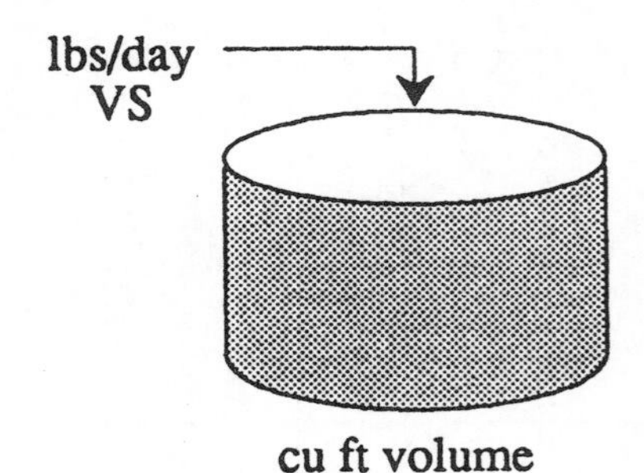

$$\text{Digester Loading Rate} = \frac{\text{VS, lbs/day}}{\text{Dig. Volume, cu ft}}$$

SUMMARY—Cont'd

11. The **digester volatile solids loading ratio** is used to determine the seed sludge required in the digester. It can also be used to determine the balance between volatile solids added to the digester and the volatile solid in the digester.

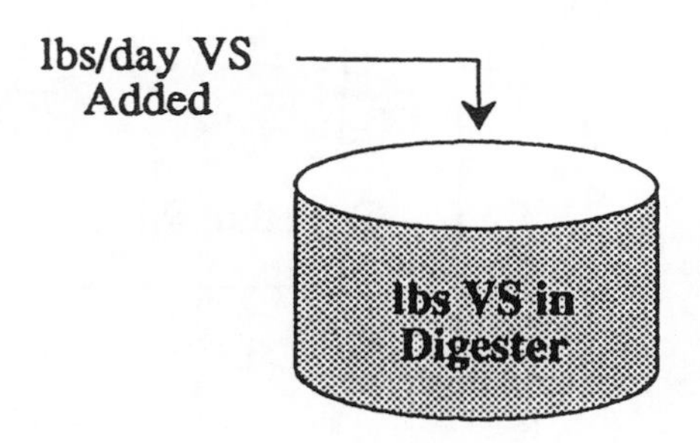

$$\text{Volatile Solids Loading Ratio} = \frac{\text{VS Added, lbs/day}}{\text{VS in Digester, lbs}}$$

12. **Population loading** is a calculation most often associated with wastewater ponds. It is the number of people served per acre of pond area.

$$\text{Population Loading} = \frac{\text{People Served by the System}}{\text{Pond Area, acres}}$$

Population equivalent is a calculation that expresses the organic content of a wastewater (BOD) in terms of an equivalent number of people using the system. This calculation assumes that for each person using the system, about 0.2 lbs/day BOD enter the system.

$$\text{Population Equivalent} = \frac{(\text{mg}/L\ \text{BOD})\ (\text{MGD flow})\ (8.34\ \text{lbs/gal})}{0.2\ \text{lbs/day BOD/person}}$$

4.1 HYDRAULIC LOADING RATE CALCULATIONS

Hydraulic loading rate is a term used to indicate the total flow, in gpd, loaded or entering each square foot of water surface area. It is the total gpd flow to the process divided by the water surface area of the tank or pond.

As shown in the diagram to the right **recirculated flows must be included** as part of the total flow (total Q) to the process.

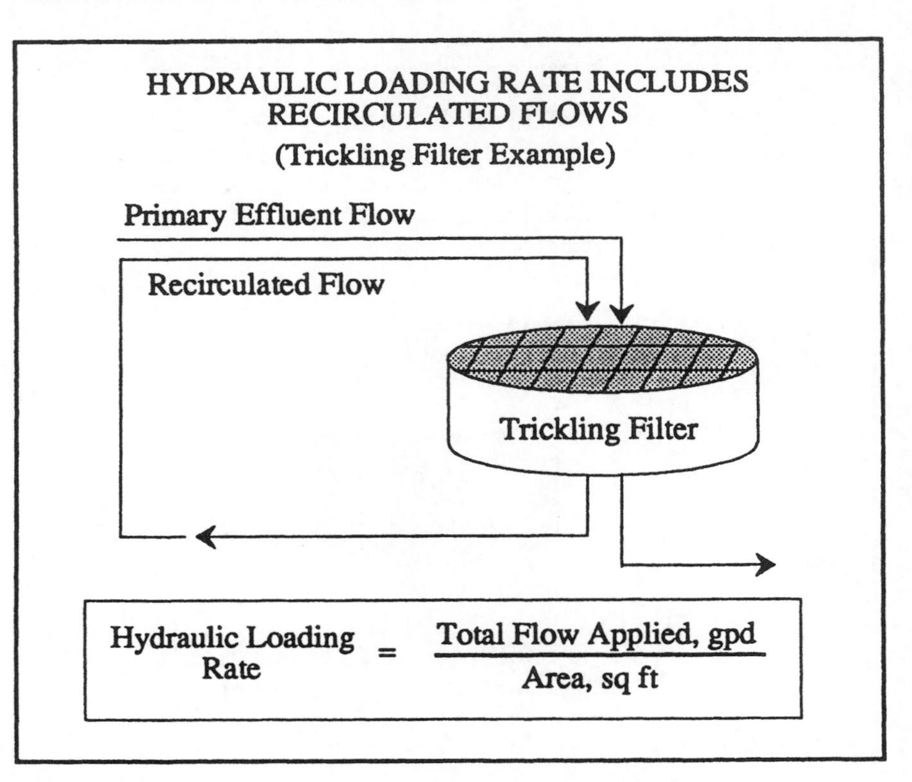

Example 1: (Hydraulic Loading)

❑ A trickling filter 80 ft in diameter treats a primary effluent flow of 1.8 MGD. If the recirculated flow is 0.3 MGD, what is the hydraulic loading on the trickling filter?

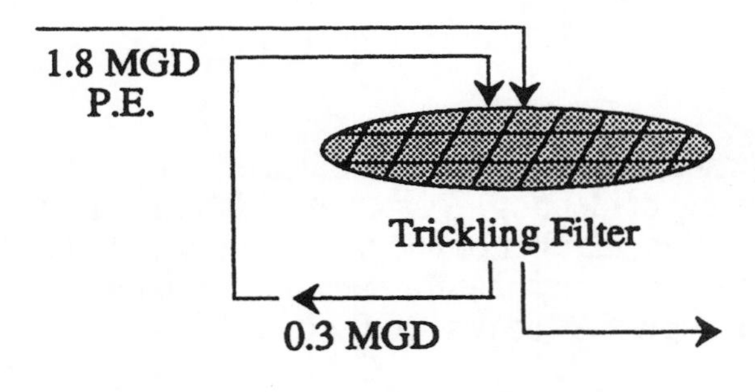

$$\text{Hydraulic Loading Rate} = \frac{\text{Flow, gpd}}{\text{Area, sq ft}}$$

$$= \frac{2{,}100{,}000 \text{ gpd}}{(0.785)\ (80 \text{ ft})\ (80 \text{ ft})}$$

$$= \boxed{418 \text{ gpd/sq ft}}$$

Example 2: (Hydraulic Loading)
❑ If 50,000 gpd are pumped to a 30 ft diameter gravity thickener, what is the hydraulic loading rate on the thickener?

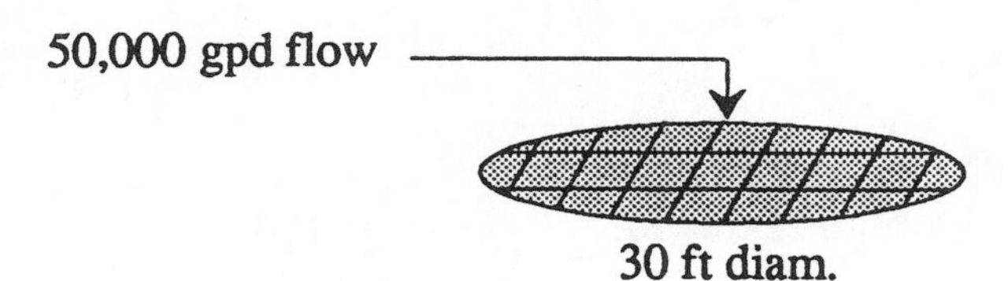

$$\text{Hydraulic Loading Rate} = \frac{\text{Flow, gpd}}{\text{Area, sq ft}}$$

$$= \frac{50{,}000 \text{ gpd}}{(0.785)\ (30 \text{ ft})\ (30 \text{ ft})}$$

$$= \boxed{71 \text{ gpd/sq ft}}$$

WHEN THERE IS NO RECIRCULATION

When there is no recirculated flow, the total flow applied is simply the flow to the unit process.

Example 3: (Hydraulic Loading)
❑ A rotating biological contactor treats a flow of 2.8 MGD. The manufacturer's data indicates a media surface area of 800,000 sq ft. What is the hydraulic loading rate on the RBC?

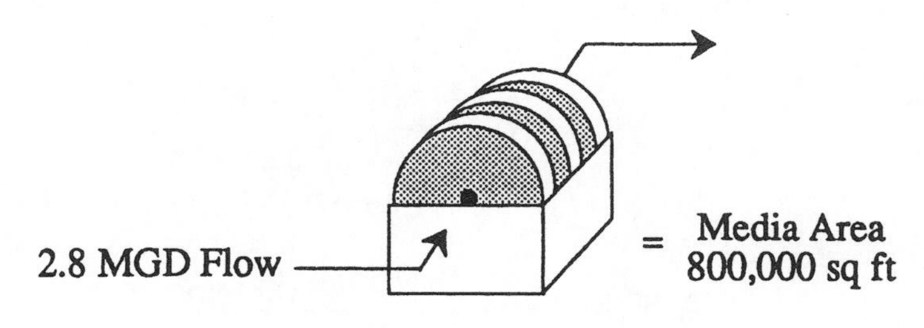

$$\text{Hydraulic Loading Rate} = \frac{\text{Flow, gpd}}{\text{Area, sq ft}}$$

$$= \frac{2{,}800{,}000 \text{ gpd}}{800{,}000 \text{ sq ft}}$$

$$= \boxed{3.5 \text{ gpd/sq ft}}$$

HYDRAULIC LOADING FOR ROTATING BIOLOGICAL CONTACTORS (RBC)

When calculating the hydraulic loading rate on a rotating biological contactor, use the **sq ft area of the media** rather than the sq ft area of the water surface, as in other hydraulic loading calculations. The RBC manufacturer provides media area information.

HYDRAULIC LOADING FOR PONDS

When calculating hydraulic loading for wastewater ponds, the answer is generally expressed as in./day rather than gpd/sq ft.

There are two ways to calculate in./day hydraulic loading, depending on how the flow to the pond is expressed.

HYDRAULIC LOADING RATE FOR PONDS, in./day

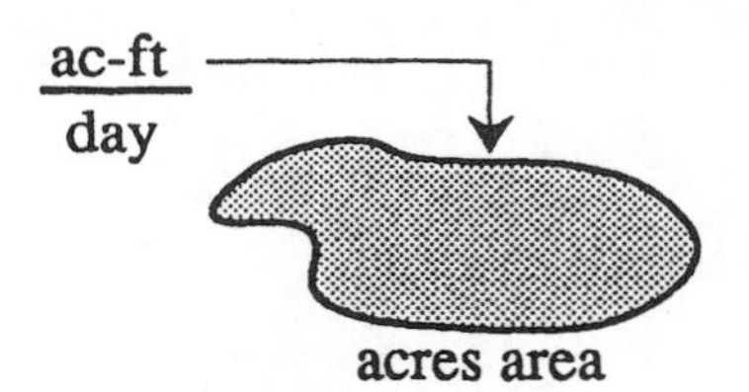

$$\text{Hydraulic Loading Rate} = \frac{\text{Flow, ac-ft/day}}{\text{Area, ac}}$$

$$= \text{ft/day}$$

Then, by multiplying by 12 in./ft, the hydraulic loading rate can be expressed in in./day.

$$\text{Hydraulic Loading Rate} = \frac{\text{in.}}{\text{day}}$$

If the flow to the pond is expressed in gpd:

1. Set up the hydraulic loading equation as usual.
2. Convert gpd flow to cubic feet per day flow (cfd or ft^3/day). This is done by dividing gpd by 7.48 gal/cu ft.*
3. Cancel terms to obtain ft/day hydraulic loading.**

   $$\frac{\text{cfd}}{\text{sq ft}} = \frac{\text{ft}^3\text{/day}}{\text{ft}^2} = \text{ft/day}$$

4. Then convert ft/day to in./day by multiplying by 12 in./ft.

Example 4: (Hydraulic Loading)

❑ A pond receives a flow of 1,980,000 gpd. If the surface area of the pond is 700,000 sq ft, what is the hydraulic loading in in./day?

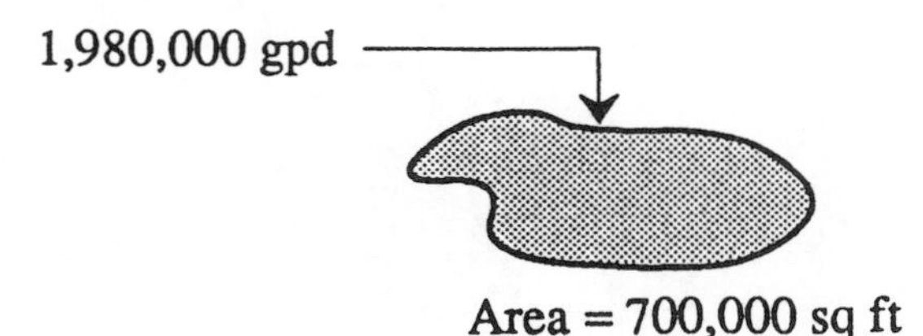

$$\text{Hydraulic Loading Rate} = \frac{\text{1,980,000 gpd}}{\text{700,000 sq ft}}$$

Convert gpd flow to ft^3/day flow (1,980,000 gpd ÷ 7.48 gal/cu ft):

$$= \frac{\text{264,706 ft}^3\text{/day}}{\text{700,000 ft}^2}$$

$$= \text{0.4 ft/day}$$

Then convert to in/day:

(0.4 ft/day) (12 in./ft) 5 in./day

* For a review of flow conversions, refer to Chapter 8 in *Basic Math Concepts*.

** To review cancellation of terms, refer to Chapter 15, Dimensional Analysis, in *Basic Math Concepts*.

Example 5: (Hydraulic Loading)
❑ A pond receives a flow of 2,400,000 gpd. If the surface area of the pond is 15 acres, what is the hydraulic loading in in/day?

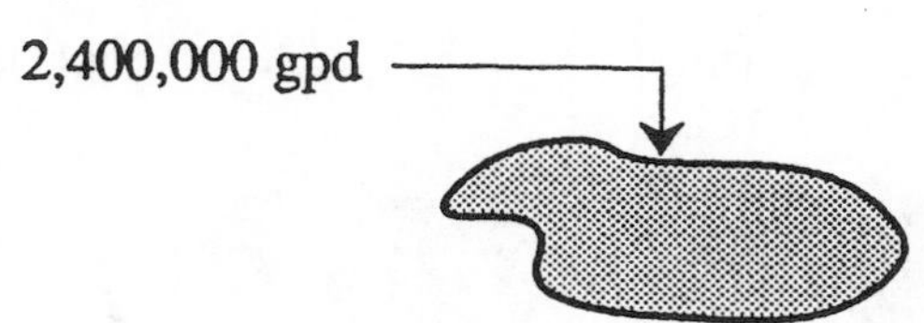

Area = 15 ac

or (15 ac) (43,560 sq ft/ac)

= 653,400 sq ft

$$\text{Hydraulic Loading Rate} = \frac{2{,}400{,}000 \text{ gpd}}{653{,}400 \text{ sq ft}}$$

Convert gpd flow to ft^3/day flow (2,400,000 gpd ÷ 7.48 gal/cu ft):

$$= \frac{320{,}856 \text{ ft}^3\text{/day}}{653{,}400 \text{ ft}^2}$$

$$= 0.5 \text{ ft/day}$$

Then convert to in/day:

(0.5 ft/day) (12 in./ft) = **6 in./day**

Example 6: (Hydraulic Loading)
❑ A 25-acre pond receives a flow of 6.2 acre-feet/day. What is the hydraulic loading on the pond in in/day?

Use the equation for hydraulic loading which includes acre-ft/day flow:

$$\text{Hydraulic Loading Rate} = \frac{6.2 \text{ ac-ft/day}}{25 \text{ ac}}$$

$$= 0.25 \text{ ft/day}$$

Then convert ft/day to in./day:

(0.25 ft/day) (12 in./ft) = **3 in./day**

If the flow to the pond is expressed in acre-feet/day:

1. Set up the hydraulic loading equation in a slightly different form. Instead of gpd/sq ft, use the form of acre-ft/day flow per acres area.

2. Canceling terms results in ft/day hydraulic loading.

$$\frac{\text{Flow, ac-ft/day}}{\text{Area, ac}} = \text{ft/day}$$

3. Then convert ft/day to in./day.

This calculation is illustrated in Example 6.

4.2 SURFACE OVERFLOW RATE CALCULATIONS

Surface overflow rate is used to determine loading on clarifiers. It is similar to hydraulic loading rate—flow per unit area. However, hydraulic loading rate measures the total water entering the process (plant flow plus recirculation) whereas **surface overflow rate measures only the water overflowing the process (plant flow only).**

As indicated in the diagram to the right, **surface overflow rate calculations do not include recirculated flows.** This is because recirculated flows are taken from the bottom of the clarifier and hence do not flow up and out of the clarifier (overflow).

Since surface overflow rate is a measure of flow (Q) divided by area (A), surface overflow is an indirect measure of the **upward velocity** of water as it overflows the clarifier:*

$$V = \frac{Q}{A}$$

This calculation is important in maintaining proper clarifier operation since settling solids will be drawn upward and out of the clarifier if surface overflow rates are too high.

Other terms used synonymously with surface overflow rate are:

- Surface Loading Rate, and
- Surface Settling Rate

SURFACE OVERFLOW RATE DOES NOT INCLUDE RECIRCULATED FLOWS

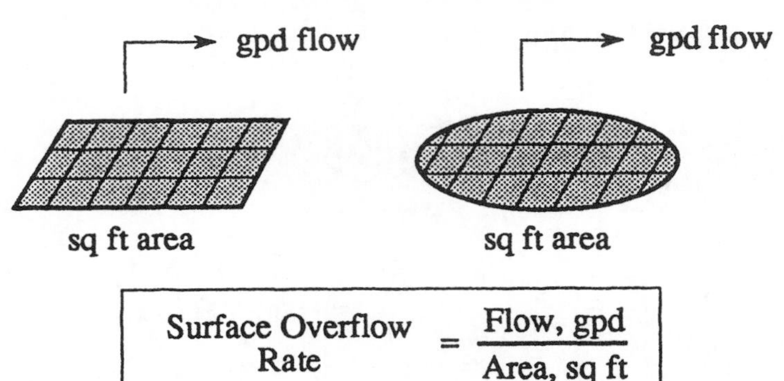

$$\text{Surface Overflow Rate} = \frac{\text{Flow, gpd}}{\text{Area, sq ft}}$$

Surface overflow rate for wastewater calculations is normally expressed as gpd/sq ft, as shown above. However this calculation for water systems is often expressed as gpm/sq ft.

$$\text{Surface Overflow Rate} = \frac{\text{Flow, gpm}}{\text{Area, sq ft}}$$

Example 1: (Surface Overflow Rate)

❑ A circular clarifier has a diameter of 60 ft. If the primary effluent flow is 2.3 MGD, what is the surface overflow rate in gpd/sq ft?

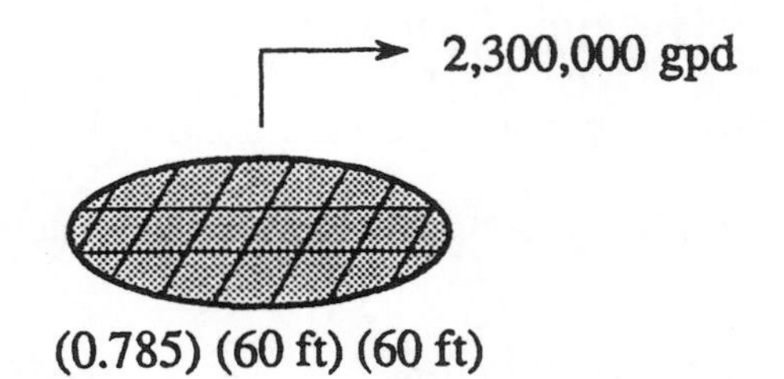

$$\text{Surface Overflow Rate} = \frac{\text{Flow, gpd}}{\text{Area, sq ft}}$$

$$= \frac{2{,}300{,}000 \text{ gpd}}{(0.785)(60 \text{ ft})(60 \text{ ft})}$$

$$= \boxed{814 \text{ gpd/sq ft}}$$

* Refer to Chapter 2 for a review of $Q = AV$ problems.

Example 2: (Surface Overflow Rate)

❑ A sedimentation basin 75 ft by 20 ft receives a flow of 1.3 MGD. What is the surface overflow rate in gpd/sq ft?

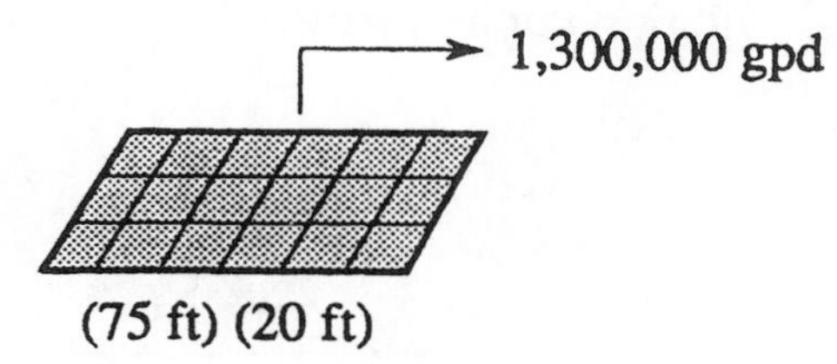

$$\text{Surface Overflow Rate} = \frac{\text{Flow, gpd}}{\text{Area, sq ft}}$$

$$= \frac{1{,}300{,}000 \text{ gpd}}{(75 \text{ ft})\ (20 \text{ ft})}$$

$$= \boxed{867 \text{ gpd/sq ft}}$$

Example 3: (Surface Overflow Rate)

❑ The flow to a sedimentation tank is 3.2 MGD. If the length of the basin is 90 ft and the width is 45 ft, what is the surface overflow rate in gpd/sq ft?

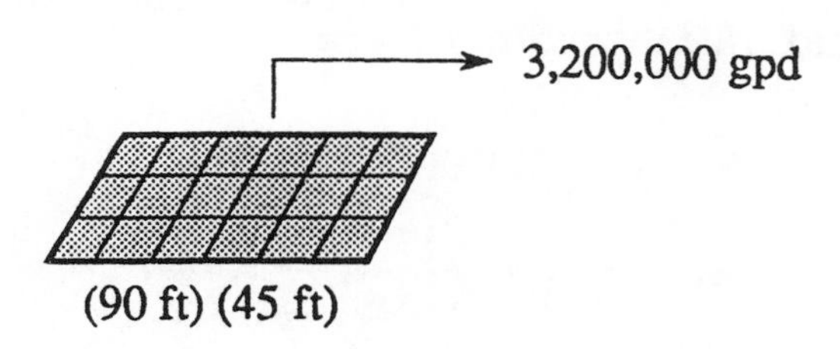

$$\text{Surface Overflow Rate} = \frac{\text{Flow, gpd}}{\text{Area, sq ft}}$$

$$= \frac{3{,}200{,}000 \text{ gpd}}{(90 \text{ ft})\ (45 \text{ ft})}$$

$$= \boxed{790 \text{ gpd/sq ft}}$$

4.3 FILTRATION RATE CALCULATIONS

The calculation of filtration rate or filter loading rate is similar to that of hydraulic loading rate. It is gpm filtered by each square foot of filter area.

$$\text{Filtration Rate} = \frac{\text{Flow, gpm}}{\text{Area, sq ft}}$$

Example 1: (Filtration Rate)

❑ A filter 20 ft by 25 ft receives a flow of 1940 gpm. What is the filtration rate in gpm/sq ft?

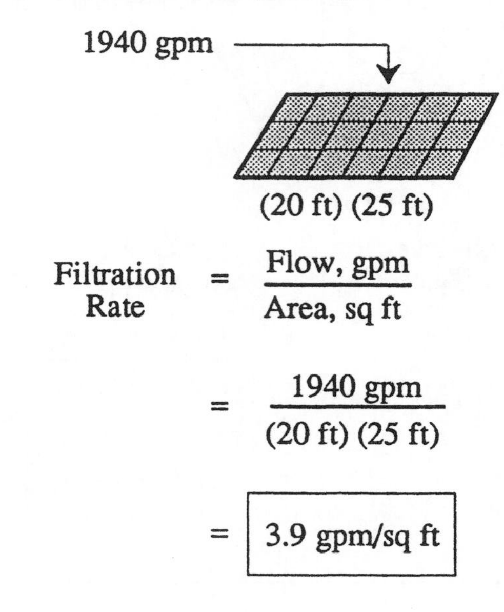

$$\text{Filtration Rate} = \frac{\text{Flow, gpm}}{\text{Area, sq ft}}$$

$$= \frac{1940 \text{ gpm}}{(20 \text{ ft})(25 \text{ ft})}$$

$$= \boxed{3.9 \text{ gpm/sq ft}}$$

Example 2: (Filtration Rate)

❑ A filter 20 ft by 35 ft treats a flow of 1530 gpm. What is the filtration rate in gpm/sq ft?

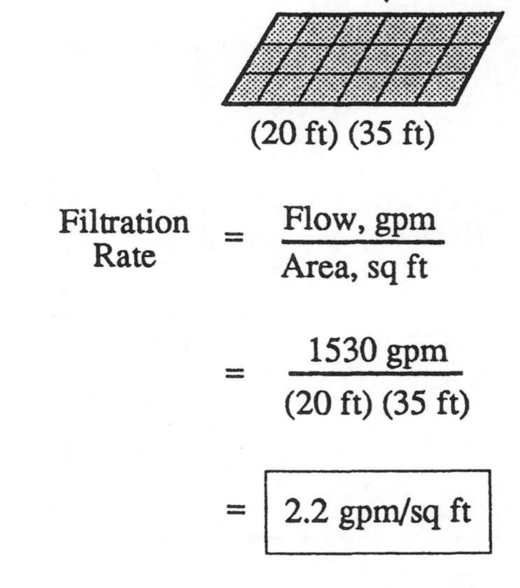

$$\text{Filtration Rate} = \frac{\text{Flow, gpm}}{\text{Area, sq ft}}$$

$$= \frac{1530 \text{ gpm}}{(20 \text{ ft})(35 \text{ ft})}$$

$$= \boxed{2.2 \text{ gpm/sq ft}}$$

Example 3: (Filtration Rate)

❑ A filter 25 ft by 30 ft treats a flow of 3.3 MGD. What is the filtration rate in gpm/sq ft?

$$\frac{3{,}300{,}000 \text{ gpd}}{1440 \text{ min/day}}$$

$= 2292$ gpm

(20 ft) (35 ft)

$$\text{Filtration Rate} = \frac{\text{Flow, gpm}}{\text{Area, sq ft}}$$

$$= \frac{2292 \text{ gpm}}{(25 \text{ ft})(30 \text{ ft})}$$

$$= \boxed{3.1 \text{ gpm/sq ft}}$$

Example 4: (Filtration Rate)

❑ A filter has a surface area of 35 ft by 25 ft. If the filter receives a flow of 2,912,000 gpd, what is the filtration rate in gpm/sq ft?

$$\frac{2{,}912{,}000 \text{ gpd}}{1440 \text{ min/day}}$$

$= 2022$ gpm

(35 ft) (25 ft)

$$\text{Filtration Rate} = \frac{\text{Flow, gpm}}{\text{Area, sq ft}}$$

$$= \frac{2022 \text{ gpm}}{(35 \text{ ft})(25 \text{ ft})}$$

$$= \boxed{2.3 \text{ gpm/sq ft}}$$

4.4 BACKWASH RATE CALCULATIONS

A filter backwash rate is a measure of the gpm flowing upward through each sq ft of filter surface area. The calculation of backwash rate is similar to filtration rate.

$$\text{Backwash Rate} = \frac{\text{Flow, gpm}}{\text{Area, sq ft}}$$

Example 1: (Backwash Rate)

❑ A filter with a surface area of 150 sq ft has a backwash flow rate of 2900 gpm. What is the filter backwash rate in gpm/sq ft?

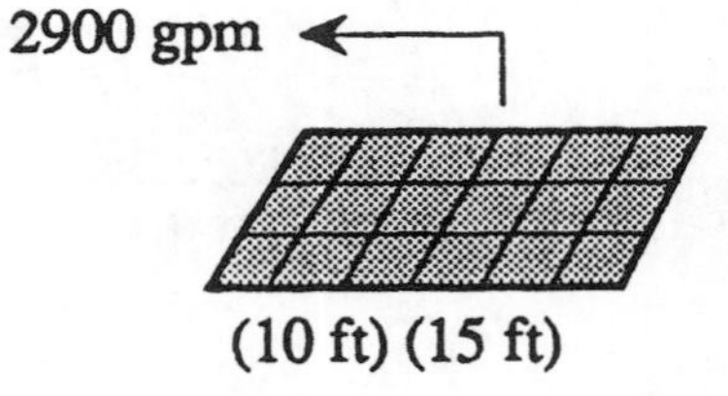

$$\text{Backwash Rate} = \frac{\text{Flow, gpm}}{\text{Area, sq ft}}$$

$$= \frac{2900 \text{ gpm}}{150 \text{ sq ft}}$$

$$= \boxed{19.3 \text{ gpm/sq ft}}$$

Example 2: (Backwash Rate)

❑ A filter 25 ft by 10 ft has a backwash rate of 3400 gpm. What is the backwash rate in gpm/sq ft?

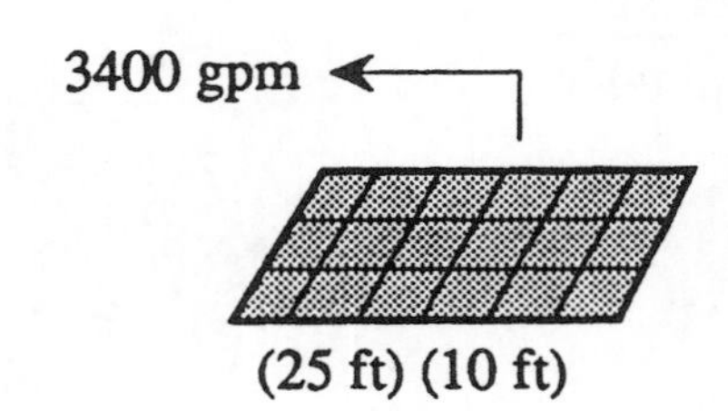

$$\text{Backwash Rate} = \frac{\text{Flow, gpm}}{\text{Area, sq ft}}$$

$$= \frac{3400 \text{ gpm}}{(25 \text{ ft})(10 \text{ ft})}$$

$$= \boxed{13.6 \text{ gpm/sq ft}}$$

Example 3: (Backwash Rate)
❑ A filter 15 ft by 15 ft has a backwash flow rate of 3150 gpm. What is the filter backwash rate in gpm/sq ft?

3150 gpm

(15 ft) (15 ft)

$$\text{Backwash Rate} = \frac{\text{Flow, gpm}}{\text{Area, sq ft}}$$

$$= \frac{3150 \text{ gpm}}{(15 \text{ ft})(15 \text{ ft})}$$

$$= \boxed{14 \text{ gpm/sq ft}}$$

Example 4: (Backwash Rate)
❑ A filter 20 ft long and 15 ft wide has a backwash flow rate of 4.64 MGD. What is the filter backwash rate in gpm/sq ft?

$$\frac{4{,}640{,}000 \text{ gpd}}{1440 \text{ min/day}} = 3222 \text{ gpm}$$

(20 ft) (15 ft)

$$\text{Backwash Rate} = \frac{\text{Flow, gpm}}{\text{Area, sq ft}}$$

$$= \frac{3222 \text{ gpm}}{(20 \text{ ft})(15 \text{ ft})}$$

$$= \boxed{10.7 \text{ gpm/sq ft}}$$

WHEN THE FLOW RATE IS EXPRESSED AS GPD

Normally the backwash flow rate is expressed as gpm. If it is expressed in any other flow rate terms, simply convert the given flow rate to gpm.* For example, if gpd flow rate is given, convert the gpd flow rate as follows:

$$\frac{\text{Flow Rate, gpd}}{1440 \text{ min/day}} = \text{Flow Rate, gpm}$$

* For a review of flow rate conversions, refer to Chapter 8 in *Basic Math Concepts*.

4.5 UNIT FILTER RUN VOLUME CALCULATIONS

The unit filter run volume (UFRV) calculation indicates the total gallons passing through each square foot of filter surface area during an entire filter run. The equation to be used in these calculations is shown to the right.

As the performance of the filter begins to deteriorate, the UFRV value will begin to decline as well.

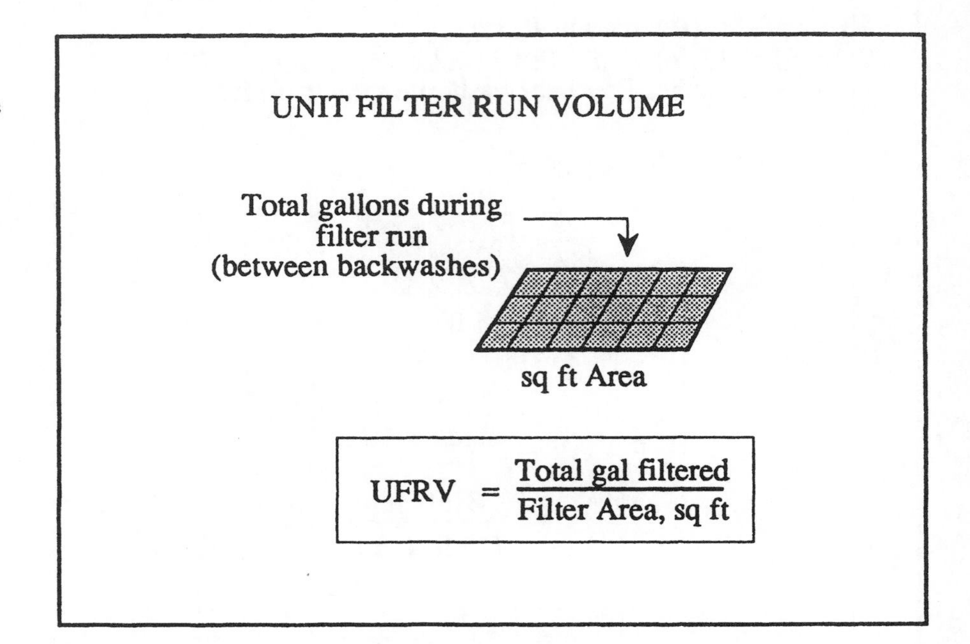

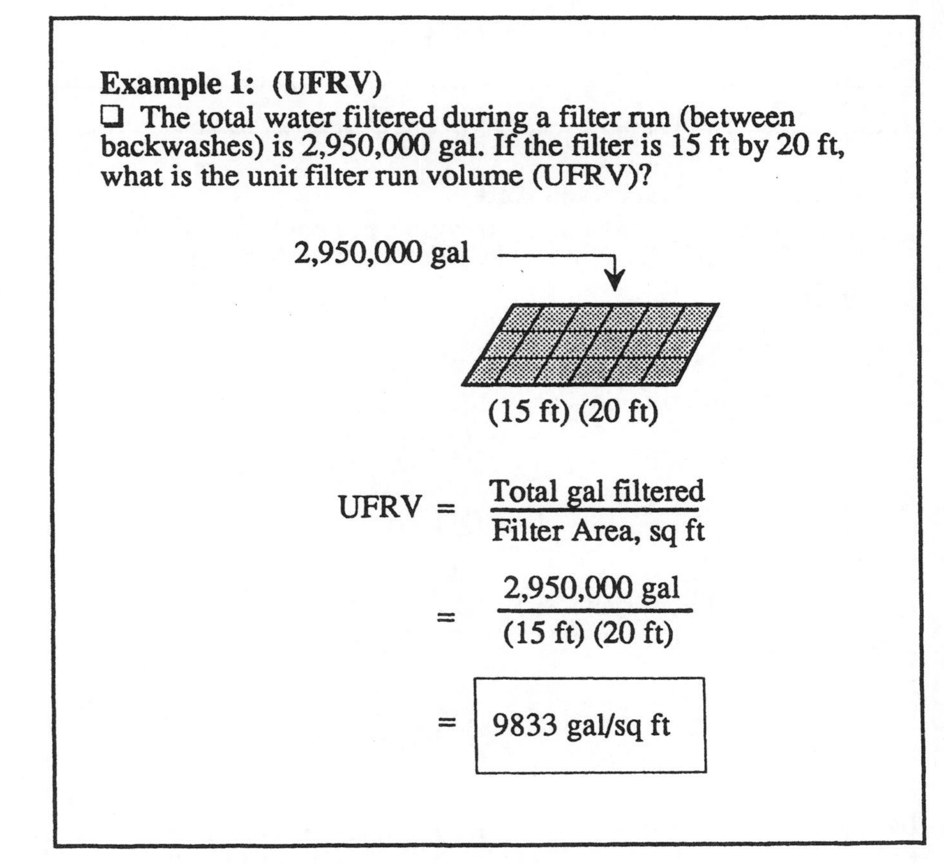

Example 2: (UFRV)

❑ The total water filtered during a filter run is 3,220,000. If the filter is 20 ft by 20 ft what is the UFRV?

3,220,000 gal

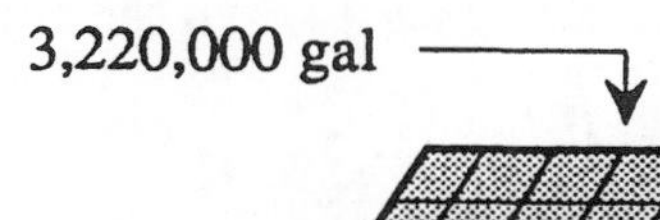

(20 ft) (20 ft)

$$\text{UFRV} = \frac{\text{Total gal filtered}}{\text{Filter Area, sq ft}}$$

$$= \frac{3{,}220{,}000 \text{ gal}}{(20 \text{ ft})(20 \text{ ft})}$$

$$= \boxed{8050 \text{ gal/sq ft}}$$

Example 3: (UFRV)

❑ The total water filtered during a filter run is 4,583,000? If the filter is 20 ft by 30 ft, what is the unit filter run volume?

4,583,000 gal

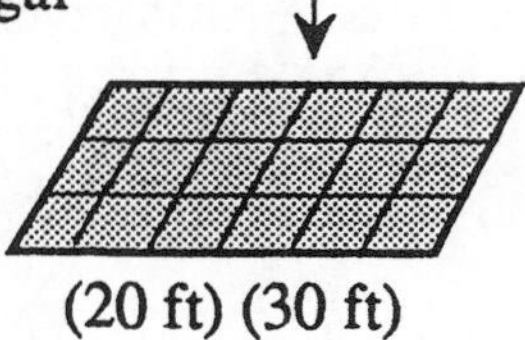

(20 ft) (30 ft)

$$\text{UFRV} = \frac{\text{Total gal filtered}}{\text{Filter Area, sq ft}}$$

$$= \frac{4{,}583{,}000 \text{ gal}}{(20 \text{ ft})(30 \text{ ft})}$$

$$= \boxed{7638 \text{ gal/sq ft}}$$

4.6 WEIR OVERFLOW RATE CALCULATIONS

Weir overflow rate is a measure of the gallons per day flowing over each foot of weir.

$$\text{Weir Overflow Rate} = \frac{\text{Flow, gpd}}{\text{Weir Length, ft}}$$

Example 1: (Weir Overflow Rate)

❑ A rectangular clarifier has a total of 100 ft of weir. What is the weir overflow rate in gpd/ft when the flow is 1.2 MGD?

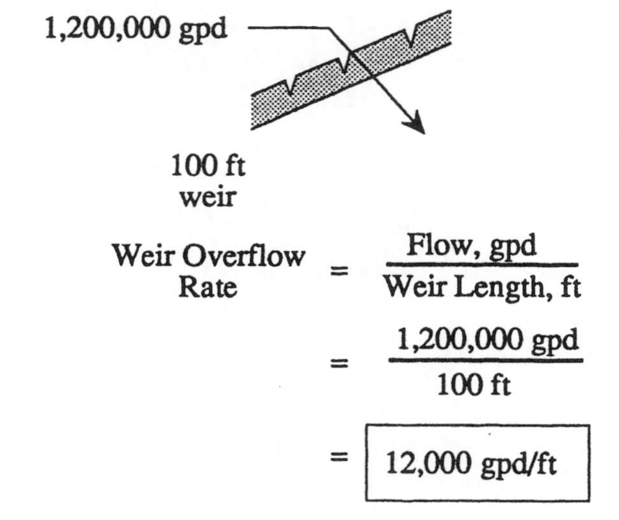

$$\text{Weir Overflow Rate} = \frac{\text{Flow, gpd}}{\text{Weir Length, ft}}$$

$$= \frac{1{,}200{,}000 \text{ gpd}}{100 \text{ ft}}$$

$$= \boxed{12{,}000 \text{ gpd/ft}}$$

CALCULATING WEIR CIRCUMFERENCE

In some calculations of weir overflow rate, you will have to calculate the total weir length given the weir diameter. To calculate the length of weir around the clarifier, you need to know the relationship between the diameter and circumference of a circle. **The distance around any circle (circumference) is about three times the distance across the circle (diameter).** In fact, the circumference is (3.14) (Diameter).* Therefore, given a diameter, the total ft of weir can be calculated as:

$$\text{Weir Length, ft} = (3.14)\ (\text{Weir Diam., ft})$$

Example 2: (Weir Overflow Rate)

❑ A circular clarifier receives a flow of 3.38 MGD. If the diameter of the weir is 80 ft, what is the weir overflow rate in gpd/ft?

The total ft of weir is not given directly in this problem. However, weir diameter is given (80 ft) and from that information we can determine the length of the weir.

3,380,000 gpd

ft weir:

$= (3.14)\ (80 \text{ ft})$

$= 251 \text{ ft}$

$$\text{Weir Overflow Rate} = \frac{\text{Flow, gpd}}{\text{Weir Length, ft}}$$

$$= \frac{3{,}380{,}000 \text{ gpd}}{251 \text{ ft}}$$

$$= \boxed{13{,}466 \text{ gpd/ft}}$$

* For a review of circumference calculations, refer to Chapter 9, Linear Measurement, in *Basic Math Concepts*..

Example 3: (Weir Overflow Rate)

❑ The flow to a circular clarifier is 2.12 MGD. If the diameter of the weir is 60 ft, what is the weir overflow rate in gpd/ft?

2,120,000 gpd

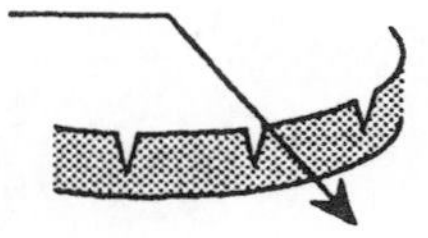

ft weir:
= (3.14) (60 ft)
= 188 ft

$$\text{Weir Overflow Rate} = \frac{\text{Flow, gpd}}{\text{Weir Length, ft}}$$

$$= \frac{2{,}120{,}000 \text{ gpd}}{188 \text{ ft}}$$

$$= \boxed{11{,}277 \text{ gpd/ft}}$$

Example 4: (Weir Overflow Rate)

❑ A rectangular sedimentation basin has a total weir length of 80 ft. If the flow to the basin is 1.3 MGD, what is the weir loading rate in gpm/ft?

$$\frac{1{,}300{,}000 \text{ gpd}}{1440 \text{ min/day}}$$

= 903 gpm

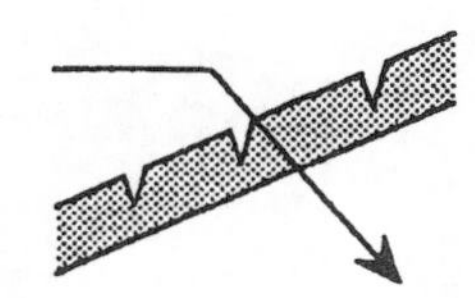

$$\text{Weir Loading Rate} = \frac{\text{Flow, gpm}}{\text{Weir Length, ft}}$$

$$= \frac{903 \text{ gpm}}{80 \text{ ft weir}}$$

$$= \boxed{11.3 \text{ gpm/ft}}$$

WEIR LOADING RATE

Weir overflow rate is a term most often associated with wastewater clarifier calculations. A similar calculation often used for water system clarifiers is weir loading rate, expressed as gpm/ft.

4.7 ORGANIC LOADING RATE CALCULATIONS

When calculating the pounds of BOD entering a wastewater process daily, you are calculating the organic load on the process—the food entering that process. Organic loading for trickling filters is calculated as lbs BOD/day per 1000 cu ft media:

$$\text{Organic Loading Rate} = \frac{\text{BOD, lbs/day}}{\text{Vol., 1000 cu ft}}$$

In most instances BOD will be expressed as a mg/*L* concentration and must be converted to lbs BOD/day.* Therefore the equation given above can be expanded as:

$$\text{O.L. Rate} = \frac{(\text{mg/}L\text{ BOD})\ (\text{MGD flow})\ (8.34\text{ lbs/gal})}{1000\text{ cu ft}}$$

Note that the "1000" in the denominator of both equations is a unit of measure ("thousand cu ft") and is **not part of the numerical calculation.**

To determine the number of 1000 cu ft, **find the thousands comma and place a decimal** at that position. In Example 1, 19,233 cu ft is 19.233 units of 1000 cu ft. So 19.233 is placed in the denominator.

Example 1: (Organic Loading Rate)

❑ A trickling filter 70 ft in diameter with a media depth of 5 feet receives a flow of 1,150,000 gpd. If the BOD concentration of the primary effluent is 230 mg/*L*, what is the organic loading on the trickling filter?

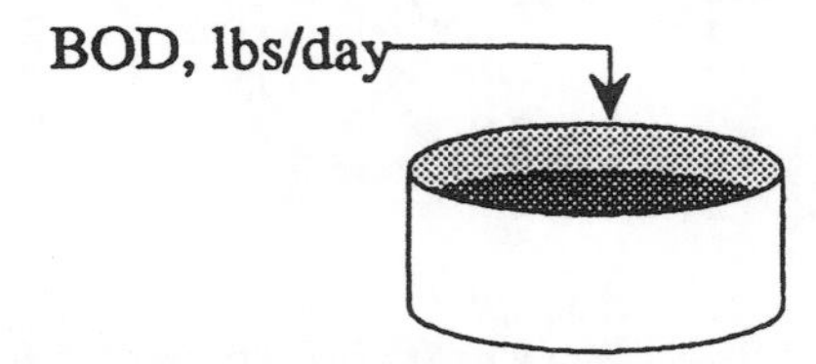

$$(0.785)\ (70\text{ ft})\ (70\text{ ft})\ (5\text{ ft}) = 19{,}233\text{ cu ft}$$

$$\text{Organic Loading Rate} = \frac{\text{BOD, lbs/day}}{\text{Volume, 1000 cu ft}}$$

$$= \frac{(230\text{ mg/}L)\ (1.15\text{ MGD})\ (8.34\text{ lbs/gal})}{19.233\quad 1000\text{-cu ft}}$$

$$= \boxed{\frac{115\text{ lbs BOD/day}}{1000\text{ cu ft}}}$$

Example 2: (Organic Loading Rate)

❑ A100-ft diameter trickling filter with a media depth of 4 ft receives a primary effluent flow of 1.65 MGD with a BOD of 105 mg/*L*. What is the organic loading on the trickling filter?

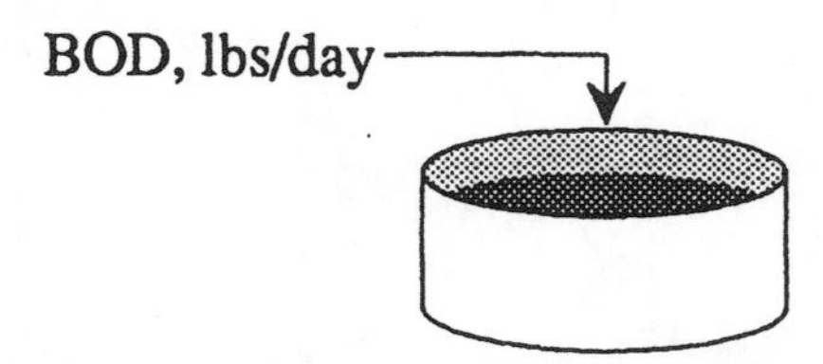

$$(0.785)\ (100\text{ ft})\ (100\text{ ft})\ (4\text{ ft}) = 31{,}400\text{ cu ft}$$

$$\text{Organic Loading Rate} = \frac{\text{BOD, lbs/day}}{\text{1000 cu ft}}$$

$$= \frac{(105\text{ mg/}L)\ (1.65\text{ MGD})\ (8.34\text{ lbs/gal})}{31.4\quad 1000\text{-cu ft}}$$

$$= \boxed{\frac{46\text{ lbs BOD/day}}{1000\text{ cu ft}}}$$

---

* For a review of milligrams per liter (mg/*L*) to pounds per day (lbs/day) BOD, refer to Chapter 3.

Example 3: (Organic Loading Rate)

❑ The flow to a 3-acre wastewater pond is 90,000 gpd. The influent BOD concentration is 125 mg/*L*. What is the organic loading to the pond?

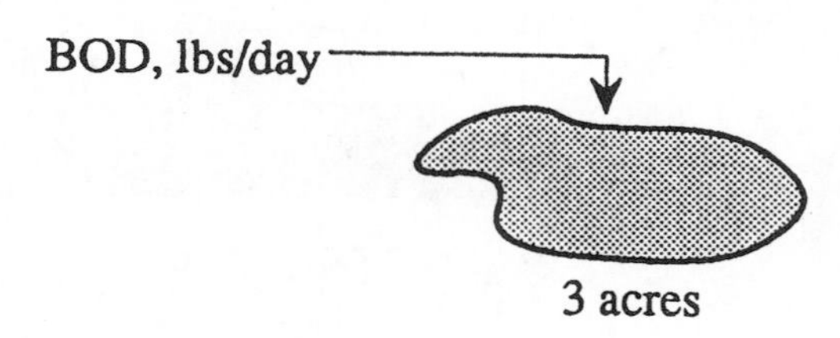

$$\text{Organic Loading Rate} = \frac{\text{BOD, lbs/day}}{\text{Area, ac}}$$

$$= \frac{(125\ \text{mg}/L)\ (0.09\ \text{MGD})\ (8.34\ \text{lbs/gal})}{3\ \text{ac}}$$

$$= \boxed{\frac{31.3\ \text{BOD/day}}{\text{ac}}}$$

ORGANIC LOADING FOR PONDS

Organic loading to ponds is generally calculated as lbs BOD/day per acre of pond surface area.

$$\text{Organic Loading Rate} = \frac{\text{BOD, lbs/day}}{\text{Area, ac}}$$

Example 4: (Organic Loading Rate)

❑ A rotating biological contactor (RBC) receives a flow of 3.6 MGD. If the soluble BOD of the influent wastewater to the RBC is 122 mg/*L*, and the surface area of the media is 600,000 sq ft, what is the organic loading rate?

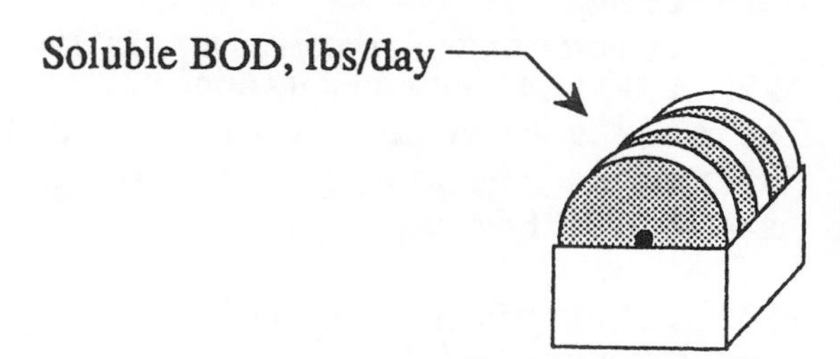

$$\text{Organic Loading Rate} = \frac{\text{Sol. BOD, lbs /day}}{\text{Area, 1000 sq ft}}$$

$$= \frac{(122\ \text{mg}/L)\ (3.6\ \text{MGD})\ (8.34\ \text{lbs/gal})}{600\ \ 1000\text{-sq ft}}$$

$$= \boxed{\frac{6.1\ \text{lbs/day Sol. BOD}}{1000\ \text{sq ft}}}$$

ORGANIC LOADING FOR ROTATING BIOLOGICAL CONTACTORS

There are two different aspects to calculating organic loading on rotating biological contactors:

1. **Soluble BOD** is used to measure organic content rather than total BOD, and

2. The calculation of organic loading is **per 1000 sq ft media** rather than 1000 cu ft media as with trickling filters.

4.8 FOOD/MICROORGANISM RATIO CALCULATIONS

In order for the activated sludge process to operate properly, there must be a balance between food entering the system (as measured by BOD or COD) and microorganisms in the aeration tank. The best F/M ratio for a particular system depends on the type of activated sludge process and the characteristics of the wastewater entering the system.

COD is sometimes used as the measure of food entering the system.* Since the COD test can be completed in only a few hours, compared with 5 days for a BOD test, the COD more accurately reflects the current food loading on the system.

Note that the F/M equation is given in two forms: the simplified equation and the expanded equation. If BOD and MLVSS data is given as lbs/day and lbs, then the simplified equation should be used. In most instances, however, BOD and MLVSS data will be given as mg/*L*, and a calculation of mg/*L* to lbs/day or lbs will be required as shown in the expanded equation.

FOOD SUPPLY (BOD OR COD) AND
MICROORGANISMS (MLVSS)
MUST BE IN BALANCE

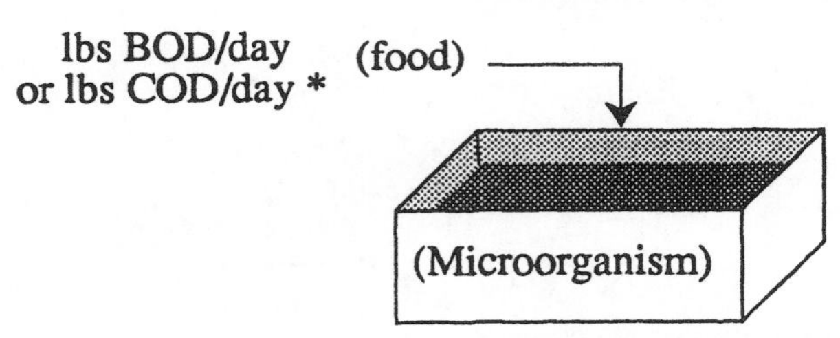

Simplified Equation:

$$F/M = \frac{\text{BOD, lbs/day}}{\text{MLVSS, lbs}}$$

Expanded Equation:

$$F/M = \frac{(\text{mg/}L\text{ BOD})(\text{MGD Flow})(8.34\text{ lbs/gal})}{(\text{mg/}L\text{ MLVSS})(\text{Aer Vol, MG})(8.34\text{ lbs/gal})}$$

Example 1: (F/M Ratio)

❑ An activated sludge aeration tank receives a primary effluent flow of 2,100,000 gpd with a BOD concentration of 158 mg/*L*. The mixed liquor volatile suspended solids is 1840 mg/*L* and the aeration tank volume is 300,000 gallons. What is the current F/M ratio?

Since BOD and MLVSS data is given in mg/*L*, the expanded equation will be needed.**

$$F/M = \frac{(\text{mg/}L\text{ BOD})(\text{MGD})(8.34\text{ lbs/gal})}{(\text{mg/}L\text{ MLVSS})(\text{Aer Vol, MG})(8.34\text{ lbs/gal})}$$

$$= \frac{(158\text{ mg/}L)(2.1\text{ MGD})(\cancel{8.34}\text{ lbs/gal})}{(1840\text{ mg/}L)(0.3\text{ MG})(\cancel{8.34}\text{ lbs/gal})}$$

$$= \boxed{0.6}$$

* COD may be used as the measure of food if there is generally a good correlation in BOD and COD characteristics of the wastewater. If not, the COD will not accurately reflect the microbiological content of the aeration tank.

** It is sometimes desirable to calculate the value in the numerator and denominator before calculating the final answer (see Example 2), since lbs BOD/day or lbs COD/day and lbs MLVSS may be required for other calculations.

Example 2: (F/M Ratio)

❑ The volume of an aeration tank is 200,000 gallons. The aeration tank receives a primary effluent flow of 2,320,000 gpd, with a COD concentration of 100 mg/*L*. If the mixed liquor volatile suspended solids is 1900 mg/*L*, what is the current F/M ratio?

Since COD and MLVSS information is given as mg/*L*, the expanded form of the equation is used in the calculation:

$$F/M = \frac{(\text{mg}/L \text{ COD})(\text{MGD flow})(8.34 \text{ lbs/gal})}{(\text{lbs MLVSS})(\text{Aer Vol, MG})(8.34 \text{ lbs/gal})}$$

$$= \frac{(100 \text{ mg}/L)(2.32 \text{ MGD})(8.34 \text{ lbs/gal})}{(1900 \text{ mg}/L)(0.2 \text{ MGD})(8.34 \text{ lbs/gal})}$$

$$= \frac{(1935 \text{ lbs COD/day})}{(3169 \text{ lbs MLVSS})}$$

$$= \boxed{0.6}$$

CALCULATING MLVSS USING THE F/M RATIO

The F/M ratio calculation can be used to calculate the desired pounds of MLVSS to be maintained in the aerator. Use the same F/M equation, fill in the given information, then solve for the unknown value (MLVSS).*

Example 3: (F/M Ratio)

❑ The desired F/M ratio at a particular activated sludge plant is 0.5 lbs BOD/1 lb mixed liquor volatile suspended solids. If the 3 MGD primary effluent flow has a BOD of 165 mg/*L* how many lbs of MLVSS should be maintained in the aeration tank?

$$F/M = \frac{\text{BOD, lbs/day}}{\text{MLVSS, lbs}}$$

Fill in the equation with the information known. Since BOD is given as mg/*L*, the lbs/day BOD must be written in expanded form.

$$0.5 = \frac{(165 \text{ mg}/L)(3 \text{ MGD})(8.34 \text{ lbs/gal})}{x \text{ lbs MLVSS}}$$

Then solve for the unknown value:*

$$x = \frac{(165)(3)(8.34)}{0.5}$$

$$= \boxed{8257 \text{ lbs MLVSS}}$$

* To review solving for the unknown value, refer to Chapter 2 in *Basic Math Concepts*.

4.9 SOLIDS LOADING RATE CALCULATIONS

The solids loading rate indicates the lbs/day solids loaded to each square foot of clarifier surface area. This calculation is used to determine solids loading on activated sludge secondary clarifiers and gravity sludge thickeners. The general solids loading rate equation is:

$$\text{Solids Loading Rate} = \frac{\text{Solids Applied, lbs/day}}{\text{Surface Area, sq ft}}$$

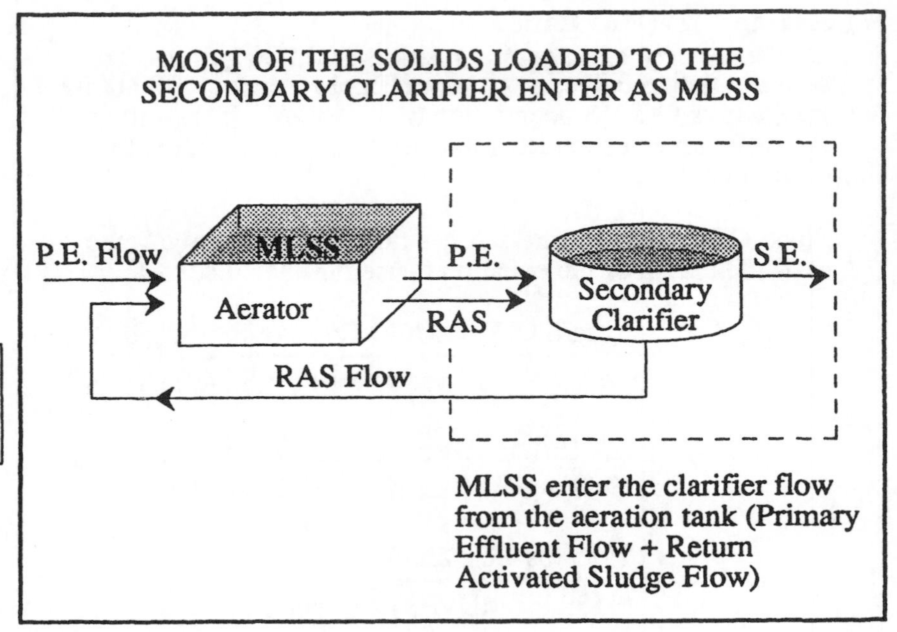

In expanded form, the equation includes the mg/L to lbs/day calculation in the numerator* and surface area calculation in the denominator:**

$$\text{S.L.R.} = \frac{(\underset{\text{mg}/L}{\text{MLSS}})\,(\underset{\text{flow}}{\text{MGD}})\,(8.34)}{(0.785)\,(D^2)}$$

The vast majority of solids coming into the secondary clarifier comes in as mixed liquor suspended solids (MLSS) from the aeration tank. A negligible amount of suspended solids enter the clarifier by the primary effluent flow. (Remember, up to 70% of the suspended solids are removed by the primary system.)

Example 1: (Solids Loading Rate)

❑ A secondary clarifier is 90 ft in diameter and receives a combined primary effluent (P.E.) and return activated sludge (RAS) flow of 4.9 MGD. If the MLSS concentration in the aeration tank is 3100 mg/*L*, what is the solids loading rate on the secondary clarifier in lbs/day/sq ft?

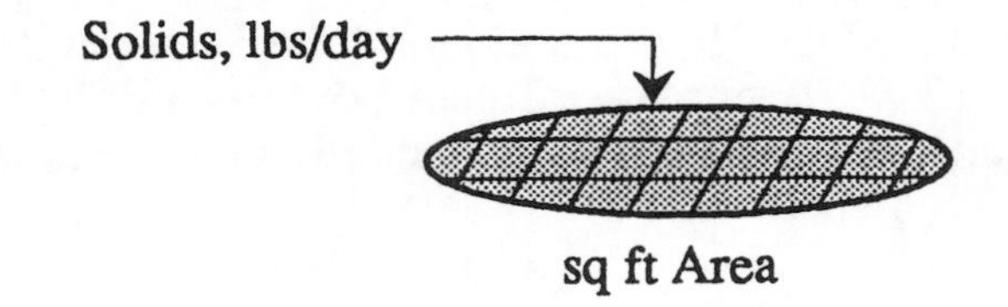

$$\text{Solids Loading Rate, lbs/day/sq ft} = \frac{(\text{MLSS mg}/L)\,(\text{MGD Flow})\,(8.34\text{ lbs/gal})}{(0.785)\,(D^2)}$$

$$= \frac{(3100\text{ mg}/L)\,(4.9\text{ MGD})\,(8.34\text{ lbs/gal})}{(0.785)\,(90\text{ ft})\,(90\text{ ft})}$$

$$= \boxed{\frac{19.9\text{ lbs solids/day}}{\text{sq ft}}}$$

---

* For a review of mg/L to lbs/day calculations refer to Chapter 3.

** Secondary clarifiers are typically circular. For a rectangular clarifier, the surface area would be $(l)\,(w)$. Area calculations are discussed in Chapter 10 of *Basic Math Concepts*.

Example 2: (Solids Loading Rate)

❑ A secondary clarifier 80 ft in diameter receives a primary effluent flow of 3.15 MGD and a return sludge flow of 0.8 MGD. If the MLSS concentration is 3650 mg/*L*, what is the solids loading rate on the clarifier in lbs/day/sq ft?

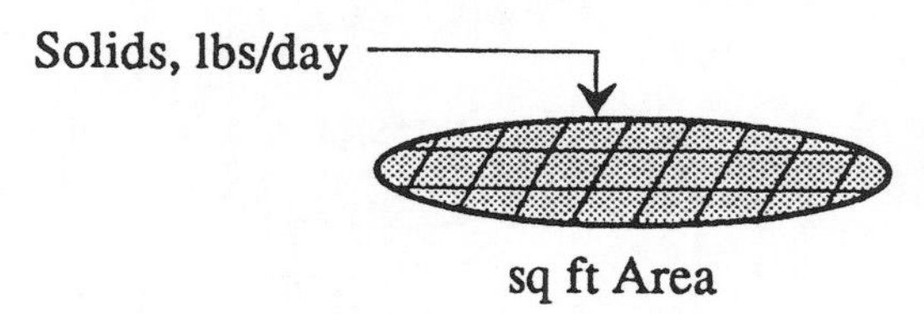

$$\text{Solids Loading Rate, lbs/day/sq ft} = \frac{\text{Solids, lbs/day}}{\text{Area, sq ft}}$$

$$= \frac{(3650 \text{ mg}/L)\ (3.95 \text{ MGD})\ (8.34 \text{ lbs/gal})}{(0.785)\ (80 \text{ ft})\ (80 \text{ ft})}$$

$$= \boxed{\frac{23.9 \text{ lbs solids/day}}{\text{sq ft}}}$$

Example 3: (Solids Loading Rate)

❑ The total flow to a 70-ft diameter clarifier is 4,600,000 gpd (P.E. + RAS flows). If the MLSS concentration is 2500 mg/*L*, what is the solids loading rate on the clarifier in lbs/day/sq ft?

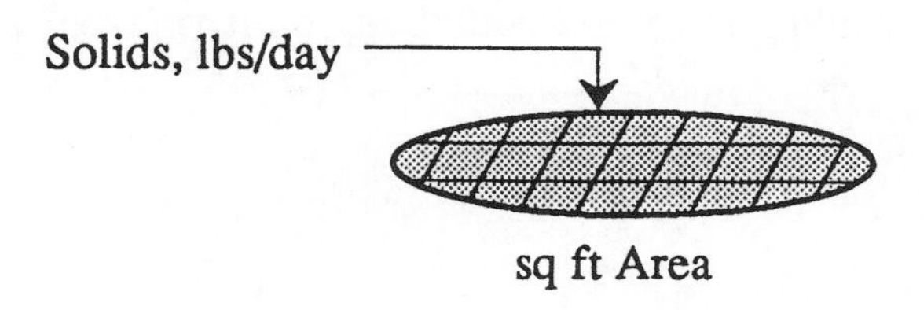

$$\text{Solids Loading Rate, lbs/day/sq ft} = \frac{\text{Solids, lbs/day}}{\text{Area, sq ft}}$$

$$= \frac{(2500 \text{ mg}/L)\ (4.6 \text{ MGD})\ (8.34 \text{ lbs/gal})}{(0.785)\ (70 \text{ ft})\ (70 \text{ ft})}$$

$$= \boxed{\frac{24.9 \text{ lbs solids/day}}{\text{sq ft}}}$$

4.10 DIGESTER LOADING RATE CALCULATIONS

Sludge is sent to a digester in order to break down or stabilize the organic portion of the sludge. Therefore, it is the organic part of the sludge (the volatile solids portion) that is of interest when calculating solids loading on a digester.

Digester loading rate is a measure of the pounds of volatile solids* entering each cubic foot of digester volume daily, as illustrated in the diagram to the right.

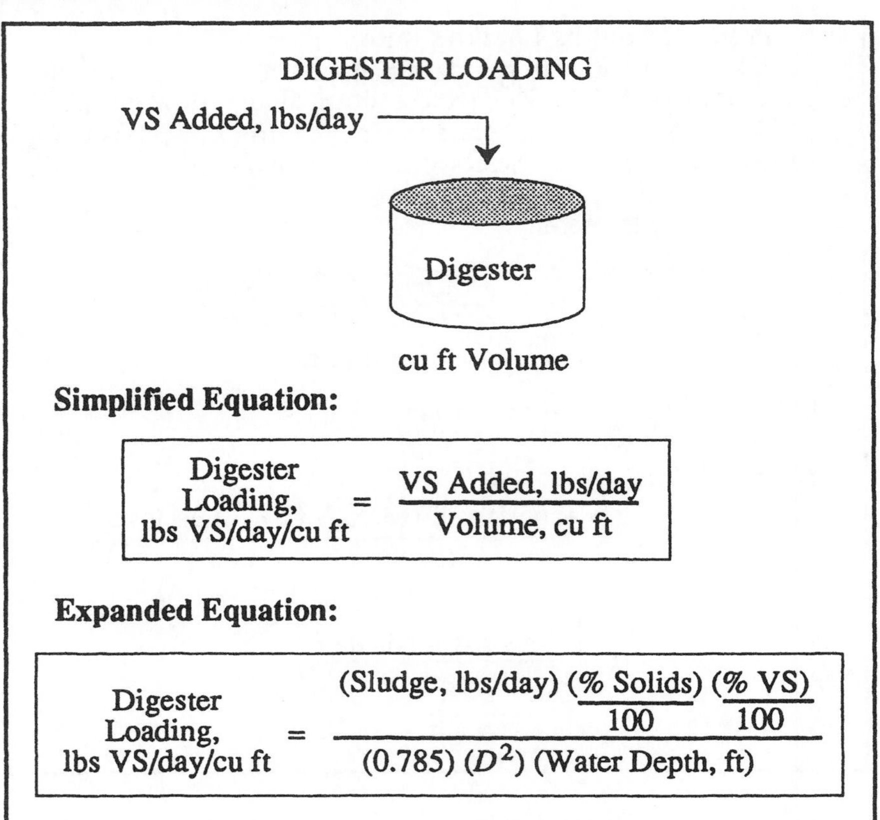

Simplified Equation:

$$\text{Digester Loading, lbs VS/day/cu ft} = \frac{\text{VS Added, lbs/day}}{\text{Volume, cu ft}}$$

Expanded Equation:

$$\text{Digester Loading, lbs VS/day/cu ft} = \frac{(\text{Sludge, lbs/day})\,\frac{(\%\ \text{Solids})}{100}\,\frac{(\%\ \text{VS})}{100}}{(0.785)\,(D^2)\,(\text{Water Depth, ft})}$$

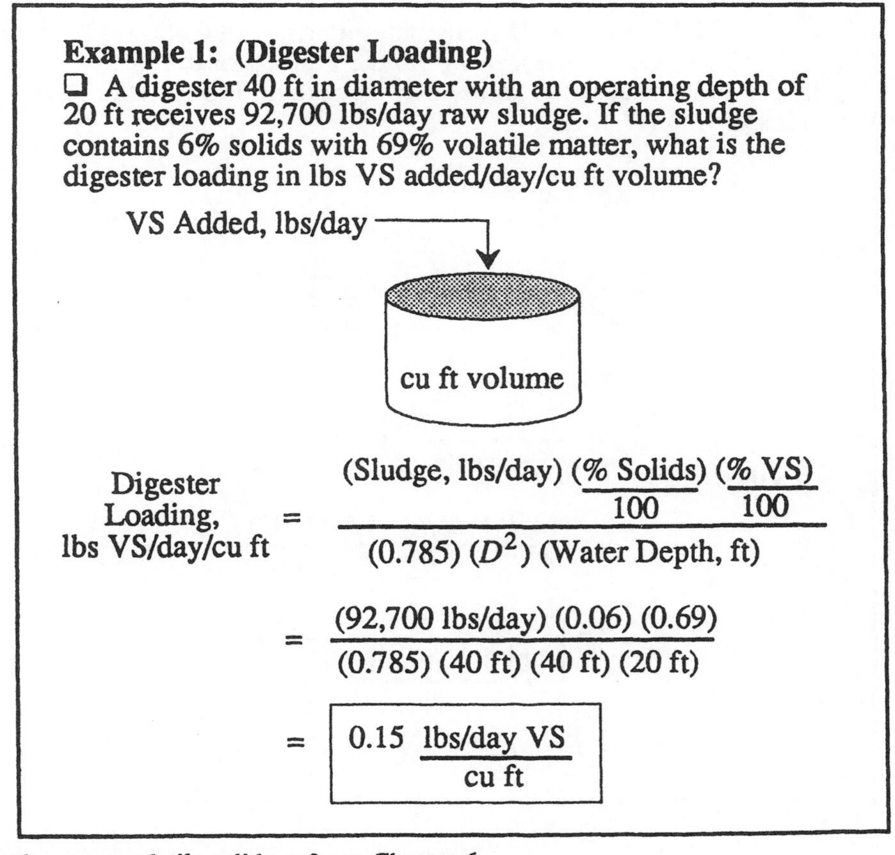

Example 1: (Digester Loading)

❑ A digester 40 ft in diameter with an operating depth of 20 ft receives 92,700 lbs/day raw sludge. If the sludge contains 6% solids with 69% volatile matter, what is the digester loading in lbs VS added/day/cu ft volume?

$$\text{Digester Loading, lbs VS/day/cu ft} = \frac{(\text{Sludge, lbs/day})\,\frac{(\%\ \text{Solids})}{100}\,\frac{(\%\ \text{VS})}{100}}{(0.785)\,(D^2)\,(\text{Water Depth, ft})}$$

$$= \frac{(92{,}700\ \text{lbs/day})\,(0.06)\,(0.69)}{(0.785)\,(40\ \text{ft})\,(40\ \text{ft})\,(20\ \text{ft})}$$

$$= 0.15\ \frac{\text{lbs/day VS}}{\text{cu ft}}$$

* For a review of calculating percent solids and percent volatile solids, refer to Chapter 6.

Example 2: (Digester Loading)

❑ A digester 40 ft in diameter operating at a depth of 18 ft receives 180,000 lbs/day sludge with 5% total solids and 72% volatile solids. What is the digester loading in lbs VS/day/cu ft?

$$\text{Digester Loading, lbs VS/day/cu ft} = \frac{(\text{Sludge, lbs/day})\left(\frac{\%\ \text{Solids}}{100}\right)\left(\frac{\%\ \text{VS}}{100}\right)}{(0.785)(D^2)(\text{Water Depth, ft})}$$

$$= \frac{(180{,}000\ \text{lbs/day})(0.05)(0.72)}{(0.785)(40\ \text{ft})(40\ \text{ft})(18\ \text{ft})}$$

$$= \boxed{0.29\ \frac{\text{lbs VS/day}}{\text{cu ft}}}$$

Example 3: (Digester Loading)

❑ A digester 50 ft in diameter operating at a depth of 20 ft receives 34,300 gpd sludge with 5.5% solids and 70% volatile solids. What is the digester loading in lbs VS/day/cu ft? (Assume the sludge weighs 8.34 lbs/gal.)

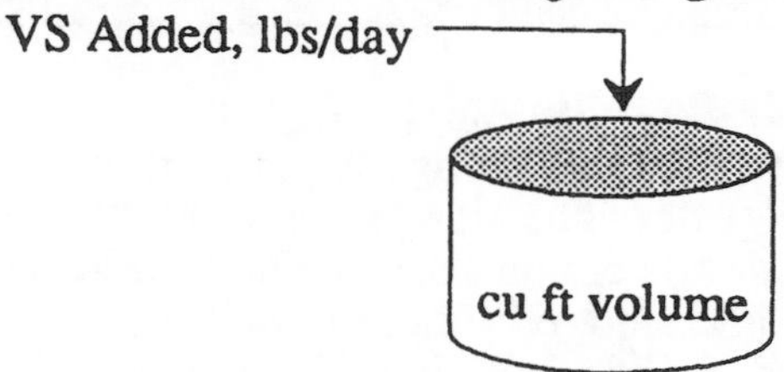

$$\text{Digester Loading, lbs VS/day/cu ft} = \frac{(\text{Sludge, gpd})(8.34\ \text{lbs/gal})\left(\frac{\%\ \text{Sol.}}{100}\right)\left(\frac{\%\ \text{VS}}{100}\right)}{(0.785)(D^2)(\text{Water Depth, ft})}$$

$$= \frac{(34{,}300\ \text{gpd})(8.34\ \text{lbs/gal})(0.055)(0.70)}{(0.785)(50\ \text{ft})(50\ \text{ft})(20\ \text{ft})}$$

$$= \boxed{0.28\ \frac{\text{lbs VS/day}}{\text{cu ft}}}$$

GIVEN GPD OR GPM SLUDGE PUMPED TO DIGESTER

In Examples 1 and 2, the sludge pumped to the digester was expressed as lbs/day. Many times, however, sludge pumped to the digester is expressed as gpd or gpm. When this is the case, convert the gpd or gpm pumping rate to lbs/day* and continue as in Examples 1 and 2. You can make the gpd to lbs/day conversion a separate calculation, or you can incorporate it into the numerator of the equation as shown in Example 3.

---

* To review flow conversions, refer to Chapter 8 in *Basic Math Concepts*..

4.11 DIGESTER VOLATILE SOLIDS LOADING RATIO CALCULATIONS

One way of expressing digester loading was described in the previous section—lbs/day volatile solids added per cu ft digester volume.

Another way to express digester loading is lbs/day volatile solids* added per lb of volatile solids under digestion (in the digester).

VOLATILE SOLIDS LOADING RATIO COMPARES VS ADDED WITH VS IN THE DIGESTER

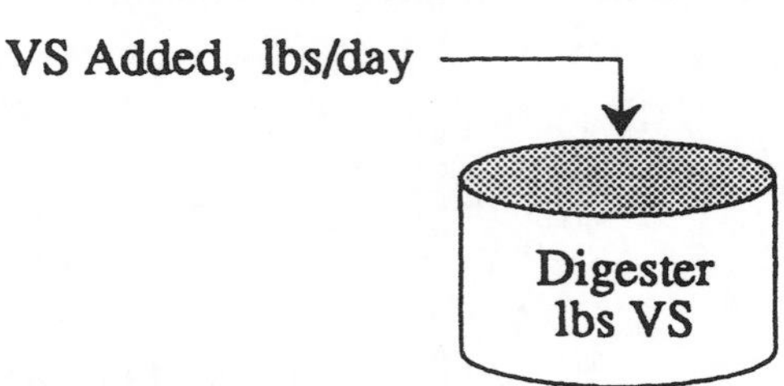

Simplified Equation:

$$\text{VS Loading Ratio} = \frac{\text{VS Added, lbs/day}}{\text{VS in Digester, lbs}}$$

Expanded Equation:**

$$\text{VS Loading Ratio} = \frac{(\text{Sludge Added, lbs/day}) \left(\frac{\%\ \text{Sol}}{100}\right) \left(\frac{\%\ \text{VS}}{100}\right)}{(\text{Sludge in Dig., lbs}) \left(\frac{\%\ \text{Sol}}{100}\right) \left(\frac{\%\ \text{VS}}{100}\right)}$$

Example 1: (Volatile Solids Loading Ratio)

❑ A total of 52,500 lbs/day sludge is pumped to a 100,000-gallon digester. The sludge being pumped to the digester has total solids content of 5% and volatile solids content of 74%. The sludge in the digester has a solids content of 6% with a 58% volatile solids content. What is the volatile solids loading on the digester in lbs VS added/day/lb VS in digester?

$$\text{VS Loading Ratio} = \frac{\text{VS Added, lbs/day}}{\text{VS in Digester, lbs}}$$

$$= \frac{(52{,}500\ \text{lbs/day}) \left(\frac{5}{100}\right) \left(\frac{74}{100}\right)}{\underbrace{(100{,}000\ \text{gal})(8.34\ \text{lbs/gal})}_{\text{This is lbs digester sludge}} \left(\frac{6}{100}\right) \left(\frac{58}{100}\right)}$$

$$= \boxed{\frac{0.067\ \text{lbs/day VS Added}}{\text{lb VS in Digester}}}$$

* For a review of calculating percent solids and percent volatile solids, refer to Chapter 6.

** The two hundreds in the numerator and denominator cancel each other out and may be omitted, if desired.

Example 2: (Volatile Solids Loading Ratio)

❑ A total of 22,850 gpd sludge is pumped to an 80,000-gal digester. This sludge has a solids content of 6% and a volatile solids concentration of 72%. The sludge in the digester has a solids content of 5.1% with a 56% volatile solids content. What is the volatile solids loading on the digester in lbs VS added/day/lb VS in digester?

$$\text{VS Loading Ratio} = \frac{\text{VS Added, lbs/day}}{\text{VS in Digester, lbs}}$$

$$= \frac{(22{,}850 \text{ lbs/day}) \left(\frac{6}{100}\right) \left(\frac{72}{100}\right)}{\underbrace{(80{,}000 \text{ gal}) (8.34 \text{ lbs/gal})}_{\textit{This is lbs digester sludge}} \left(\frac{5.1}{100}\right) \left(\frac{56}{100}\right)}$$

$$= \boxed{\frac{0.052 \text{ lbs/day VS Added}}{\text{lb VS in Digester}}}$$

Example 3: (Volatile Solids Loading Ratio)

❑ A total of 63,000 gpd sludge is pumped to the digester. The sludge has 4% solids with a volatile solids content of 74%. If the desired VS loading ratio is 0.08 lbs VS added/lb VS in digester, how many lbs VS should be in the digester for this volatile solids load?

$$\text{VS Loading Ratio} = \frac{\text{VS Added, lbs/day}}{\text{VS in Digester, lbs}}$$

$$0.08 = \frac{\overbrace{(63{,}000 \text{ gpd}) (8.34 \text{ lbs/day})}^{\textit{This is lbs/day sludge added}} \left(\frac{4}{100}\right) \left(\frac{74}{100}\right)}{x \text{ lbs VS in Digester}}$$

Then solve for x:

$$x = \frac{(63{,}000) (8.34) (0.04) (0.74)}{0.08}$$

$$x = \boxed{194{,}405 \text{ lbs VS in Digester}}$$

CALCULATING OTHER UNKNOWN VALUES

Volatile solids loading ratio calculations have three variables: VS loading ratio, lbs VS added/day, and lbs VS in the digester.

Given a **desired** volatile solids loading ratio, you can calculate the desired lbs of volatile solids in the digester. This type of calculation is used for determining seed sludge requirements for startup of a digester. Example 3 illustrates this calculation.

4.12 POPULATION LOADING AND POPULATION EQUIVALENT

POPULATION LOADING

Population loading is a calculation associated with wastewater treatment by ponds. Population loading is an indirect measure of both water and solids loading to a system. It is calculated as the number of persons served per acre of pond:

$$\text{Population Loading} = \frac{\text{persons}}{\text{acre}}$$

Example 1: (Population Loading)

❑ A 3.5 acre wastewater pond serves a population of 1500. What is the population loading on the pond?

$$\text{Population Loading} = \frac{\text{persons}}{\text{acre}}$$

$$= \frac{1500 \text{ persons}}{3.5 \text{ acres}}$$

$$= \boxed{429 \frac{\text{persons}}{\text{acre}}}$$

Example 2: (Population Loading)

❑ A wastewater pond serves a population of 4000. If the pond is 16 acres, what is the population loading on the pond?

$$\text{Population Loading} = \frac{\text{persons}}{\text{acre}}$$

$$= \frac{4000 \text{ persons}}{16 \text{ acres}}$$

$$= \boxed{250 \frac{\text{persons}}{\text{acre}}}$$

Example 3: (Population Equivalent)

❑ A 0.4-MGD wastewater flow has a BOD concentration of 1800 mg/*L* BOD. Using an average of 0.2 lbs/day BOD/person, what is the population equivalent of this wastewater flow?

$$\text{Population Equivalent} = \frac{\text{BOD, lbs/day}}{\text{lbs BOD/day/person}}$$

Convert mg/*L* BOD to lbs/day BOD* then divide by 0.2 lbs BOD/day/person:

$$\text{Population Equivalent} = \frac{(1800 \text{ mg}/L)(0.4 \text{ MGD})(8.34 \text{ lbs/gal})}{0.2 \text{ lbs BOD/day/person}}$$

$$= \boxed{30{,}024 \text{ people}}$$

Example 4: (Population Equivalent)

❑ A 100,000 gpd wastewater flow has a BOD content of 2800 mg/*L*. Using an average of 0.2 lbs/day BOD/person, what is the population equivalent of this flow?

$$\text{Population Equivalent} = \frac{\text{BOD, lbs/day}}{\text{lbs BOD/day/person}}$$

$$= \frac{(2800 \text{ mg}/L)(0.1 \text{ MGD})(8.34 \text{ lbs/gal})}{0.2 \text{ lbs BOD/day/person}}$$

$$= \boxed{11{,}676 \text{ people}}$$

POPULATION EQUIVALENT

Industrial or commercial wastewater generally has a higher organic content than domestic wastewater. Population equivalent calculations equate these concentrated flows with the number of people that would produce a domestic wastewater of that strength. For a domestic wastewater system, each person served by the system contributes about 0.17 or 0.2 lbs BOD/day . To determine the population equivalent of a wastewater flow, divide the lbs BOD/day content by the lbs BOD/day contributed per person (e.g. 0.2 lbs BOD/day).

* For a review of mg/*L* to lbs/day calculations, refer to Chapter 3.

NOTES:

5 Detention and Retention Times Calculations

SUMMARY

1. **Detention time** indicates the amount of time a given flow of water is retained by a unit process. It is calculated as the tank volume divided by the flow rate:

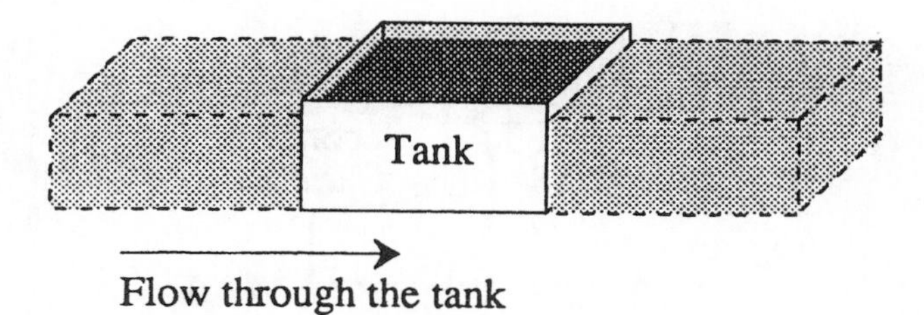

$$\text{Detention Time} = \frac{\text{Volume of Tank, gal}}{\text{Flow, gal/time}}$$

2. **Sludge age** is a measure of the average time a suspended solids particle remains under aeration. Sludge age (sometimes called Gould sludge age) is based on the **pounds of solids added daily** to the activated sludge process.

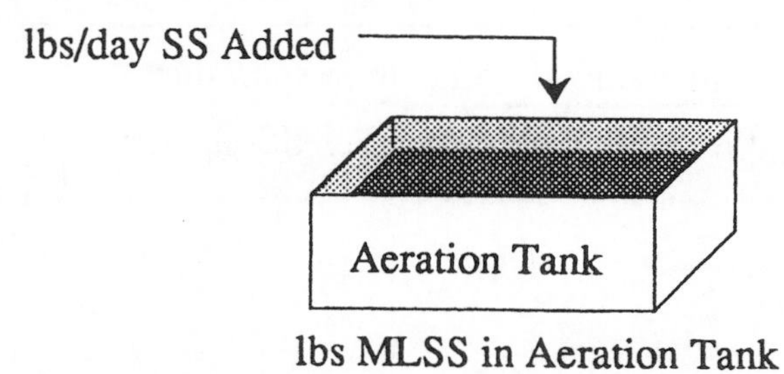

Simplified Equation:

$$\underset{\text{days}}{\text{Sludge Age}} = \frac{\text{MLSS, lbs}}{\text{SS Added, lbs/day}}$$

Expanded Equation:

$$\underset{\text{days}}{\text{Sludge Age}} = \frac{(\text{MLSS mg}/L)\ (\text{Aer. Vol., MG})\ (8.34\ \text{lbs/gal})}{(\text{P. E. SS, mg}/L)\ (\text{Flow, MGD})\ (8.34\ \text{lbs/gal})}$$

In the previous chapter we focused on calculations that measure the water and solids loading on a system. Now we will examine calculations that measure **how long the water and solids are retained in the system**. Three calculations will be discussed:

- Detention Time or Fill Time
- Sludge Age
- Solids Retention Time (SRT) (also called Mean Cell Residence Time, MCRT)

SUMMARY—Cont'd

3. **Solids Retention Time**, SRT (also called Mean Cell Residence Time, MCRT) is another measure of the length of time a suspended solid particle remains under aeration. However, the SRT calculation is **based on the pounds of solids leaving the activated sludge process** rather than the pounds of solids added, as with sludge age.

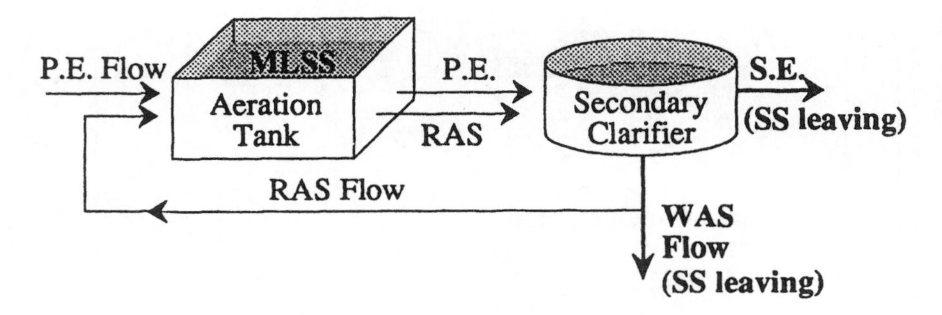

Simplified Equation:

$$\text{Solids Retention Time} = \frac{\text{Suspended Solids in System, lbs**}}{\text{Suspended Solids Leaving System, lbs/day}}$$

Or

$$\text{Solids Retention Time} = \frac{\text{Suspended Solids in System lbs**}}{\text{WAS SS, lbs/day + S.E.* SS, lbs/day}}$$

Expanded Equation:

$$\text{SRT, days} = \frac{(\text{MLSS mg}/L)\,(\text{Aer. Vol. + Fin. Clar. Vol., MG})\,(8.34\text{ lbs/gal})}{\underset{\text{mg}/L}{(\text{WAS SS})}\,\underset{\text{MGD}}{(\text{WAS Flow})}\,\underset{\text{lbs/gal}}{(8.34)} + \underset{\text{mg}/L}{(\text{S.E. SS})}\,\underset{\text{MGD}}{(\text{Plant Flow})}\,\underset{\text{lbs/gal}}{(8.34)}}$$

---

* S.E. is an abbreviation for Secondary Effluent. P.E. refers to Primary Effluent.

** There are four ways to account for system solids in the SRT calculation (numerator). One commonly used calculation of system solids is given in the SRT equation above. The other three methods are discussed in Chapter 12.

NOTES:

5.1 DETENTION TIME CALCULATIONS

There are two basic ways to consider detention time:

1. Detention time is the length of time required for a given flow rate to pass through a tank.

2. Detention time may also be considered as the length of time required to fill a tank at a given flow rate.

In each case, the calculation of detention time is the same:

$$\text{Detention Time} = \frac{\text{Volume of Tank, gal}}{\text{Flow Rate, gal/time}}$$

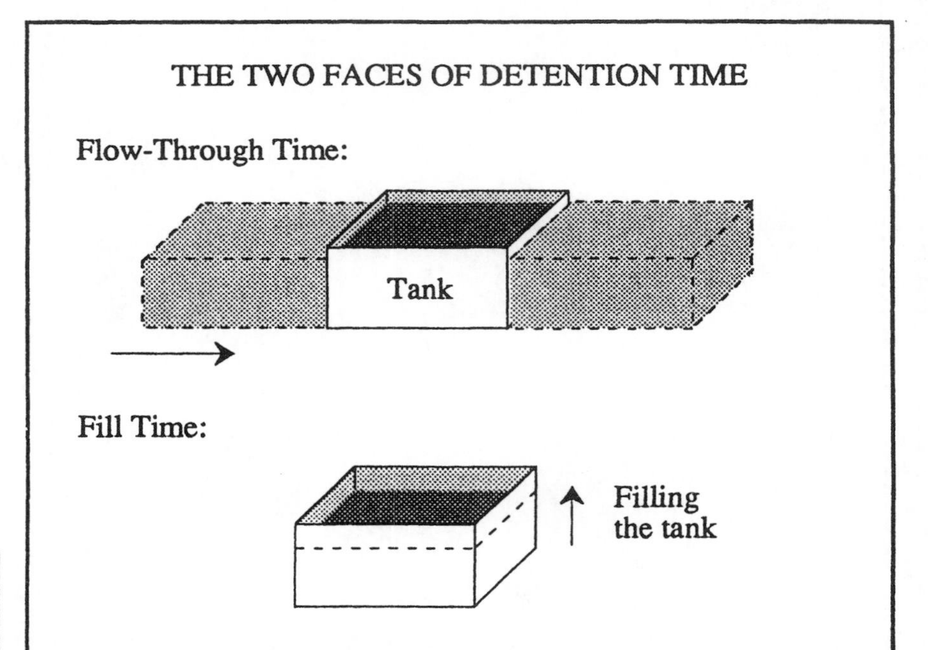

MATCHING UNITS

There are many possible ways of writing the detention time equation, depending on the time unit desired (seconds, minutes, hours, days) and the expression of volume and flow rate.

When calculating detention time, it is essential that the time and volume units used in the equation are consistent with each other, as illustrated to the right.

BE SURE THE TIME AND VOLUME UNITS MATCH

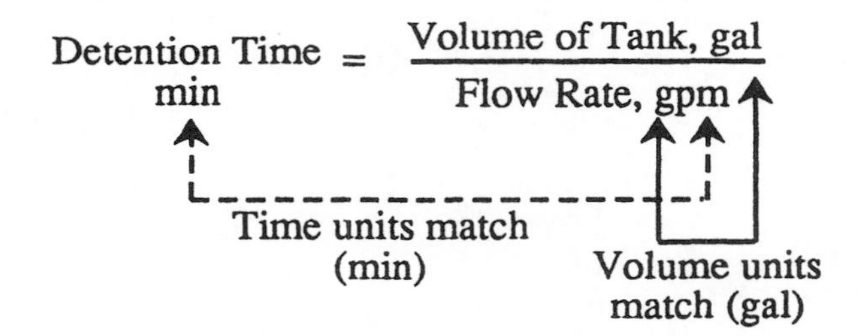

Other examples of detention time equations where time and volume units match include:

$$\text{Detention Time, sec} = \frac{\text{Volume of Tank, cu ft}}{\text{Flow Rate, cfs}}$$

$$\text{Detention Time, hrs} = \frac{\text{Volume of Tank, gal}}{\text{Flow Rate, gph}}$$

$$\text{Detention Time, days} = \frac{\text{Volume of Pond, ac-ft}}{\text{Flow Rate, ac-ft/day}}$$

Example 1: (Detention Time)
❑ The flow to a sedimentation tank 80 ft long, 30 ft wide and 10 ft deep is 3.7 MGD. What is the detention time in the tank in hours?

(80 ft) (30 ft) (10 ft) (7.48 gal/cu ft) = 179,520 gal Volume

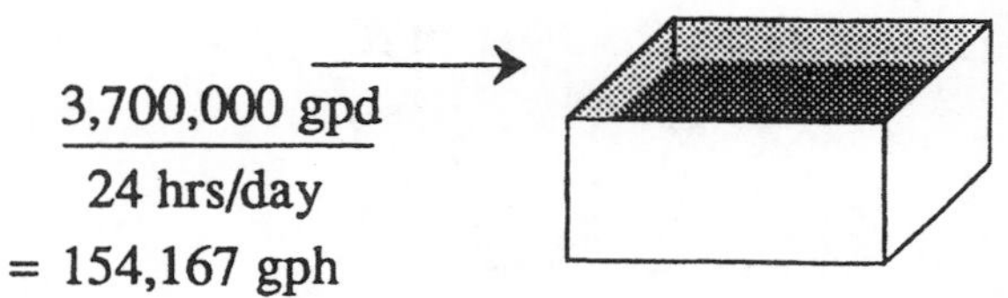

$$\frac{3{,}700{,}000 \text{ gpd}}{24 \text{ hrs/day}} = 154{,}167 \text{ gph}$$

First, write the equation so that volume and time units match. Then fill in the equation and solve for the unknown.

$$\text{Detention Time, hrs} = \frac{\text{Volume of Tank, gal}}{\text{Flow Rate, gph}}$$

$$= \frac{179{,}520 \text{ gal Volume}}{154{,}167 \text{ gph}}$$

$$= \boxed{1.2 \text{ hours}}$$

Example 2: (Detention Time)
❑ A flocculation basin is 8 ft deep, 15 ft wide, and 40 ft long. If the flow through the basin is 2.2 MGD, what is the detention time in minutes?

(40 ft) (15 ft) (8 ft) (7.48 gal/cu ft) = 35,904 gal Volume

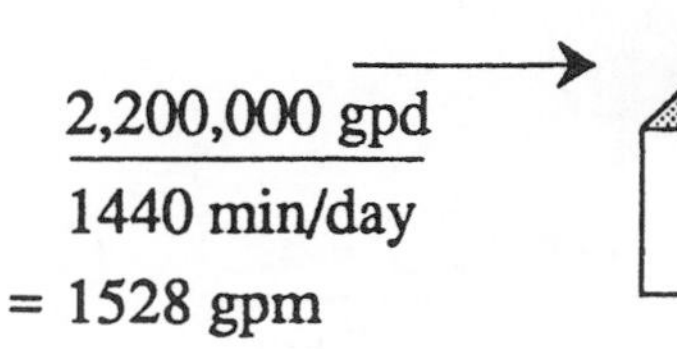

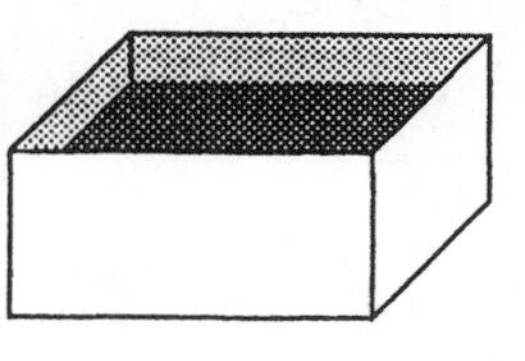

$$\frac{2{,}200{,}000 \text{ gpd}}{1440 \text{ min/day}} = 1528 \text{ gpm}$$

$$\text{Detention Time, min} = \frac{\text{Volume of Tank, gal}}{\text{Flow Rate, gpm}}$$

$$= \frac{35{,}904 \text{ gal Volume}}{1528 \text{ gpm}}$$

$$= \boxed{23 \text{ minutes}}$$

DETENTION TIME AS FLOW THROUGH A TANK

In calculating unit process detention times, you are calculating the length of time it takes the water to flow through that unit process. Detention times are normally calculated for the following basins or tanks:

- Flash mix chambers (sec)
- Flocculation basins (min)
- Sedimentation tanks or clarifiers (hrs),
- Wastewater ponds (days),
- Oxidation ditches (hrs).

There are two key points to remember when calculating detention time:

1. **Tank volume is the numerator (top) of the fraction** and flow rate is the denominator (bottom) of the fraction. Many times students have a difficult time remembering which term belongs in the numerator and which in the denominator. As a memory aid, remember that "V," the victor, is always on top.

2. **Time and volume units must match.** If detention time is desired in minutes, then the flow rate used in the calculation should have the same time frame (cfm or gpm, depending on whether tank volume is expressed as cubic feet or gallons). If detention time is desired in hours, then the flow rate used in the calculation should be cfh or gph.

DETENTION TIME FOR PONDS

Detention time for a pond may be calculated using one of two equations, depending on how the flow rate is expressed:

$$\text{Detention Time, days} = \frac{\text{Pond Volume, gal}}{\text{Flow Rate, gpd}}$$

Or

$$\text{Detention Time, days} = \frac{\text{Pond Volume, ac-ft}}{\text{Flow Rate, ac-ft/day}}$$

For a better understanding of the relative sizes of MGD and ac-ft/day, remember that 1 MGD is equivalent to about 3 ac-ft/day flow.

Examples 3 and 4 illustrate the use of both detention time equations.

Example 3: (Detention Time)

❑ A waste treatment pond is operated at a depth of 5 feet. The average width of the pond is 375 ft and the average length is 610 ft. If the flow to the pond is 570,000 gpd, what is the detention time in days?

$$(610 \text{ ft})(375 \text{ ft})(5 \text{ ft})(7.48 \text{ gal/cu ft}) = 8{,}555{,}250 \text{ gal Volume}$$

570,000 gpd →

$$\text{Detention Time days} = \frac{\text{Volume of Pond, gal}}{\text{Flow Rate, gpd}}$$

$$= \frac{8{,}555{,}250 \text{ gal Volume}}{570{,}000 \text{ gpd}}$$

$$= \boxed{15 \text{ days}}$$

Example 4: (Detention Time)

❑ A waste treatment pond is operated at a depth of 6 feet. The volume of the pond is 54 ac-ft. If the flow to the pond is 2.7 ac-ft/day, what is the detention time in days?

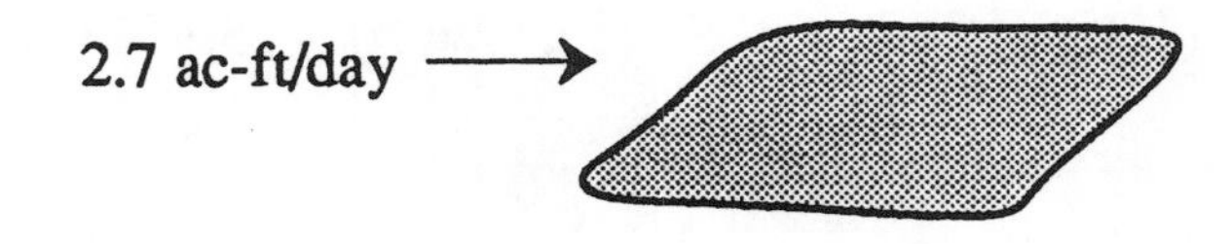

$$\text{Detention Time days} = \frac{\text{Volume of Pond, ac-ft}}{\text{Flow Rate, ac-ft/day}}$$

$$= \frac{54 \text{ ac-ft Volume}}{2.7 \text{ ac-ft/day}}$$

$$= \boxed{20 \text{ days}}$$

Example 5: (Detention Time)

❑ A basin 4 ft square is to be filled to the 3 ft level. If the flow to the tank is 3 gpm, how long will it take to fill the tank (in hours)?

(4 ft) (4 ft) (3 ft) (7.48 gal/cu ft) = 359 gal Volume

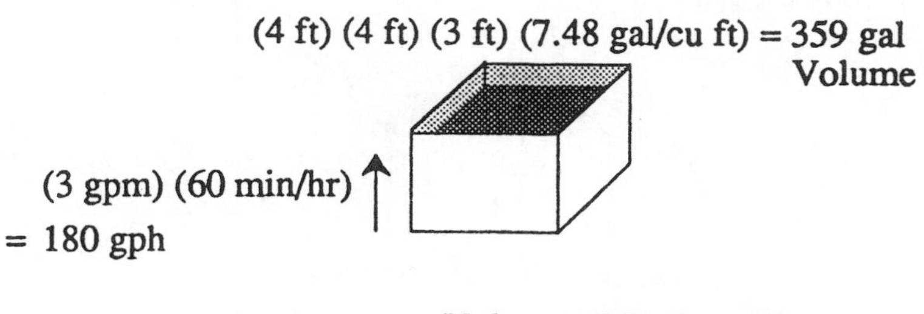

(3 gpm) (60 min/hr) = 180 gph

$$\text{Fill Time, hrs} = \frac{\text{Volume of Tank, gal}}{\text{Flow Rate, gph}}$$

$$= \frac{\text{359 gal Volume}}{\text{180 gph}}$$

$$= \boxed{\text{2 hrs}}$$

Example 6: (Detention Time)

❑ A tank has a diameter of 5 ft with an overflow depth at 4 ft. The current water level is 2.8 ft. Water is flowing into the tank at a rate of 4.1 gpm. At this rate, how long will it take before the tank overflows (in min)?

The volume of the tank remaining to be filled is 5 ft in diameter and 1.2 ft deep (4 ft – 2.8 ft = 1.2 ft). Therefore, the fill volume is:

(0.785) (5 ft) (5 ft) (1.2 ft) (7.48 gal/cu ft) = 176 gal Vol.

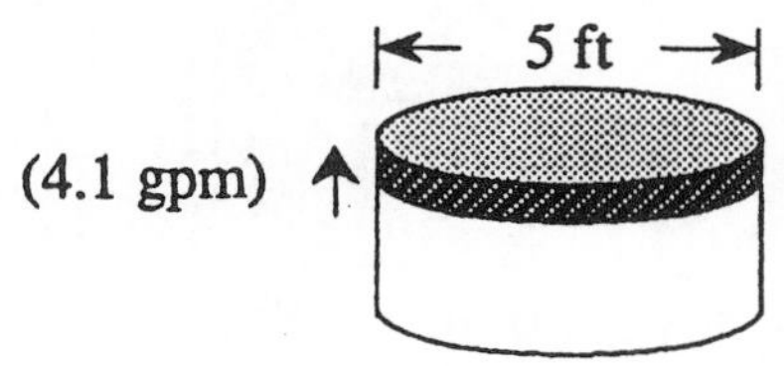

$$\text{Time Until Overflow, min} = \frac{\text{Volume of Tank, gal}}{\text{Flow Rate, gpm}}$$

$$= \frac{\text{176 gal Volume}}{\text{4.1 gpm}}$$

$$= \boxed{\text{43 min until overflow}}$$

DETENTION TIME AS FILL TIME

Another way to think of detention time is the time required to fill a tank or basin at a given flow rate. Regardless of whether you consider detention time flow time through a tank, or fill time, the calculation is precisely the same:

$$\text{Detention Time} = \frac{\text{Volume of Tank, gal}}{\text{Flow Rate, gal/time}}$$

In some equations, the word *fill time* is used rather than detention time.

$$\text{Fill Time} = \frac{\text{Volume of Tank, gal}}{\text{Flow Rate, gal/time}}$$

In each equation listed above the volume can be given as cubic feet, if desired (cu ft and cu ft/time).

The fill time calculation can also be used to determine the **time remaining before a tank overflows**, as illustrated in Example 6. Such a calculation can be critical during equipment failure conditions.

5.2 SLUDGE AGE CALCULATIONS

Sludge age refers to the average number of days a particle of suspended solids remains under aeration. It is a calculation used to maintain the proper amount of activated sludge in the aeration tank.

When considering sludge age, in effect you are asking, "how many days of suspended solids are in the aeration tank?" If you know how many pounds of suspended solids enter the aeration tank daily and you can determine how many total pounds of suspended solids are in the aeration tank, then you can calculate how many days of solids are in the aeration tank. For example, if 2000 lbs SS enter the aeration tank daily and the aeration tank contains 10,000 lbs of suspended solids, then 5 days of solids are in the aeration tank—a sludge age of 5 days.

Notice the similarity of this calculation with that of detention time—sludge age is **solids retained** calculated using units of lbs and lbs/day; detention time is **water retained**, using units of gal and gal/time or cu ft and cu ft/time:

$$\text{Sludge Age, days} = \frac{\text{SS in Tank, lbs}}{\text{SS Added, lbs/day}}$$

$$\text{Detention Time, min} = \frac{\text{Volume of Tank, gal}}{\text{Flow Rate, gpm}}$$

SLUDGE AGE IS BASED ON SUSPENDED SOLIDS <u>ENTERING</u> THE AERATION TANK*

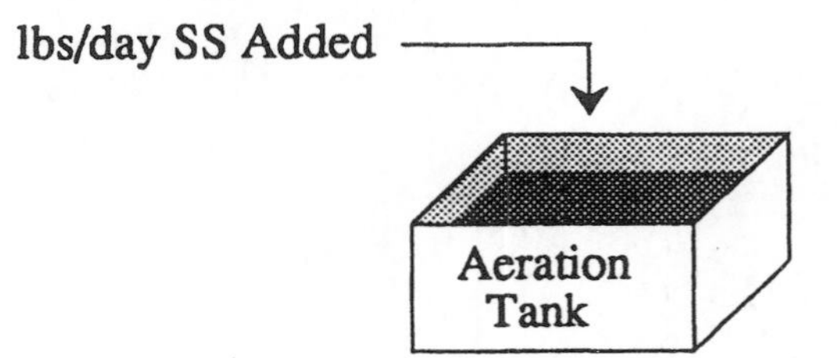

Simplified Equation:

$$\text{Sludge Age, days} = \frac{\text{MLSS, lbs}}{\text{SS Added, lbs/day}}$$

Expanded Equation:

$$\text{Sludge Age, days} = \frac{(\text{MLSS mg}/L)(\text{Aer. Vol., MG})(8.34 \text{ lbs/gal})}{(\text{P. E. SS, mg}/L)(\text{Flow, MGD})(8.34 \text{ lbs/gal})}$$

Example 1: (Sludge Age)

❑ An aeration tank has a total of 13,000 lbs of mixed liquor suspended solids. If a total of 2540 lbs/day suspended solids enter the aeration tank in the primary effluent flow, what is the sludge age in the aeration tank?

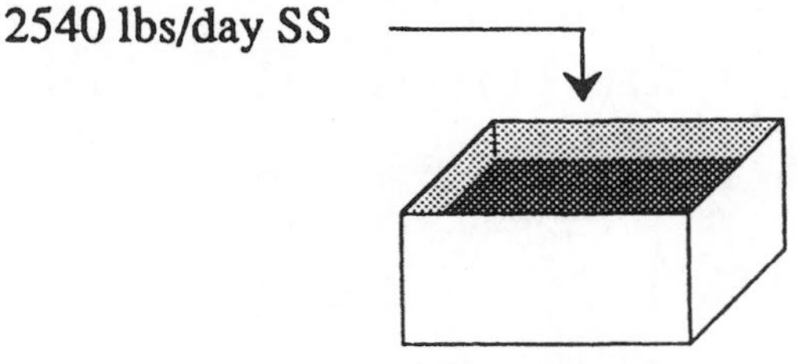

$$\text{Sludge Age, days} = \frac{\text{MLSS, lbs}}{\text{SS Added, lbs/day}}$$

$$= \frac{13{,}000 \text{ lbs}}{2540 \text{ lbs/day}}$$

$$= \boxed{5.1 \text{ days}}$$

---

* Sludge age based on solids <u>entering</u> the aeration tank is sometimes referred to as Sludge Age (Gould) to distinguish it from the calculation of Solids Retention Time (or Mean Cell Residence Time), which is based on solids <u>leaving</u> the aeration tank.

Example 2: (Sludge Age)

❏ An aeration tank contains 500,000 gallons of wastewater. The MLSS is 2200 mg/*L*. If the primary effluent flow is 3.7 MGD with a suspended solids concentration of 72 mg/*L*, what is the sludge age?

(72 mg/*L*) (3.7 MGD) (8.34 lbs/gal) = 2222 lbs/day SS
SS

lbs MLSS in Aeration Tank:

(2200 mg/*L*) (0.5 MG) (8.34 lbs/gal) = 9174 lbs MLSS
MLSS

$$\text{Sludge Age, days} = \frac{\text{MLSS, lbs}}{\text{SS Added, lbs/day}}$$

$$= \frac{9174 \text{ lbs MLSS}}{2222 \text{ lbs/day SS}}$$

$$= \boxed{4.1 \text{ days}}$$

Example 3: (Sludge Age)

❏ An aeration tank is 80 ft long, 20 ft wide with wastewater to a depth of 15 ft. The mixed liquor suspended solids concentration is 2800 mg/*L*. If the primary effluent flow is 1.6 MGD with a suspended solids concentration of 65 mg/*L*, what is the sludge age in the aeration tank?

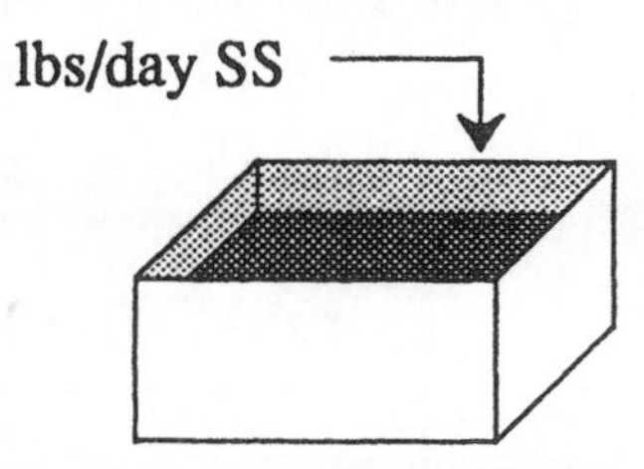

<u>Aeration Tank Volume</u>

(80 ft) (20 ft) (15 ft) (7.48 gal/cu ft) = 179,520 gal

$$\text{Sludge Age, days} = \frac{(2800 \text{ mg/}L)(0.18 \text{ MG})(\cancel{8.34} \text{ lbs/gal})}{(65 \text{ mg/}L)(1.6 \text{ MGD})(\cancel{8.34} \text{ lbs/gal})}$$

$$= \boxed{4.8 \text{ days}}$$

USING SLUDGE AGE TO CALCULATE OTHER UNKNOWNS

The sludge age equation can be used to calculate:

- Desired lbs MLSS, and
- Desired mg/*L* SS

In Examples 1-3, sludge age was the unknown variable. In Examples 4-6, the sludge age equation is used to calculate the desired pounds of MLSS in the aeration tank.

Example 4: (Sludge Age)

❑ A sludge age of 5 days is desired. Assume 1200 lbs/day suspended solids enter the aeration tank in the primary effluent. To maintain the desired sludge age, how many lbs of MLSS must be maintained in the aeration tank?

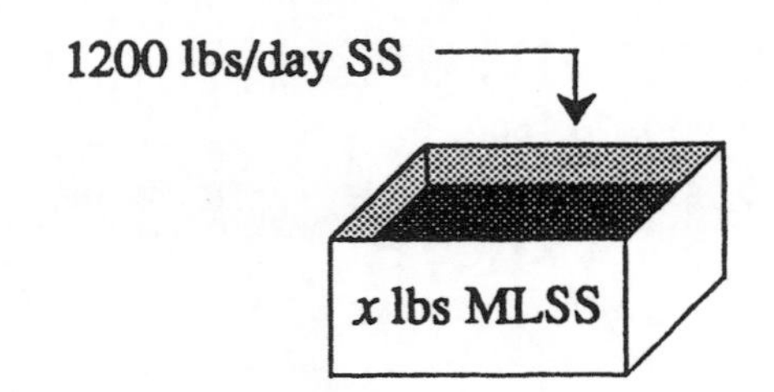

$$\text{Sludge Age, days} = \frac{\text{MLSS, lbs}}{\text{SS Added, lbs/day}}$$

$$5 \text{ days} = \frac{x \text{ lbs MLSS}}{1200 \text{ lbs/day SS Added}}$$

$$(5)\,(1200) = x$$

$$\boxed{6000 \text{ lbs}} = x$$

Example 5: (Sludge Age)

❑ A sludge age of 5.2 days is desired for an aeration tank 100 ft long, 40 ft wide, with a liquid level of 15 ft. If 1950 lbs/day suspended solids enter the aeration tank in the primary effluent flow, how many lbs of MLSS must be maintained in the aeration tank to maintain the desired sludge age?

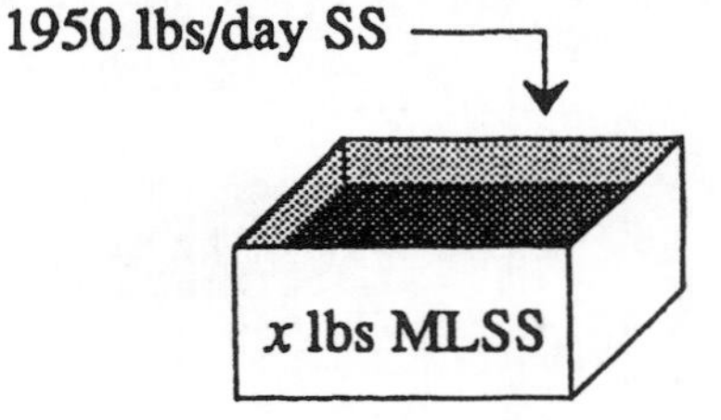

$$\text{Sludge Age, days} = \frac{\text{MLSS, lbs}}{\text{SS Added, lbs/day}}$$

$$5.2 \text{ days} = \frac{x \text{ lbs MLSS}}{1950 \text{ lbs/day SS Added}}$$

$$(5.2)\,(1950) = x$$

$$\boxed{10{,}140 \text{ lbs}} = x$$

Example 6: (Sludge Age)

❑ The 1.3-MGD primary effluent flow to an aeration tank has a suspended solids concentration of 65 mg/*L*. The aeration tank volume is 180,000 gallons. If a sludge age of 6 days is desired, what is the desired MLSS concentration?

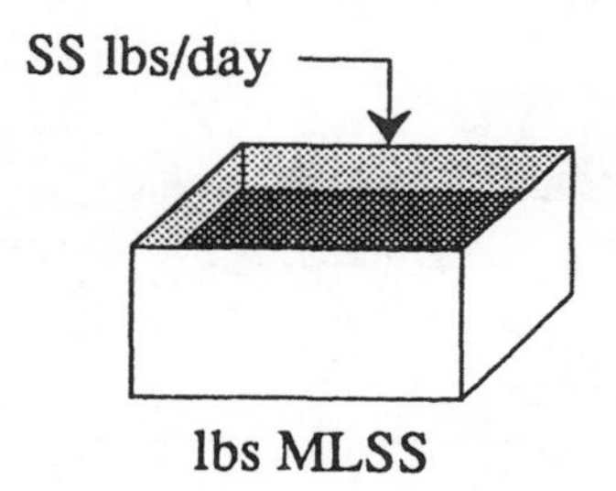

$$6 \text{ days} = \frac{(x \text{ mg/}L \text{ MLSS})\ (0.18 \text{ mg})\ (\cancel{8.34} \text{ lbs/gal})}{(65 \text{ mg/}L)\ (1.3 \text{ MGD})\ (\cancel{8.34} \text{ lbs/gal})}$$

After dividing out the 8.34 factor from the numerator and denominator, solve for x:

$$x = \frac{(65)\ (1.3)\ (6)}{0.18}$$

$$x = \boxed{2817 \text{ mg/}L \text{ MLSS}}$$

Example 7: (Sludge Age)

❑ A total of 5500 lbs of MLSS are desired in the aeration tank. What is the desired MLSS concentration (in mg/*L*) if the aeration tank volume is 250,000 gallons?

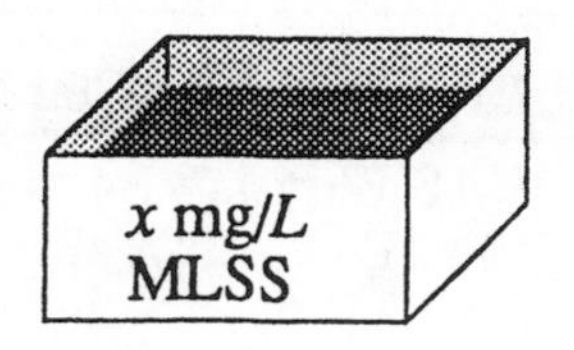

$$(x \text{ mg/}L \text{ MLSS})\ (0.25 \text{ MG})\ (8.34) = 5500 \text{ lbs MLSS}$$

$$x = \frac{5500}{(0.25)\ (8.34)}$$

$$x = \boxed{2638 \text{ mg/}L \text{ MLSS}}$$

CALCULATING MLSS CONCENTRATION GIVEN POUNDS SOLIDS AND AERATION TANK VOLUME

The sludge age equation can be used in some cases, such as in Example 6, to calculate the actual or desired MLSS concentration. However, many times sludge age is not mentioned at all and yet MLSS concentration must be determined.

When converting from mg/*L* to lbs MLSS or vice versa, the following equation is used:*

$$(\text{mg/}L)\ (\text{Vol, MG})\ (\underset{\text{lbs/gal}}{8.34}) = \text{lbs}$$

Note that this is the same equation used in the sludge age numerator to calculate lbs SS (lbs MLSS) in the aeration tank :

$$(\underset{\text{MLSS}}{\text{mg/}L})\ (\underset{\text{MG}}{\text{Aer.Vol.}})\ (\underset{\text{lbs/gal}}{8.34}) = \underset{\text{MLSS}}{\text{lbs}}$$

* mg/*L* to lbs/day calculations are discussed in Chapter 3.

5.3 SOLIDS RETENTION TIME CALCULATIONS

Solids Retention Time (SRT), also called Mean Cell Residence Time (MCRT), is a calculation very similar to the sludge age calculation. There are two principal differences in calculating SRT:

1. **The SRT calculation is based on suspended solids leaving the system.** (Sludge age is based on suspended solids entering the system.)

2. **There are four different methods that may be used to calculate lbs MLSS (system solids).**** (In sludge age calculations, only the MLSS concentration and aeration tank volume are used in calculating system solids.)

Examples 1 and 2 illustrate the calculation of SRT, using the combined volume method of estimating system solids.

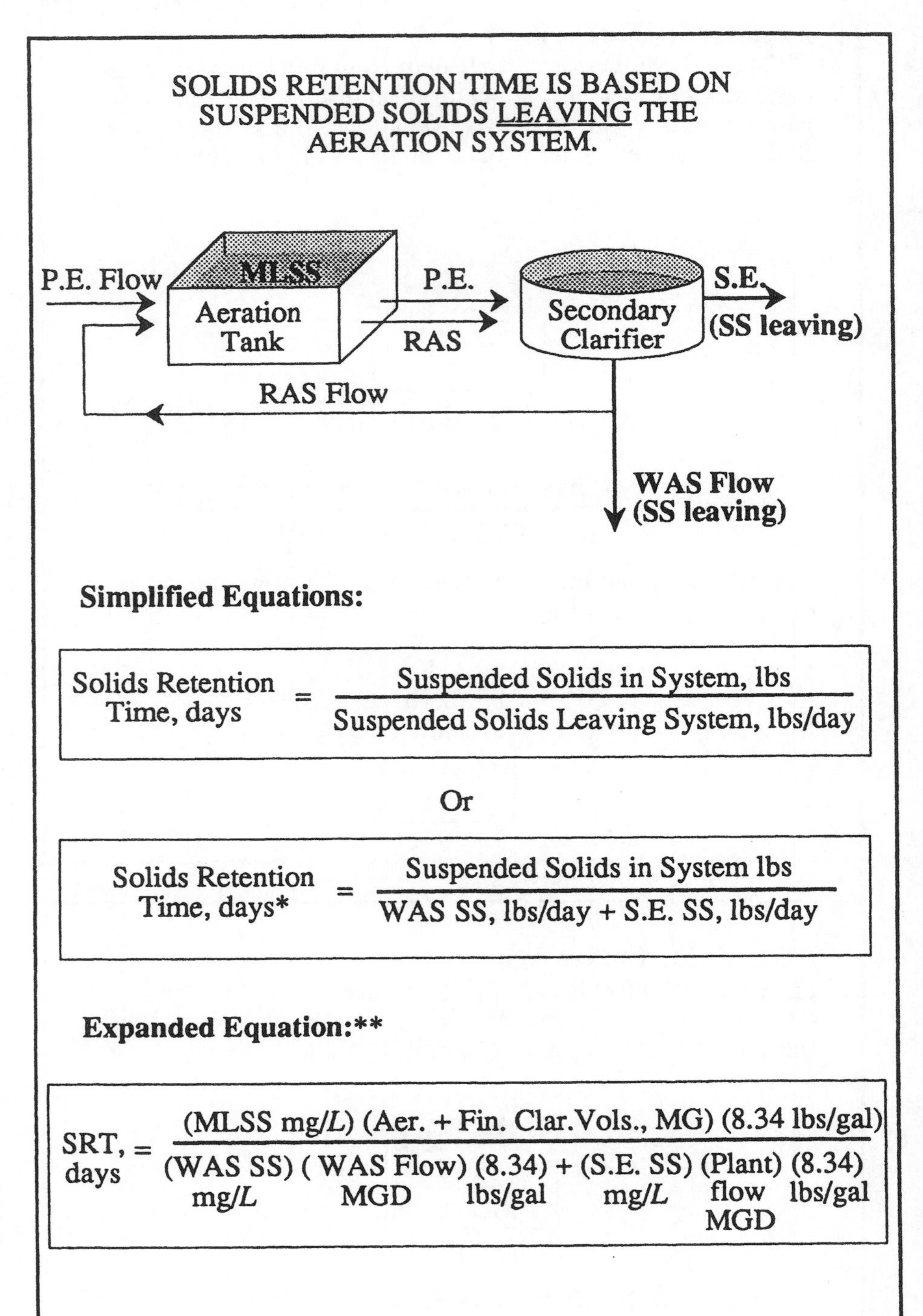

* S.E. is Secondary Effluent.

** There are four ways to account for system solids in the SRT calculation (numerator). The other three methods are discussed in Chapter 12.

Example 1: (Solids Retention Time)

❑ An activated sludge system has a total of 9930 lbs of mixed liquor suspended solids. The suspended solids leaving the final clarifier in the effluent is calculated to be 290 lbs/day. The lbs suspended solids wasted from the final clarifier is 1050 lbs/day. What is the solids retention time, in days?

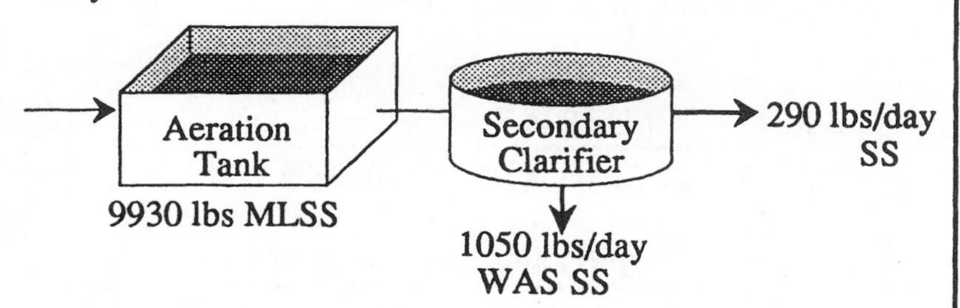

$$\text{SRT days} = \frac{\text{MLSS in System, lbs}}{\text{WAS SS, lbs/day} + \text{S.E. SS, lbs/day}}$$

$$= \frac{9930 \text{ lbs MLSS}}{1050 \text{ lbs/day} + 290 \text{ lbs/day}}$$

$$= \boxed{7.4 \text{ days}}$$

Example 2: (Solids Retention Time)

❑ An aeration tank has a volume of 330,000 gal. The final clarifier has a volume of 150,000 gallons. The MLSS concentration in the aeration tank is 2900 mg/*L*. If a total of 1520 lbs SS/day are wasted and 400 lbs SS/day are in the secondary effluent, what is the solids retention time for the activated sludge system?

$$\text{SRT days} = \frac{\text{MLSS in System, lbs}}{\text{WAS SS, lbs/day} + \text{S.E. SS, lbs/day}}$$

$$= \frac{(2900 \text{ mg}/L)\ (0.48 \text{ MGD})\ (8.34 \text{ lbs/gal})}{1520 \text{ lbs/day} + 400 \text{ lbs/day}}$$

$$= \frac{11{,}609 \text{ lbs MLSS}}{1920 \text{ lbs/day SS leaving}}$$

$$= \boxed{6 \text{ days}}$$

Example 3: (Solids Retention Time)

❑ Determine the solids retention time (SRT) given the following data:

Aer. Tank Vol. 1.1 MG
Fin. Clar. Vol. 0.4 MG
P.E. Flow 3.9 MGD
WAS Pumping Rate 80,000 gpd
MLSS 2500 mg/*L*
WAS SS 6150 mg/*L*
S.E. SS 15 mg/*L*

$$\text{SRT, days} = \frac{\text{MLSS in System, lbs}}{\text{SS Leaving System, lbs/day}}$$

$$= \frac{(2500\ \text{mg}/L)\ (1.5\ \text{MG})\ (8.34\ \text{lbs/gal})}{\underset{\text{mg}/L}{(6150)}\ \underset{\text{MGD}}{(0.08)}\ \underset{\text{lbs/gal}}{(8.34)} + \underset{\text{mg}/L}{(15)}\ \underset{\text{MGD}}{(3.9)}\ \underset{\text{lbs/gal}}{(8.34)}}$$

$$= \frac{31{,}275\ \text{lbs MLSS}}{4103\ \text{lbs SS} + 488\ \text{lbs SS}}$$

$$= \boxed{6.8\ \text{days}}$$

Example 4: (Solids Retention Time)

❑ Calculate the solids retention time (SRT) given the following data:

Aer. Tank Vol. 300,000 gal
Fin. Clar. Vol. 120,000 gal
P.E. Flow 2.2 MGD
WAS Pumping Rate 19,000 gpd
MLSS 2400 mg/*L*
WAS SS 5900 mg/*L*
S.E. SS 20 mg/*L*

$$\text{SRT, days} = \frac{\text{MLSS in System, lbs}}{\text{SS Leaving System, lbs/day}}$$

$$= \frac{(2400\ \text{mg}/L)\ (0.42\ \text{MG})\ (8.34\ \text{lbs/gal})}{\underset{\text{mg}/L}{(5900)}\ \underset{\text{MGD}}{(0.019)}\ \underset{\text{lbs/gal}}{(8.34)} + \underset{\text{mg}/L}{(20)}\ \underset{\text{MGD}}{(2.2)}\ \underset{\text{lbs/gal}}{(8.34)}}$$

$$= \frac{8407\ \text{lbs MLSS}}{935\ \text{lbs/day SS} + 367\ \text{lbs/day SS}}$$

$$= \boxed{6.5\ \text{days}}$$

Example 5: (Solids Retention Time)

❑ The volume of an aeration tank is 320,000 gal and the final clarifier is 130,000 gal. The desired SRT for a plant is 7 days. The primary effluent flow is 2 MGD and the WAS pumping rate is 25,000 gpd. If the WAS SS is 5400 mg/*L* and the secondary effluent SS is 18 mg/*L*, what is the desired MLSS mg/*L*?

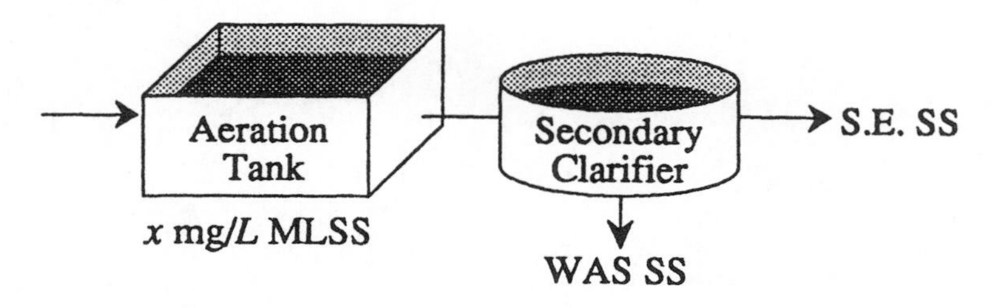

$$\text{SRT, days} = \frac{\text{MLSS in System, lbs}}{\text{SS Leaving System, lbs/day}}$$

$$7 \text{ days} = \frac{(x \text{ mg/}L \text{ MLSS})(0.45 \text{ MG})(8.34 \text{ lbs/gal})}{\underset{\text{mg/}L}{(5400)}\ \underset{\text{MGD}}{(0.025)}\ \underset{\text{lbs/gal}}{(8.34)} + \underset{\text{mg/}L}{(18)}\ \underset{\text{MGD}}{(2)}\ \underset{\text{lbs/gal}}{(8.34)}}$$

$$7 \text{ days} = \frac{(x)(0.45)(8.34)}{1126 + 300}$$

$$7 = \frac{(x)(0.45)(8.34)}{1426}$$

$$\frac{(1426)(7)}{(0.45)(8.34)} = x$$

$$\boxed{2660 \text{ mg/}L \text{ MLSS}} = x$$

CALCULATING OTHER UNKNOWN FACTORS

The SRT calculation incorporates many variables:

- SRT
- MLSS, mg/*L*
- Aeration Tank and Clarifier Volumes
- WAS SS, mg/*L*
- WAS Pumping Rate, MGD
- S.E. SS, mg/*L*
- Plant Flow

In Examples 1-4, SRT was the unknown variable. Other variables can also be unknown, as illustrated in Example 5. Regardless of which variable is unknown, use the same equation, fill in the given information, then solve for the unknown value.

NOTES:

6 Efficiency and Other Percent Calculations

SUMMARY

1. **Unit process efficiency calculations** refer to the **percent removal** of a water or wastewater constituent such as suspended solids (SS) or biochemical oxygen demand (BOD).

Simplified Equation:

$$\% \text{ Removed} = \frac{\text{Part Removed}}{\text{Total}} \times 100$$

SS Removal Efficiency:

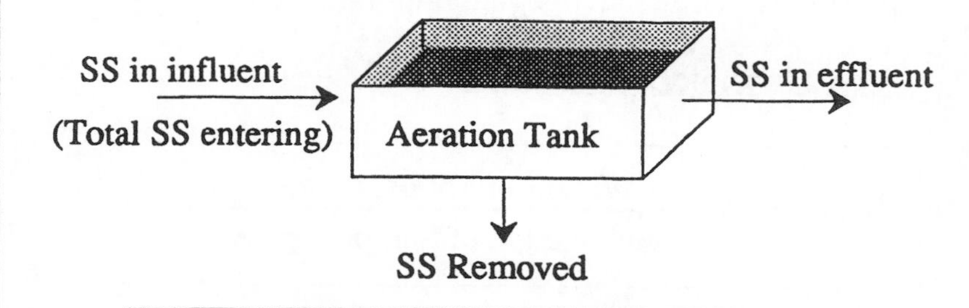

$$\% \text{ SS Removed} = \frac{\text{SS Removed, mg/}L}{\text{SS Total, mg/}L} \times 100$$

BOD Removal Efficiency:

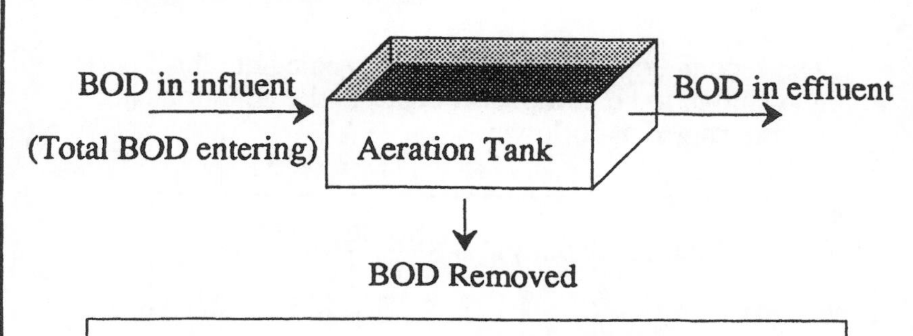

$$\% \text{ BOD Removed} = \frac{\text{BOD Removed, mg/}L}{\text{BOD Total, mg/}L} \times 100$$

This chapter focuses on various percent calculations in water and wastewater math. The underlying concept in each of these calculations is percent:*

$$\% = \frac{\text{Part}}{\text{Whole}} \times 100$$

Note that **for every percent sign, %, included in an equation, there should be a 100 factor** in the equation as well—either directly under the percent sign or on the opposite side of the equation in the same relative location. (If the percent sign is in the numerator, the 100 factor on the opposite side of the equation will also be in the numerator. And if the percent sign is in the denominator, the 100 factor on the opposite side of the equation will also be in the denominator. This can be verified by moving the 100 factor according to the diagonal rule of movement.)**

The equation above shows the 100 factor on the opposite side of the equation. Written on the same side of the equation it would be:

$$\frac{\%}{100} = \frac{\text{Part}}{\text{Whole}}$$

This concept of the location of the100 factor is important since you will encounter equations using both means of expression.

* To review percent calculations, refer to Chapter 5 in *Basic Math Concepts*.

** For a review of moving terms from one side of the equation to the other, refer to Chapter 2 in *Basic Math Concepts*.

SUMMARY—Cont'd

Sludge percent calculations include four different calculations (problem types 2-5 below):

2. **Percent Solids and Sludge Pumping Rate**—In making these calculations it is important to distinguish between the terms "solids" and "sludge". Solids refers to dry solids; whereas, sludge refers to water and solids.*

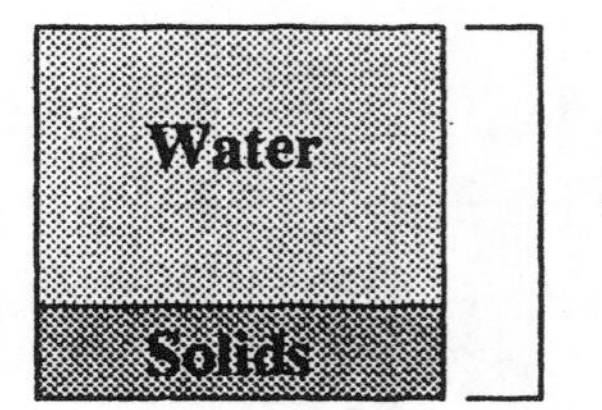

The two most common sludge percent calculations are percent solids and sludge pumping rate.

Using laboratory data:

$$\% \text{ Solids} = \frac{\text{Weight of Solids, grams}}{\text{Total Weight of Sample, grams}} \times 100$$

Using plant data:

$$\% \text{ Solids} = \frac{\text{Solids, lbs/day}}{\text{Sludge, lbs/day}} \times 100$$

The second equation can be used to calculate lbs/day or gpd sludge to be pumped. If desired, the equation can be rearranged as follows:

$$(\text{Sludge, lbs/day}) = \frac{\text{Solids, lbs/day}}{\frac{\% \text{ Solids}}{100}}$$

* The diagram of sludge and solids is for illustration purposes only and is not to scale. Primary sludge contains about 3-5% solids, and secondary sludges about 1-3% solids. Sludge is actually a mixture of solids and water. The diagram above shows the portions of water and solids if they were separated.

SUMMARY—Cont'd

3. **Mixing Different Percent Solids Sludges**
When mixing sludges with different percent solids, use the following equation to calculate the percent solids concentration of the resulting sludge:

$$\text{\% Solids of Sludge Mixture} = \frac{\text{Solids in Mixture, lbs/day}}{\text{Sludge Mixture, lbs/day}} \times 100$$

4. **Percent Volatile Solids**—Two equations represent the two most common volatile solids calculations—the first using laboratory data, the second using plant data.*

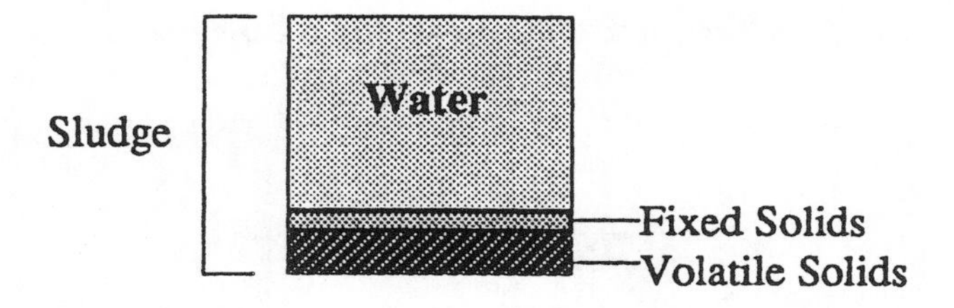

$$\text{\% Volatile Solids} = \frac{\text{Weight Volatile Solids, grams}}{\text{Weight of Total Solids, grams}} \times 100$$

$$\text{\% Volatile Solids} = \frac{\text{Volatile Solids, lbs}}{\text{Total Solids, lbs}} \times 100$$

5. **Seed Sludge (Based on % Digester Volume)**
In Chapter 4, required digester seed sludge was calculated on the basis of a volatile solids loading ratio. Another way to calculate seed sludge required is based on a percent of the digester volume.

$$\text{\% Seed Sludge} = \frac{\text{Seed Sludge, gal}}{\text{Total Digester Capacity, gal}} \times 100$$

SUMMARY—Cont'd

Chemical dosage percent calculations include solution strength problems and solution mixture problems.

6. **Solution Strength**
   The strength of a solution is a measure of the amount of chemical dissolved in the solution.

$$\% \text{ Strength} = \frac{\text{Weight of Chemical}}{\text{Weight of Solution}} \times 100$$

7. **Mixing Different Percent Strength Solutions**

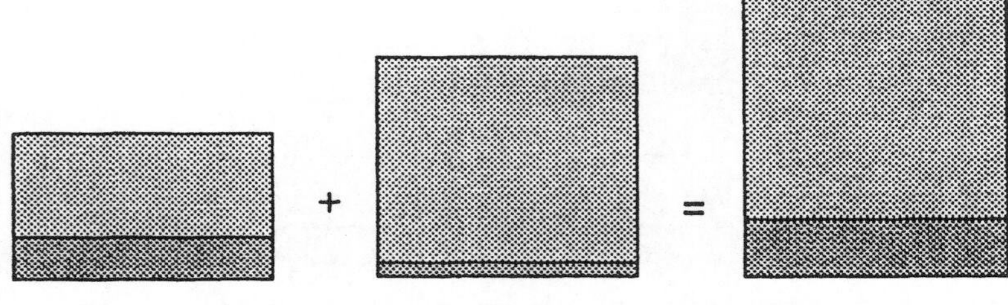

10% Strength Solution

1% Strength Solution

Solution Mixture (% Strength somewhere between 10% and 1% depending on the quantity contributed by each.)

Simplified Equation:

$$\begin{matrix}\% \text{ Strength} \\ \text{of Mixture}\end{matrix} = \frac{\text{Chemical in Mixture, lbs}}{\text{Solution Mixture, lbs}} \times 100$$

Expanded Equation:

$$\begin{matrix}\% \text{ Strength} \\ \text{of Mixture}\end{matrix} = \frac{\begin{matrix}\text{lbs Chem. from} \\ \text{Solution 1}\end{matrix} + \begin{matrix}\text{lbs Chem. from} \\ \text{Solution 2}\end{matrix}}{\text{lbs Solution 1} + \text{lbs Solution 2}} \times 100$$

SUMMARY—Cont'd

8. **Pump and motor efficiency calculations** are based on horsepower input and output.

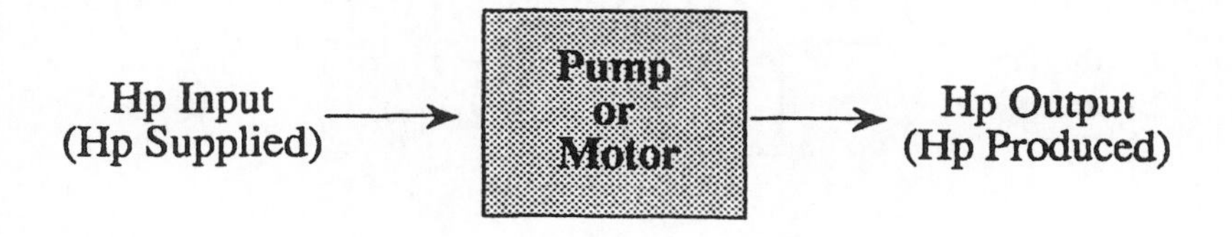

$$\% \text{ Efficiency} = \frac{\text{Hp Output}}{\text{Hp Input}} \times 100$$

6.1 UNIT PROCESS EFFICIENCY CALCULATIONS

The efficiency of a treatment process is its effectiveness in removing various constituents from the water or wastewater. Suspended solids, BOD and COD removal are the most common calculations of unit process efficiency.

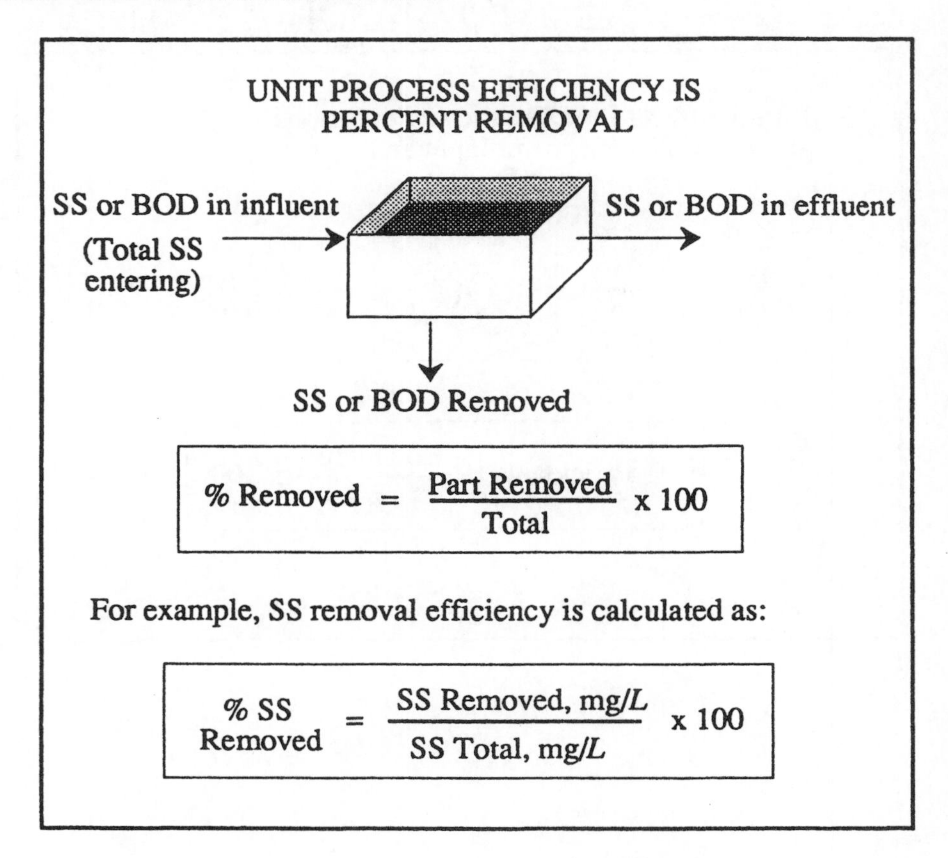

Example 1: (Unit Process Efficiency)

❑ The suspended solids entering a trickling filter is 190 mg/*L*. If the suspended solids in the trickling filter effluent is 22 mg/*L*, what is suspended solids removal efficiency of the trickling filter?

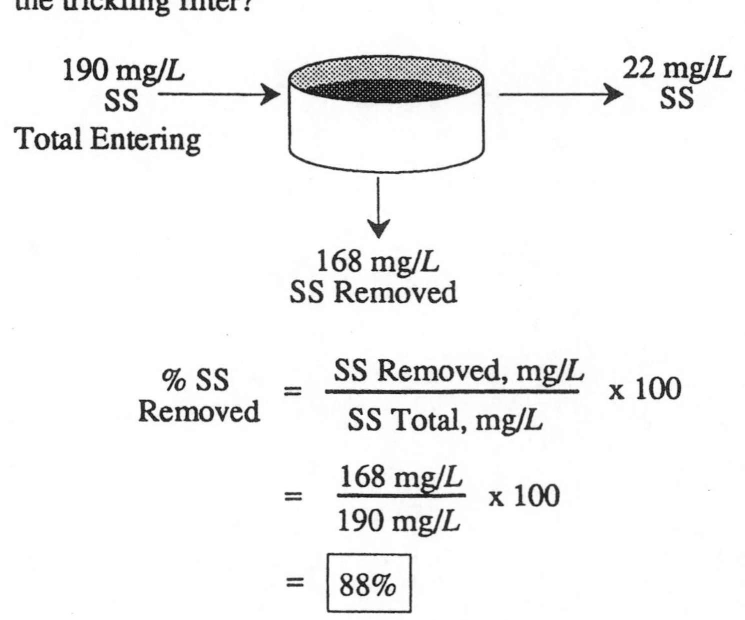

Example 2: (Unit Process Efficiency)

❑ The influent of a primary clarifier has a BOD content of 250 mg/*L*. If the clarifier effluent has a BOD content of 120 mg/*L*, what is the BOD removal efficiency?

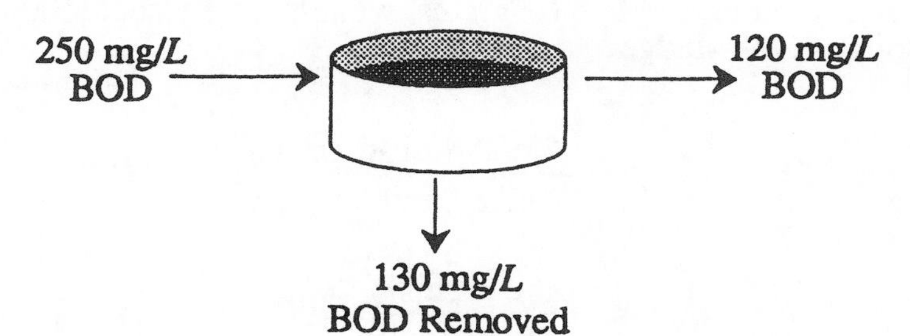

$$\%\ \text{BOD Removed} = \frac{\text{BOD Removed, mg}/L}{\text{BOD Total, mg}/L} \times 100$$

$$= \frac{130\ \text{mg}/L}{250\ \text{mg}/L} \times 100$$

$$= \boxed{52\%}$$

Example 3: (Unit Process Efficiency)

❑ The suspended solids entering a primary clarifier is 220 mg/*L*. The suspended solids concentration of the primary clarifier effluent is 99 mg/*L*. What is the suspended solids removal efficiency?

$$\%\ \text{SS Removed} = \frac{\text{SS Removed, mg}/L}{\text{SS Total, mg}/L} \times 100$$

$$= \frac{121\ \text{mg}/L\ \text{SS}}{220\ \text{mg}/L\ \text{SS}} \times 100$$

$$= \boxed{55\%}$$

6.2 PERCENT SOLIDS AND SLUDGE PUMPING RATE CALCULATIONS

PERCENT SOLIDS

Sludge is composed of water and solids. The vast majority of sludge is water—usually in the range of 93 to 97%.

To determine the solids content of a sludge, a sample of sludge is dried overnight in an oven at about 103°- 105° C. **The solids that remain after drying represent the total solids content of the sludge.** This solids content may be expressed as a percent or as a mg/*L* concentration.* Two equations are used to calculate percent solids, depending on whether lab data or plant data is used in the calculation. In both cases, the calculation is **on the basis of solids and sludge weight.**

$$\% \text{ Solids} = \frac{\text{Total Solids, g}}{\text{Sludge Sample, g}} \times 100$$

$$\% \text{ Solids} = \frac{\text{Solids, lbs}}{\text{Sludge, lbs}} \times 100$$

Example 1: (Percent Solids)

❑ The total weight of a sludge sample is 15 grams. (Sludge sample only, not the dish.) If the weight of the solids after drying is 0.62 grams, what is the percent total solids of the sludge?

$$\% \text{ Solids} = \frac{\text{Total Solids, grams}}{\text{Sludge Sample, grams}} \times 100$$

$$= \frac{0.62 \text{ grams}}{15 \text{ grams}} \times 100$$

$$= \boxed{4.1\%}$$

Example 2: (Percent Solids)

❑ A total of 3000 gallons of sludge is pumped to a digester. If the sludge has a 6% solids content, how many lbs/day solids are pumped to the digester?

First, write the % solids equation and fill in the given information:

$$\% \text{ Solids} = \frac{\text{Solids, lbs/day}}{\text{Sludge, lbs/day}} \times 100$$

$$6 = \frac{x \text{ lbs/day Solids}}{(3000 \text{ gal})(8.34 \text{ lbs/gal})} \times 100$$

$$\frac{(3000)(8.34)(6)}{100} = x$$

$$\boxed{\text{1501 lbs/day Solids}} = x$$

---

* 1% solids = 10,000 mg/*L*. For a review of % to mg/*L* conversions, refer to Chapter 8 in *Basic Math Concepts*.

Example 3: (Percent Solids)

❑ A total of 10,000 lbs/day SS are removed from a primary clarifier and pumped to a sludge thickener. If the sludge has a solids content of 3%, how many lbs/day sludge are pumped to the thickener?

$$\% \text{ Solids} = \frac{\text{Solids, lbs/day}}{\text{Sludge, lbs/day}} \times 100$$

$$3 = \frac{10{,}000 \text{ lbs/day Solids}}{x \text{ lbs/day Sludge}} \times 100$$

$$x = \frac{(10{,}000)\,(100)}{3}$$

$$= \boxed{333{,}333 \text{ lbs/day Sludge}}$$

Example 4: (Percent Solids)

❑ It is anticipated that 200 lbs/day SS will be pumped from the primary clarifier of a new plant. If the primary clarifier sludge has a solids content of 5%, how many gpd sludge will be pumped from the clarifier? (Assume the sludge weighs 8.34 lbs/gal.)

First calculate lbs/day sludge to be pumped using the % solids equation, then convert lbs/day sludge to gpd sludge:

$$\text{Sludge, lbs/day} = \frac{\text{Solids, lbs/day}}{\frac{\% \text{ Solids}}{100}}$$

$$x \text{ lbs/day Sludge} = \frac{200 \text{ lbs/day Solids}}{0.05}$$

$$x = 4000 \text{ lbs/day Sludge}$$

Converting lbs/day sludge to gpd sludge:

$$\frac{4000 \text{ lbs/day Sludge}}{8.34 \text{ lbs/gal}} = \boxed{480 \text{ gpd Sludge}}$$

CALCULATING OTHER UNKNOWN VARIABLES

The three variables in percent solids calculations are percent solids, lbs/day solids, and lbs/day sludge. In Example 1, percent solids was the unknown variable. Examples 2-4 illustrate calculations when other variables are unknown. In solving these problems, write the equation as usual, fill in the known information, then solve for the unknown value.*

SLUDGE TO BE PUMPED

As mentioned above, one of the variables in the % solids calculation is lbs/day sludge. Example 3 illustrates a calculation of lbs/day sludge. Example 4 illustrates the calculation of gpd sludge to be pumped.

Because lbs/day sludge is calculated relatively frequently, the % solids equation is often rearranged as follows:

$$\text{Sludge, lbs/day} = \frac{\text{Solids, lbs/day}}{\frac{\% \text{ Solids}}{100}}$$

* Refer to Chapter 2 in *Basic Math Concepts* for a review of solving for the unknown value.

6.3 MIXING DIFFERENT PERCENT SOLIDS SLUDGES CALCULATIONS

When sludges with different percent solids content are mixed, the resulting sludge has a percent solids content somewhere **between** the solids contents of the original sludges. For example, if a 4% primary sludge is mixed with a 1% secondary sludge, the resulting sludge might have a solids content of about 2 or 3%. The actual percent solids content will depend on how much (lbs) of each sludge is mixed together. If, in the example, most of the sludge is from the secondary sludge (1% solids) and very little from the primary sludge (4% solids), then the resulting sludge would be closer to a 1% sludge (perhaps a 1.5% sludge). If, on the other hand, most of the sludge is primary sludge and very little is secondary sludge, then the resulting sludge mixture might have a solids content closer to 4%—such as 3 or 3.5%.

The actual solids content of a mixture of two or more sludges depends on the pounds of sludge contributed from each source.

As with the sludge thickening equation, remember that if the thickened sludge has a density greater than 8.34 lbs/gal, it must be used instead of 8.34 lbs/gal.*

WHEN SLUDGES ARE MIXED THE MIXTURE HAS A % SOLIDS CONTENT BETWEEN THE TWO ORIGINAL % SOLIDS VALUES

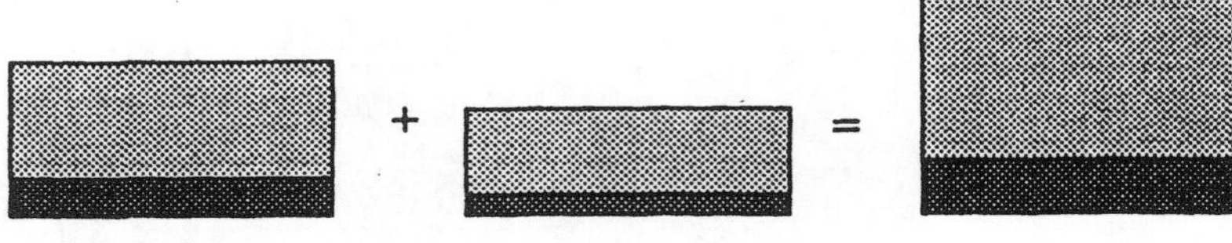

Simplified Equation:

$$\text{\% Solids of Sludge Mixture} = \frac{\text{Solids in Mixture, lbs/day}}{\text{Sludge Mixture, lbs/day}} \times 100$$

Expanded Equation:

$$\text{\% Solids of Sludge Mixture} = \frac{\text{Prim. Sol., lbs/day} + \text{Sec. Sol., lbs/day}}{\text{Prim. Sludge, lbs/day} + \text{Sec. Sludge, lbs/day}} \times 100$$

Example 8: (Mixing Sludges)

❑ A primary sludge flow of 5000 gpd (5% solids) is mixed with a thickened secondary sludge flow of 3500 gpd (3% solids). What is the percent solids content of the mixed sludge flow?

$$\text{\% Solids of Sludge Mixture} = \frac{\text{Prim. Sl. Sol., lbs/day} + \text{Sec. Sl. Sol., lbs/day}}{\text{Prim. Sludge, lbs/day} + \text{Sec. Sludge, lbs/day}} \times 100$$

$$= \frac{\underset{\text{Prim. Slud.}}{(5000 \text{ gpd})}\ \underset{\text{lbs/gal}}{(8.34)}\ \left(\frac{5}{100}\right) + \underset{\text{Sec. Slud.}}{(3500 \text{ gpd})}\ \underset{\text{lbs/gal}}{(8.34)}\ \left(\frac{3}{100}\right)}{\underset{\text{Prim. Sludge}}{(5000 \text{ gpd})\ (8.34)} + \underset{\text{Sec. Sludge}}{(3500 \text{ gpd})\ (8.34)}} \times 100$$

$$= \frac{2085 \text{ lbs/day Prim Sol.} + 876 \text{ lbs/day Sec. Sol.}}{41{,}700 \text{ lbs/day Prim Slud.} + 29{,}190 \text{ lbs/day Sec. Slud.}} \times 100$$

$$= \frac{2961 \text{ lbs/day Solids}}{70{,}890 \text{ lbs/day Sludge}} \times 100$$

$$= \boxed{4.2\% \text{ Solids}}$$

* Refer to Chapter 7 for a review of density and specific gravity.

Example 9: (Mixing Sludges)

❑ Primary and thickened secondary sludges are to be mixed and sent to the digester. The 5700 gpd primary sludge has a solids content of 5.5%; the 4200 gpd thickened secondary sludge has a solids content of 3.8%. What would be the percent solids content of the mixed sludge?

$$\text{\% Solids of Sludge Mixture} = \frac{\text{Prim. Sl. Sol., lbs/day} + \text{Sec. Sl. Sol., lbs/day}}{\text{Prim. Sludge, lbs/day} + \text{Sec. Sludge, lbs/day}} \times 100$$

$$= \frac{\underset{\text{Prim. Sl.}}{(5700\text{ gpd})}\ \underset{\text{lbs/gal}}{(8.34)}\ \left(\frac{5.5}{100}\right) + \underset{\text{Sec. Sl.}}{(4200\text{ gpd})}\ \underset{\text{lbs/gal}}{(8.34)}\ \left(\frac{3.8}{100}\right)}{\underset{\text{Prim. Sludge}}{(5700\text{ gpd})}\ \underset{\text{lbs/gal}}{(8.34)} + \underset{\text{Sec. Sludge}}{(4200\text{ gpd})}\ \underset{\text{lbs/gal}}{(8.34)}} \times 100$$

$$= \frac{2615\text{ lbs/day Prim Sol.} + 1331\text{ lbs/day Sec Sol.}}{47{,}538\text{ lbs/day Prim. Slud.} + 35{,}028\text{ lbs/day Sec. Slud.}} \times 100$$

$$= \frac{3946\text{ lbs/day Solids}}{82{,}566\text{ lbs/day Sludge}} \times 100$$

$$= \boxed{4.8\%\text{ Solids}}$$

Example 10: (Mixing Sludges)

❑ A primary sludge flow of 6800 gpd (3.8% solids) is mixed with a thickened secondary sludge flow of 4500 gpd (7% solids). What is the percent solids of the combined sludge flow?

$$\text{\% Solids of Sludge Mixture} = \frac{\text{Prim. Sl. Sol., lbs/day} + \text{Sec. Sl. Sol., lbs/day}}{\text{Prim. Sludge, lbs/day} + \text{Sec. Sludge, lbs/day}} \times 100$$

$$= \frac{(6800\text{ gpd})\ \underset{\text{lbs/gal}}{(8.34)}\ \left(\frac{3.8}{100}\right) + (4500)\ \underset{\text{lbs/gal}}{(8.34)}\ \left(\frac{7}{100}\right)}{(6800\text{ gpd})\ (8.34\text{ lbs/day}) + (4500)\ (8.34\text{ lbs/day})} \times 100$$

$$= \frac{2155\text{ lbs/day} + 2627\text{ lbs/day}}{56{,}712\text{ lbs/day} + 37{,}530\text{ lbs/day}} \times 100$$

$$= \frac{4782\text{ lbs/day Solids}}{94{,}242\text{ lbs/day Sludge}} \times 100$$

$$= \boxed{5.1\%\text{ Solids}}$$

6.4 PERCENT VOLATILE SOLIDS

Sludge solids are comprised of organic matter (from plant or animal sources) and inorganic matter (material from mineral sources, such as sand and grit). The organic matter is called **volatile solids**, the inorganic matter is called **fixed solids**. Together, the volatile solids and fixed solids make up the **total solids**.

When calculating percent solids (also called percent total solids) and percent volatile solids, it is essential to focus on the general concept of percent:

$$\text{Percent} = \frac{\text{Part}}{\text{Whole}} \times 100$$

As illustrated in the diagrams to the right, **when calculating percent solids**, the "part" of interest is the weight of the total solids; the "whole" is the weight of the sludge:

$$\text{\% Solids} = \frac{\text{Wt. of Solids}}{\text{Wt. of Sludge}} \times 100$$

When calculating percent volatile solids, the "part" of interest is the weight of the volatile solids; the "whole" is the weight of total solids:

$$\text{\% VS} = \frac{\text{Wt. of VS}}{\text{Wt. of Tot. Solids}} \times 100$$

The calculation of volatile solids using laboratory data is described more thoroughly in Chapter 18.

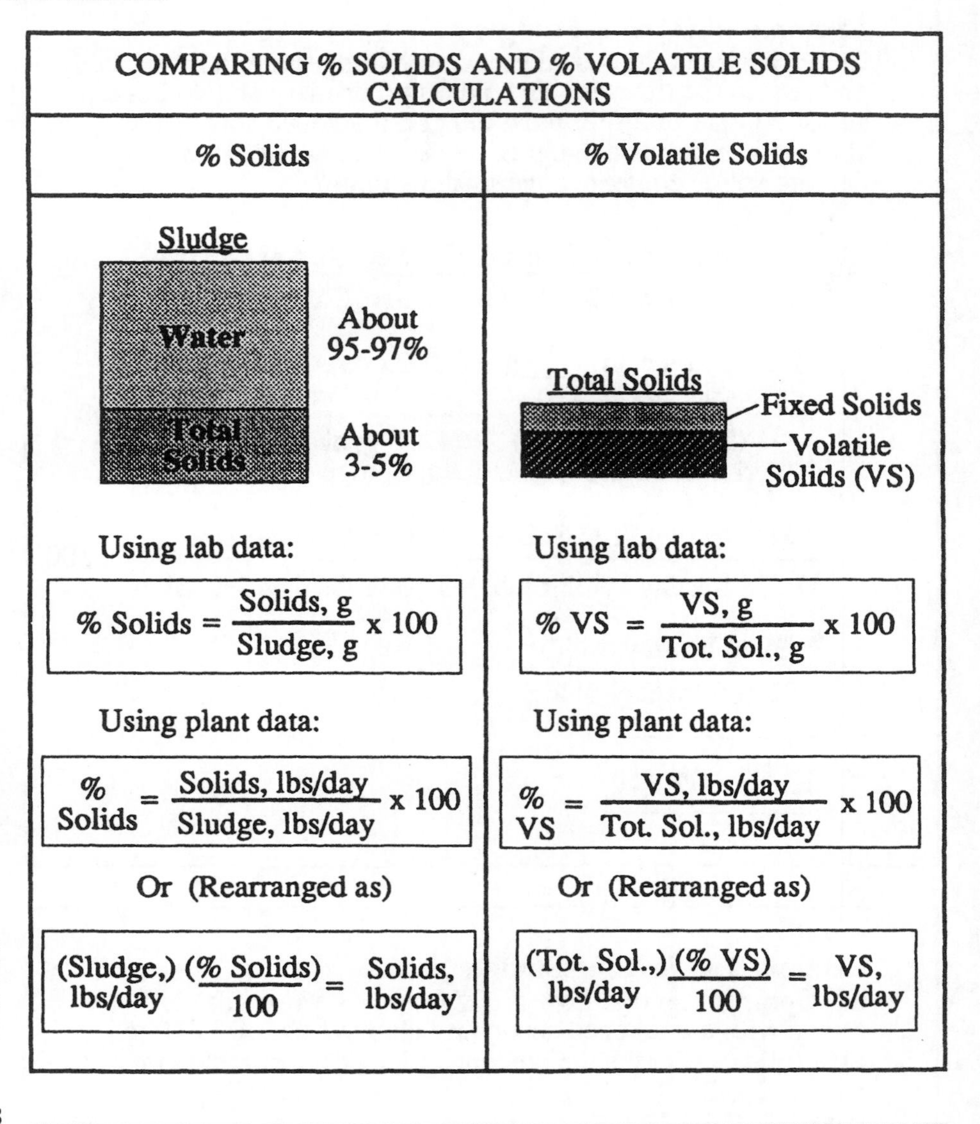

Example 11: (% Volatile Solids)

❑ If 1250 lbs/day solids are sent to the digester, with a volatile solids content of 72%, how many lbs/day volatile solids are sent to the digester?

Either the % Vol solids equation or the rearranged equation may be used to calculate the lbs/day volatile solids.

$$(\text{Total Solids, lbs/day}) \frac{(\text{\% VS})}{100} = \text{Vol. Sol. lbs/day}$$

$$(1250 \text{ lbs/day}) \frac{(72)}{100} = \boxed{\text{900 lbs/day Vol. Solids}}$$

Example 12: (% Volatile Solids)

❑ A total of 3000 gpd sludge is to be pumped to the digester. If the sludge has a 6% solids content with 70% volatile solids, how many lbs/day volatile solids are pumped to the digester?

$$\underset{\text{lbs/day}}{(\text{Sludge})}\left(\frac{\%\ \text{Solids}}{100}\right)\left(\frac{\%\text{Vol. Sol.}}{100}\right) = \underset{\text{lbs/day}}{\text{Vol. Sol.}}$$

Since sludge is given in gpd, 8.34 lbs/gal must be added to the equation to convert gpd sludge to lbs/day sludge:*

lbs/day Sludge

$$\underset{\text{gpd}}{(\text{Sludge})}\ \underset{\text{lbs/gal}}{(8.34)}\left(\frac{\%\ \text{Solids}}{100}\right)\left(\frac{\%\ \text{Vol. Sol.}}{100}\right) = \underset{\text{lbs/day}}{\text{Vol. Sol.}}$$

$$\underset{\text{gpd}}{(3000)}\ \underset{\text{lbs/gal}}{(8.34)}\left(\frac{6}{100}\right)\left(\frac{70}{100}\right) = \boxed{\text{1051 lbs/day Vol. Sol.}}$$

CALCULATING VOLATILE SOLIDS GIVEN SLUDGE DATA

Sometimes you will have lbs/day sludge information and will want to calculate lbs/day volatile solids. When this is the case, you must include the % solids factor in the equation as well, shown in the equation below. In effect, you are calculating lbs/day solids first, (using the % Solids factor), then the lbs/day Volatile Solids (using the % Volatile Solids factor):

$$\underset{\text{lbs/day}}{(\text{Sludge})}\left(\frac{\%\ \text{Sol.}}{100}\right)\left(\frac{\%\ \text{VS}}{100}\right) = \underset{\text{lbs/day}}{\text{VS}}$$

Example 13: (% Volatile Solids)

❑ A sludge with 5% solids has a volatile solids content of 68%. If 1200 lbs/day volatile solids are pumped to the digester (a) how many lbs/day sludge are pumped to the digester? and (b) how many gpd sludge are pumped to the digester? (Assume the sludge weighs 8.34 lbs/gal.)

(a)

$$\underset{\text{lbs/day}}{(\text{Sludge})}\left(\frac{\%\ \text{Sol.}}{100}\right)\left(\frac{\%\ \text{Vol. Sol.}}{100}\right) = \underset{\text{lbs/day}}{\text{Vol. Sol.}}$$

$$(x\ \text{lbs/day Sludge})\left(\frac{5}{100}\right)\left(\frac{68}{100}\right) = \text{1200 lbs/day Vol. Sol.}$$

$$(x)\,(0.05)\,(0.68) = 1200$$

$$x = \frac{(1200)}{(0.05)\,(0.68)}$$

$$= \boxed{\text{35,294 lbs/day Sludge}}$$

(b) Convert lbs/day to gpd:

$$\frac{\text{35,294 lbs/day}}{\text{8.34 lbs/gal}} = \boxed{\text{4232 gpd Sludge}}$$

SOLVING FOR OTHER UNKNOWN FACTORS

The equations shown above can be used to solve for any one of the three or four variables shown. Use the same equation, fill-in the known information, and solve for the unknown variable. Example 13 illustrates this type of calculation.

* For a review of flow conversions, refer to Chapter 8 in *Basic Math Concepts.*

6.5 PERCENT SEED SLUDGE CALCULATIONS

There are many methods to determine seed sludge required to start a new digester. One method, discussed in Chapter 4, is to use a volatile solids loading ratio—lbs volatile solids added per lb volatile solids in the digester.

Another method is to calculate seed sludge required based on the volume of the digester. This method is not quite as sensitive to the volatile solids balance in the seed sludge and incoming sludge. Examples 1-4 illustrate this calculation.

Although most digesters have cone-shaped bottoms, for simplicity, it is assumed that the side water depth represents the average digester depth.

Example 1: (% Seed Sludge)

❑ A digester has a volume of 350,000 gallons. If the digester seed sludge is to be 22% of the digester volume, how many gallons of seed sludge will be required?*

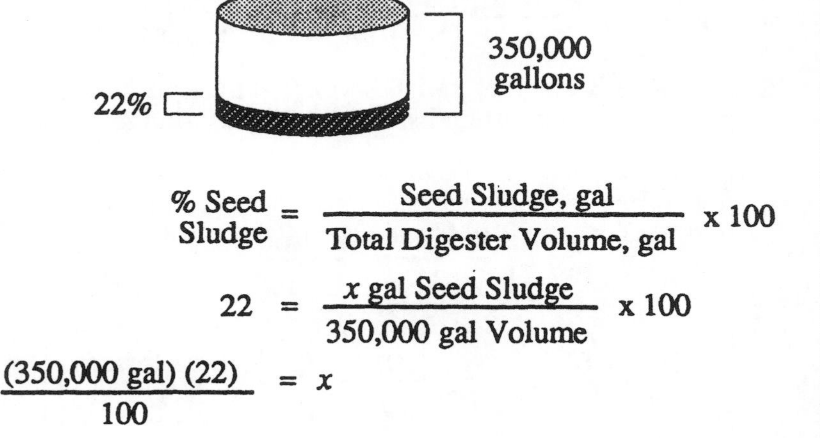

$$\text{\% Seed Sludge} = \frac{\text{Seed Sludge, gal}}{\text{Total Digester Volume, gal}} \times 100$$

$$22 = \frac{x \text{ gal Seed Sludge}}{350{,}000 \text{ gal Volume}} \times 100$$

$$\frac{(350{,}000 \text{ gal})(22)}{100} = x$$

$$\boxed{\begin{array}{c}77{,}000 \text{ gal} \\ \text{Seed Sludge}\end{array}} = x$$

Example 2: (% Seed Sludge)

❑ A 40-ft diameter digester has a typical water depth of 20 ft. If the seed sludge to be used is 20% of the tank volume, how many gallons of seed sludge will be required?

40 ft

20 ft

20%

$$\text{\% Seed Sludge} = \frac{\text{Seed Sludge, gal}}{\text{Total Digester Volume, gal}} \times 100$$

$$20 = \frac{x \text{ gal Seed Sludge}}{(0.785)(40 \text{ ft})(40 \text{ ft})(20 \text{ ft})(7.48 \text{ gal/cu ft})} \times 100$$

$$\frac{(0.785)(40 \text{ ft})(40 \text{ ft})(20 \text{ ft})(7.48)(20)}{100} = x$$

$$\boxed{\begin{array}{c}37{,}580 \text{ gal} \\ \text{Seed Sludge}\end{array}} = x$$

* For a review of volume calculations, refer to Chapter 11 in *Basic Math Concepts*.

Example 3: (% Seed Sludge)

❑ A digester 50 ft in diameter has a side water depth of 20 ft. If the digester seed sludge is to be 25% of the digester volume, how many gallons of seed sludge will be required?

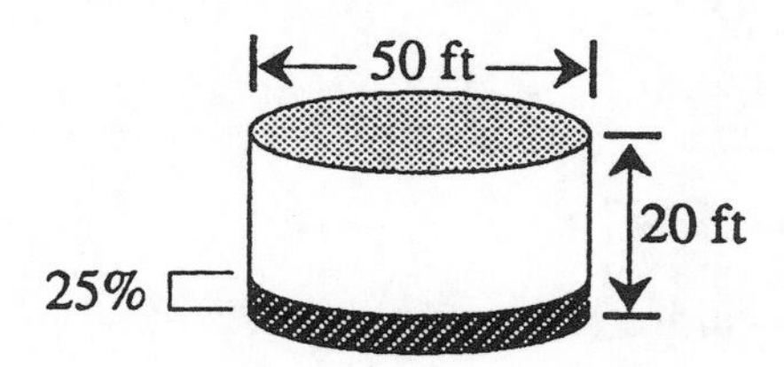

$$\%\ \text{Seed Sludge} = \frac{\text{Seed Sludge, gal}}{\text{Total Digester Volume, gal}} \times 100$$

$$25 = \frac{x \text{ gal Seed Sludge}}{(0.785)\ (50\ \text{ft})\ (50\ \text{ft})\ (20\ \text{ft})\ (7.48\ \text{gal/cu ft})} \times 100$$

$$\frac{(0.785)\ (50\ \text{ft})\ (50\ \text{ft})\ (20\ \text{ft})\ (7.48)\ (25)}{100} = x$$

$$\boxed{\text{73,398 gal Seed Sludge}} = x$$

Example 4: (% Seed Sludge)

❑ A 40-ft diameter digester has a typical side water depth of 18 ft. If 45,700 gallons of seed sludge are to be used in starting up the digester, what percent of the digester volume will be seed sludge?

40 ft
18 ft
x %

$$\%\ \text{Seed Sludge} = \frac{\text{Seed Sludge, gal}}{\text{Total Digester Volume, gal}} \times 100$$

$$x = \frac{\text{45,700 gal Seed Sludge}}{(0.785)\ (40\ \text{ft})\ (40\ \text{ft})\ (18\ \text{ft})\ (7.48\ \text{gal/cu ft})} \times 100$$

$$x = \boxed{27\%}$$

CALCULATING OTHER UNKNOWN FACTORS

There are three variables in percent seed sludge calculations: percent seed sludge, gallons seed sludge, and total gallons digester volume.

In Examples 1-3, the unknown factor was seed sludge gallons. However, the same equation can be used to calculate either one of the other two variables. Example 4 is one such calculation.

6.6 PERCENT STRENGTH OF A SOLUTION CALCULATIONS

PERCENT STRENGTH USING DRY CHEMICALS

The strength of a solution is a measure of the amount of chemical (solute) dissolved in the solution. Since percent is calculated as "part over whole,"

$$\% = \frac{\text{Part}}{\text{Whole}} \times 100$$

percent strength is calculated as **part chemical**, in lbs, divided by the **whole solution**, in lbs:

$$\%\ \text{Strength} = \frac{\text{Chem., lbs}}{\text{Sol'n, lbs}} \times 100$$

The denominator of the equation, lbs solution, includes both chemical (lbs) and water (lbs). Therefore, the equation can be written in expanded form as:

$$\%\ \text{Strength} = \frac{\text{Chem., lbs}}{\text{Water, lbs} + \text{Chem., lbs}} \times 100$$

As the two equations above illustrate, **the chemical added must be expressed in pounds**. If the chemical weight is expressed in ounces (as in Example 1) or grams (as in Example 2), it must first be converted to pounds (to correspond with the other lbs terms in the equation) before percent strength is calculated.

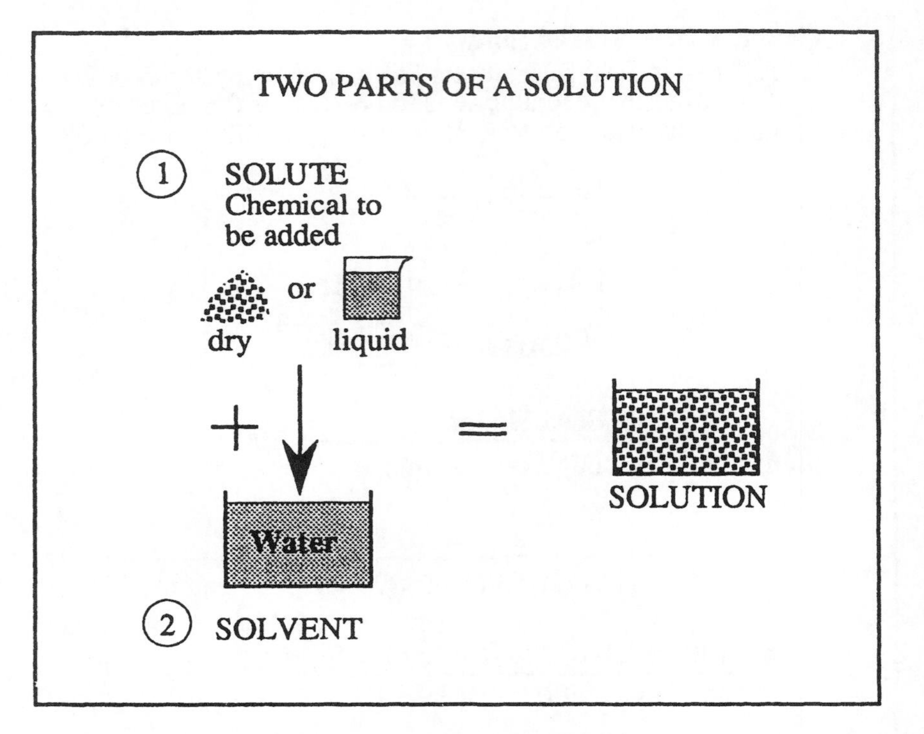

Example 1: (Percent Strength)

❑ If a total of 8 ounces of dry polymer are added to 10 gallons of water, what is the percent strength (by weight) of the polymer solution?

Before calculating percent strength, the ounces chemical must be converted to lbs chemical:*

$$\frac{8 \text{ ounces}}{16 \text{ ounces/pound}} = 0.5 \text{ lbs chemical}$$

Now calculate percent strength:

$$\%\ \text{Strength} = \frac{\text{Chemical, lbs}}{\text{Water, lbs} + \text{Chemical, lbs}} \times 100$$

$$= \frac{0.5 \text{ lbs Chemical}}{(10 \text{ gal})(8.34 \text{ lbs/gal}) + 0.5 \text{ lbs}} \times 100$$

$$= \frac{0.5 \text{ lbs Chemical}}{83.9 \text{ lbs Solution}} \times 100$$

$$= \boxed{0.6\%}$$

---

* To review ounces to pounds conversions refer to Chapter 8 in *Basic Math Concepts*.

Example 2: (Percent Strength)

❑ If 100 grams of dry polymer are dissolved in 5 gallons of water, what percent strength is the solution? (1 g = 0.0022 lbs)

First, convert grams chemical to pounds chemical. Since 1 gram equals 0.0022 lbs, 100 grams is 100 times 0.0022 lbs:

$$(100 \text{ grams Chemical})(0.0022 \text{ lbs/gram}) = 0.22 \text{ lbs Chemical}$$

Now calculate percent strength of the solution:

$$\% \text{ Strength} = \frac{\text{lbs Chemical}}{\text{lbs Water} + \text{lbs Chemical}} \times 100$$

$$= \frac{0.22 \text{ lbs Chemical}}{(5 \text{ gal})(8.34 \text{ lbs/gal}) + 0.22 \text{ lbs}} \times 100$$

$$= \frac{0.22 \text{ lbs}}{41.92 \text{ lbs}} \times 100$$

$$= \boxed{0.52\%}$$

WHEN GRAMS CHEMICAL ARE USED

The chemical (solute) to be used in making a solution may be measured in grams rather than pounds or ounces. When this is the case, convert grams of chemical to pounds of chemical before calculating percent strength. The following conversion equations may be used:

$$\boxed{1\ g = 0.0022 \text{ lbs}}$$

Or

$$\boxed{1 \text{ lb} = 454 \text{ grams}^{**}}$$

Example 3: (Percent Strength)

❑ How many pounds of dry polymer must be added to 25 gallons of water to make a 1% polymer solution?

First, write the equation as usual and fill in the known information. Then solve for the unknown.*

$$\% \text{ Strength} = \frac{\text{lbs Chemical}}{\text{lbs Water} + \text{lbs Chemical}} \times 100$$

$$1 = \frac{x \text{ lbs Chemical}}{(25 \text{ gal})(8.34 \text{ lbs/gal}) + x \text{ lbs Chemical}} \times 100$$

$$1 = \frac{100\,x}{208.5 + x}$$

$$1\,(208.5 + x) = 100\,x$$

$$208.5 = 100\,x - 1\,x$$

$$208.5 = 99\,x$$

$$\boxed{2.1 \text{ lbs Chem.}} = x$$

SOLVING FOR OTHER UNKNOWN VARIABLES

In the percent strength equation there are three variables: percent strength, lbs chemical and lbs water. In Examples 1 and 2, the unknown value was percent strength. However, the same equation can be used to determine either of the other two variables. Example 3 illustrates this type of calculation.

Note that gallons water can also be the unknown variable in percent strength calculations. First set lbs water as the unknown variable in the equation. Then when you have calculated the lbs water required, you can then convert lbs water to gallons water using the 8.34 lbs/gal factor.

* To review solving for the unknown value, refer to Chapter 2 in *Basic Math Concepts*.

** If the box method of conversions is used (see Chapter 8 in *Basic Math Concepts*), both numbers in the conversion equation must be greater than one.

6.7 MIXING DIFFERENT PERCENT STRENGTH SOLUTIONS CALCULATIONS

There are two types of solution mixture calculations. In one type of calculation, two solutions of different strengths are mixed, with no particular target solution strength. The calculation involves determining the percent strength of the solution mixture. These calculations are similar to the sludge mixture problems described in Section 6.3.

The second type of solution mixture calculation includes a desired or target strength. This calculation is described in Chapter 14, Section 5.

WHEN DIFFERENT % STRENGTH SOLUTIONS ARE MIXED

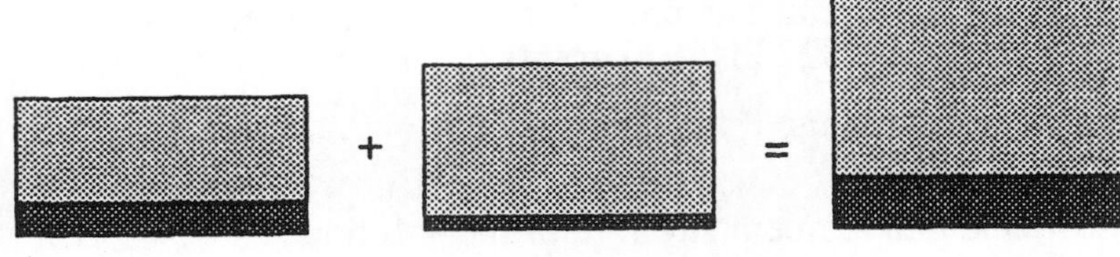

Simplified Equation:

$$\text{\% Strength of Mixture} = \frac{\text{lbs Chemical in Mixture}}{\text{lbs Solution Mixture}} \times 100$$

Expanded Equations:

$$\text{\% Strength of Mixture} = \frac{\text{lbs Chem. from Solution 1} + \text{lbs Chem. from Solution 2}}{\text{lbs Solution 1} + \text{lbs Solution 2}} \times 100$$

$$\text{\% Strength of Mixture} = \frac{(\text{Sol'n 1 lbs})\left(\frac{\text{\% Strength}}{100}\right) + (\text{Sol'n 2 lbs})\left(\frac{\text{\% Strength}}{100}\right)}{\text{lbs Solution 1} + \text{lbs Solution 2}} \times 100$$

Example 7: (Solution Mixtures)

❑ If 20 lbs of a 10% strength solution are mixed with 50 lbs of 1% strength solution, what is the percent strength of the solution mixture?

$$\text{\% Strength of Mixture} = \frac{\text{lbs Chem. from Solution 1} + \text{lbs Chem. from Solution 2}}{\text{lbs Solution 1} + \text{lbs Solution 2}} \times 100$$

$$= \frac{(20 \text{ lbs})\left(\frac{10}{100}\right) + (50 \text{ lbs})\left(\frac{1}{100}\right)}{20 \text{ lbs} + 50 \text{ lbs}} \times 100$$

$$= \frac{2 \text{ lbs} + 0.5 \text{ lbs}}{70 \text{ lbs}} \times 100$$

$$= \frac{2.5 \text{ lbs}}{70 \text{ lbs}} \times 100$$

$$= \boxed{3.6\%}$$

Example 8: (Solution Mixtures)

❑ If 5 gallons of an 8% strength solution are mixed with 40 gallons of a 0.5% strength solution, what is the percent strength of the solution mixture? (Assume the 8% solution weighs 9.5 lbs/gal and the 0.5% solution weighs 8.34 lbs/gal.)

$$\text{\% Strength of Mixture} = \frac{\text{lbs Chem. from Solution 1} + \text{lbs Chem. from Solution 2}}{\text{lbs Solution 1} + \text{lbs Solution 2}}$$

$$= \frac{(5\text{ gal})(9.5\text{ lbs/gal})\left(\frac{8}{100}\right) + (40\text{ gal})(8.34\text{ lbs/gal})\left(\frac{0.5}{100}\right)}{(5\text{ gal})(9.5\text{ lbs/gal}) + (40\text{ gal})(8.34\text{ lbs/gal})} \times 100$$

$$= \frac{3.8\text{ lbs Chem.} + 1.7\text{ lbs Chem.}}{47.5\text{ lbs Soln 1} + 333.6\text{ lbs Soln 2}} \times 100$$

$$= \frac{5.5\text{ lbs Chemical}}{381.1\text{ lbs Solution}} \times 100$$

$$= \boxed{1.4\%\text{ Strength}}$$

USE DIFFERENT DENSITY FACTORS WHEN APPROPRIATE

Percent strength should be expressed in terms of **pounds chemical per pounds solution.** Therefore, when solutions are expressed in terms of gallons, the gallons should be expressed as pounds before continuing with the percent strength calculation. It is important to consider what density factor should be used to convert from gallons to pounds. If the solution has a density the same as water, 8.34 lbs/gal would be used. If, however, the solution has a higher density, such as for some polymer solutions, then a higher density factor should be used. When the density is unknown, sometimes it is possible to weigh the chemical solution to determine its density.

Example 9: (Solution Mixtures)

❑ If 15 gallons of a 10% strength solution are added to 50 gallons of 0.8% strength solution, what is the percent strength of the solution mixture? (Assume the 10% strength solution weighs 10.2 lbs/gal and the 0.8% strength solution weighs 8.8 lbs/gal.)

$$\text{\% Strength of Mixture} = \frac{\text{lbs Chem. from Solution 1} + \text{lbs Chem. from Solution 2}}{\text{lbs Solution 1} + \text{lbs Solution 2}} \times 100$$

$$= \frac{(15\text{ gal})(10.2\text{ lbs/gal})\left(\frac{10}{100}\right) + (50\text{ gal})(8.8\text{ lbs/gal})\left(\frac{0.8}{100}\right)}{(15\text{ gal})(10.2\text{ lbs/gal}) + (50\text{ gal})(8.8\text{ lbs/gal})} \times 100$$

$$= \frac{15.3\text{ lbs Chem.} + 3.5\text{ lbs Chem.}}{153\text{ lbs Soln 1} + 440\text{ lbs Soln 2}} \times 100$$

$$= \frac{18.8\text{ lbs Chemical}}{593\text{ lbs Solution}} \times 100$$

$$= \boxed{3.2\%\text{ Strength}}$$

6.5 PUMP AND MOTOR EFFICIENCY CALCULATIONS

Pump and motor efficiencies are a measure of horsepower output compared with the horsepower input to the pump or motor. Since percent is a calculation of "part over whole,"

$$\% = \frac{\text{Part}}{\text{Whole}} \times 100$$

in these efficiency calculations the "part" is represented by the hp output, and the "whole" is represented by the total hp supplied (or hp input), as shown in the general efficiency equation to the right.

PUMP AND MOTOR EFFICIENCIES ARE CALCULATIONS OF **PERCENT HORSEPOWER OUTPUT**

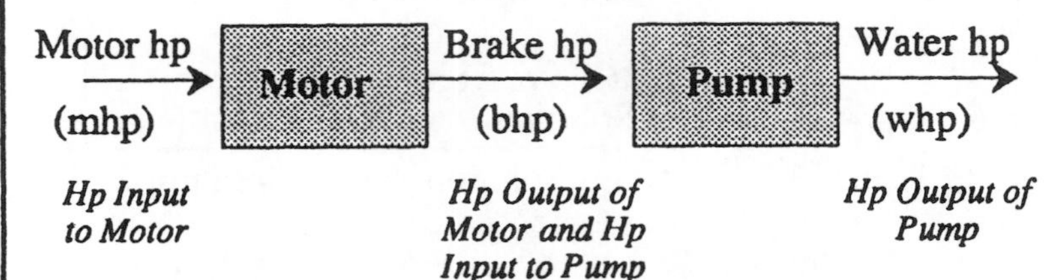

Hp Input to Motor — *Hp Output of Motor and Hp Input to Pump* — *Hp Output of Pump*

General Efficiency Equation:

$$\%\ \text{Hp Output} = \frac{\text{Hp Output}}{\text{Total hp Input}} \times 100$$

Motor and Pump Efficiency Equations:

These equations can be written using the terms Hp Input and Hp Output or mhp and bhp, as shown below:

$$\%\ \text{Motor Efficiency} = \frac{\text{Hp Output}}{\text{Hp Input}} \times 100 \quad \underline{\text{Or}} \quad \frac{\text{bhp}}{\text{mhp}} \times 100$$

$$\%\ \text{Pump Efficiency} = \frac{\text{Hp Output}}{\text{Hp Input}} \times 100 \quad \underline{\text{Or}} \quad \frac{\text{whp}}{\text{bhp}} \times 100$$

Overall Efficiency Equation:

$$\%\ \text{Overall Efficiency} = \frac{\text{Hp Output}}{\text{Hp Input}} \times 100 \quad \underline{\text{Or}} \quad \frac{\text{whp}}{\text{mhp}} \times 100$$

Example 1: (Pump and Motor Efficiency)

❑ The brake horsepower of a pump is 18 hp. If the water horsepower is 14 hp, what is the efficiency of the pump?

18 bhp → **Pump** → 14 whp

$$\%\ \text{Pump Efficiency} = \frac{\text{Hp Output}}{\text{Hp Input}} \times 100$$

$$= \frac{14 \text{ hp}}{18 \text{ hp}} \times 100$$

$$= \boxed{78\%}$$

Example 2: (Pump and Motor Efficiency)

❑ If the motor horsepower is 25 hp and the brake horsepower is 22 hp, what is the efficiency of the motor?

25 mhp → Motor → 22 bhp

$$\text{\% Motor Efficiency} = \frac{\text{Hp Output}}{\text{Hp Input}} \times 100$$

$$= \frac{22 \text{ hp}}{25 \text{ hp}} \times 100$$

$$= \boxed{88\%}$$

Example 3: (Pump and Motor Efficiency)

❑ The brake horsepower is 13.5 hp. If the motor is 90% efficient, what is the motor horsepower?

x mhp → Motor → 13.5 bhp

$$90 = \frac{13.5 \text{ bhp}}{x \text{ mhp}} \times 100$$

$$x = \frac{13.5}{90} \times 100$$

$$= \boxed{15 \text{ mhp}}$$

CALCULATING OTHER UNKNOWN VALUES

Pump and motor efficiency calculations have three variables: efficiency, hp output, and hp input. In Examples 1 and 2, efficiency was the unknown term. However, **any one of the variables can be the unknown term** as long as data is given for the other two. Example 3 illustrates this type of calculation.

Example 4: (Pump and Motor Efficiency)

❑ A total of 20 hp is supplied to a motor. If the wire-to-water efficiency of the pump and motor is 65%, what will the whp be?

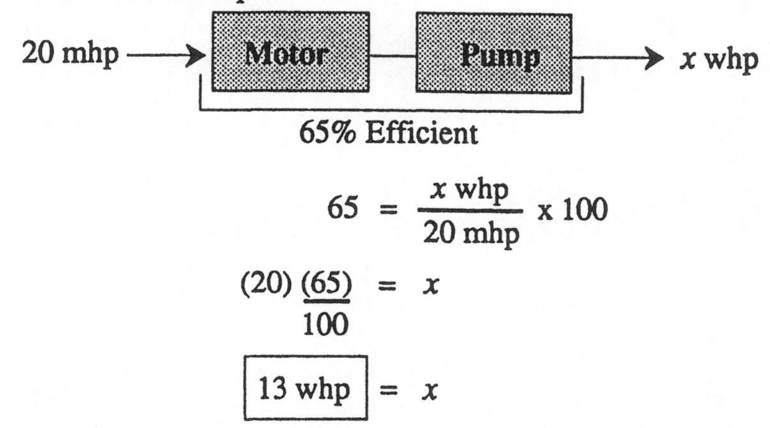

$$65 = \frac{x \text{ whp}}{20 \text{ mhp}} \times 100$$

$$\frac{(20)(65)}{100} = x$$

$$\boxed{13 \text{ whp}} = x$$

WIRE-TO-WATER EFFICIENCY

Wire-to-water efficiency is another name for **overall efficiency of the pump and motor**. In other words, the pump and motor are considered one unit. The hp input to the "unit" is the motor horsepower (mhp); the hp output to the "unit" is water horsepower (whp).

NOTES:

7 Pumping Calculations

SUMMARY

1. Density and Specific Gravity

The density of a substance is mass per volume (cu ft).

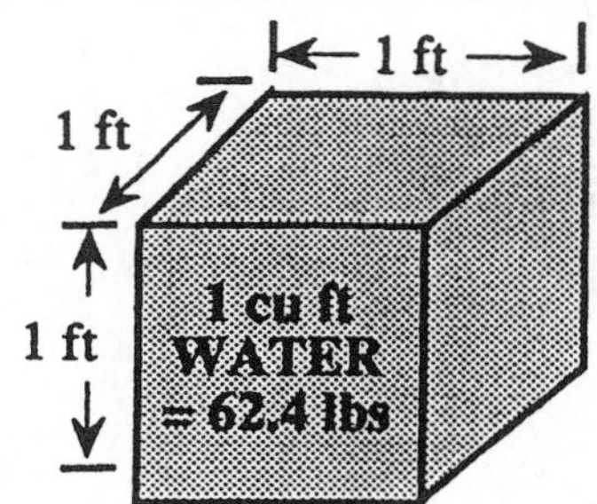

Density of Water = 62.4 $\frac{\text{lbs}}{\text{cu ft}}$

$$\text{Density of any Substance} = \frac{\text{lbs}}{\text{cu ft}}$$

The density of liquids is commonly expressed as pounds per gallon:

$$\text{Density of a Liquid} = \frac{\text{lbs of Liquid}}{\text{1 gal of Liquid}}$$

The specific gravity of a liquid is the density of the liquid compared to the density of water.

$$\text{Specific Gravity of a Liquid} = \frac{\text{Density of the Liquid}}{\text{Density of Water}}$$

If the specific gravity of a liquid is known, the density of that liquid may be calculated as:*

$$(\text{Spec. Grav. of a liquid})(\text{Density of Water, 62.4 lbs/gal}) = \text{Density of the Liquid, lbs/gal}$$

The specific gravity of a gas is the density of the gas compared to the density of air.

$$\text{Specific Gravity of a Gas} = \frac{\text{Density of the Gas}}{\text{Density of Air}}$$

Pumping calculations include a variety of different types of problems including:

- Density and Specific Gravity,
- Pressure and Force,
- Head and Head Loss,
- Horsepower, and
- Pump Capacity.

* Note that this equation is simply the specific gravity equation with the terms rearranged.

SUMMARY

2. Pressure and Force

Pressure exerted by solid objects depends on contact area.

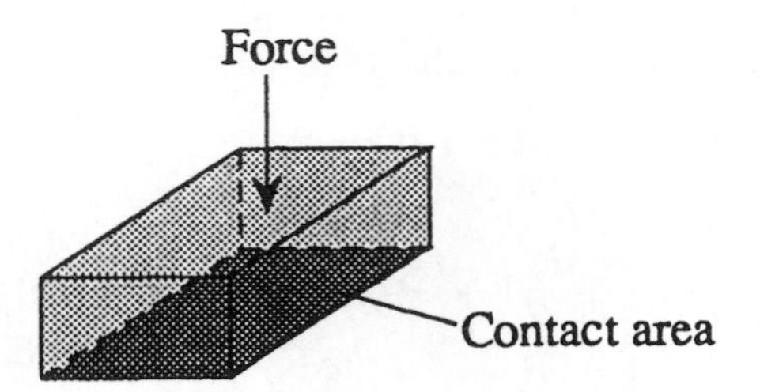

$$\text{Pressure} = \frac{\text{Force}}{\text{Area}}$$

Pressure exerted by a liquid depends on both liquid depth and density.

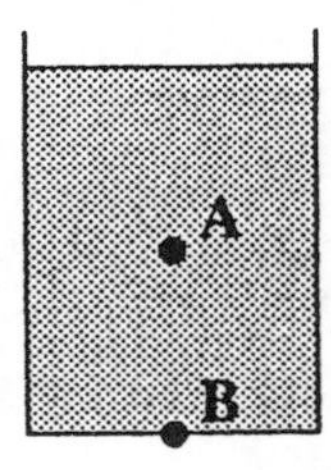

The pressure at Point B is greater than the pressure at Point A

$$\text{Pressure} = (\text{depth})\,(\text{Density})$$

or

$$\text{Pressure} = dD$$

Since d and D may be confused, the depth is generally expressed as height, h:

$$\text{Pressure} = hD$$

SUMMARY—Cont'd

2. Pressure and Force—Cont'd

The total force on a surface is the sum of all unit pressures against it. The total force against the bottom of a tank is:

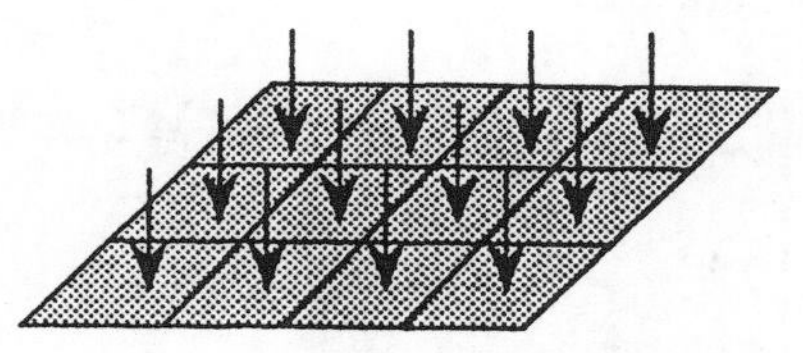

Total Force, lbs	=	(Pressure) lbs/sq ft	(Area) sq ft

or

Total Force, lbs	=	(Pressure) lbs/sq in.	(Area) sq in.

The total force against the side of a tank is:

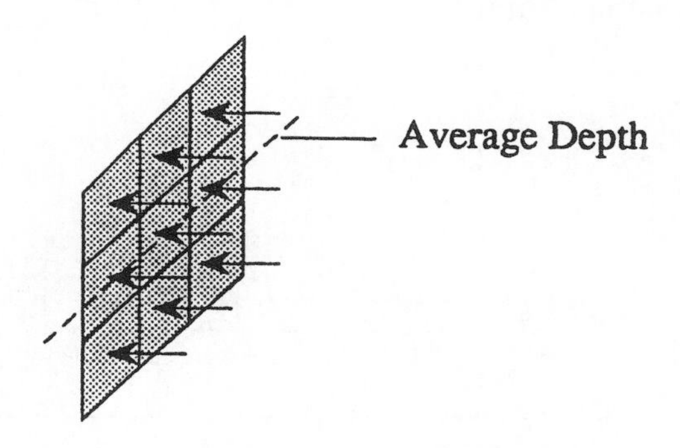

Total Force, lbs	=	(Pressure at Average) Depth, lbs/sq ft	(Total Area) sq ft

or

Total Force, lbs	=	(Pressure at Average) Depth, lbs/sq in.	(Total Area) sq in.

SUMMARY—Cont'd

2. Pressure and Force—Cont'd

The center of force on the side of a tank filled with water is located at a point two-thirds of the way down from the water surface:

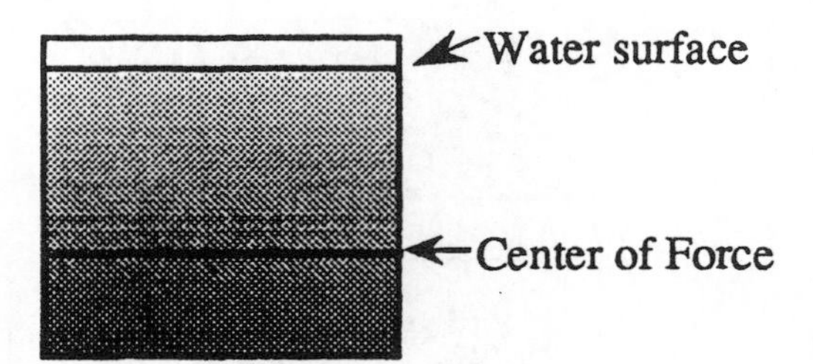

Front View of Wall

$$\text{Center of Force} = \frac{(2)}{3}(\text{Depth of Water})$$

A hydraulic press operates on the principle of total force. The pressure applied at the smaller piston (Point A) is transferred through the liquid to the larger piston (Point B).

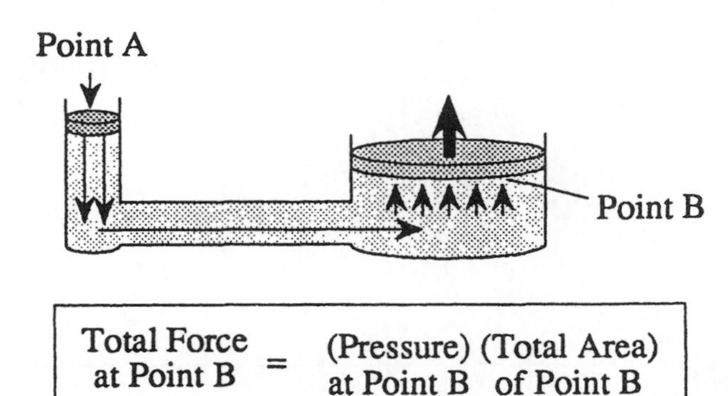

$$\text{Total Force at Point B} = (\text{Pressure at Point B})(\text{Total Area of Point B})$$

Gage pressures do not include atmospheric pressure. Absolute pressure includes both gage and atmospheric pressures.

$$\text{Absolute Pressure, psi} = \text{Gage Pressure, psi} + \text{Atmos. Pressure, psi}$$

SUMMARY—Cont'd

3. Head and Head Loss

When the two water surfaces are located above the pump, static head is the difference in water surface elevations:

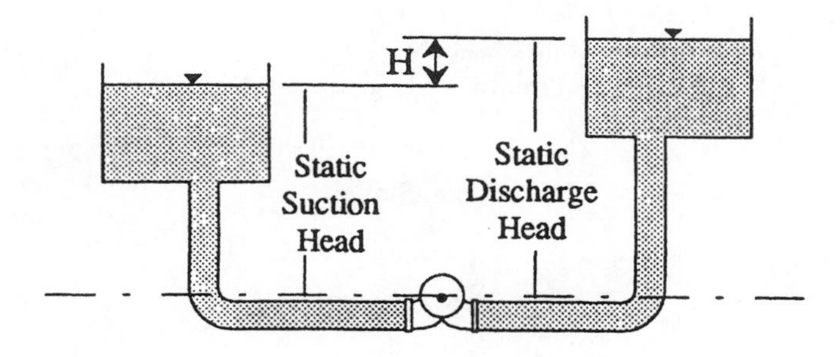

$$\text{Total Static Head, ft} = \text{Higher Elevation, ft} - \text{Lower Elevation, ft}$$

When the water surface on the suction side of the pump is below the pump centerline, the two distances must be added:

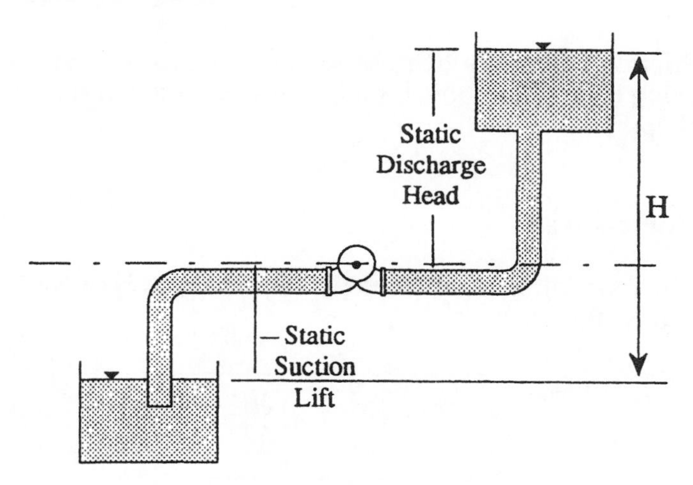

$$\text{Total Static Head, ft} = \text{Height Above Pump Centerline, ft} + \text{Distance Below Pump Centerline, ft}$$

SUMMARY—Cont'd

3. Head and Head Loss—Cont'd

Dynamic head is the static head plus friction and minor head losses:

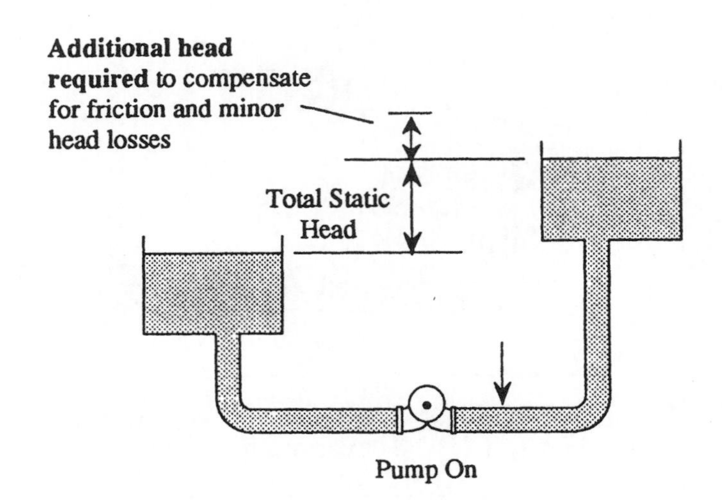

$$\text{Total Dynamic Head, ft} = \text{Total Static Head, ft} + \text{Head Losses ft}$$

Friction and minor head losses may be determined using hydraulics tables, such as those shown in this chapter.

4. Horsepower

The foundation of all horsepower problems is power—ft-lbs/min:

$$\text{ft} \times \frac{\text{lbs}}{\text{min}} = \frac{\text{ft-lbs}}{\text{min}}$$

$$\text{hp} = \frac{\text{ft-lbs/min}}{33{,}000 \text{ ft-lbs/min/hp}}$$

SUMMARY—Cont'd

4. Horsepower—Cont'd

Motor, brake, and water horsepower are terms used to indicate where the horsepower is measured.

mhp → **Motor** → bhp → **Pump** → whp

(Motor: Partial Loss of hp) (Pump: Partial Loss of hp)

The equations for motor, brake, and water horsepower are:

$$\text{Brake hp} = \frac{\text{Water hp}}{\frac{\text{Pump Effic.}}{100}}$$

$$\text{Motor hp} = \frac{\text{Brake hp}}{\frac{\text{Motor Effic.}}{100}}$$

$$\text{Motor hp} = \frac{\text{Water hp}}{\left(\frac{\text{Motor Effic.}}{100}\right)\left(\frac{\text{Pump Effic.}}{100}\right)}$$

Each of the three equations above may be rearranged as follows:

$$\text{Water hp} = (\text{Brake hp})\left(\frac{\text{Pump Effic.}}{100}\right)$$

$$\text{Brake hp} = (\text{Motor hp})\left(\frac{\text{Motor Effic.}}{100}\right)$$

$$\text{Water hp} = (\text{Motor hp})\left(\frac{\text{Motor Effic.}}{100}\right)\left(\frac{\text{Pump Effic.}}{100}\right)$$

SUMMARY—Cont'd

4. Horsepower—Cont'd

When pumping liquids with a specific gravity different than that of water, the specific gravity factor must be added to the ft-lbs/min calculation:

$$(\text{ft})(\text{lbs/min})(\text{spec. grav}) = \begin{array}{c}\text{ft-lbs/min} \\ \text{for different} \\ \text{liquid}\end{array}$$

To calculate pumping costs, first calculate the kilowatt-hours (kWh) power consumption:

$$\underset{\text{draw}}{(\text{kW})}\underset{\text{Operation}}{(\text{Hrs of Pump})} = \begin{array}{c}\text{kWh} \\ \text{power} \\ \text{consumed}\end{array}$$

Then determine pumping cost:

$$\underset{\text{Power Use}}{(\text{kWh})}\ \underset{\text{Use}}{(\text{Cost/kWh})} = \begin{array}{c}\text{Total} \\ \text{Cost}\end{array}$$

5. Pump Capacity

These calculations are based on a volume of wastewater or sludge pumped during a specific time period. The two general types of pumping rate calculations include:

- Pumping into or out of a tank, and
- Positive displacement calculations

General Equation:

$$\begin{array}{c}\text{Pumping} \\ \text{Rate, gpm}\end{array} = \frac{\text{gallons pumped}}{\text{minutes}}$$

SUMMARY—Cont'd

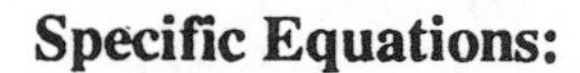

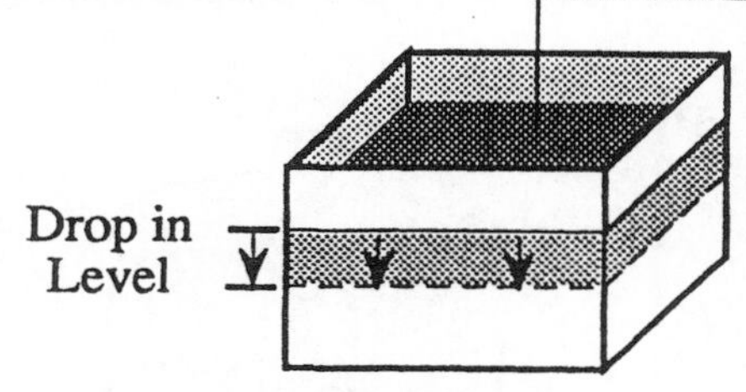

When Influent Valve Is Closed—

$$\text{Pumping Rate, gpm} = \frac{\text{gallons pumped}}{\text{minutes}}$$

$$\text{Pumping Rate, gpm} = \frac{(\text{Length, ft})(\text{Width, ft})(\text{Drop, ft})(7.48\ \text{gal/cu ft})}{\text{minutes}}$$

When Influent Valve Is Open—

The water level will drop if the pump is pumping at a rate greater than the influent flow:

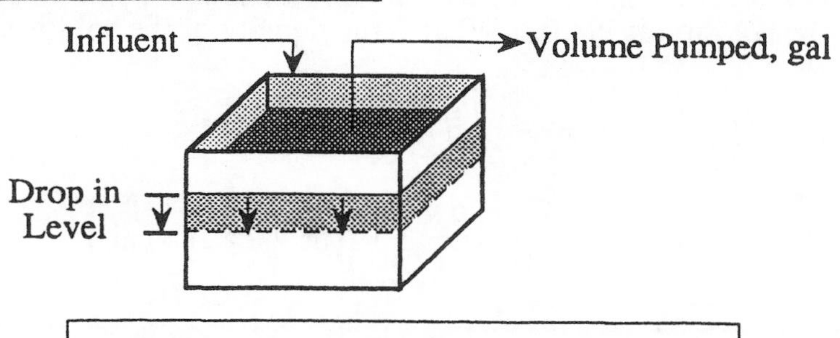

$$\text{Pumping Rate, gpm} = \text{Influent Flow, gpm} + \text{Drop in Level, gpm}$$

The water level will rise if the pump is pumping at a rate less than the influent flow:

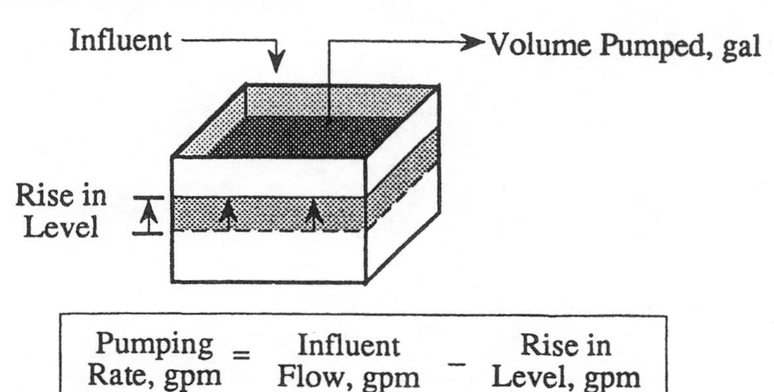

$$\text{Pumping Rate, gpm} = \text{Influent Flow, gpm} - \text{Rise in Level, gpm}$$

7.1 DENSITY AND SPECIFIC GRAVITY

DENSITY

The "lightness" or "heaviness" of an object is the layman's term for what scientists refer to as the density of an object. For example, petrified wood is heavy when compared to volcanic ash, and lead is heavy compared to aluminum. **Such comparisons presume a given volume.** That is, any given volume of lead is heavier than the same volume of aluminum. Without keeping volume constant, no comparison between objects or substances may be made.

The **density** of a substance is therefore the amount of matter or "mass" in a given volume of that substance. It is normally measured in lbs/cu ft. Tables listing the densities of a variety of substances are available in chemical and engineering handbooks. A listing of a few substances is given below.

Substance	Density (lbs/cu ft)
Water	62.4
Seawater	64
Gasoline	44
Aluminum	170
Lead	700
Wood, pine	30

In the water and wastewater field, the density of water and other liquids is commonly measured in lbs/gal. This is simply another expression of mass per unit volume.

DENSITY IS MEASURED IN LBS/CU FT OR LBS/GAL

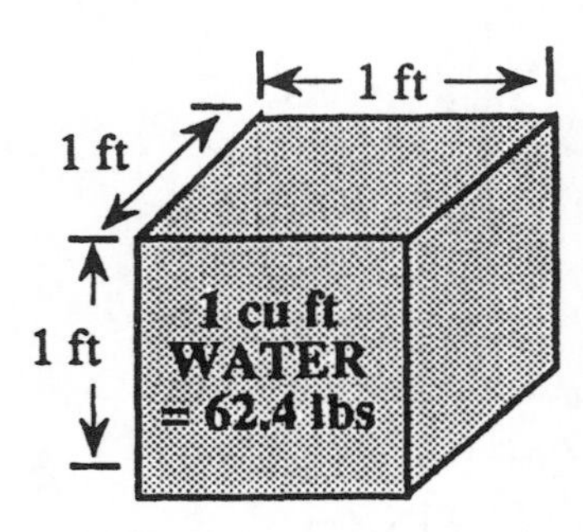

The density of water is 62.4 lbs/cu ft

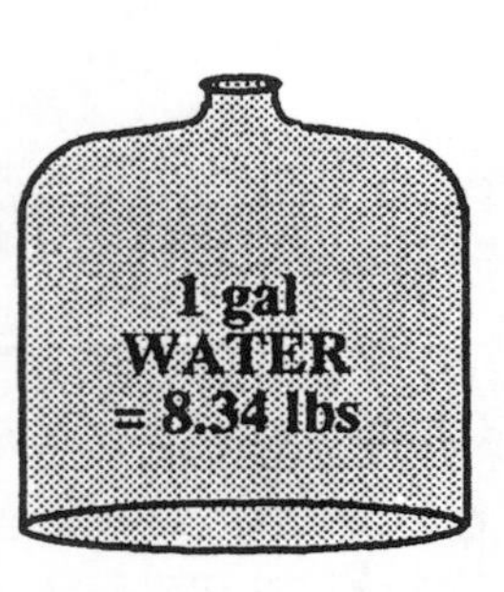

The density of water is 8.34 lbs/gal

Example 1: (Density and Specific Gravity)

❑ A gallon of solution is weighed. After the weight of the container is subtracted, it is determined that the weight of the solution is 9.8 lbs. What is the density of the solution?

$$\text{Density} = \frac{\text{lbs of Solution}}{\text{gal of Solution}}$$

$$= \boxed{\frac{9.8 \text{ lbs}}{1 \text{ gal}}} \quad \text{or} \quad \boxed{9.8 \text{ lbs/gal}}$$

Example 2: (Density and Specific Gravity)

❑ Suppose that only a half a gallon had been weighed in the example above, with a resulting weight of 4.9 lbs. How would density be calculated?

$$\text{Density} = \frac{\text{lbs of Solution}}{\text{gal of Solution}}$$

$$= \frac{4.9 \text{ lbs}}{0.5 \text{ gal}}$$

$$= \boxed{9.8 \text{ lbs/gal}}$$

Example 3: (Density and Specific Gravity)

❑ The density of a substance is given as 76.3 lbs/cu ft. What is this density expressed in lbs/gal?

To make a conversion in densities, first sketch the box diagram:

lbs/gal ← 7.48 — lbs/cu ft

Converting from lbs/cu ft to lbs/gal, the move is from a larger box to a smaller box. Therefore, division is indicated:

$$\frac{76.3 \text{ lbs/cu ft}}{\frac{7.48 \text{ lbs/cu ft}}{1 \text{ lb/gal}}} = \boxed{10.2 \text{ lbs/gal}}$$

Example 4: (Density and Specific Gravity)

❑ The density of a solution is 9.6 lbs/gal. What is the density expressed as lbs/cu ft?

First sketch the box diagram:

lbs/gal — 7.48 → lbs/cu ft

Converting from lbs/gal to lbs/cu ft involves a move from the smaller box to the larger box. Therefore, multiplication is indicated:

$$(9.6 \text{ lbs/gal})\frac{(7.48 \text{ lbs/cu ft})}{1 \text{ lbs/gal}} = \boxed{71.8 \text{ lbs/cu ft}}$$

ESTIMATING THE DENSITY OF A SUBSTANCE

You can estimate the density of a substance by weighing a known volume of it. For example, to estimate the density of sludge being pumped, weigh a gallon sample of it. (Be sure to subtract the weight of the container.) You will then have the lbs/gal density of that sludge. The accuracy of your estimate, of course, depends on whether the sludge sample is a representative sample.

LBS/CU FT AND LBS/GAL EXPRESSIONS OF DENSITY

Occasionally you may know the density of a substance expressed in lbs/cu ft but need to know the density in lbs/gal, or vice versa. To make such a conversion, you may use the following box diagram:* (based on the conversion equation, 1 lb/gal = 7.48 lbs/cu ft)

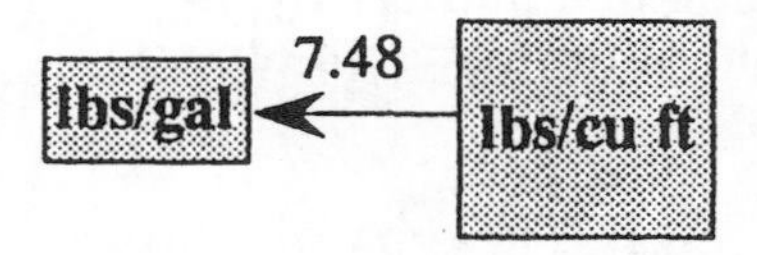

One aspect of this diagram is very different from other box diagrams. In other box diagram conversions, the same quantity is simply expressed in different units. For example, a quantity of water expressed as gallons is re-expressed in terms of cubic feet or pounds. The quantity of water has not changed, only how it is described.

In these density conversions, however, you are converting the weight of one quantity of water to the equivalent weight of a different quantity of water. Remember that the smaller box is associated with the smaller quantity of water (note gallons in the denominator of the smaller box), and the larger box is associated with the larger quantity of water (note cubic feet in the denominator of the larger box).

---

* For a review of the "box diagram" method of conversions, refer to Chapter 8 in *Basic Math Concepts*.

SPECIFIC GRAVITY

Two expressions of density have been mentioned thus far—lbs/cu ft and lbs/gal. There is a third measurement, a metric measurement, of density: grams per cubic centimeter (g/cm^3). With three different ways to express density, comparison of one density to another can be difficult. This problem is resolved by the use of specific gravity.

The density of water was established as the "standard" and all other densities are then compared to that of water. **The specific gravity of any liquid* is therefore the ratio or comparison of the density of a substance to the density of water.** Practically speaking, the specific gravity of a liquid may be determined by weighing a given volume of that liquid and then dividing that number by the weight of the same volume of water.

The specific gravity of water is one, since comparing the density of water to the density of water results in the following calculation:

$$\frac{62.4 \text{ lbs/cu ft}}{62.4 \text{ lbs/cu ft}} = 1.0$$

The specific gravity of seawater (with a density of 64 lbs/cu ft) would be:

$$\frac{64 \text{ lbs/cu ft}}{62.4 \text{ lbs/cu ft}} = 1.03$$

Any substance with a density greater than that of water will have a specific gravity greater than 1.0. And any substance with a density less than that of water will have a specific gravity less than 1.0.**

SPECIFIC GRAVITY IS A COMPARISON (OR RATIO) OF DENSITIES

The densities of all liquids are compared to the density of water.

$$\text{Specific Gravity of Liquids} = \frac{\text{Density of the Liquid}}{\text{Density of Water}}$$

The densities of all gases are compared to the density of air.

$$\text{Specific Gravity of Gases} = \frac{\text{Density of the Gas}}{\text{Density of Air}}$$

Example 5: (Density and Specific Gravity)

❑ Using a density of 44 lbs/cu ft for gasoline, what is the specific gravity of gasoline?

The specific gravity of gasoline is the comparison, or ratio, of the density of gasoline to that of water:

$$\text{Specific Gravity of Gasoline} = \frac{\text{Density of Gasoline}}{\text{Density of Water}}$$

$$= \frac{44 \text{ lbs/cu ft}}{62.4 \text{ lbs/cu ft}}$$

$$= \boxed{0.71}$$

(Note: both the density and specific gravity of gasoline indicate that gasoline will float in water.)

---

* The specific gravity of gases is based on a standard of air rather than water. These densities are very dependent on pressure and temperature. Therefore density listings of gases normally include pressure and temperature readings.

** The density (and specific gravity) of a substance indicates whether it will sink or float in water. If its density is greater than that of water, it will sink; if less, it will float.

Example 6: (Density and Specific Gravity)

❑ You wish to determine the specific gravity of a solution. After weighing a gallon of solution and subtracting the weight of the container, the solution is found to weigh 9.17 lbs. What is the specific gravity of the solution?

To determine the specific gravity of the solution, its density must be first be determined. Since one gallon of the solution weighed 9.17 lbs, the density is 9.17 lbs/gal.

Now compare the density of the solution to that of water to determine specific gravity:

$$\text{Specific Gravity} = \frac{\text{Density of the Solution}}{\text{Density of Water}}$$

$$= \frac{9.17 \text{ lbs/gal}}{8.34 \text{ lbs/gal}}$$

$$= \boxed{1.1}$$

DETERMINING THE SPECIFIC GRAVITY OF A SUBSTANCE

To determine the specific gravity of a substance, you must first determine its density (described on the previous two pages). Then the density of that substance is compared to the density of water. Example 6 illustrates such a problem.

Example 7: (Density and Specific Gravity)

❑ The specific gravity of a liquid is 0.95. What is the density of that liquid?(Density of water = 8.34 lbs/gal).

The specific gravity of a liquid can be used to determine its density. Multiply the specific gravity times the density of water:

$$\text{(Spec. Grav.) of the Liq.} \; \text{(Density) of water} = \text{Density of the Liquid}$$

$$(0.95)(8.34 \text{ lbs/gal}) = \boxed{7.9 \text{ lbs/gal}}$$

WHEN SPECIFIC GRAVITY IS KNOWN AND DENSITY IS UNKNOWN

If you know the specific gravity of any substance, you can always determine its density by multiplying the specific gravity by the density of water:

(Specific) Gravity of a Liquid	(Density) of Water, always 8.34 lbs/gal	=	Density of the Liquid, lbs/gal

Example 7 illustrates this type of calculation.

7.2 PRESSURE AND FORCE

Force is a push or pull measured in terms of weight, such as pounds or kilograms. The force on the bottom of an open tank, for example, is a measure of the weight of the water above it. The deeper the water, the more force on the bottom of the tank.

Although the force exerted against the entire tank bottom is an important calculation, we will first focus on another calculation related to force—that of pressure.

Pressure is a measure of the force or weight pushing against a specified area, usually a square inch or a square foot. Thus, pressure is normally expressed in pounds per square inch (lbs/sq in. or psi) or pounds per square foot (lbs/sq ft). The general equation used in calculations of pressure is:

$$\text{Pressure} = \frac{\text{Force}}{\text{Area}}$$

PRESSURE DEPENDS ON CONTACT AREA

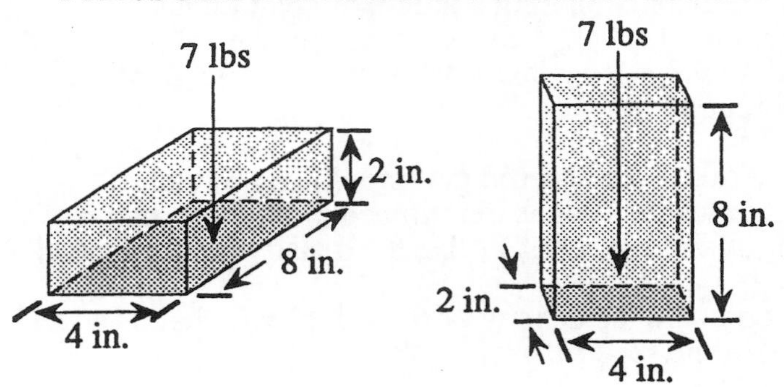

A brick is set on a table, as shown. If the 7 lbs of force (weight) is spread over the 32 sq in. of its side;* the pressure against the table at the point of contact is:

$$\frac{7 \text{ lbs}}{32 \text{ sq in.}} = \boxed{0.2 \text{ lbs/sq in.}}$$

A brick is placed on a table, as shown. Now the 7 lbs of force (weight) is spread over the 8 sq in. of its bottom. The pressure against the table at the point of contact is:

$$\frac{7 \text{ lbs}}{8 \text{ sq in.}} = \boxed{0.9 \text{ lbs/sq in.}}$$

Example 1: (Pressure and Force)

❑ The object shown below weighs 30 lbs. What is the lbs/sq in. pressure at the surface of contact?

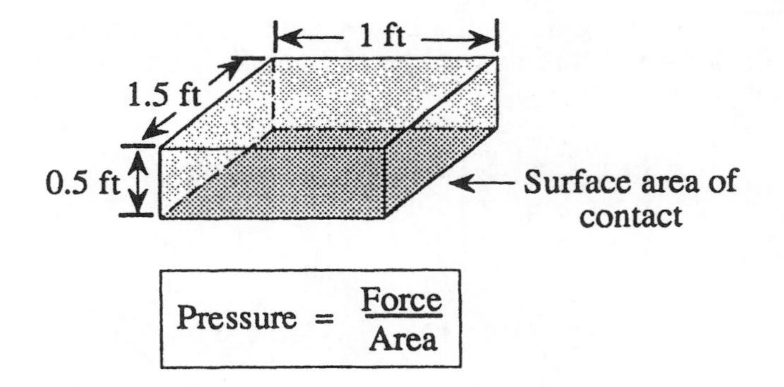

$$\text{Pressure} = \frac{\text{Force}}{\text{Area}}$$

Because lbs/sq in. pressure is desired, the dimensions will be expressed in inches rather than ft:

$$\text{Pressure} = \frac{30 \text{ lbs}}{(12 \text{ in.})(18 \text{ in.})}$$

$$= \boxed{0.14 \text{ lbs/sq in.}}$$

---

* For a review of area calculations, refer to Chapter 10, "Area Measurements", in *Basic Math Concepts*.

Example 2: (Pressure and Force)
❑ Compare the pressures, in lbs/sq ft, on the contact area for the two positions of the object shown below. The object weighs 300 lbs.

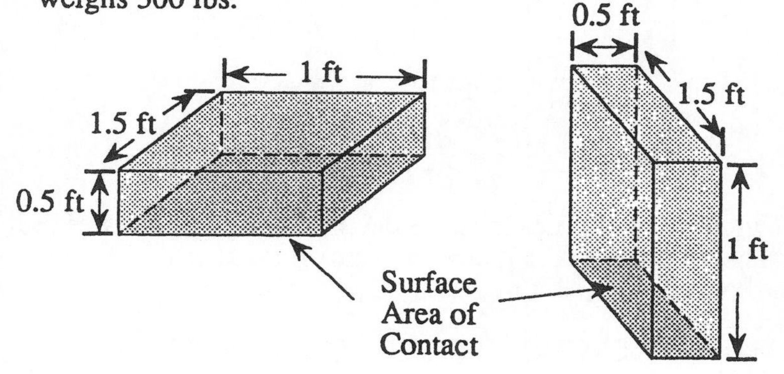

$$\text{Pressure} = \frac{\text{Force}}{\text{Area}} = \frac{300 \text{ lbs}}{(1.5 \text{ ft})(1 \text{ ft})} = \boxed{200 \text{ lbs/sq ft}}$$

$$\text{Pressure} = \frac{\text{Force}}{\text{Area}} = \frac{300 \text{ lbs}}{(0.5 \text{ ft})(1.5 \text{ ft})} = \boxed{400 \text{ lbs/sq ft}}$$

Example 3: (Pressure and Force)
❑ An object rests on the floor. The pressure at the surface of contact is 0.5 lbs/sq in.. If the object is placed on another side that has only one-third the contact area, what is the new lbs/sq in. pressure?

Since the area and pressure are inversely related, with a decrease in contact area there will be an increase in pressure. From this, we know that the new pressure should be greater than 0.5 lbs/sq in.:

Area decreased to 1/3 ⟶ Pressure increased 3 times

$$(0.5 \text{ lbs/sq in.})(3) = \boxed{1.5 \text{ lbs/sq in.}}$$

PRESSURE AND AREA ARE INVERSELY RELATED

As the pressure equation indicates, pressure reflects both the force exerted and the contact area. If the force is increased, the numerator is larger, resulting in a larger pressure reading. And if the force is decreased, the resulting pressure reading is also decreased. This is called a **direct proportion**—as force increases, pressure increases; and as force decreases, pressure decreases.

But what is the effect of changes in area? Assuming a constant force, if the area increases, the denominator of the fraction increases, resulting in a smaller pressure reading. Likewise, if the area is decreased, the denominator is thereby decreased, resulting in a larger pressure reading. This is called an **inverse or indirect proportion***—as one increases, the other decreases, and vice versa.

Apart from the mathematics, this reasoning makes sense. As the area decreases, the same weight is distributed over a smaller area. Therefore each square inch receives a greater force.

Assuming the force remains constant, if the area of contact is cut in half, the pressure is increased twofold. If the area is reduced to one-quarter of its size, the pressure is increased fourfold. Note the relationship between these changes:

Area Decreased to	Pressure Increased by Factor of
1/2 size	2
1/4 size	4

Other numbers apply as well—1/8 and 8, etc. This inverse relationship also occurs when the area is increased from its original size.

Area Increased to	Pressure Decreased by Factor of
2 times size	1/2
4 times size	1/4

Example 3 illustrates a calculation using inverse relationships.

* Inverse or indirect proportions are described in Chapter 7, "Ratios and Proportions", in *Basic Math Concepts*.

LIQUID PRESSURE

In a liquid at rest, such as in a container, tank, or reservoir, the pressure at any one point is exerted in all directions, not just toward the bottom contact surface. It is exerted toward the sides and top as well.

The amount of pressure at any point in the water depends on two factors:

- Depth (measured vertically) and
- Density.

The greater the water depth, the greater the pressure. At deeper and deeper levels of water, there is an increased weight of water above. As shown in the diagram at the top of the facing page, the pressure at 1 ft depth is 0.433 psi, while the pressure at 2.31 ft depth is 1 psi.*

Pressure is also dependent on the density of the liquid. For example, at point A_1 in the box to the right, the pressure would be less than that at point A_2, because the liquid is less dense at A_1 than at A_2. The equation used to determine pressure in a liquid is shown to the right.

Note how this equation results in the same units as the general equation for pressure:

$$P = h D$$

$$P = \frac{(\text{ft})(\text{lbs})}{\text{ft}^3}$$

$$P = \frac{\text{lbs}}{\text{ft}^2} \begin{matrix} \leftarrow (\text{force}) \\ \leftarrow (\text{area}) \end{matrix}$$

PRESSURE DEPENDS ON DEPTH

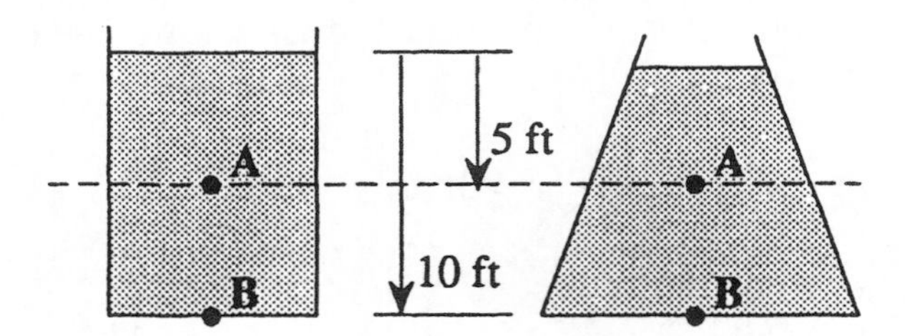

Twice the depth means twice the pressure, regardless of the shape of the container. The pressure at Point B is twice that at Point A.

PRESSURE DEPENDS ON DENSITY

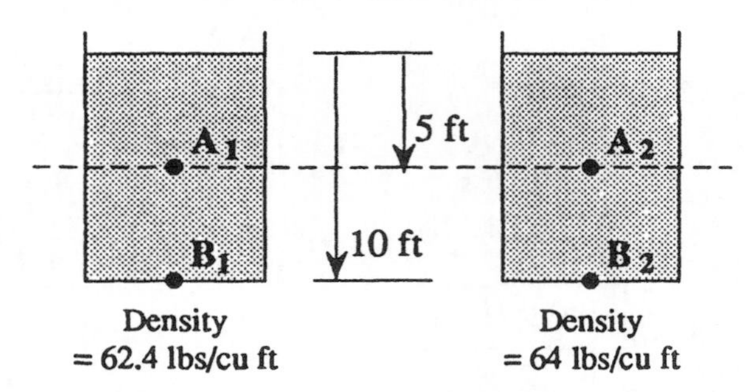

The pressures at A_1 and B_1 are less than corresponding pressures at A_2 and B_2. This is due to differences in the density of the liquids.

THE PRESSURE EQUATION INCLUDES BOTH DEPTH AND DENSITY

Pressure = (depth) (Density)

Due to possible confusion with abbreviation of terms, depth is replaced by height, *h*. Using *D* for density, the equation is written as:

Pressure = hD

Example 4: (Pressure and Force)

❑ What is the pressure (in lbs/sq ft) at a point 8 feet below the surface of the water? (The density of water is 62.4 lbs/cu ft.)

$$\begin{aligned} \text{Pressure} &= hD \\ &= (8 \text{ ft})(62.4 \text{ lbs/cu ft}) \\ &= \boxed{499 \text{ lbs/sq ft}} \end{aligned}$$

* This is gage pressure, not absolute pressure. These terms are described later in this section.

PRESSURE AND PSI

1 ft = 0.433 psi

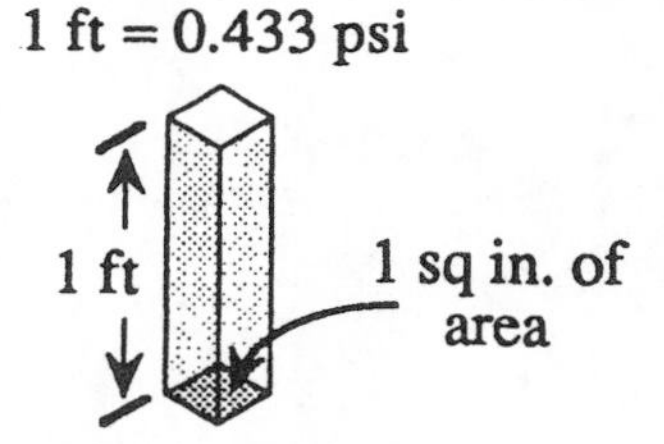

The weight bearing down on each square inch of area is about a half pound (0.433 lbs) for each foot of depth.

1 psi = 2.31 ft

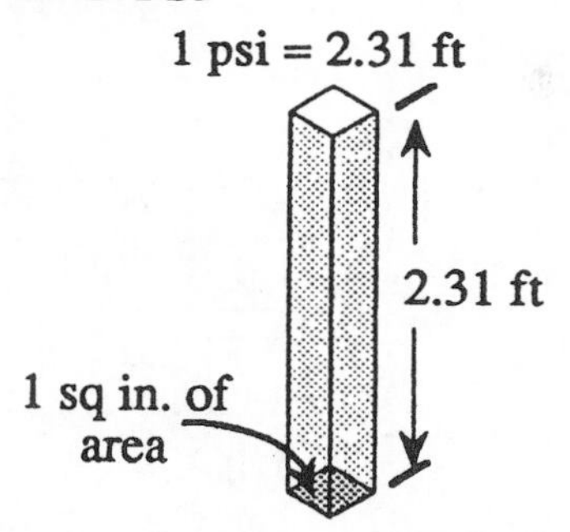

At a depth of 2.31 ft, there is one pound of weight bearing down on each square inch of area.

Example 5: (Pressure and Force)

❑ What is the pressure (in psi) at a point 12 ft below the surface?

Using the equation 1 psi = 2.31 ft, draw the box diagram:

psi ← 2.31 — ft

Converting from feet to psi, the move is from a larger box to a smaller box. Division by 2.31 is therefore indicated:

$$\frac{12 \text{ ft}}{2.31 \text{ ft/psi}} = \boxed{5.2 \text{ psi}}$$

Example 6: (Pressure and Force)

❑ At a point 3 ft below the liquid surface, what is the pressure in psi? (The specific gravity of the liquid is 0.95.)

First calculate psi as usual, using the box diagram:

psi ← 2.31 — ft

Division is indicated:

$$\frac{3 \text{ ft}}{2.31 \text{ ft/psi}} = \boxed{1.3 \text{ psi}}$$

The specific gravity may now be taken into account. A specific gravity less than that of water will result in a smaller psi reading for the same water depth:

$$\text{Pressure at Different Sp. Grav.} = (\text{psi})(\text{specific gravity})$$
$$= (1.3 \text{ psi})(0.95)$$
$$= \boxed{1.2 \text{ psi}}$$

USING PSI AND FT

Since the density of water is a given in most water and wastewater calculations, the $P = hD$ equation can be shortened. The pressure at 1 ft depth is always 62.4 lbs/sq ft or 0.433 lbs/sq in.*

$$\boxed{1 \text{ ft} = 0.433 \text{ psi}}$$

When the "box method" of conversion is used, however, both numbers of the equation must be greater than one.** The equation shown above may be easily converted to the desired form by dividing both sides of the equation by 0.433, as follows:

$$\frac{1 \text{ ft}}{0.433} = \frac{\cancel{0.433} \text{ psi}}{\cancel{0.433}}$$

$$\boxed{2.31 \text{ ft} = 1 \text{ psi}}$$

Now rearrange the equation so the one is on the left side of the equation:

$$\boxed{1 \text{ psi} = 2.31 \text{ ft}}$$

This equation can be used in all conversions between feet and psi. It is recommended that this equation be memorized. Example 5 illustrates a conversion between feet and psi.

PRESSURE AND SPECIFIC GRAVITY

Specific gravity can be used to determine pressures within liquids of different densities. First, calculate the pressure in psi using the equation given above (1 psi = 2.31 ft). Then multiply the psi result by the specific gravity of the liquid. Example 6 illustrates this calculation.

* To verify this $P = (1 \text{ ft})(62.4 \text{ lbs/cu ft}) = 62.4 \text{ lbs/sq ft}$. To convert to lbs/sq in., divide by 144 sq in./sq ft. For a discussion of square terms conversions, refer to Chapter 8 in *Basic Math Concepts*.

** The box method of conversions is described in Chapter 8 in *Basic Math Concepts*.

TOTAL FORCE

The total force of water against the side or bottom of a tank or wall is determined by multiplying the pressure at that depth times the entire area:

Total Force = (Pressure)(Area)

If the pressure is given as lbs/sq ft, the area must be expressed as square feet. And if the pressure is given in lbs/sq in., the area must be expressed as square inches.

$$\text{Total Force} = (\text{Pressure, lbs/sq ft})(\text{Area, sq ft})$$

$$\text{Total Force} = (\text{Pressure, lbs/sq in.})(\text{Area, sq in.})$$

Calculating the total force against the bottom of a tank is a straight - forward calculation—simply determine the pressure at the bottom of the tank (based on the vertical distance beneath the surface of the water) and multiply this pressure by the area of the bottom.

Calculating the total force against the side of a tank is a little different—which pressure value should be used? The pressure at the water surface is zero*, and the pressures increase toward the bottom of the tank. The pressure to be used in these calculations is the **average pressure**. The average pressure occurs at half of the water depth, or what might be termed the **average depth**. Average depth may be determined as follows:

$$\text{Average Depth, ft} = \frac{\text{Total Depth, ft}}{2}$$

Example 8 illustrates this type of calculation.

TOTAL FORCE AGAINST THE BOTTOM OF A TANK

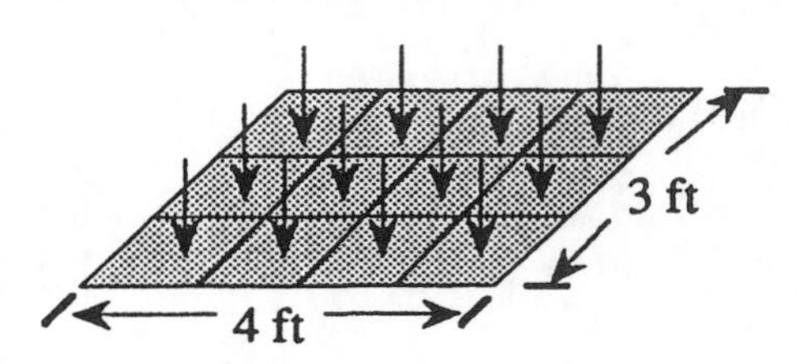

$$\text{Total Force} = (\text{Pressure, lbs/sq ft})(\text{Area, sq ft})$$

TOTAL FORCE AGAINST THE SIDE OF A TANK

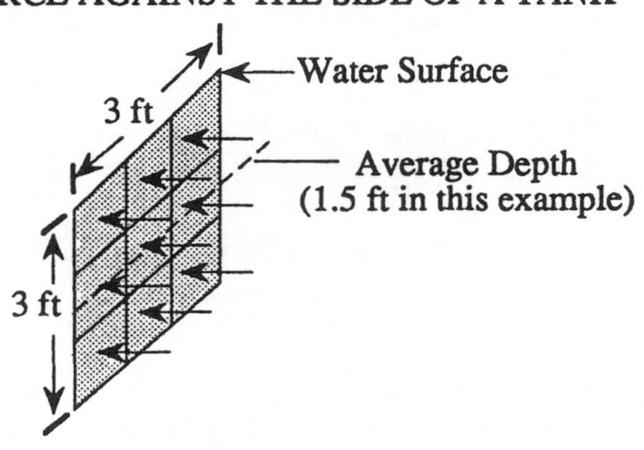

$$\text{Total Force} = (\text{Pressure at Average Depth, lbs/sq ft})(\text{Total Area, sq ft})$$

Example 7: (Pressure and Force)

❑ What is the total force against the bottom of a tank 30 ft long and 10 ft wide? The water depth is 8 ft.

First, calculate the pressure at the bottom of the tank. At 8 ft depth, the pressure is:

psi ← 2.31 — ft

$$\frac{8 \text{ ft}}{2.31 \text{ ft/psi}} = 3.46 \text{ psi}$$

Then calculate total force. Since pressure is given in psi, the dimensions must be reported in inches (30 ft x 12 in/ft = 360 in.; 10 ft x 12 in./ft = 120 in.)

$$\begin{aligned}\text{Total Force} &= (\text{Pressure, psi})(\text{Area, sq in.})\\ &= (3.46 \text{ psi})(360 \text{ in.})(120 \text{ in.})\\ &= \boxed{149{,}472 \text{ lbs}}\end{aligned}$$

---

* Referring to gage pressure. Absolute and gage pressures are described later in this section.

Example 8: (Pressure and Force)

❑ What is the total force exerted on the side of a tank if the tank is 16 ft wide and the water depth is 12 ft?

Since the water depth is 12 ft, the halfway point (or average water depth) is 6 ft. To calculate total force, the pressure at the average depth must be calculated. The pressure at a depth of 1 ft = 62.4 lbs/sq ft, so the pressure at 6 ft would be:

$$(6)(62.4 \text{ lbs/sq ft}) = \underset{\text{average pressure}}{374 \text{ lbs/sq ft}}$$

Now the total force can be calculated:

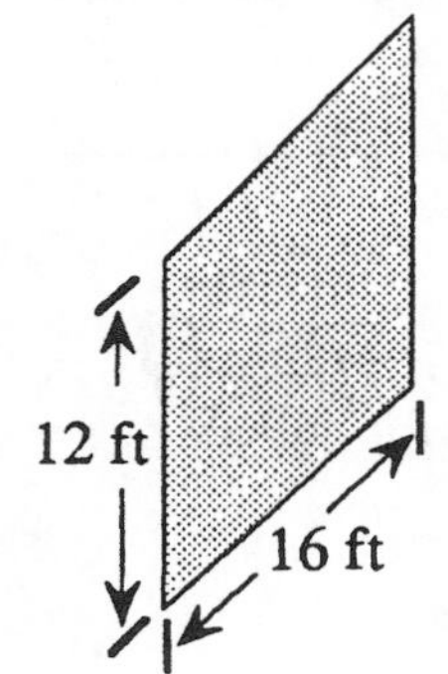

$$\underset{\text{Force}}{\text{Total}} = \underset{\text{Depth, lbs/sq ft}}{(\text{Pressure at Average})} \; \underset{\text{sq ft}}{(\text{Total Area})}$$

$$= (374 \text{ lbs/sq ft})(12 \text{ ft})(16 \text{ ft})$$

$$= \boxed{71{,}808 \text{ lbs}}$$

An equation sometimes given for the total force on the side of a tank is:

$$F = (31.2)(H^2)(L),$$

where F = force in lbs, H = head in ft, and L = length of wall in ft. This equation is simply an abbreviation of the typical total force equation, as shown below: (note the water depth is written as H, ft of head)

$$\underset{\text{lbs}}{\text{Total Force,}} = \underset{\text{lbs/sq ft}}{(\text{Pressure})}\underset{\text{sq ft}}{(\text{Area})}$$

$$= \underset{\text{height}}{(\text{Aver.})}(\text{Den.})\,(\text{Area})$$

$$= \frac{(H)}{2}(62.4)(H)(L)$$

$$= (31.2)(H^2)(L)$$

TOTAL FORCE VS CENTER OF FORCE

Although the **total force** on a side wall is calculated using the average depth (or average pressure, since these are related), the center of force is not located at the halfway line—it is located along a line **two-thirds of the way down from the water surface.**

This is because the pressure against the wall increases with depth **(forming what is called a "triangular load")**. As shown in the diagram to the left, more of the force against the wall is located nearer the bottom. To calculate the center of force on a wall, simply multiply the water depth by 2/3:

$$\underset{\text{of Force}}{\text{Center}} = \frac{(2)}{3}\underset{\text{Water}}{(\text{Depth of})}$$

THE CENTER OF FORCE IS LOCATED ALONG A LINE 2/3 FROM THE SURFACE

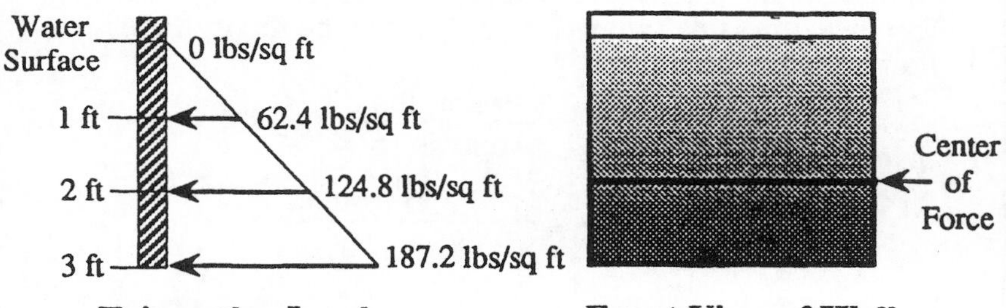

Triangular Load (Side View of Wall)

Front View of Wall

In Example 8, the total force against the wall is 71,808 lbs. Where is the center of force located?

$$\underset{\text{of Force}}{\text{Center}} = \frac{(2)}{3}\underset{\text{Water}}{(\text{Depth of})}$$

$$= \frac{(2)}{3}(12 \text{ ft})$$

$$= \boxed{8 \text{ ft}}$$

HYDRAULIC PRESS

The operation of the **hydraulic press or hydraulic jack** is based on two primary principles:

- Force applied to a liquid is distributed equally within that liquid, and
- Total Force = (Pressure)(Area)

As illustrated in the graphic to the right, the force applied to the smaller cylinder is distributed evenly throughout the liquid. The larger cylinder has a greater surface area, so the total force applied is magnified several times.

To calculate the total force on the large cylinder, you must know the pressure against it (which is the same as that applied to the smaller cylinder) and the area. Examples 9 and 10 illustrate hydraulic press calculations.

THE HYDRAULIC PRESS OPERATES ON THE PRINCIPLE OF TOTAL FORCE

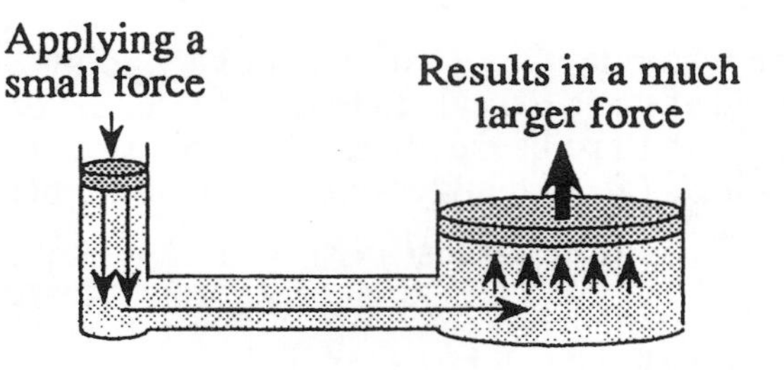

Example 9: (Pressure and Force)

❑ The force applied to the small cylinder (12-in. diameter) of a hydraulic jack is 35 lbs. If the diameter of the large cylinder is 3 ft, what is the total lifting force?

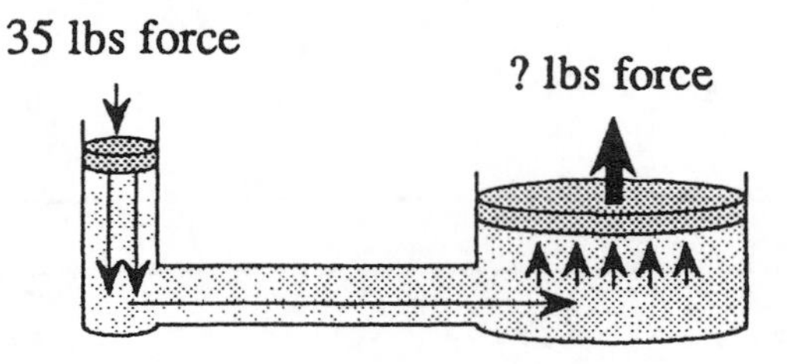

First calculate the pressure applied to the small cylinder.* The pressure is calculated in lbs/sq ft since the cylinder diameters are given in ft:

$$\text{Pressure} = \frac{\text{Force, lbs}}{\text{Area, sq ft}}$$

$$= \frac{35 \text{ lbs}}{(0.785)(1 \text{ ft})(1 \text{ ft})}$$

$$= 45 \text{ lbs/sq ft}$$

The total force at the large cylinder can now be calculated:

$$\text{Total Force} = (\text{Pressure})\ (\text{Area})$$

$$= (45 \text{ lbs/sq ft})\ (0.785)\ (3 \text{ ft})\ (3 \text{ ft})$$

$$= \boxed{318 \text{ lbs}}$$

* For a review of circular area calculations, refer to Chapter 10 in *Basic Math Concepts.*

Example 10: (Pressure and Force)

❑ The force applied to the small cylinder of a hydraulic jack is 50 lbs. The diameter of the small cylinder is 18 inches. If the diameter of the large cylinder is 40 inches, what is the total lifting force?

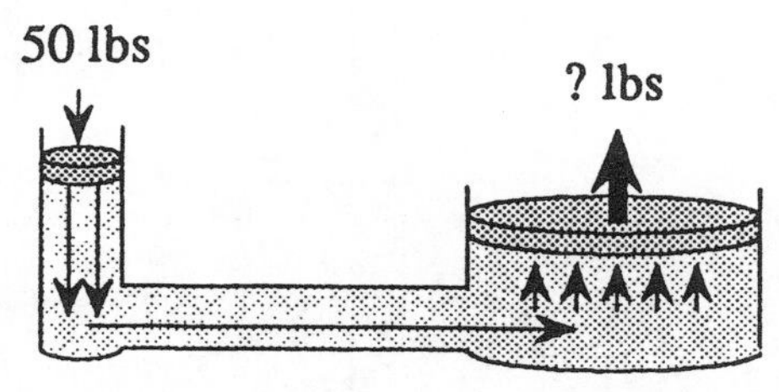

First calculate the pressure applied to the small cylinder. Since the diameters of the cylinders are given in inches, the pressure will be calculated as lbs/sq in.:

$$\text{Pressure} = \frac{\text{Force, lbs}}{\text{Area, sq in.}}$$

$$= \frac{50 \text{ lbs}}{(0.785)(18 \text{ in.})(18 \text{ in.})}$$

$$= 0.2 \text{ lbs/sq in.}$$

The same pressure is transmitted to the larger cylinder. The total force at the large cylinder is calculated as:

$$\text{Total Force} = (\text{Pressure})(\text{Area})$$

$$= \left(0.2 \frac{\text{lbs}}{\text{sq in.}}\right)(0.785)(40 \text{ in.})(40 \text{ in.})$$

$$= \boxed{251.2 \text{ lbs}}$$

Example 11: (Pressure and Force)

❑ A gage reading is 25 psi. What is the absolute pressure at the gage? (Assume sea level atmospheric pressure.)

$$\begin{matrix}\text{Absolute} \\ \text{Pressure, psi}\end{matrix} = \begin{matrix}\text{Gage Pressure} \\ \text{psi}\end{matrix} + \begin{matrix}\text{Atmospheric} \\ \text{Pressure, psi}\end{matrix}$$

$$= 25 \text{ psi} + 14.7 \text{ psi}$$

$$= \boxed{39.7 \text{ psi}}$$

GAGE VS. ABSOLUTE PRESSURES

When water pressures are measured in a tank or pipeline, they are measured by gages. These measurements are therefore called **gage pressures.** Gage pressures do not include all the pressures acting in the tank or pipeline—**they do not include atmospheric pressure.**

Atmospheric pressure is generally not considered in water and wastewater calculations, because atmospheric pressure is exerted both inside and outside the tank or pipeline. At sea level, the atmospheric pressure is 14.7 psi. The name given to the total pressure, including both gage and atmospheric pressures, is **absolute pressure.** Absolute pressure is calculated by adding the gage pressure and atmospheric pressure:

$$\begin{matrix}\text{Absolute} \\ \text{Press.,} \\ \text{psi}\end{matrix} = \begin{matrix}\text{Gage} \\ \text{Press.} \\ \text{psi}\end{matrix} + \begin{matrix}\text{Atmos.} \\ \text{Press.,} \\ \text{psi}\end{matrix}$$

Example 11 illustrates a calculation involving gage and absolute pressures.

7.3 HEAD AND HEAD LOSS

HEAD TERMINOLOGY

When describing the various types of head against which a pump must operate, several different terms may be used, depending on the side of the pump and whether or not the pump is operating.

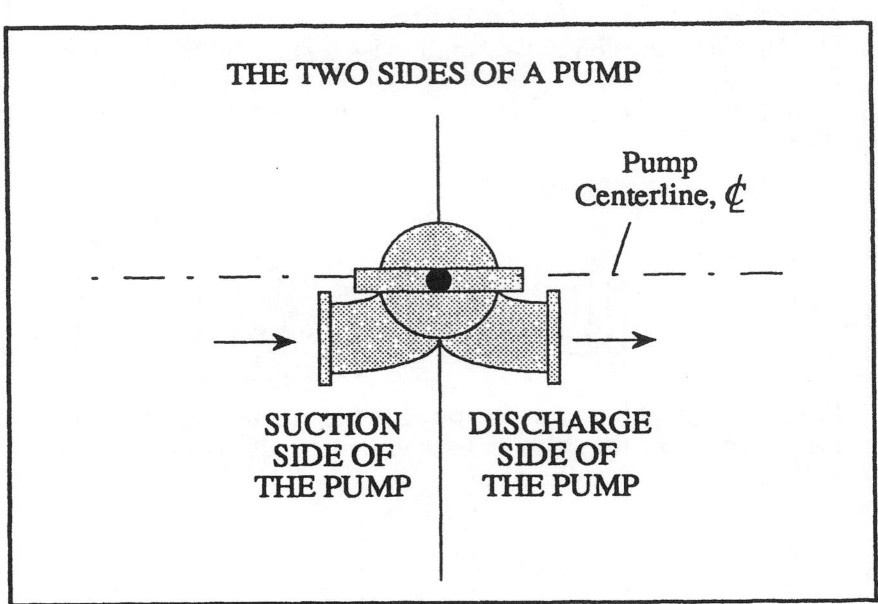

SUCTION AND DISCHARGE HEADS

The terms **suction** and **discharge** indicate two different sides of the pump. As shown in the diagram to the right, the suction side of a pump is the inlet or low pressure side of the pump. The discharge side of a pump is the outlet or high pressure side of the pump.

Heads measured on the suction side of a pump are called **suction heads.** Heads measured on the discharge side of a pump are called **discharge heads.**

SUCTION HEAD, SUCTION LIFT, AND DISCHARGE HEAD

When the water feeding the pump is **above the pump,** this is called a **suction head:**

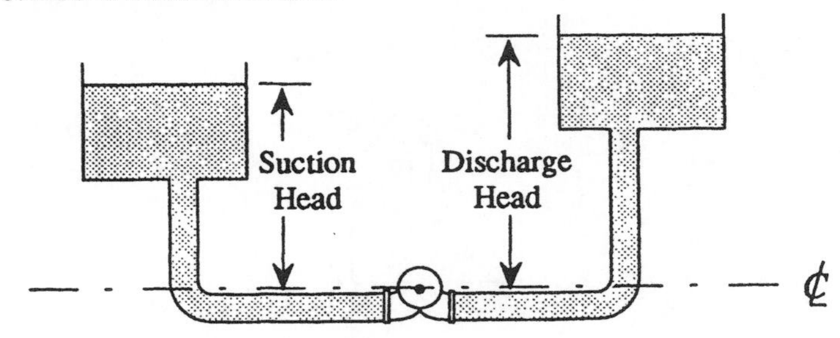

When the water feeding the pump is **below the pump,** this is called a **suction lift:**

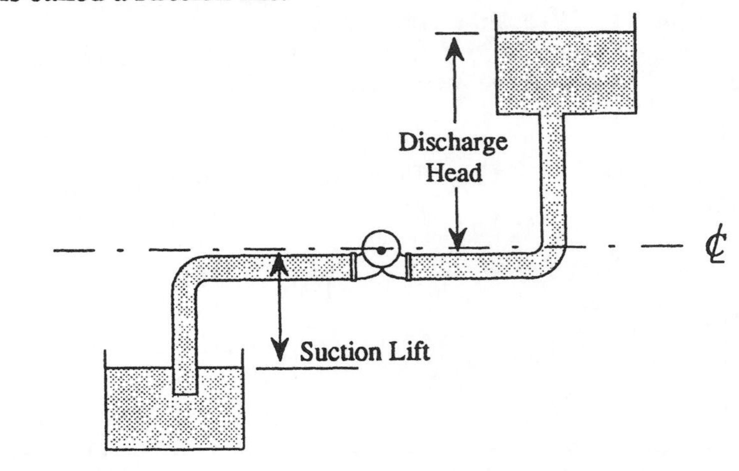

TOTAL STATIC HEAD IS THE VERTICAL DISTANCE BETWEEN THE TWO FREE WATER SURFACES

When the two water surfaces are located above the pump, the static suction head offsets part of the static discharge head. **The total static head is therefore the difference in height between the two heads:**

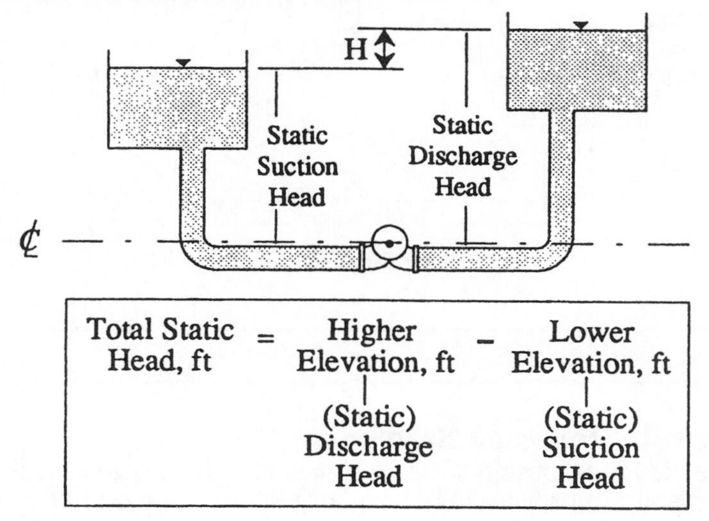

When the water surface on the suction side of the pump is below the pump centerline, there is no offsetting head on the suction side. **The total static head is therefore the sum of the two heads:**

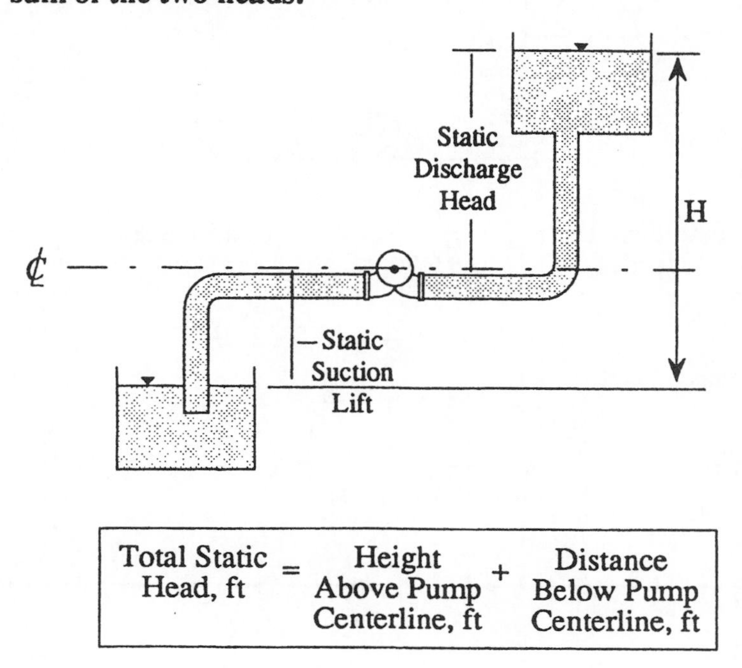

STATIC HEAD

The total head against which a pump must operate is determined principally by two calculations:

- Total static head, and
- Friction and minor head losses

Head measurements taken on either side of the pump when the pump is off are called **static heads.** Both the static suction head and static discharge head are considered when determining the **total static head,** as illustrated in the diagrams to the left.

In simplified terms, however, total static head is a measure of the vertical distance between the two water surface elevations. It can be measured using elevations, vertical measurements, or gage pressures.

TOTAL DYNAMIC HEAD (TDH)

In addition to static head, the pump must work against **friction and minor head losses.** These are head losses resulting from friction as the water rubs against the pipeline and from friction and changes in direction as the water moves through valves and orifices.

For example, if the water is to be lifted 50 ft from Point A to Point B (as shown in the diagram to the right) and head losses are equal to 5 ft, the pump must produce a total of 55 ft of head to lift the water from Point A to Point B.

The total head against which the pump must operate is called the total dynamic head (TDH).

$$\text{Total Dynamic Head, ft} = \text{Total Static Head, ft} + \text{Head Losses, ft}$$

DYNAMIC HEAD IS THE STATIC HEAD PLUS FRICTION AND MINOR HEAD LOSSES

Additional head required to compensate for friction and minor headlosses

5 ft

Total Static Head

50 ft

A

B

Pump On

Example 1: (Head and Head Loss)

❑ The elevation of two water surfaces are given below. If the friction and minor head losses equal 9 ft, what is the total dynamic head (in ft)?

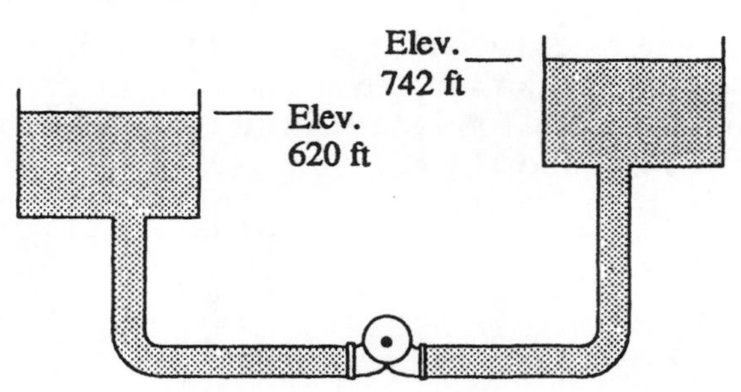

$$\text{Total Dynamic Head, ft} = \text{Total Static Head, ft} + \text{Head Losses ft}$$

$$= (742 \text{ ft} - 620 \text{ ft}) + 9 \text{ ft}$$

$$= 122 \text{ ft} + 9 \text{ ft}$$

$$= \boxed{131 \text{ ft TDH}}$$

Example 2: (Head and Head Loss)

❑ The pump inlet and outlet pressure gage readings are given below. (Pump is off.) If the friction and minor head losses are equal to 12 ft, what is the total dynamic head (in ft)?

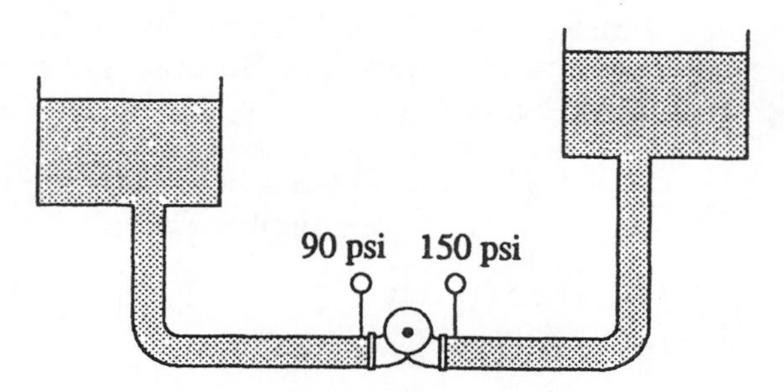

First calculate the static head in psi, then convert head in psi to head in ft. The static head is the difference in inlet and outlet pump pressures: 150 psi – 90 psi = 60 psi.

Next, convert 60 psi to ft*: (1 psi = 2.31 ft)

psi ← 2.31 — ft

(60 psi) (2.31 ft/psi) = **139 ft Static Head**

Then calculate the total dynamic head:

TDH = Static Head + Head Losses

= 139 ft + 12 ft

= **151 ft**

If the pressure is expressed in psi, it can be converted to feet, if desired.* Example 2 illustrates such a calculation.

Example 3: (Head and Head Loss)

❑ Readings taken from the inlet and outlet pressure gages of a pump while it is in operation are as follows: 87 psi and 143 psi, respectively. What is the TDH, in ft?

The difference in pressure readings is 143 psi – 87 psi = 56 psi. Convert this difference in psi to ft:

(56 psi)(2.31 ft/psi) = **129 ft TDH**

When a pump is operating, it operates against the heads and head losses described above. Inlet and outlet gage pressures **taken when the pump is operating,** therefore, are a good estimate of the total dynamic head. Example 3 illustrates this calculation.

* Refer to Section 7.2 for a discussion of psi and ft conversions.

FRICTION HEAD LOSS

Friction head losses within a pipeline depend on several factors:

- Velocity or rate of flow,
- Diameter of pipe,
- Length of pipe, and
- Pipe roughness.

Friction losses are determined by using tables such as that shown on the facing page. Values on this table are calculated using the Hazen-Williams formula, which includes a roughness coefficient, *C*. The smoother the pipe, the higher the *C*-value.

The table on the facing page is based on a *C*-value of 100. To determine friction loss:

1. Enter the table at the known pipe diameter. (Diameters are shown across the top of the table.)
2. Follow the column down until it is opposite the given flow rate, gpm.
3. Read the corresponding friction loss. The loss is given in feet and represents the friction loss for every 100-ft section of pipe.
4. Calculate the total friction loss for the pipeline by multiplying the friction loss by the number of 100-ft segments of pipe.

A wide variety of pipes in use today have *C*-values from 130 to 140 and greater. Reference books are available that include friction loss tables for various other *C*-values.

Note that these tables **do not apply when pumping sludges.**

Example 4: (Head and Head Loss)

❑ A 6-inch diameter pipe has a *C*-value of 100. When the flow rate is 240 gpm, what is the friction loss for a 2000-ft length of pipe?

Enter the top column at "6-in. Pipe". Come down the column to the entry across from 240 gpm. The friction loss shown is 0.87 ft per 100-ft segments.

2000. → There are 20 100-ft segments of pipe

↑ hundreds place decimal

Total Friction Loss, ft = (0.87 ft)(20 segments of 100-ft)

= **17.4 ft**

Example 5: (Head and Head Loss)

❑ Flow through an 8-inch pipeline is 1300 gpm. The *C*-value is 100. What is the friction loss through a 3500-ft section of pipe ?

Enter the table at the 8-inch diameter heading and follow the column down until you are across from the 1300 gpm value. The friction loss indicated is 4.85 ft per 100-ft segment of pipeline.

3500. → There are 35 100-ft segments of pipe

↑ hundreds place decimal

Total Friction Loss, ft = (4.85)(35 segments of 100-ft)

= **169.8 ft**

FRICTION LOSS IN FEET PER 100-FT LENGTH OF PIPE
(Based on Williams & Hazen Formula Using $C = 100$)

	½-in. Pipe		¾-in. Pipe		1-in. Pipe		1¼-in. Pipe		1½-in. Pipe		2-in. Pipe		2½-in. Pipe		3-in. Pipe		4-in. Pipe		5-in. Pipe		6-in. Pipe		
US gal per min	Vel. ft per sec	Loss in ft	Vel. ft per sec	Loss in ft	Vel. ft per sec	Loss in ft	Vel. ft per sec	Loss in ft	Vel. ft per sec	Loss in ft	Vel. ft per sec	Loss in ft	Vel. ft per sec	Loss in ft	Vel. ft per sec	Loss in ft	Vel. ft per sec	Loss in ft	Vel. ft per sec	Loss in ft	Vel. ft per sec	Loss in ft	US gal per min
2	2.10	7.4	1.20	1.9																			2
4	4.21	27.0	2.41	7.0	1.49	2.14	.86	.57	.63	.26													4
6	6.31	57.0	3.61	14.7	2.23	4.55	1.29	1.20	.94	.56	.61	.20											6
8	8.42	98.0	4.81	25.0	2.98	7.8	1.72	2.03	1.26	.95	.82	.33	.52	.11									8
10	10.52	147.0	6.02	38.0	3.72	11.7	2.14	3.05	1.57	1.43	1.02	.50	.65	.17	.45	.07							10
12			7.22	53.0	4.46	16.4	2.57	4.3	1.89	2.01	1.23	.79	.78	.23	.54	.10							12
15			9.02	80.0	5.60	25.0	3.21	6.5	2.36	3.00	1.53	1.08	.98	.36	.68	.15							15
18			10.84	108.2	6.69	35.0	3.86	9.1	2.83	4.24	1.84	1.49	1.18	.50	.82	.21							18
20			12.03	136.0	7.44	42.0	4.29	11.1	3.15	5.20	2.04	1.82	1.31	.61	.91	.25	.51	.06					20
25					9.30	64.0	5.36	16.6	3.80	7.30	2.55	2.73	1.63	.92	1.13	.38	.64	.09					25
30					11.15	89.0	6.43	23.0	4.72	11.0	3.06	3.84	1.96	1.29	1.36	.54	.77	.13	.49	.04			30
35					13.02	119.0	7.51	31.2	5.51	14.7	3.57	5.10	2.29	1.72	1.59	.71	.89	.17	.57	.06			35
40					14.88	152.0	8.58	40.0	6.30	18.8	4.08	6.6	2.61	2.20	1.82	.91	1.02	.22	.65	.08			40
45							9.65	50.0	7.08	23.2	4.60	8.2	2.94	2.80	2.04	1.15	1.15	.28	.73	.09			45
50							10.72	60.0	7.87	28.4	5.11	9.9	3.27	3.32	2.27	1.38	1.28	.34	.82	.11	.57	.04	50
55							11.78	72.0	8.66	34.0	5.62	11.8	3.59	4.01	2.45	1.58	1.41	.41	.90	.14	.62	.05	55
60							12.87	85.0	9.44	39.6	6.13	13.9	3.92	4.65	2.72	1.92	1.53	.47	.98	.16	.68	.06	60
65							13.92	99.7	10.23	45.9	6.64	16.1	4.24	5.4	2.89	2.16	1.66	.53	1.06	.19	.74	.076	65
70							15.01	113.0	11.02	53.0	7.15	18.4	4.58	6.2	3.18	2.57	1.79	.63	1.14	.21	.79	.08	70
75							16.06	129.0	11.80	60.0	7.66	20.9	4.91	7.1	3.33	3.00	1.91	.73	1.22	.24	.85	.10	75
80							17.16	145.0	12.59	68.0	8.17	23.7	5.23	7.9	3.63	3.28	2.04	.81	1.31	.27	.91	.11	80
85							18.21	163.8	13.38	75.0	8 68	26.5	5.56	8.1	3.78	3.54	2.17	.91	1.39	.31	.96	.12	85
90							19.30	180.0	14.71	84.0	9.19	29.4	5.88	9.8	4.09	4.08	2.30	1.00	1.47	.34	1.02	.14	90
95									14.95	93.0	9.70	32.6	6.21	10.8	4.22	4.33	2.42	1.12	1.55	.38	1.08	.15	95
100									15.74	102.0	10.21	35.8	6.54	12.0	4.54	4.96	2.55	1.22	1.63	.41	1.13	.17	100
110									17.31	122.0	11.23	42.9	7.18	14.5	5.00	6.0	2.81	1.46	1.79	.49	1.25	.21	110
120									18.89	143.0	12.25	50.0	7.84	16.8	5.45	7.0	3.06	1.17	1.96	.58	1.36	.24	120
130									20.46	166.0	13.28	58.0	8.48	18.7	5.91	8.1	3.31	1.97	2.12	.67	1.47	.27	130
	8" Pipe																						
140	.90	.08							22.04	190.0	14.30	67.0	9.15	22.3	6.35	9.2	3.57	2.28	2.29	.76	1.59	.32	140
150	.96	.09									15.32	76.0	9.81	25.5	6.82	10.5	3.82	2.62	2.45	.88	1.70	.36	150
160	1.02	.10									16.34	86.0	10.46	29.0	7.26	11.8	4.08	2.91	2.61	.98	1.82	.40	160
170	1.08	.11									17.36	96.0	11.11	34.1	7.71	13.3	4.33	3.26	2.77	1.08	1.92	.45	170
180	1.15	.13									18.38	107.0	11.76	35.7	8.17	14.0	4.60	3.61	2.94	1.22	2.04	.50	180
190	1.21	.14									19.40	118.0	12.42	39.6	8.63	15.5	4.84	4.01	3.10	1.35	2.16	.55	190
200	1.28	.15									20.42	129.0	13.07	43.1	9.08	17.8	5.11	4.4	3.27	1.48	2.27	.62	200
			10" Pipe																				
220	1.40	.18	.90	.06							22.47	154.0	14.38	52.0	9.99	21.3	5.62	5.2	3.59	1.77	2.50	.73	220
240	1.53	.22	.98	.07							24.51	182.0	15.69	61.0	10.89	25.1	6.13	6.2	3.92	2.08	2.72	.87	240
260	1.66	.25	1.06	.08							26.55	211.0	16.99	70.0	11.80	29.1	6.64	7.2	4.25	2.41	2.95	1.00	260
280	1.79	.28	1.15	.09									18.30	81.0	12.71	33.4	7.15	8.2	4.58	2.77	3.18	1.14	280
300	1.91	.32	1.22	.11									19.61	92.0	13.62	38.0	7.66	9.3	4.90	3.14	3.40	1.32	300
320	2.05	.37	1.31	.12									20.92	103.0	14.52	42.8	8.17	10.5	5.23	3.54	3.64	1.47	320
340	2.18	.41	1.39	.14									22.22	116.0	15.43	47.9	8.68	11.7	5.54	3.97	3.84	1.62	340
					12" Pipe																		
360	2 30	.45	1.47	.15									23.53	128.0	16.34	53.0	9.19	13.1	5.87	4.41	4.08	1.83	360
380	2.43	.50	1.55	.17	1.08	.069							24.84	142.0	17.25	59.0	9.69	14.0	6.19	4.86	4.31	2.00	380
400	2.60	.54	1.63	.19	1.14	.075							26.14	156.0	18.16	65.0	10.21	16.0	6.54	5.4	4.55	2.20	400
							14" Pipe																
450	2.92	.68	1.84	.23	1.28	.95									20.40	78.0	11.49	19.8	7.35	6.7	5.11	2.74	450
500	3.19	.82	2.04	.28	1.42	.113	1.04	.06							22.70	98.0	12.77	24.0	8.17	8.1	5.68	2.90	500
550	3.52	.97	2.24	.33	1.56	.135	1.15	.07							24.96	117.0	14.04	28.7	8.99	9.6	6.25	3.96	550
600	3.84	1.14	2.45	.39	1.70	.159	1.25	.08							27.23	137.0	15.32	33.7	9.80	11.3	6.81	4.65	600
650	4.16	1.34	2.65	.45	1.84	.19	1.37	.09									16.59	39.0	10.62	13.2	7.38	5.40	650
700	4.46	1.54	2.86	.52	1.99	.22	1.46	.10									17.87	44.9	11.44	15.1	7.95	6.21	700
750	4.80	1.74	3.06	.59	2.13	.24	1.58	.11									19 15	51.0	12.26	17.2	8.50	7.12	750
800	5.10	1.90	3.26	.66	2.27	.27	1.67	.13									20.42	57.0	13.07	19.4	9.08	7.96	800
									16" Pipe														
850	5.48	2.20	3.47	.75	2.41	.31	1.79	.14	1.36	.08							21.70	64.0	13.89	21.7	9.65	8.95	850
900	5.75	2.46	3.67	.83	2.56	.34	1.88	.16	1.44	.084							22.98	71.0	14.71	24.0	10.20	10.11	900
											20" Pipe												
950	6.06	2.87	3.88	.91	2.70	.38	2.00	.18	1.52	.095									15.52	26.7	10.77	11.20	950
1000	6.38	2.97	4.08	1.03	2.84	.41	2.10	19	1.60	.10	1.02	.04							16.34	29.2	11.34	12.04	1000
1100	7.03	3.52	4.49	1.19	3.13	.49	2.31	.23	1.76	.12	1.12	.04							17.97	34.9	12.48	14.55	1100
1200	7 66	4.17	4.90	1.40	3.41	.58	2.52	.27	1.92	.14	1.23	.05							19.61	40.9	13.61	17.10	1200
1300	8.30	4 85	5.31	1.62	3.69	.67	2.71	.32	2.08	.17	1.33	.06									14.72	18.4	1300
1400	8.95	5.50	5.71	1.87	3.98	.78	2.92	.36	2.24	.19	1.43	.064									15.90	22.60	1400
1500	9.58	6.24	6.12	2 13	4.26	.89	3.15	.41	2.39	21	1.53	.07									17.02	25.60	1500
1600	10.21	7.00	6 53	2.39	4.55	.98	3.34	47	2.56	24	1.63	.08									18.10	26.9	1600
													24" Pipe										
1800	11.50	8.78	7.35	2.95	5.11	1.21	3.75	.58	2 87	.30	1.84	.10	1.28	.04									1800
2000	12.78	10.71	8.16	3.59	5.68	1.49	4 17	.71	3.19	.37	2.04	.12	1.42	.05									2000
2200	14.05	12.78	8.98	4.24	6.25	1.81	4.59	.84	3.51	.44	2.25	.15	1.56	.06									2200
															30" Pipe								
2400	15.32	14.2	9.80	5.04	6.81	2.08	5.00	.99	3.83	.52	2.45	.17	1.70	.07	1.09	.02							2400
2600			10 61	5.81	7.38	2.43	5 47	1.17	4.15	.60	2.66	.20	1.84	.08	1 16	.027							2600
2800			11.41	6.70	7.95	2.75	5.84	1.32	4.47	.68	2.86	.23	1.98	.09	1.27	.03							2800
3000			12.24	7.62	8.52	3.15	6.01	1.49	4.79	.78	3.08	.27	2.13	.10	1.37	.037							3000
3200			13 05	7.8	9.10	3.51	6.68	1.67	5.12	88	3.27	.30	2.26	.12	1.46	.041							3200
3500			14 30	10 08	9.95	4.16	7 30	1.97	5.59	1 04	3.59	.35	2.49	.14	1.56	.047							3500
3800			15 51	13.4	10.80	4 90	7 98	2.36	6.07	1.20	3.88	.41	2.69	.17	1 73	.05							3800
4200					11.92	5.88	8.76	2.77	6.70	1.44	4.29	.49	2.99	.20	1.91	.07							4200
4500					12 78	6.90	9.45	3.22	7.18	1.64	4 60	.56	3.20	.22	2.04	.08							4500
5000					14 20	8.40	10.50	3.92	8.01	2 03	5.13	.68	3.54	.27	2.26	.09							5000
5500							11.55	4.65	8.78	2.39	5.64	.82	3.90	.33	2.50	.11							5500
6000							12 60	5.50	9.58	2.79	6.13	.94	4.25	.38	2.73	13							6000
6500							13.65	6.45	10.39	3 32	6.64	1.10	4.61	.45	2.96	.15							6500
7000							14 60	7.08	11.18	3.70	7 15	1.25	4 97	.52	3.18	.17							7000
8000									12.78	4.74	8.17	1.61	5.68	.66	3.64	.23							8000
9000									14 37	5.90	9.20	2.01	6 35	81	4 08	.28							9000
10000									15.96	7.19	10.20	2.44	7.07	.98	4.54	.33							10000
12000											12.25	3 41	8.50	1.40	5.46	.48							12000

Reprinted from *Basic Science Concepts and Applications*, by permission. Copyright © 1980, American Water Works Association.

MINOR HEAD LOSS

Minor head losses are a result of water moving through valves and orifices, causing rapid changes in velocity and direction of flow.

These losses may be estimated using the nomograph* given on the facing page.

To read the nomograph, follow these steps:

1. Place one end of a straightedge at the known pipe diameter on the right scale.
2. Align the other end of the straight edge with the point designated for the desired fitting or orifice on the left scale.
3. Draw a line from the left scale to the right scale and read the head loss value on the middle scale.

The values given in the table represent **equivalent length of straight pipe**. These values are to **be added to actual pipe length** for calculation of friction losses, described on the previous page.

Minor losses are normally just that—minor, when compared to friction loss values. However, the smaller the runs of pipe, the more significant are these minor head losses.

Examples 6 and 7 illustrate the use of the nomograph in determining minor friction losses.

Example 6: (Head and Head Loss)

❑ Determine the "equivalent length of pipe" for a flow through a swing check valve, fully open, if the diameter of the pipe is 6 inches.

Align one end of the straightedge with the 6-inch (nominal diameter) mark on the left side of the scale shown to the right.

Align the other end of the straightedge with the point indicated for the "Swing Check Valve, Fully Open."

Draw a line from the left scale to the right scale.

The "equivalent length" value can be read on the middle scale:

40 ft equivalent length of pipe

Example 7: (Head and Head Loss)

❑ What is the head loss through a gate valve (1/2 closed) for a 10-inch diameter pipeline?

Align one end of the straightedge with 10-inches (nominal diameter) on the scale to the right.

Align the other end of the straightedge with the point corresponding with "Gate Valve—1/2 Closed" on the scale to the left.

Draw a line between the two outside scales and read the head loss value from the middle scale. The approximate reading is:

170 ft equivalent length of pipe

* For a review of reading nomographs, refer to Chapter 12 in *Basic Math Concepts*.

RESISTANCE OF VALVES AND FITTINGS TO FLOW

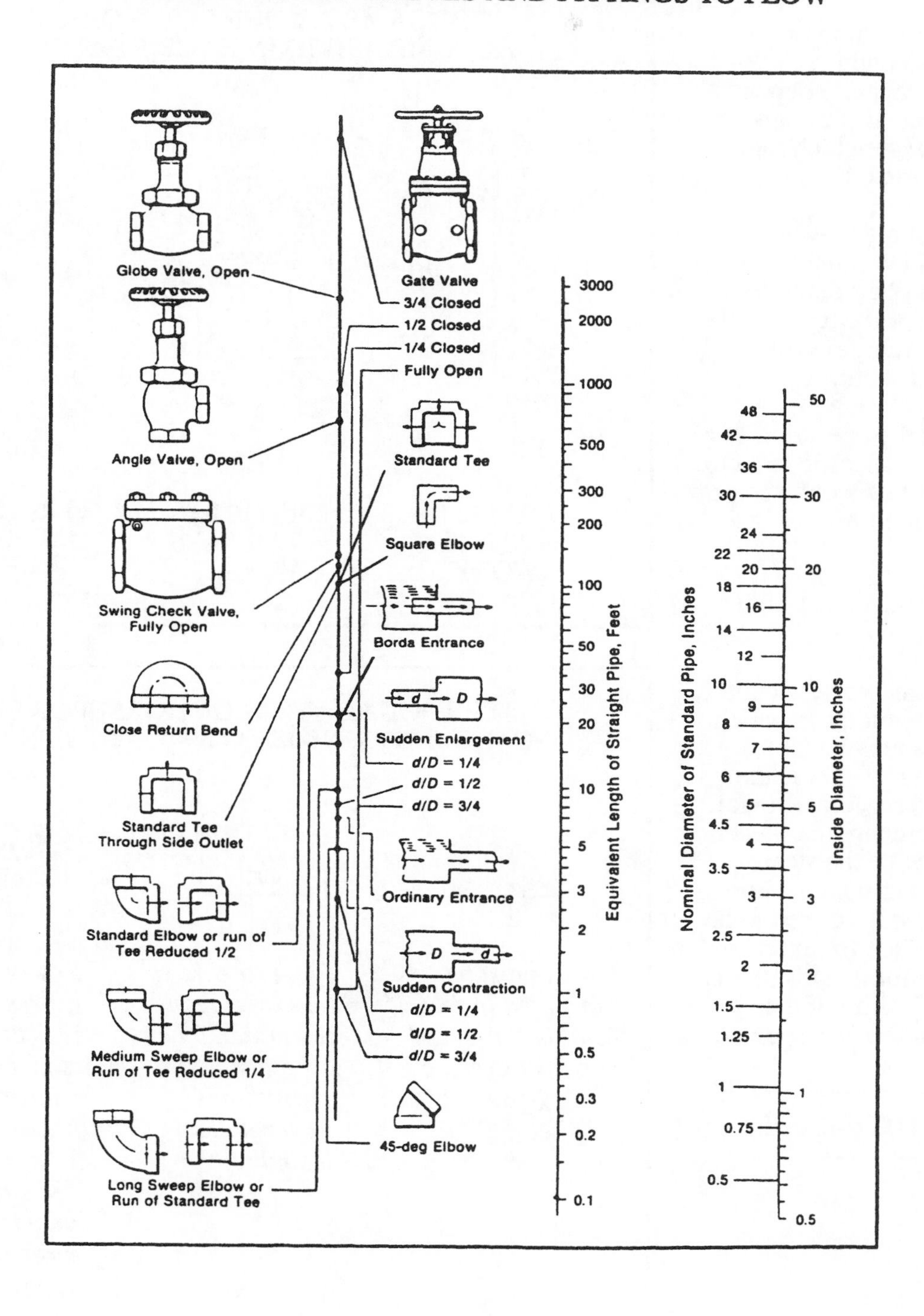

Reprinted with permission of Crane Valves

7.4 HORSEPOWER

The selection of a pump or combination of pumps with an adequate pumping capacity depends upon the flow rate desired and the effective height* to which the flow must be pumped.

Calculations of horsepower and head are made in conjunction with many treatment plant operations. The basic concept from which the horsepower calculation is derived is the concept of work.

Work involves the operation of a force (lbs) over a specific distance (ft). The **amount of work** accomplished is measured in foot-pounds:

(ft)(lbs) = ft-lbs

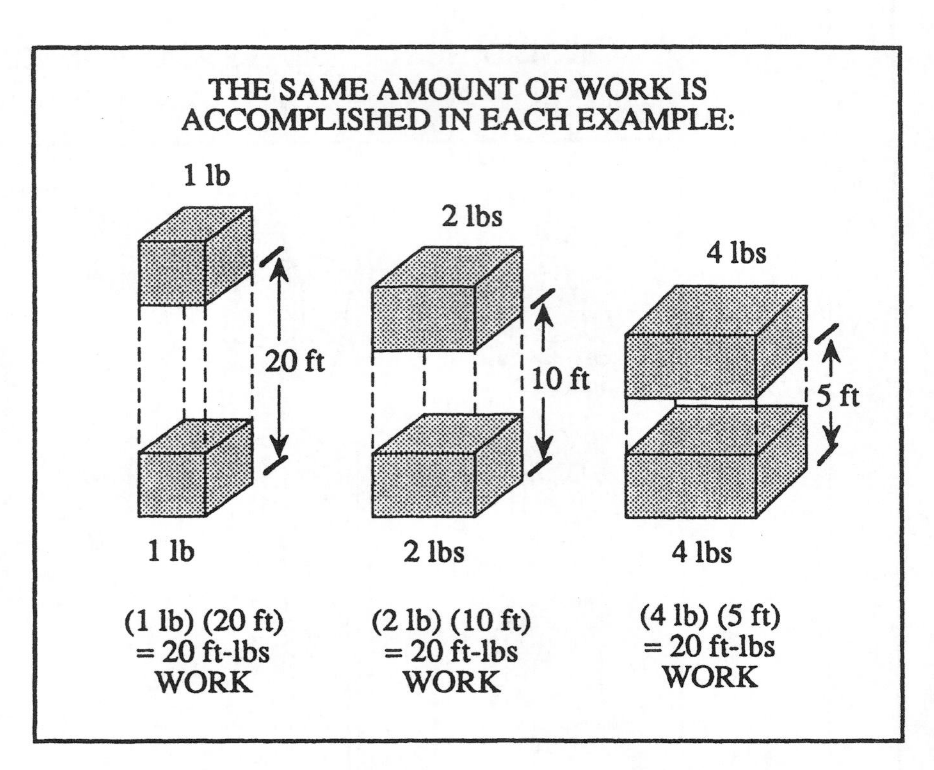

The **rate of doing work** is called **power**. The time factor in which the work occurs now becomes important. James Watt was the first to use the term **horsepower**. He used it to compare the power of a horse to that of the steam engine. The rate at which a horse could work was determined to be about 550 ft-lbs/sec (or expressed as 33,000 ft-lbs/min). This rate has become the definition of the standard unit called horsepower:

1 hp = 33,000 ft-lbs/min

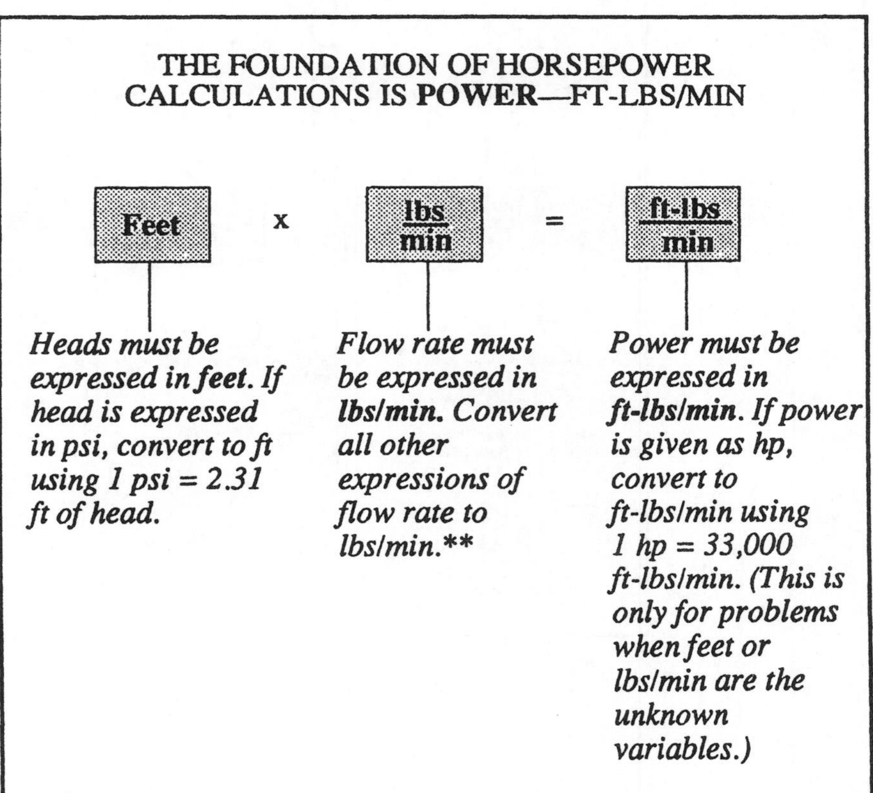

---

* "Effective height" refers to the feet of head against which the pump must pump.

** To review flow rate conversions, refer to Chapter 8 in *Basic Math Concepts*.

Example 1: (Horsepower)

❑ A pump must pump 2000 gpm against a total head of 20 ft. What horsepower is required for this work?

First calculate ft - lbs/min required:

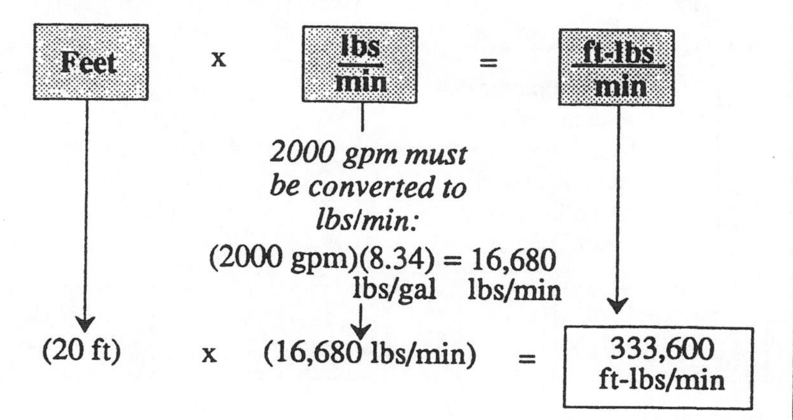

Then convert ft-lbs/min to hp:

$$\frac{333{,}600 \text{ ft-lbs/min}}{33{,}000 \text{ ft-lbs/min/hp}} = \boxed{10 \text{ hp}}$$

Example 2: (Horsepower)

❑ A flow of 8 MGD must be pumped against a total dynamic head (TDH) of 25 ft. What horsepower is required for this work?

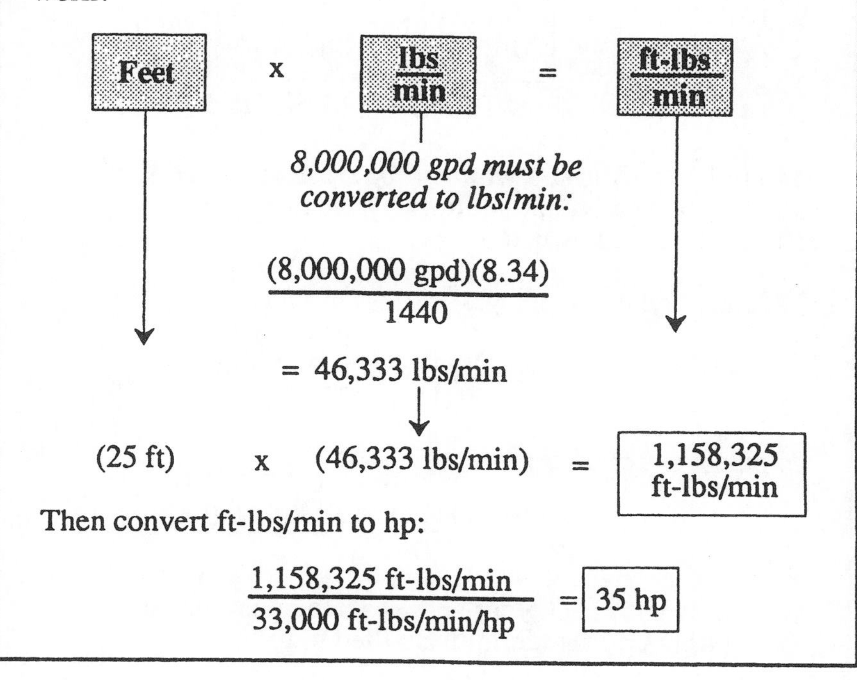

Then convert ft-lbs/min to hp:

$$\frac{1{,}158{,}325 \text{ ft-lbs/min}}{33{,}000 \text{ ft-lbs/min/hp}} = \boxed{35 \text{ hp}}$$

CALCULATING HORSEPOWER

When calculating horsepower requirements:

1. Determine the ft-lbs/min power required:

$$\boxed{(\text{ft})(\text{lbs/min}) = \text{ft-lbs/min}}$$

2. Once ft-lbs/min power has been calculated, horsepower can be determined using the equation 1 hp = 33,000 ft-lbs/min.

$$\boxed{\text{hp} = \frac{\text{ft-lbs/min}}{33{,}000 \text{ ft-lbs/min/hp}}}$$

AN ALTERNATE EQUATION

An equation frequently given for horsepower calculations is:

$$\boxed{\text{whp} = \frac{\underset{\text{gpm}}{(\text{flow rate})}\ \underset{\text{ft}}{(\text{total head})}}{3960}}$$

This equation is derived from the horsepower equation described above:

$$\text{hp} = \frac{\text{ft-lbs/min}}{33{,}000 \text{ ft-lbs/min/hp}}$$

It is then adjusted to reflect gpm flow rate, rather than lbs/min flow rate. The advantage of this equation is that gpm may be used directly, without conversions. The disadvantage is that it is somewhat "cut off from its roots":—the concept of power, ft-lbs/min. Because of this, there is often a lack in flexible application of the equation. It tends to become an equation memorized but not fully understood.

In the examples thus far, we have calculated the horsepower required to accomplish a particular pumping job. Due to motor and pump inefficiencies, however, more horsepower must be supplied in order to deliver the desired horsepower. To illustrate, in Example 2 it was calculated that 35 hp would be needed to pump 8 MGD against a TDH of 25 ft. Due to pump and motor inefficiencies, however, about 50 hp would have to be supplied to the motor in order to deliver the 35 hp from the pump.

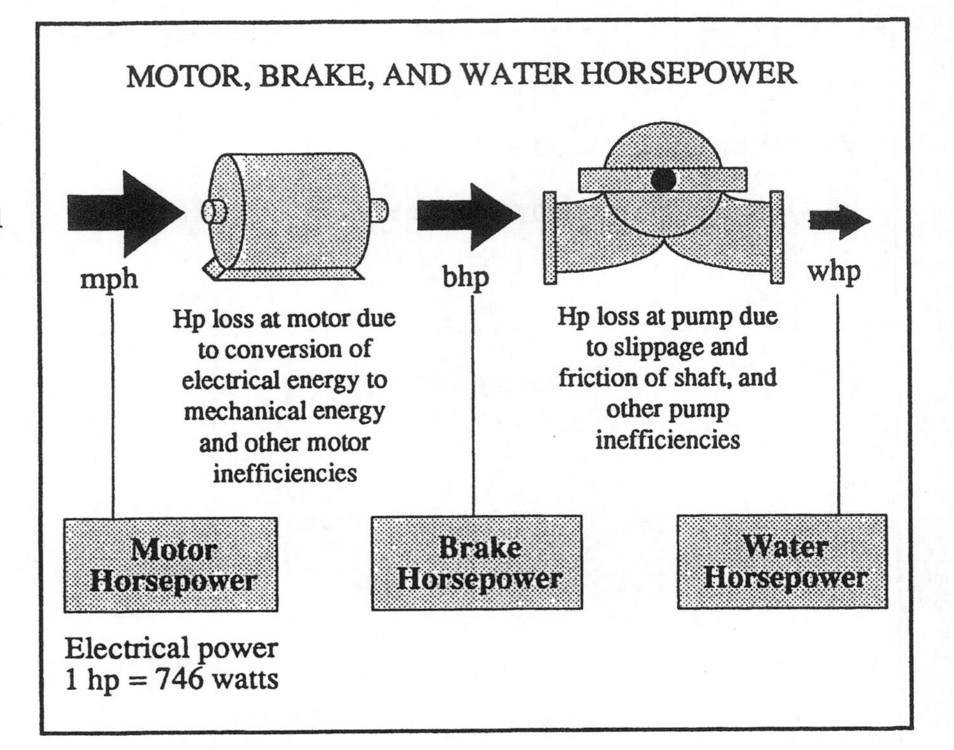

HORSEPOWER TERMINOLOGY

Three different horsepower terms are used to distinguish the type of horsepower being referred to in any particular calculation:

- Motor horsepower,
- Brake horsepower, and
- Water horsepower.

Motor horsepower (mhp) refers to the horsepower supplied to the motor in the form of electrical current. Some of this horsepower is lost due to the conversion of electrical energy to mechanical energy. The efficiency of most motors ranges from 80-95%, and is listed in manufacturer's literature.

Brake horsepower (bhp) refers to the horsepower supplied to the pump from the motor. As the power moves through the pump, additional horsepower is lost, resulting from slippage and friction of the shaft and other factors. Pump efficiencies generally range between 50-85%.

Example 3: (Horsepower)

❑ If 12 hp is supplied to a motor (mhp), what is the bhp and whp if the motor is 90% efficient and the pump is 85% efficient?

It is helpful to diagram the information given and desired:

12 mhp ⟶ **Motor** (90% Effic.) —? bhp→ **Pump** (85% Effic.) —? whp→

Mhp is 12 hp. Bhp and whp will be smaller numbers. To calculate bhp and mhp, multiply by the motor and pump efficiencies, as indicated below.

Calculate brake horsepower:

$$(12 \text{ mhp}) \frac{90}{100} = \boxed{10.8 \text{ bhp}}$$

Calculate water horsepower:

$$(10.8 \text{ bhp}) \frac{85}{100} = \boxed{9.2 \text{ whp}}$$

Note: Always check your answers. Note that bhp and whp are smaller numbers than mhp.

Example 4: (Horsepower)
❑ 40 hp is supplied to a motor. How many horsepower will be available for actual pumping loads, if the motor is 92% efficient and the pump is 85% efficient?

In this problem, the unknown hp is whp ("available for actual pumping loads"). First diagram the problem:

40 mhp → **Motor** (92% Effic.) → **Pump** (85% Effic.) → ? whp

To calculate whp, **multiply** mhp by both the motor and pump efficiencies:

$$(40 \text{ mhp}) \frac{(92)}{100} \frac{(85)}{100} = \boxed{31.3 \text{ whp}}$$

Example 5: (Horsepower)
❑ A total of 28 hp is required for a particular pumping application. If the pump efficiency is 75% and the motor efficiency is 85%, what horsepower must be supplied to the motor?

In this problem, the unknown term is mhp. Remember that the answer must be a **larger number** than the whp (28 hp).

First diagram the information in the problem:

? mhp → **Motor** (85% Effic.) → **Pump** (75% Effic.) → 28 whp

To calculate mhp, divide by pump and motor efficiencies:

$$\frac{28 \text{ whp}}{(0.85 (0.75)} = \boxed{43.9 \text{ mhp}}$$

Note that the mhp calculated is in fact a larger number than whp.

Water horsepower (whp) refers to the actual horsepower available to pump the water. Examples 1 and 2 in this section were actually calculations of water horsepower.

When making calculations of motor, brake, and water horsepower, it is important to remember that motor horsepower will be the largest number, brake horsepower next largest, followed by water horsepower. Knowing the relative sizes of these terms will help you know if your answers are reasonable.

The conversion between horsepower terms can be calculated as follows:

- When converting from a smaller term to a larger term* (such as whp to bhp, bhp to mhp, or whp to mhp), divide by the efficiency of the pump or motor:**

$$\text{Brake hp} = \frac{\text{Water hp}}{\text{Pump Effic.}}$$

$$\text{Motor hp} = \frac{\text{Brake hp}}{\text{Motor Effic.}}$$

$$\text{Motor hp} = \frac{\text{Water hp}}{(\text{Motor Effic.})(\text{Pump Effic.})}$$

- When converting from a larger term to a smaller term (such as mhp to bhp, bhp to whp, or mhp to whp), multiply by the efficiency of the pump or motor:**

$$\text{Brake hp} = (\text{Motor hp})(\text{Motor Effic.})$$

$$\text{Water hp} = (\text{Brake hp})(\text{Pump Effic.})$$

$$\text{Water hp} = (\text{Motor hp})(\text{Motor Effic.})(\text{Pump Effic.})$$

---

* Normally, when converting from a smaller term to a larger term, multiplication is indicated. However, that is only true when multiplying by a number greater than one. When a number less than one (such as pump or motor efficiency) is multiplied times a number , the answer is a smaller number; when a number less than one is used to divide a number, the resulting answer is larger number.

** Efficiency is written as %/100. For example 80% efficiency is written as 80/100. This can be simplified as 0.80.

HORSEPOWER AND SPECIFIC GRAVITY

In Examples 1-5, the horsepower calculations were based on pumping water. If another liquid is to be pumped, the specific gravity* of the liquid must be considered.

The specific gravity of a liquid is an indication of its density, or generally its weight, compared to that of water.

To account for differences in specific gravity, include the specific gravity factor when calculating ft-lbs/min pumping requirements:

(ft)(lbs/min)(sp. gr.) = ft-lbs/min for different liquid

Example 6 illustrates such a calculation.

Example 6: (Horsepower)

❑ A pump must pump against a total dynamic head of 50 ft at a flow rate of 1300 gpm. The liquid to be pumped has a specific gravity of 1.3. What is the water horsepower requirement for this pumping application?

Water horsepower is essentially a calculation of ft-lbs/min. A specific gravity factor must be included in this calculation:

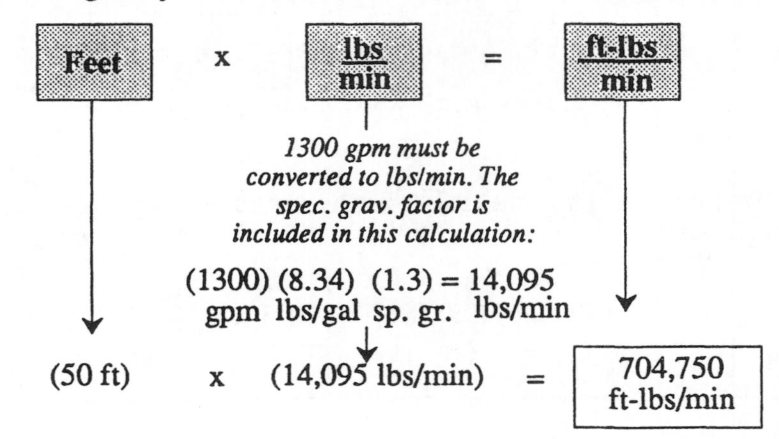

Now convert ft-lbs/min to hp:

$$\frac{704{,}750 \text{ ft-lbs/min}}{33{,}000 \text{ ft-lbs/min/hp}} = \boxed{21.4 \text{ whp}}$$

MHP AND KILOWATT REQUIREMENTS

Motor horsepower requirements can be converted to watts and then kilowatts requirements using the following equation:**

1 hp = 746 watts

Once watts requirements are determined, kilowatts are easily determined by a metric system conversion.

Example 7: (Horsepower)

❑ The motor horsepower requirement has been calculated to be 35 hp. How many kilowatts electric power does this represent?

First calculate the watts required using the equation 1 hp = 746 watts. The box method of conversions may be used:

hp —746→ watts

Multiplication by 746 is indicated:

$$(35 \text{ hp})(746 \text{ watts/hp}) = 26{,}110 \text{ watts}$$

Then:

$$\frac{26{,}110 \text{ watts}}{1000 \text{ watts/kW}} = \boxed{26.1 \text{ kW}}$$

---

* Specific gravity is discussed in Section 7.1 of this chapter.

** This equation is preferred to the equation 1 hp = 0.746 kW, since the box method of conversion works only if both numbers are greater than one. Refer to Chapter 8 in *Basic Math Concepts*.

Example 8: (Horsepower)
❑ 22 mhp is required for a pumping application. If the cost of power is \$0.0526/kWh, and the pump is in operation 24 hrs/day, what is the daily pump cost?

To calculate kWh pump operation, you must know the kW power requirements of the motor and hours of operation. First convert 22 mhp to kW: (1 hp = 746 watts)

$$(22 \text{ mhp})\ (746 \text{ watts/hp}) = 16{,}412 \text{ watts}$$

$$\text{or} = 16.4 \text{ kW}$$

The kWh of power consumption can now be determined:

$$(16.4 \text{ kW})\ (24 \text{ hrs/day}) = \boxed{393.6 \text{ kWh daily}}$$

Now complete the cost calculation:

$$(393.6 \text{ kWh/day})\ (\$0.0526/\text{kWh}) = \boxed{\$20.70 \text{ daily}}$$

Example 9: (Horsepower)
❑ The motor horsepower requirement has been calculated to be 50 mhp. During the week, the pump is in operation a total of 148 hours. Using a power cost of \$0.09439/kWh, what would be the power cost that week for the pumping?

First convert 50 mhp to kW so that kWh can be calculated: (1 hp = 746 watts)

$$(50 \text{ mhp})\ (746 \text{ watts/hp}) = 37{,}300 \text{ watts}$$

$$\text{or} = 37.3 \text{ kW}$$

Next calculate kWh of power consumed:

$$(37.3 \text{ kW})\ (148 \text{ hrs}) = \boxed{5520 \text{ kWh}}$$

Then powers costs may be calculated:

$$(5520 \text{ kWh/wk})\ \left(\frac{\$0.09439}{\text{kWh}}\right) = \boxed{\$521.03 \text{ cost for the week}}$$

PUMPING COST CALCULATIONS

Pumps costs are determined on the basis of two primary considerations:

- Kilowatt-hours of pump operation, and
- Power cost per kilowatt-hour.

Kilowatt-hours of pump operation are determined by multiplying power drawn by the pump (kW) by the hours of operation (hrs):

$$\boxed{(\text{kW})\ (\text{hrs}) = \text{kWh used}}$$

Once the kilowatt-hours of power use has been determined, then determine the cost of that power use using the cost factor:

$$\boxed{(\text{kWh Power Use})\ (\text{Cost/kWh use}) = \text{Total Cost}}$$

Examples 8 and 9 are pumping cost calculations.

7.5 PUMP CAPACITY

PUMP CAPACITY TESTING

Pump capacity may be determined by timing the pumping into or out of a tank of known size. Assuming water is not entering or leaving from any source other than the pump being tested, then—

- **When pumping into a tank** the rise in water level will correspond with the pumping rate.
- **When pumping out of a tank**, the drop in water level will correspond with the pumping rate.

To calculate the pumping capacity or rate, determine the gallons rise or fall and divide by the time of the pump test:

$$\text{Pumping Rate, gpm} = \frac{\text{gal rise or fall}}{\text{minutes of test}}$$

Or

(In expanded form for a rectangular tank)

$$\text{Pumping Rate, gpm} = \frac{(l)\,(w)\,(d)\,(7.48\text{ gal/cu ft})}{\text{minutes}}$$

WHEN PUMPING INTO AN EMPTY TANK, **THE RISE IN WATER LEVEL** INDICATES PUMPING RATE

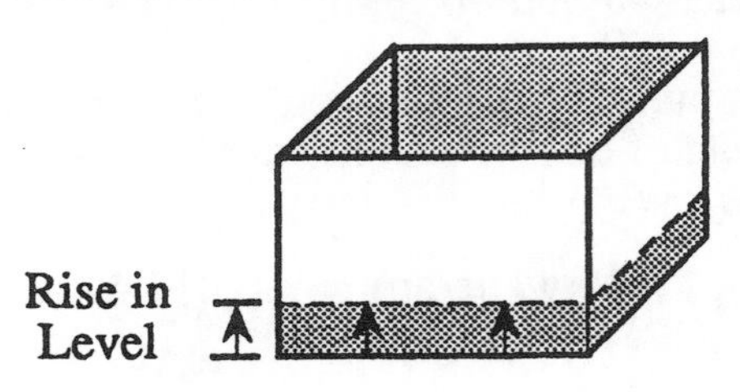

WHEN PUMPING FROM A TANK, (WITH INFLUENT VALVE CLOSED) **THE DROP IN WATER LEVEL** INDICATES PUMPING RATE

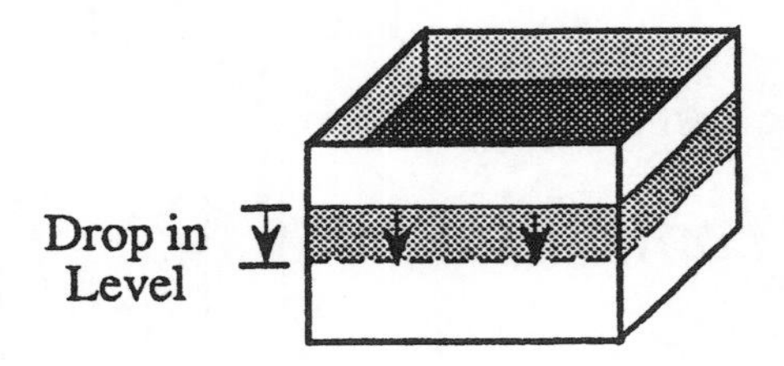

Example 1: (Pump Capacity)

❑ A wet well is 15 ft long and 12 ft wide. The influent valve to the wet well is closed. If a pump lowers the water level 1.25 ft during a 5-minute pumping test, what is the gpm pumping rate?

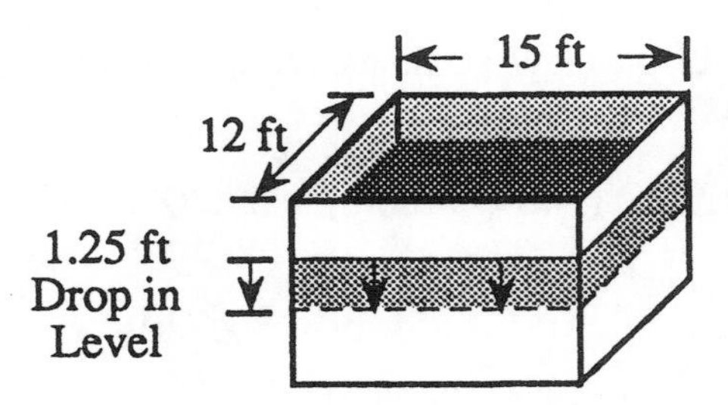

$$\text{Pumping Rate, gpm} = \frac{(\text{Length, ft})\,(\text{Width, ft})\,(\text{Drop, ft})\,(7.48\text{ gal/cu ft})}{\text{Test time, min}}$$

$$= \frac{(15\text{ ft})\,(12\text{ ft})\,(1.25\text{ ft})\,(7.48\text{ gal/cu ft})}{5\text{ minutes}}$$

$$= \boxed{337\text{ gpm}}$$

Example 2: (Pump Capacity)

❑ A pump is discharged into a 55-gallon barrel. If it takes 35 seconds to fill the barrel, what is the pumping rate in gpm?

Since gallons <u>per minute</u> are desired, first, convert 35 seconds to minutes,

$$\frac{35 \text{ sec}}{60 \text{ sec/min}} = 0.58 \text{ min}$$

then calculate the gpm pumping rate. The equation using tank dimensions is not needed since the gallons pumped (55 gallons) is already known. The general equation may be used:

55 gallons "rise"

$$\text{Pumping Rate, gpm} = \frac{\text{Rise, gallons}}{\text{Test time, min}}$$

$$= \frac{55 \text{ gallons}}{0.58 \text{ minutes}}$$

$$= \boxed{95 \text{ gpm}}$$

Example 3: (Pump Capacity)

❑ A wet well pump is rated at 300 gpm. A pump test is conducted for 3 minutes. What is the actual gpm pumping rate if the wet well is 10 ft long and 8 ft wide and the water level drops 1.33 ft during the pump test?

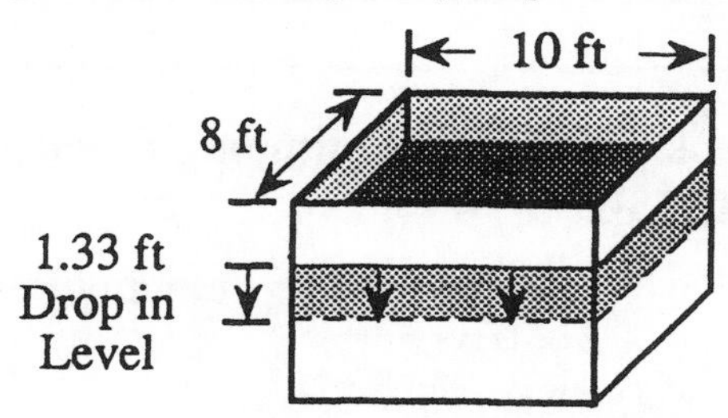

$$\text{Pumping Rate, gpm} = \frac{(\text{Length, ft})(\text{Width, ft})(\text{Drop, ft})(7.48 \text{ gal/cu ft})}{\text{Test time, min}}$$

$$= \frac{(10 \text{ ft})(8 \text{ ft})(1.33 \text{ ft})(7.48 \text{ gal/cu ft})}{3 \text{ minutes}}$$

$$= \boxed{265 \text{ gpm}}$$

CAPACITY TESTING WHEN INFLUENT VALVE IS OPEN

In the previous two pages, pump capacity was determined for two conditions:

- Pumping into a tank with known dimensions, and
- Pumping from a tank with the influent valve closed.

However, it is possible to determine pumping capacity (or pumping rate) even when the influent valve is open. Examples 4 and 5 illustrate how to calculate pumping rate out of a tank when there is influent entering the tank.

WHEN INFLUENT VALVE IS OPEN INFLUENT FLOW (GPM) MUST BE INCLUDED IN THE CALCULATION

When the water level remains the same, the pumping rate is equal to the influent rate:

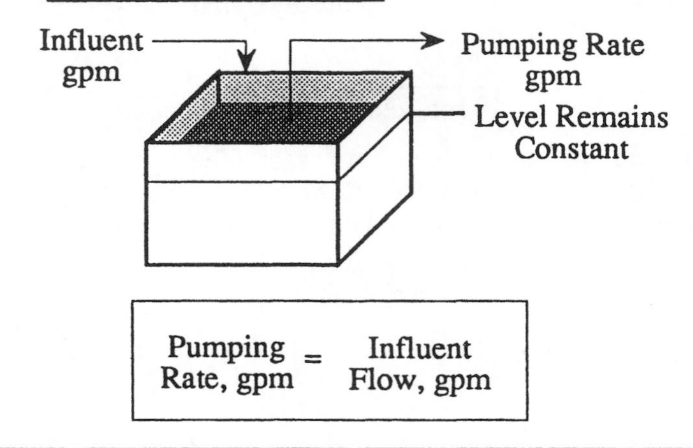

When the water level drops, the pumping rate is greater than the influent rate:

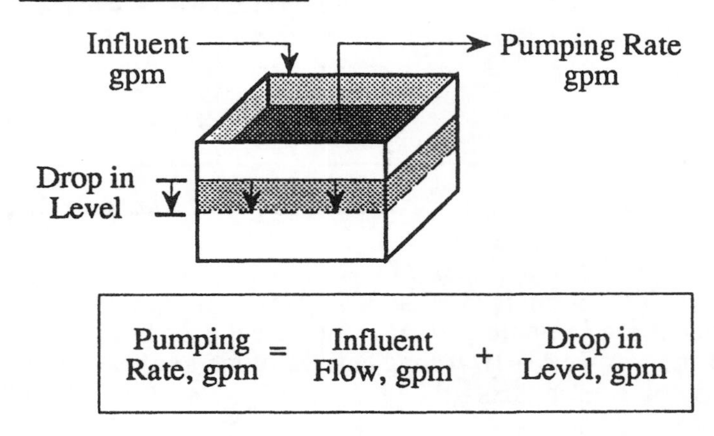

When the water level rises, the pumping rate is less than the influent rate:

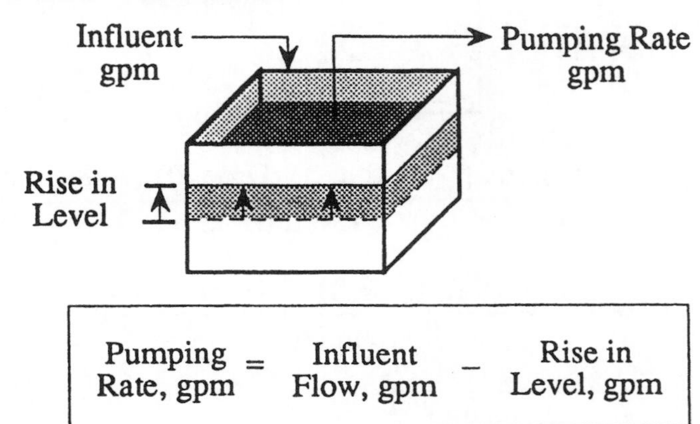

Example 4: (Pump Capacity)

❑ A tank is 6 ft wide and 10 ft long. During a 3-minute pumping test, the influent valve remains open. If the water level drops 6 inches during the pump test, what is the pumping rate in gpm? The influent flow is 0.8 MGD.

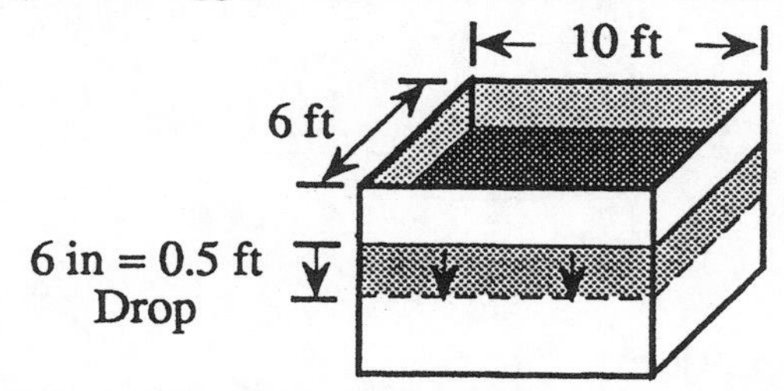

First calculate the gpm corresponding to the drop in water level: (6 in = 0.5 ft)

$$\text{Drop, gpm} = \frac{(10\text{ ft})\ (6\text{ ft})\ (0.5\text{ ft})\ (7.48)\text{ gallons}}{3\text{ minutes}}$$

$$= 75\text{ gpm}$$

Now calculate pumping rate:

$$\text{Pumping Rate, gpm} = \text{Influent Flow, gpm} + \text{Drop in Level, gpm}$$

$$= \frac{800{,}000\text{ gpd}}{1440\text{ min/day}} + 75\text{ gpm}$$

$$= 556\text{ gpm} + 75\text{ gpm}$$

$$= \boxed{631\text{ gpm}}$$

Example 5: (Pump Capacity)

❑ A pump test is conducted for 5 minutes while influent flow continues. During the test, the water level rises 3 inches. If the tank is 10 ft by 10 ft and the influent flow is 750,000 gpd, what is the pumping rate in gpm?

First calculate the gpm corresponding to the rise in level: (3 inches = 0.25 ft)

$$\text{Rise, gpm} = \frac{(10\text{ ft})\ (10\text{ ft})\ (0.25\text{ ft})\ (7.48)\text{ gallons}}{5\text{ minutes}}$$

$$= 37\text{ gpm}$$

Now calculate the pumping rate:

$$\text{Pumping Rate, gpm} = \text{Influent Flow, gpm} - \text{Rise in Level, gpm}$$

$$= \frac{750{,}000\text{ gpd}}{1440\text{ min/day}} - 37\text{ gpm}$$

$$= 521\text{ gpm} - 37\text{ gpm}$$

$$= \boxed{484\text{ gpm}}$$

CAPACITY FOR POSITIVE DISPLACEMENT PUMPS

One of the most common types of sludge pumps is the piston pump.* This type of pump operates on the principle of positive displacement. This means that it displaces, or pushes out, a volume of sludge equal to the volume of the piston. The length of the piston, called the stroke, can be adjusted (lengthened or shortened) to increase or decrease the gpm sludge delivered by the pump. Normally, the piston pump is operated no faster than about 50 gpm.

EACH STROKE OF A PISTON PUMP "DISPLACES" OR PUSHES OUT SLUDGE

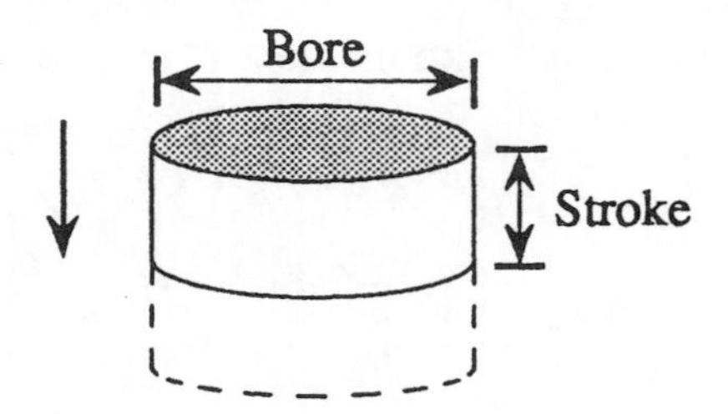

Simplified Equation:

$$\text{Volume of Sludge Pumped (gal/min)} = \frac{\text{(Gallons pumped)}}{\text{Stroke}} \frac{\text{(No. of Strokes)}}{\text{Minute}}$$

Expanded Equation:

$$\text{Volume of Sludge Pumped (gal/min)} = \left[(0.785)(D^2)\,\underset{\text{Length}}{(\text{Stroke})}\,\underset{\text{gal/cu ft}}{(7.48)}\right]\left[\underset{\text{Strokes/min}}{\text{No. of}}\right]$$

Example 6: (Pump Capacities)

❑ A piston pump discharges a total of 0.8 gallons per stroke (or revolution). If the pump operates at 20 revolutions per minute, what is the gpm pumping rate? (Assume the piston is 100% efficient and displaces 100% of its volume each stroke)

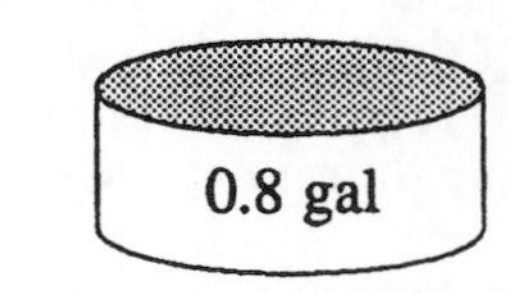

$$\text{Vol. of Sludge Pumped} = \frac{\text{(Gallons pumped)}}{\text{Stroke}} \frac{\text{(No. of Strokes)}}{\text{Minute}}$$

$$= \frac{(0.8 \text{ gal})}{\text{stroke}} \frac{(20 \text{ strokes})}{\text{min}}$$

$$= \boxed{16 \text{ gpm}}$$

* This type pump is also known as a plunger-type pump or positive displacement pump.

Example 7: (Pump Capacities)

❑ A sludge pump has a bore of 8 inches and a stroke length of 3 inches. If the pump operates at 50 strokes (or revolutions) per minute, how many gpm are pumped? (Assume the piston is 100% efficient and displaces 100% of its volume each stroke.)

$$\frac{8 \text{ in.}}{12 \text{ in./ft}} = 0.67 \text{ ft}$$

0.67 ft

$$\frac{3 \text{ in.}}{12 \text{ in./ft}} = 0.25 \text{ ft}$$

$$\text{Vol. of Sludge Pumped} = \frac{(\text{Gallons pumped})}{\text{Stroke}} \frac{(\text{No. of Strokes})}{\text{Minute}}$$

$$= \left[(0.785)(D^2)(\text{Stroke Length})(7.48 \text{ gal/cu ft})\right]\left[\text{Strokes/min}\right]$$

$$= \left[(0.785)(0.67 \text{ ft})(0.67 \text{ ft})(0.25 \text{ ft})\left(7.48 \frac{\text{gal}}{\text{cu ft}}\right)\right]\left[50 \frac{\text{Strokes}}{\text{min}}\right]$$

$$= \left(0.66 \frac{\text{gal}}{\text{stroke}}\right)\left(50 \frac{\text{strokes}}{\text{min}}\right)$$

$$= \boxed{33 \text{ gpm}}$$

Example 8: (Pump Capacities)

❑ A sludge pump has a bore of 6 inches and a stroke setting of 3 inches. The pump operates at 45 revolutions per minute. If the pump operates a total of 80 minutes during a 24-hour period, what is the gpd pumping rate? (Assume the piston is 100% efficient.)

$$\frac{6 \text{ in.}}{12 \text{ in./ft}} = 0.5 \text{ ft}$$

0.5 ft

$$\frac{3 \text{ in.}}{12 \text{ in./ft}} = 0.25 \text{ ft}$$

First calculate the gpm pumping rate:

$$\text{Vol. Pumped gpm} = \frac{(\text{Gallons pumped})}{\text{Stroke}} \frac{(\text{No. of Strokes})}{\text{Minute}}$$

$$= \left[(0.785)(0.5 \text{ ft})(0.5 \text{ ft})(0.25 \text{ ft})\left(7.48 \frac{\text{gal}}{\text{cu ft}}\right)\right]\left[45 \frac{\text{Strokes}}{\text{min}}\right]$$

$$= \left(0.37 \frac{\text{gal}}{\text{stroke}}\right)\left(45 \frac{\text{strokes}}{\text{min}}\right)$$

$$= 16.7 \text{ gpm}$$

Then convert gpm to gpd pumping rate, based on total minutes pumped during 24-hours:

$$(16.7 \text{ gpm})\left(80 \frac{\text{min}}{\text{day}}\right) = \boxed{1336 \text{ gpd}}$$

CALCULATING GPD PUMPED

There are two methods to determine gpd pumping rate:

- Calculate the gpm pumping rate, then multiply by the total minutes operation during the 24-hour period:

$$\text{Pumping Rate, gpd} = (\text{Pumping Rate, gpm})(\text{Total min pumping in 24 hrs})$$

- Calculate the gallons pumped each revolution, then multiply by the total revolutions during the 24-hour period:

$$\text{Pumping Rate, gpd} = \frac{(\text{Gallons})}{\text{Revolution}} \frac{(\text{Total Revol.})}{\text{day}}$$

NOTES:

8 Wastewater Collection and Preliminary Treatment

SUMMARY

1. Wet Well Capacity

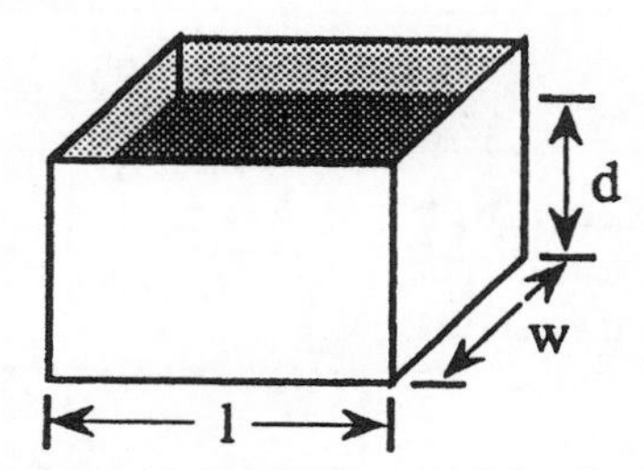

$$\text{Volume, cu ft} = (\text{length, ft})\,(\text{width, ft})\,(\text{depth, ft})$$

$$\text{Volume, gal} = (\text{length, ft})\,(\text{width, ft})\,(\text{depth, ft})\,(7.48,\ \text{gal/cu ft})$$

2. Wet Well Pumping Rate

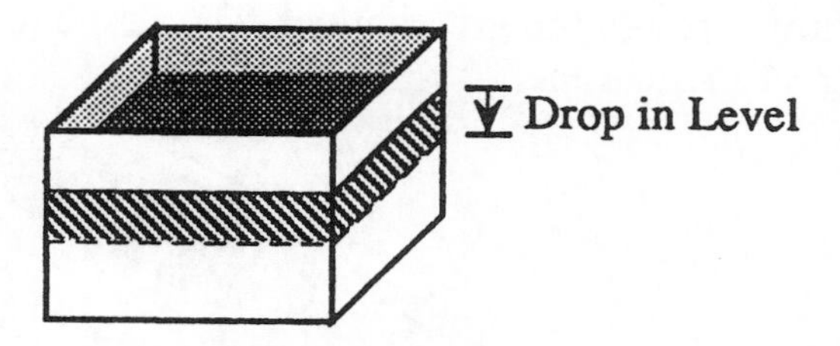

Simplified Equation:

$$\text{Pumping Rate, gpm} = \frac{\text{Volume pumped, gal}}{\text{Duration of Test, min}}$$

Expanded Equation:

$$\text{Pumping Rate, gpm} = \frac{(\text{length})\,(\text{width})\,(\text{depth})\,(7.48)}{\text{Duration of Test, min}}$$

(length in ft, width in ft, depth in ft, 7.48 in gal/cu ft)

SUMMARY

3. Screenings Removed

$$\text{Screenings Removed (cu ft/day)} = \frac{\text{Screenings, cu ft}}{\text{day}}$$

Or

$$\text{Screenings Removed (cu ft/MG)} = \frac{\text{Screenings, cu ft}}{\text{MG Flow}}$$

4. Screenings Pit Capacity

$$\text{Screening Pit Capacity, days} = \frac{\text{Screening Pit Vol., cu ft}}{\text{Screenings Removed, cu ft/day}}$$

5. Grit Channel Velocity

Two equations may be used to estimate the velocity of flow through the grit channel. The first equation is the $Q=AV$ equation.

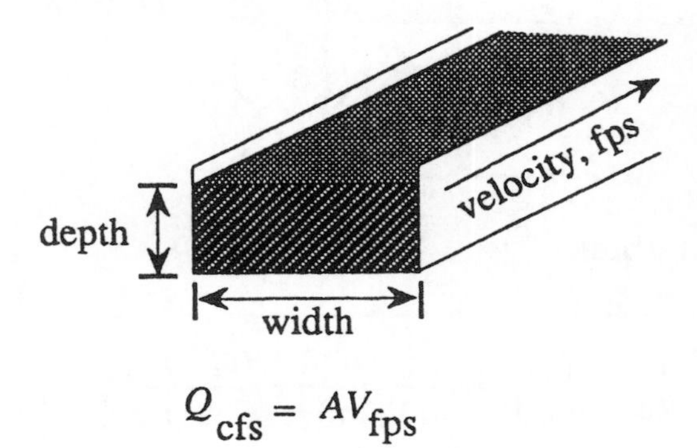

$$Q_{cfs} = AV_{fps}$$

$$\underset{\text{cfs}}{Q} = \underset{\text{ft}}{(\text{width})}\ \underset{\text{ft}}{(\text{depth})}\ \underset{\text{fps}}{(\text{velocity})}$$

SUMMARY—Cont'd

The second equation is necessary when using a float or dye to determine velocity.

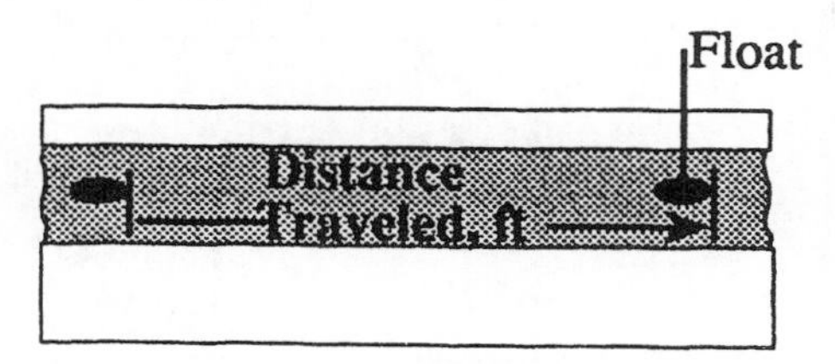

$$\text{Velocity} = \frac{\text{Distance traveled, ft}}{\text{Seconds of Test}}$$

$$\text{Velocity} = \frac{\text{ft}}{\text{sec}}$$

6. **Grit Removal**

$$\text{Grit Removal cu ft/MG} = \frac{\text{Grit Volume, cu ft}}{\text{Flow, MG}}$$

7. **Flow measurement** for a particular moment can be determined using the $Q=AV$ equation. Metering devices often have instrumentation to read and record flow rates. However, charts, graphs and nomographs can also be used to estimate flow rates.

8.1 WET WELL CAPACITY

In conveying wastewater from outlying areas to the treatment plant by gravity flow, it is occasionally necessary to install a lift station or pump station. The pump may lift the flow to a higher sewer in which the wastewater may again flow by gravity, or it may lift the wastewater directly into the treatment plant for gravity flow through the plant.

Because a minimum velocity of two feet per second is normally necessary to prevent the settling of solids in the sewer, pumping stations are sometimes used to increase the velocity of flow in a sewer.

WET WELL CAPACITY IS A VOLUME CALCULATION

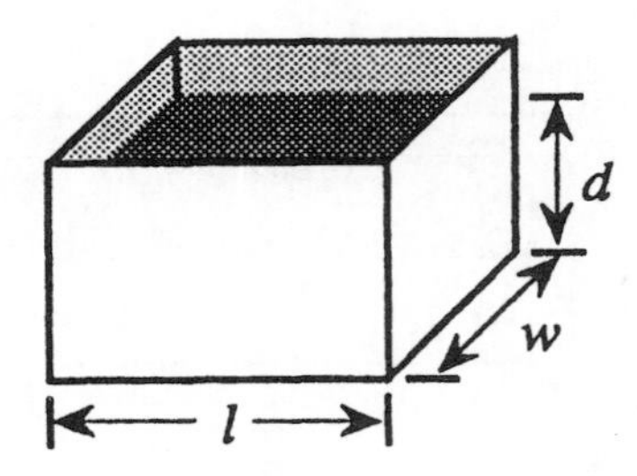

Volume, cu ft = (length, ft) (width, ft) (depth, ft)

Volume, gal = (length, ft) (width, ft) (depth, ft) (7.48, gal/cu ft)

Example 1: (Wet Well Capacity)

❑ A wet well is 13 ft long, 10 ft wide and 10 ft deep. What is the gallon capacity of the wet well?

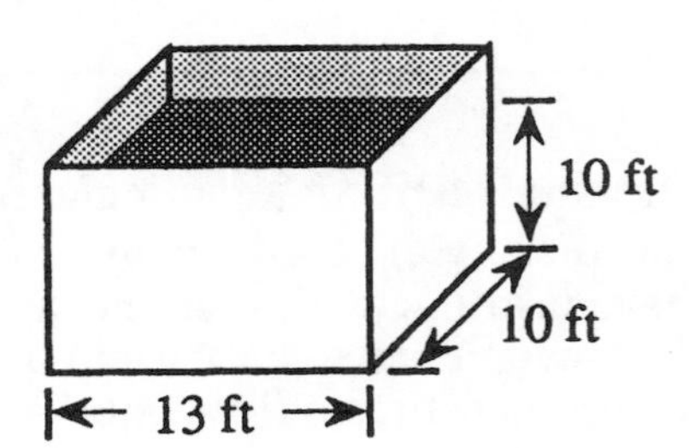

Volume, gal = (length, ft) (width, ft) (depth, ft) (7.48, gal/cu ft)

= (13 ft) (10 ft) (10 ft) (7.48, gal/cu ft)

= **9724 gallons**

Example 2: (Wet Well Capacity)

❑ A wet well is 8 ft long, 6 ft wide and 6 ft deep. what is the gallon capacity of the wet well?

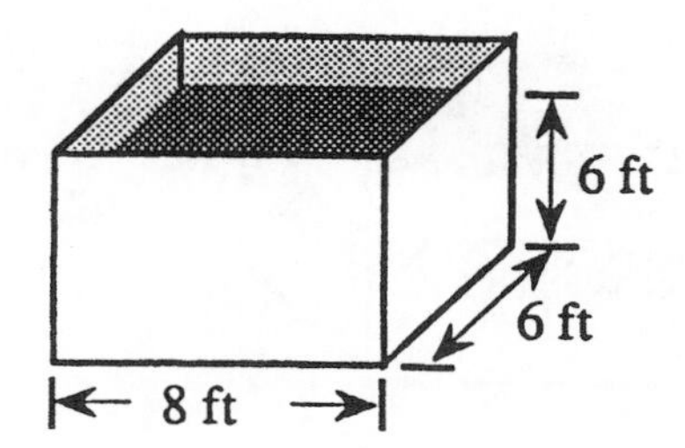

$$\text{Volume, gal} = (\text{length, ft})\ (\text{width, ft})\ (\text{depth, ft})\ (7.48,\ \text{gal/cu ft})$$

$$= (8\ \text{ft})\ (6\ \text{ft})\ (6\ \text{ft})\ (7.48,\ \text{gal/cu ft})$$

$$= \boxed{2154\ \text{gallons}}$$

Example 3: (Wet Well Capacity)

❑ A wet well 12 ft long and 10 ft wide contains wastewater to a depth of 3.7 ft. How many gallons of wastewater are in the wet well?

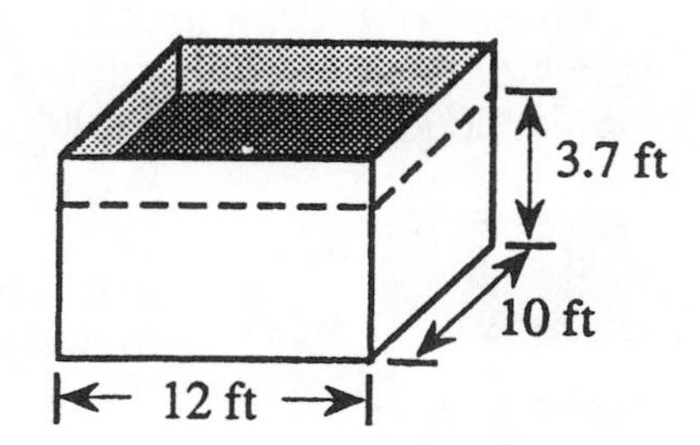

$$\text{Volume, gal} = (\text{length, ft})\ (\text{width, ft})\ (\text{depth, ft})\ (7.48,\ \text{gal/cu ft})$$

$$= (12\ \text{ft})\ (10\ \text{ft})\ (3.7\ \text{ft})\ (7.48,\ \text{gal/cu ft})$$

$$= \boxed{3321\ \text{gallons}}$$

8.2 WET WELL PUMPING RATE

Depending upon head, pump wear or general usage, the **actual gpm delivered** by a pump will vary more or less from the **gpm pump design** or **pump rating**. The pump may be calibrated by conducting a brief pumping test.

With the flow into the wet well stopped during the pumping test, allow the pump to begin pumping the water from the wet well. Measure the drop in water at the end of the test (2-5 minutes). The pumping rate can then be calculated using the equations shown to the right.

TO CALCULATE PUMPING RATE CALCULATE "VOLUME DROPPED"

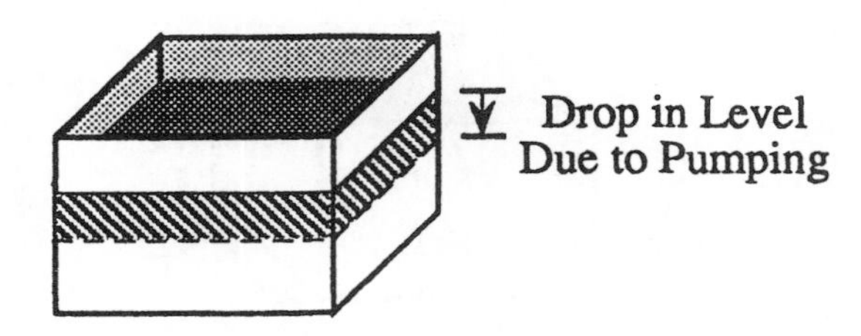

Simplified Equation:

$$\text{Pumping Rate, gpm} = \frac{\text{Volume Pumped, gal}}{\text{Duration of Test, min}}$$

Expanded Equation:

$$\text{Pumping Rate, gpm} = \frac{(\text{length})\ (\text{width})\ (\text{depth})\ (7.48)}{\text{Duration of Test, min}}$$

(length, ft; width, ft; depth, ft; 7.48 gal/cu ft)

Example 1: (Wet Well Pumping Rate)

❑ A wet well is 12 ft by 10 ft. With no influent to the well, a pump lowers the water level 1.2 ft during a 4-minute pumping test. What is the gpm pumping rate?

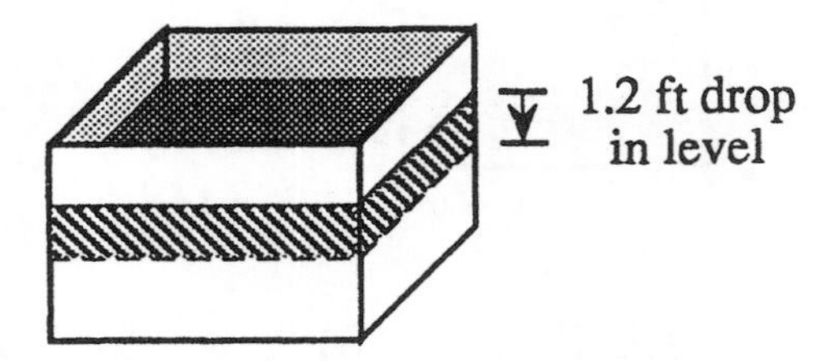

$$\text{Pumping Rate, gpm} = \frac{(\text{length})\ (\text{width})\ (\text{depth})\ (7.48)}{\text{Duration of Test, min}}$$

(length, ft; width, ft; depth, ft; 7.48 gal/cu ft)

$$= \frac{(12\text{ ft})\ (10\text{ ft})\ (1.2\text{ ft})\ (7.48\text{ gal/cu ft})}{4\text{ min}}$$

$$= \boxed{269\text{ gpm}}$$

Example 2: (Wet Well Pumping Rate)

❑ A wet well is 8 ft by 10 ft. During a 2-minute pumping test (with no influent to the wet well), the water level dropped 5 inches. What is the gpm pumping rate?

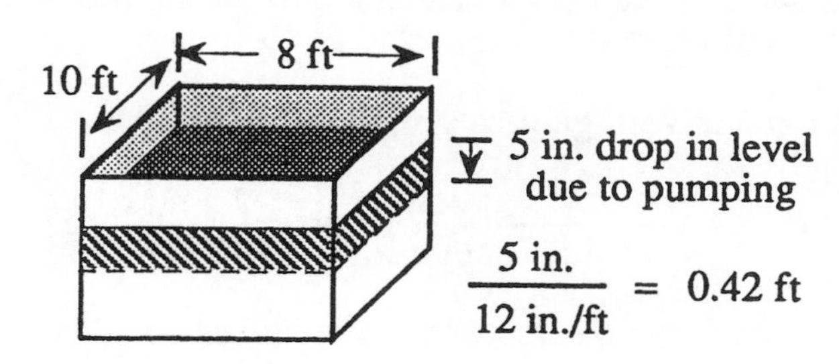

$$\frac{5 \text{ in.}}{12 \text{ in./ft}} = 0.42 \text{ ft}$$

$$\text{Pumping Rate, gpm} = \frac{\underset{\text{ft}}{(\text{length})}\ \underset{\text{ft}}{(\text{width})}\ \underset{\text{ft}}{(\text{depth})}\ \underset{\text{gal/cu ft}}{(7.48)}}{\text{Duration of Test, min}}$$

$$= \frac{(8 \text{ ft})\ (10 \text{ ft})\ (0.42 \text{ ft})\ (7.48 \text{ gal/cu ft})}{2 \text{ min}}$$

$$= \boxed{126 \text{ gpm}}$$

GIVEN INCHES DROP IN LEVEL

When the drop in water level is measured in inches, the inches must first be converted to feet before using the pumping rate equation. Example 2 illustrates such a calculation.

Example 3: (Wet Well Pumping Rate)

❑ The depth of wastewater in a wet well is sufficiently low to allow shutting off all pumps. With a rod and a stopwatch, you are able to determine that the water level rises 1.5 ft in 2 minutes 30 seconds. The pumps are restarted. What is the gpm influent rate to the 8 ft long, 8 ft wide wet well?

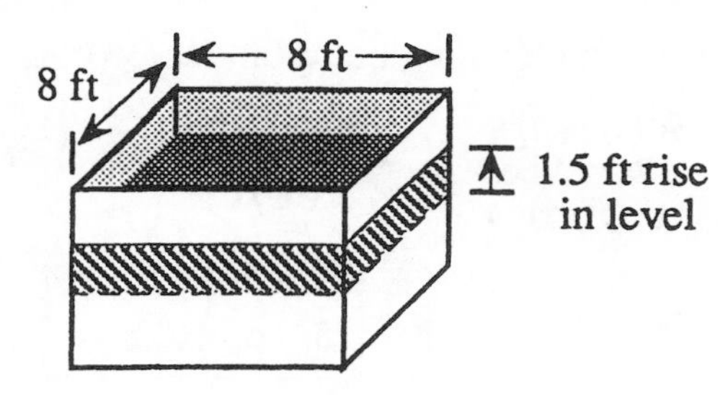

$$\text{Pumping Rate, gpm} = \frac{\underset{\text{ft}}{(\text{length})}\ \underset{\text{ft}}{(\text{width})}\ \underset{\text{ft}}{(\text{Rise in Level})}\ \underset{\text{gal/cu ft}}{(7.48)}}{\text{Duration of Test, min}}$$

$$= \frac{(8 \text{ ft})\ (8 \text{ ft})\ (1.5 \text{ ft})\ (7.48 \text{ gal/cu ft})}{2.5 \text{ min}}$$

$$= \boxed{287 \text{ gpm}}$$

CALCULATING INFLUENT RATE

This same basic concept can be used to estimate the flow rate of the water entering the wet well. An example of this type of calculation is given in Example 3.

8.3 SCREENINGS REMOVED

Screens are used in pretreatment to remove large debris such as rags, cans, cardboard, etc., from the wastewater flow. A range of 0.5 to 12 cu ft screenings per million gallons of flow may be removed from a plant. The withdrawal rate may remain somewhat stable for a plant, depending on sources of wastewater flow.

In order to plan properly for screenings disposal, it is important to keep a record of the amount of screenings removed from the wastewater flow. Two methods commonly used to calculate the volume of screenings withdrawn are:

$$\text{Screenings Removed (cu ft/day)} = \frac{\text{Screenings, cu ft}}{\text{days}}$$

$$\text{Screenings Removed (cu ft/MG)} = \frac{\text{Screenings, cu ft}}{\text{Flow, MG}}$$

Examples 1 and 2 illustrate the first calculation, and Examples 3 and 4 illustrate the second calculation.

Example 1: (Screenings Removed)

❑ A total of 55 gallons of screenings are removed from the wastewater flow during a 24-hour period. What is the screenings removal reported as cu ft/day?

First, convert gallons screenings to cu ft:*

$$\frac{55 \text{ gallons}}{7.48 \text{ gal/cu ft}} = \begin{matrix}7.4 \text{ cu ft}\\ \text{Screenings}\end{matrix}$$

Now calculate screenings removed as cu ft/day:

$$\text{Screenings Removed (cu ft/day)} = \frac{7.4 \text{ cu ft}}{1 \text{ day}}$$

$$= \boxed{7.4 \text{ cu ft/ day}}$$

Example 2: (Screenings Removed)

❑ During one week, a total of 290 gallons of screenings were removed from the wastewater screens. What is the average screenings removal in cu ft/day?

First, gallons screenings must be converted to cu ft screenings:

$$\frac{290 \text{ gallons}}{7.48 \text{ gal/cu ft}} = \begin{matrix}38.8 \text{ cu ft}\\ \text{Screenings}\end{matrix}$$

Now the screenings removal calculation can be completed:

$$\text{Screenings Removed (cu ft/day)} = \frac{38.8 \text{ cu ft}}{7 \text{ days}}$$

$$= \boxed{5.5 \text{ cu ft/ day}}$$

---

* For a review of gal to cu ft conversions, refer to Chapter 8 in *Basic Math Concepts*.

Example 3: (Screenings Removed)

❑ The flow at a treatment plant is 3.6 MGD. If a total of 5.5 cu ft screenings are removed during the 24-hour period, what is the screenings removal reported as cu ft/MG?

$$\text{Screenings Removed (cu ft/MG)} = \frac{\text{Screenings, cu ft}}{\text{Flow, MG}}$$

$$= \frac{5.5 \text{ cu ft}}{3.6 \text{ MG}}$$

$$= \boxed{1.5 \text{ cu ft/ MG}}$$

Example 4: (Screenings Removed)

❑ On a particular day a treatment plant receives a flow of 3,850,000 gpd. If 80 gallons of screenings are removed that day, what is the screenings removal expressed as cu ft/MG?

First, convert gallons screening to cu ft:

$$\frac{80 \text{ gallons}}{7.48 \text{ gal/cu ft}} = 10.7 \text{ cu ft}$$

Now calculate the desired screening rate. Remember that the flow must be expressed in **million gallons**:

$$\text{Screenings Removed (cu ft/MG)} = \frac{\text{Screenings, cu ft}}{\text{Flow, MG}}$$

$$= \frac{10.7 \text{ cu ft}}{3.85 \text{ MG}}$$

$$= \boxed{2.8 \text{ cu ft/ MG}}$$

8.4 SCREENINGS PIT CAPACITY

Screenings pit capacity calculations are actually detention time calculations.* Remember that detention time may be considered the time required to flow through a tank or the **time required to fill a tank or basin at a given flow rate.** In screenings pit capacity problems, the time required to fill a screenings pit is being calculated.

Both the general fill time equation and the specific equation to be used in screenings pit capacity problems are given below.

$$\text{Fill Time} = \frac{\text{Volume of Tank, gal}}{\text{Flow Rate, gal/time}}$$

$$\text{Screenings Pit Fill Time, days} = \frac{\text{Volume of Pit, cu ft}}{\text{Screenings Removed, cu ft/day}}$$

Example 1: (Screenings Pit Capacity)

❑ A screenings pit has a capacity of 400 cu ft. (The pit is actually larger than 400 cu ft to accommodate soil for covering.) If an average of 3.4 cu ft of screenings are removed daily from the wastewater flow, in how many days will the pit be full?

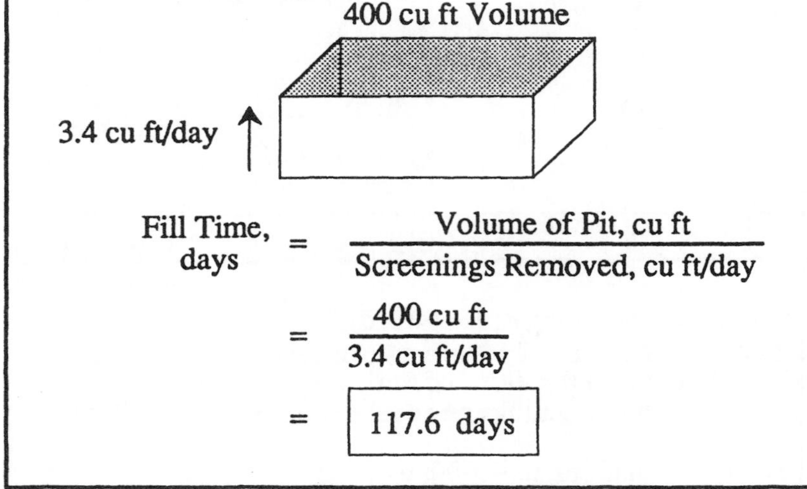

$$\text{Fill Time, days} = \frac{\text{Volume of Pit, cu ft}}{\text{Screenings Removed, cu ft/day}}$$

$$= \frac{400 \text{ cu ft}}{3.4 \text{ cu ft/day}}$$

$$= \boxed{117.6 \text{ days}}$$

Example 2: (Screenings Pit Capacity)

❑ A plant has been averaging a screenings removal of 2 cu ft/MG. If the average daily flow is 1.6 MGD how many days will it take to fill the pit with an available capacity of 150 cu ft?

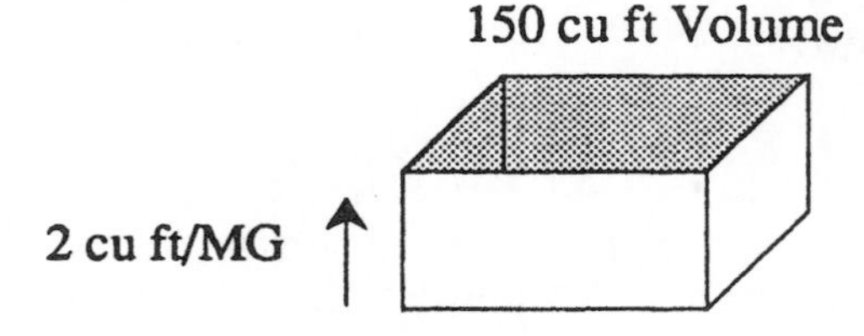

The filling rate must first be expressed as cu ft/day:

$$\left(\frac{2 \text{ cu ft}}{\text{MG}}\right)(1.6 \text{ MGD}) = 3.2 \text{ cu ft/day}$$

Now calculate fill time:

$$\text{Fill Time, days} = \frac{150 \text{ cu ft}}{3.2 \text{ cu ft/day}}$$

$$= \boxed{46.9 \text{ days}}$$

---

* For a review of detention time calculations, refer to Chapter 5.

Example 3: (Screenings Pit Capacity)

❑ A screenings pit has a capacity of 10 cu yds available for screenings. If the plant removes an average of 2.4 cu ft of screenings per day, in how many days will the pit be filled?

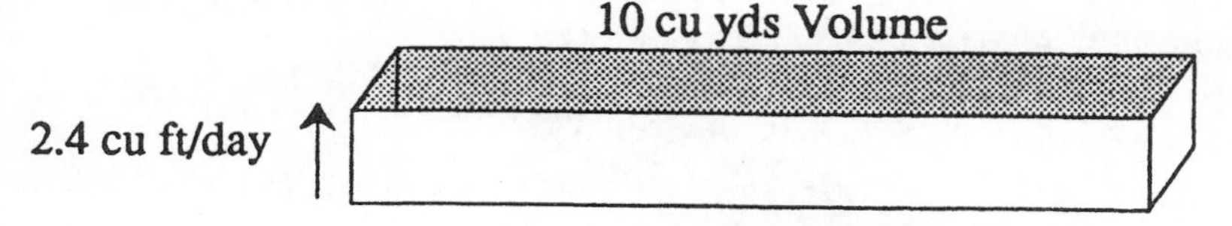

Since the filling rate is expressed as cu ft/day, the pit volume must be expressed as cu ft:**

$$(10 \text{ cu yds}) \left(27 \frac{\text{cu ft}}{\text{cu yds}}\right) = 270 \text{ cu ft}$$

Now calculate fill time:

$$\text{Fill Time, days} = \frac{\text{Volume of Pit, cu ft}}{\text{Screenings Removed, cu ft/day}}$$

$$= \frac{270 \text{ cu ft}}{2.4 \text{ cu ft/day}}$$

$$= \boxed{112.5 \text{ days}}$$

Example 4: (Screenings Pit Capacity)

❑ Suppose you want to have a screenings pit capacity of 90 days (not including dirt for cover). If the screenings removal rate is 4 cu ft/day, what will the available volume of the screenings pit have to be (cu ft)?

$$\text{Fill Time, days} = \frac{\text{Volume of Pit, cu ft}}{\text{Screenings Removed, cu ft/day}}$$

$$90 \text{ days} = \frac{x \text{ cu ft}}{4 \text{ cu ft/day}}$$

$$(90)(4) = x$$

$$\boxed{360 \text{ cu ft}} = x$$

CALCULATING OTHER UNKNOWN VALUES

In Examples 1-3 the unknown variable was fill time. However, other variables such as the required pit volume or the filling rate can also be the unknown variable. Example 4 illustrates this type of calculation.

** To review cu yd to cu ft conversions, refer to Chapter 8 in *Basic Math Concepts*.

8.5 GRIT CHANNEL VELOCITY

The desired velocity in sewers is approximately 2 fps at peak flow, because this velocity will normally prevent solids from settling from the lines. However, when the flow reaches the grit channel, **the velocity is decreased to about 1 fps to permit settling of the heavy inorganic solids**. These solids are removed early in the treatment process because they produce an unnecessary load on biological processes and greater stress on mechanical equipment.

There are two methods used to estimate the velocity of flow in a grit channel—the *Q=AV* method and the float or dye method.

VELOCITY USING *Q=AV*

The *Q=AV* equation may be used to estimate the velocity of flow in a channel or pipeline. Write the equation, filling in the known data, then solve for the unknown factor (velocity in this case).

Be sure that the volume and time expressions match on both sides of the equation. For instance, if the velocity is desired in ft/sec, then the flow rate should be converted to cu ft/sec before beginning the *Q=AV* calculation. Examples 1 and 2 illustrate a velocity estimate using the *Q=AV* equation.

Example 1: (Grit Channel Velocity)

❑ A grit channel is 2.5 ft wide, with water flowing to a depth of 14 inches. If the flow meter indicates a flow rate of 1450 gpm, what is the velocity of flow through the channel?(fps)

$\frac{14 \text{ in.}}{12 \text{ in./ft}} = 1.17 \text{ ft}$

2.5 ft

x fps

Convert gpm to cfs:

$$\frac{1450 \text{ gpm}}{(7.48 \frac{\text{gal}}{\text{cu ft}})(60 \frac{\text{sec}}{\text{min}})} = 3.2 \text{ cfs}$$

$$Q_{cfs} = AV_{fps}$$

$$3.2 \text{ cfs} = (2.5 \text{ ft})(1.17 \text{ ft})(x \text{ fps})$$

$$\frac{3.2}{(2.5)(1.17)} = x$$

$$\boxed{1.1 \text{ fps}} = x$$

Example 2: (Grit Channel Velocity)

❑ The total flow through both channels of a grit channel is 5.9 cfs. If each channel is 2 ft wide and water is flowing to a depth of 15 inches, what is the velocity of flow through the channels? (fps)

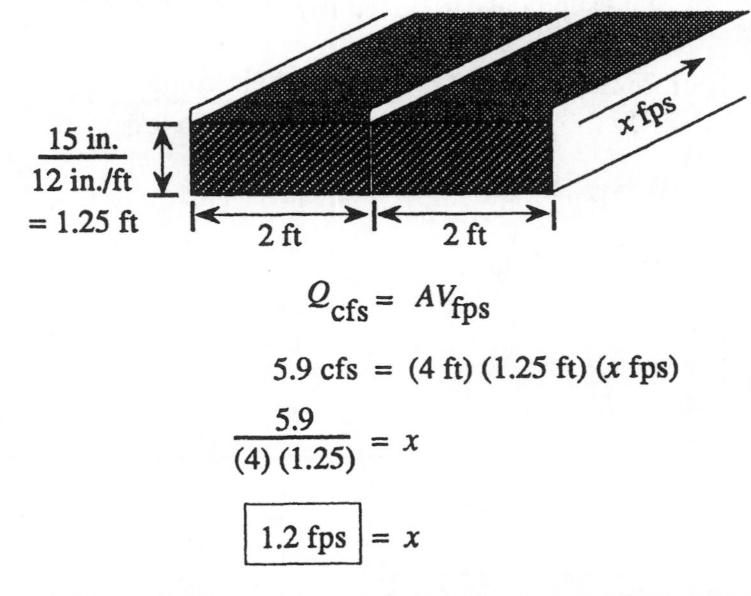

$$Q_{cfs} = AV_{fps}$$

$$5.9 \text{ cfs} = (4 \text{ ft})(1.25 \text{ ft})(x \text{ fps})$$

$$\frac{5.9}{(4)(1.25)} = x$$

$$\boxed{1.2 \text{ fps}} = x$$

Example 3: (Grit Channel Velocity)

❑ A stick travels 35 ft in 30 seconds in a grit channel. What is the estimated velocity in the channel (ft/sec)?

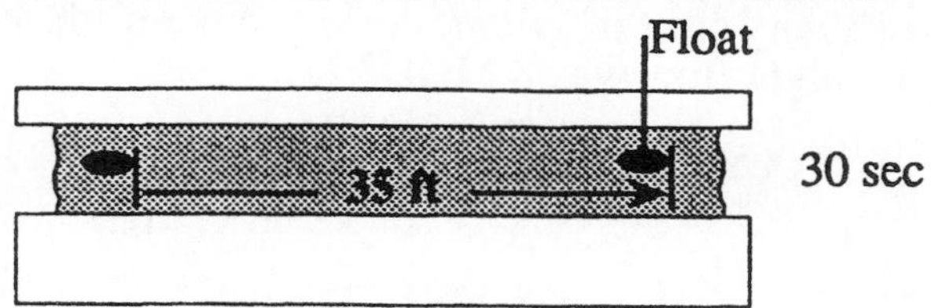

$$\text{Velocity, ft/sec} = \frac{\text{Distance, ft}}{\text{Time, sec}}$$

$$= \frac{35 \text{ ft}}{30 \text{ sec}}$$

$$= \boxed{1.2 \text{ ft/sec}}$$

Example 4: (Grit Channel Velocity)

❑ A stick is placed in a grit channel and flows 19 ft in 15 seconds. What is the estimated velocity in the channel (ft/sec)?

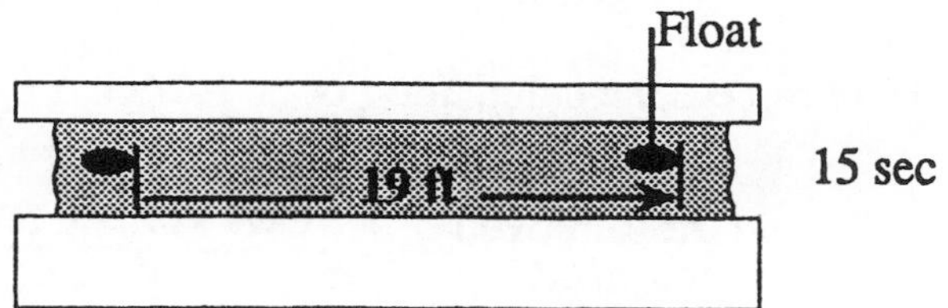

$$\text{Velocity, ft/sec} = \frac{\text{Distance, ft}}{\text{Time, sec}}$$

$$= \frac{19 \text{ ft}}{15 \text{ sec}}$$

$$= \boxed{1.3 \text{ ft/sec}}$$

VELOCITY USING THE FLOAT OR DYE METHOD

Velocities can also be estimated by the use of a float placed in the water. By timing the distance traveled by the float, the velocity can be determined:

$$\boxed{\text{Velocity, ft/sec} = \frac{\text{Distance, ft}}{\text{Time, sec}}}$$

Remember that these velocity calculations are estimates only. Since a float or stick tends to move along with the faster surface waters, the calculated velocity can be as much as 10 or 15 percent faster than the actual average flow rate.

Other flow patterns and currents through the channel can also affect the flow velocity of the float or stick, even slowing it somewhat.

8.6 GRIT REMOVAL

Sanitary wastewater systems normally average 1 to 4 cubic feet of grit per million gallons of flow. Combined wastewater systems—systems which accept both sanitary wastes and storm flow—average from 4 to 15 cubic feet per million gallons of flow, with higher ranges during periods of heavy rains.

Because grit is normally disposed of by burial, it is important for planning purposes that accurate records be kept of grit removal. Most often the data is reported as cubic feet of grit removed per million gallons of flow:

$$\frac{\text{Grit Removed}}{\text{cu ft/MG}} = \frac{\text{Grit Vol., cu ft}}{\text{Flow, MG}}$$

Example 1: (Grit Removal)

❑ A treatment plant removes 12 cu ft of grit in one day. How many cu ft of grit are removed per million gallons if the plant flow was 8 MGD?

$$\text{Grit Removed, cu ft/MG} = \frac{\text{Grit Vol., cu ft}}{\text{Flow, MG}}$$

$$= \frac{12 \text{ cu ft}}{8 \text{ MG}}$$

$$= \boxed{1.5 \text{ cu ft/MG}}$$

Example 2: (Grit Removal)

❑ The total daily grit removal for a plant is 270 gallons. If the plant flow is 12.3 MGD, how many cubic feet of grit are removed per MG flow?

First convert gallons grit removed to cu ft:*

$$\frac{270 \text{ gallons}}{7.48 \text{ gal/cu ft}} = 36 \text{ cu ft}$$

Now complete the calculation of cu ft/MG:

$$\text{Grit Removed, cu ft/MG} = \frac{\text{Grit Vol., cu ft}}{\text{MG flow}}$$

$$= \frac{36 \text{ cu ft}}{12.3 \text{ MG}}$$

$$= \boxed{2.9 \text{ cu ft/MG}}$$

* Refer to Chapter 8 in *Basic Math Concepts* for a review of gal to cu ft conversions.

Example 3: (Grit Removal)

❑ The average grit removal at a particular treatment plant is 2.1 cu ft/MG. If the monthly average daily flow is 4.5 MGD, how many cu yds of grit would be removed from the wastewater flow during one month? (Assume the month has 30 days.)

First calculate the cu ft grit removed from the average daily flow:

$$\frac{(2.1 \text{ cu ft})}{\text{MG}}(4.5 \text{ MGD}) = 9.45 \text{ cu ft each day}$$

Then calculate the anticipated grit removed for the month:

$$\frac{(9.45 \text{ cu ft})}{\text{day}}(30 \text{ days}) = 283.5 \text{ cu ft}$$

And convert cu ft grit removed to cu yds grit:

$$\frac{283.5 \text{ cu ft}}{27 \text{ cu ft/cu yds}} = \boxed{10.5 \text{ cu yds}}$$

Example 4: (Grit Removal)

❑ The monthly average grit removal is 3 cu ft/MG. If the monthly average flow is 2,800,000 gpd, how many cu yds must be available for grit disposal if the disposal pit is to have a 90-day capacity?

First calculate the grit generated each day:

$$\frac{(3 \text{ cu ft})}{\text{MG}}(2.8 \text{ MGD}) = 8.4 \text{ cu ft each day}$$

The cu ft grit generated for 90 days would be:

$$\frac{(8.4 \text{ cu ft})}{\text{day}}(90 \text{ days}) = 756 \text{ cu ft}$$

Converting cu ft grit to cu yds grit:

$$\frac{756 \text{ cu ft}}{27 \text{ cu ft/cu yds}} = \boxed{28 \text{ cu yds}}$$

FORECASTING DISPOSAL NEEDS

Over a given time, the average grit removal rate at a plant (at least a seasonal average) can be determined and used for planning purposes. Often grit removal is calculated as cubic yards, since excavation is normally expressed in terms of cubic yards.*

$$\boxed{\frac{\text{Total Grit, cu ft}}{27 \text{ cu ft/cu yds}} = \text{cu yds grit}}$$

* Cu ft to cu yds conversions are described in Chapter 8 of *Basic Math Concepts*.

8.7 FLOW MEASUREMENT

FLOW MEASUREMENT USING $Q=AV$

The flow rate through grit channels is normally measured by some type of flow metering device as described in the following pages. However, the flow rate for any particular moment can also be determined by using the $Q=AV$ equation.

The flow rate (Q) is equal to the cross-sectional area (A) of the channel or pipeline multiplied by the velocity (V) through the channel.

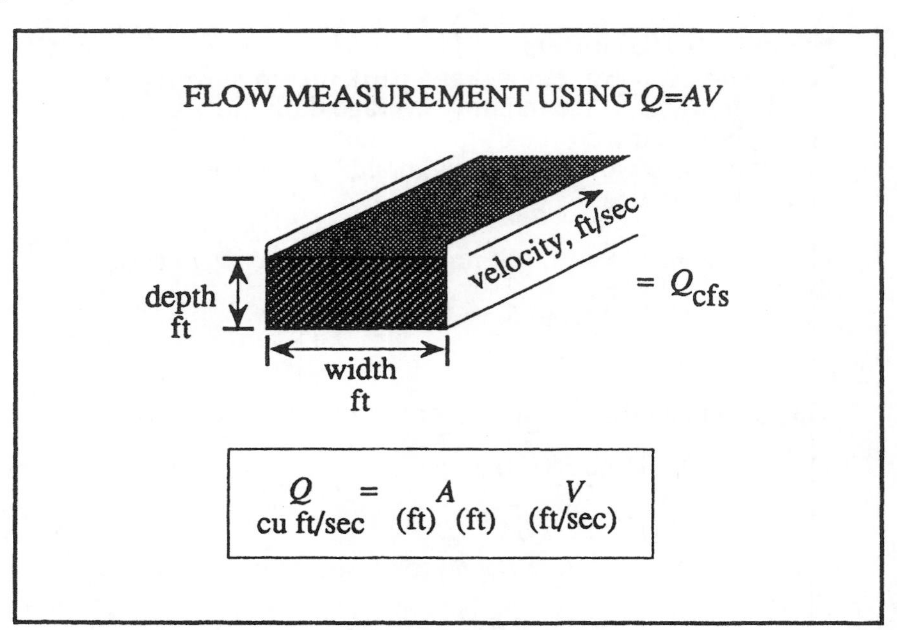

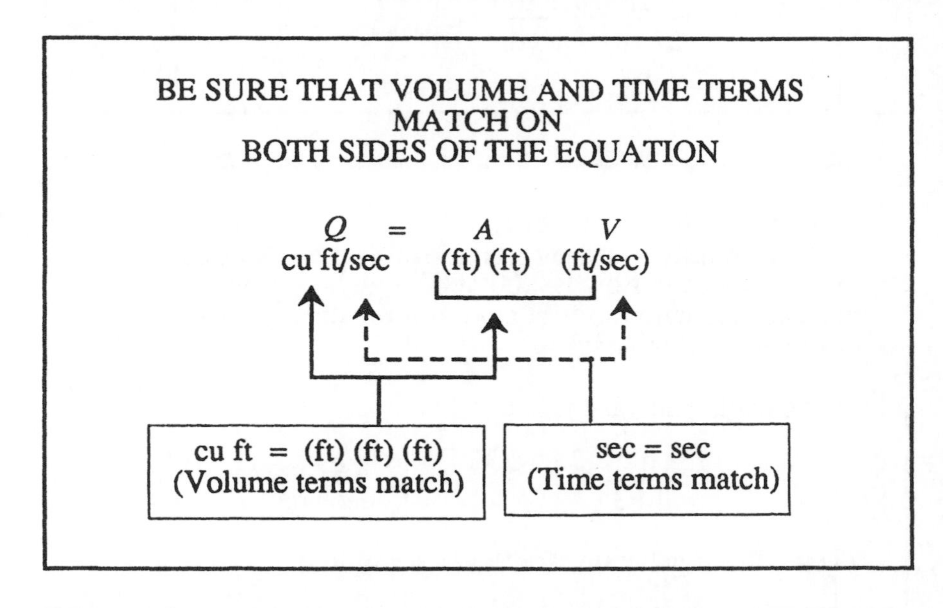

Example 1: (Flow Measurement)

❑ A grit channel 2 ft wide has water flowing to a depth of 1.5 ft. If the velocity through the channel is 1.2 fps, what is the cfs flow rate through the channel?

$$Q_{cfs} = AV_{fps}$$

$$= (2\text{ ft})\ (1.5\text{ ft})\ (1.2\text{ fps})$$

$$= \boxed{3.6\text{ cfs}}$$

Example 2: (Flow Measurement)

❑ A grit channel 30 inches wide has water flowing to a depth of 1 ft. If the velocity of the water is 1.3 fps, what is the cfs flow in the channel?

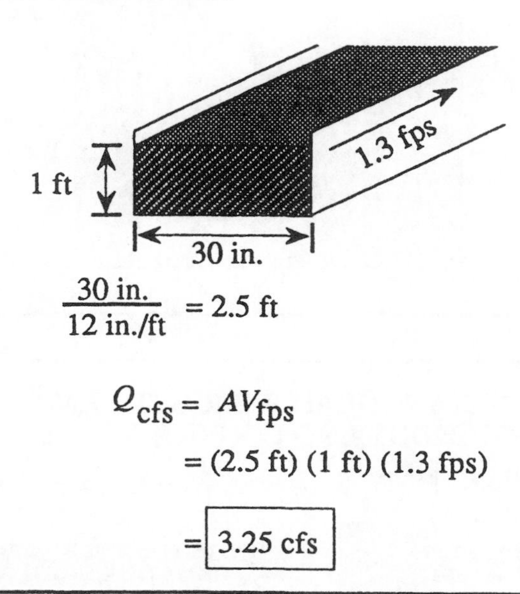

$$\frac{30 \text{ in.}}{12 \text{ in./ft}} = 2.5 \text{ ft}$$

$$Q_{cfs} = AV_{fps}$$

$$= (2.5 \text{ ft})(1 \text{ ft})(1.3 \text{ fps})$$

$$= \boxed{3.25 \text{ cfs}}$$

DIMENSIONS SHOULD BE EXPRESSED AS FEET

The dimensions in a $Q=AV$ calculation should always be expressed in feet because (ft) (ft) (ft) = cu ft. Therefore when dimensions are given as inches, first convert all dimensions to feet before beginning the $Q=AV$ calculation.

Example 3: (Flow Measurement)

❑ A grit channel 2 ft wide has water flowing at a velocity of 0.9 fps. If the depth of water is 1.25 ft, what is the gpd flow rate through the channel? (Assume the flow is steady and continuous.)

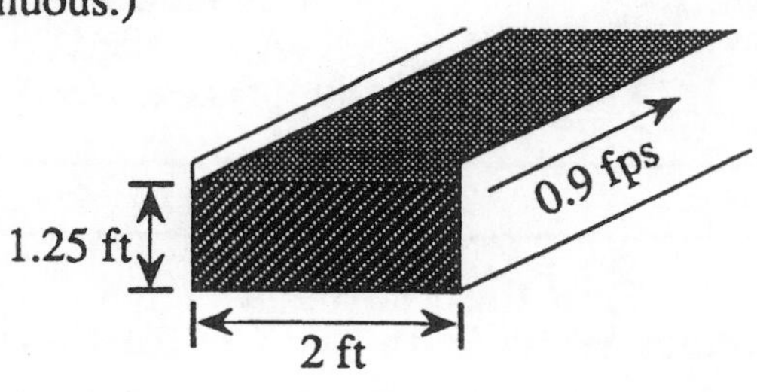

$$Q_{cfs} = AV_{fps}$$

$$= (2 \text{ ft})(1.25 \text{ ft})(0.9 \text{ fps})$$

$$= 2.25 \text{ cfs}$$

Now convert cfs flow rate to gpd:

$$\frac{(2.25 \text{ cu ft})}{\text{sec}} \frac{(60 \text{ sec})}{\text{min}} \frac{(7.48 \text{ gal})}{\text{cu ft}} \frac{(1440 \text{ min})}{\text{day}} = 1{,}454{,}112 \text{ gpd}$$

CALCULATING GPM OR GPD FLOW

If gpm or gpd flow rate is desired, simply calculate the cfs flow rate, as illustrated in Examples 1 and 2, then convert cfs flow to gpm or gpd flow rate.*

* To review flow conversions, refer to Chapter 8 in *Basic Math Concepts*.

FLOW MEASUREMENT USING WEIRS

Several devices may be used to determine flow rate including weirs, flumes, venturi meters, and magnetic flow meters. Tables, graphs, and nomographs can be used to determine flow rates through these devices. In day-to-day operation, instrumentation is normally used to read and record flows continuously.

The most common type of weirs are V-notch and rectangular weirs, so named because of the shape of the opening through which the water flows. Cipolletti weirs are weirs with a trapezoidal opening.

To read discharge charts and tables for V-notch weirs, you will need to know two things:

- the angle of the weir (common angles are 22-1/2°, 45°, 60° and 90°), and
- the feet of head (the height of the water in the V).

To read discharge charts and tables for rectangular weirs, you will need to know three things:

- whether the rectangular weir has end contractions or not,
- the length of the weir crest, and
- the feet of head (the height of water in the weir).

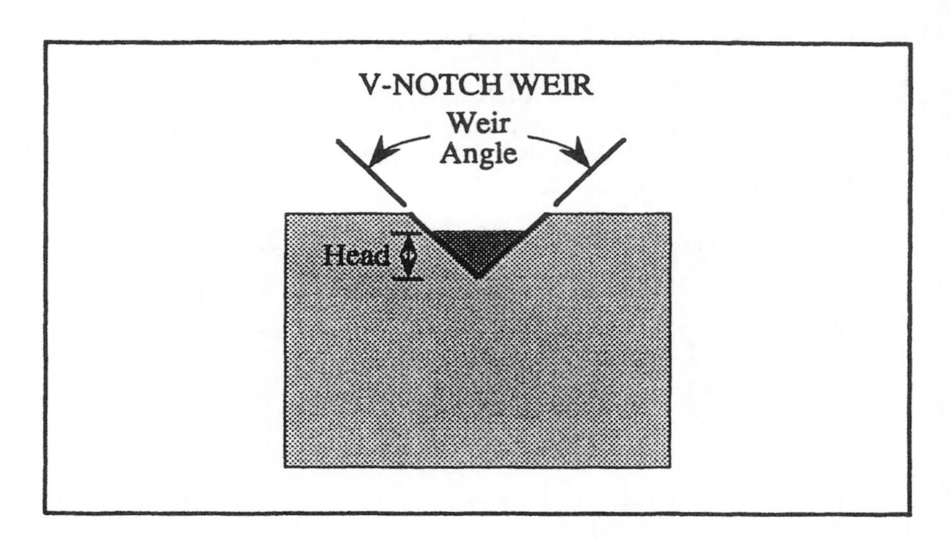

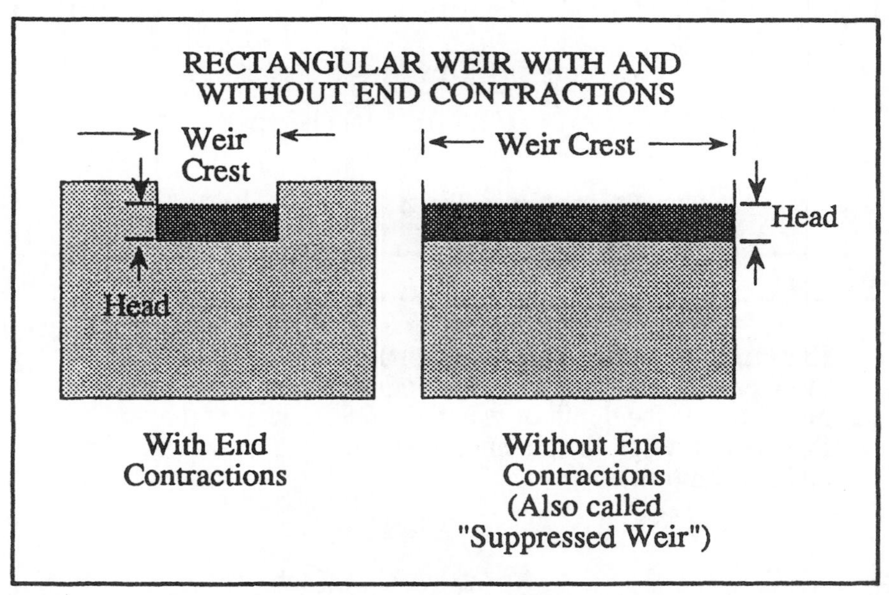

Flow Rate Through Rectangular Weirs With End Contractions

Head, ft	Length of Weir Crest in Feet					
	1		1-1/2		2	
	cfs	MGD	cfs	MGD	cfs	MGD
0.11	0.119	0.077	0.178	0.116	0.241	0.155
0.12	0.135	0.087	0.204	0.132	0.273	0.177
0.13	0.152	0.098	0.230	0.149	0.308	0.199
0.14	0.169	0.110	0.256	0.166	0.343	0.222

cfs $= 3.33(L - 0.2\,H)H^{3/2}$

FLOW THROUGH A 90° V-NOTCH WEIR—$Q=2.5H^{5/2}$

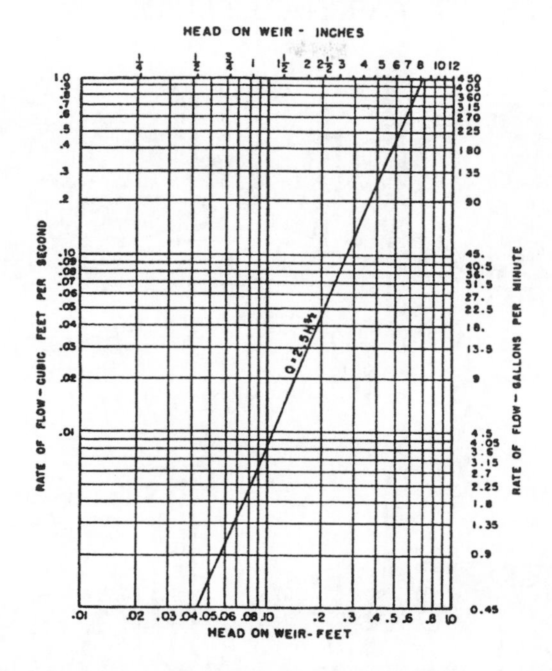

Source: New York Manual of Instruction for Water Treatment Plant Operators

Example 4: (Flow Measurement)
❑ Use the flow table to determine the MGD flow rate through a rectangular weir with end contractions if the feet of head indicated at the staff gage is 0.14 ft and the length of the weir crest is 1.5 ft.

First find 0.14 ft in the "Head" column. Then move to the right until you come under the column 1 1/2. The MGD reading is 0.166 MGD

0.166 MGD

Example 5: (Flow Measurement)
❑ The head on a V-notch weir is 0.5 ft. What is the cfs flow through the weir? (Use the graph above to determine flow rate.)*

The bottom of the chart indicates flow rate in feet. Find 0.5 ft and follow the line upward to the diagonal line. Then move horizontally to the left scale. The line falls halfway between the 0.4 and 0.5. So the reading is about 0.45 cfs.

0.45 cfs

* For a discussion of reading scales, refer to Chapter 12 in *Basic Math Concepts*.

FLOW MEASUREMENT USING PARSHALL FLUMES

One of the most commonly used flow measuring devices is the Parshall flume. A Parshall flume is a specially designed constriction in an open channel. The water flows into the converging section, then drops down through the throat section and out into the diverging section. Parshall flumes are classified according to throat width, *W*.

Parshall flumes are often favored over weirs for flow measurement since the flumes are self-cleaning (there are no places for debris to get lodged) and there is very little head loss in this type of flow measurement device.

Flow measurement for a Parshall flume is based on the depth of water (head) measured at H_a in the converging section. To use the flow charts or graphs, you must know whether there are free flow conditions or submerged flow conditions in the flume.

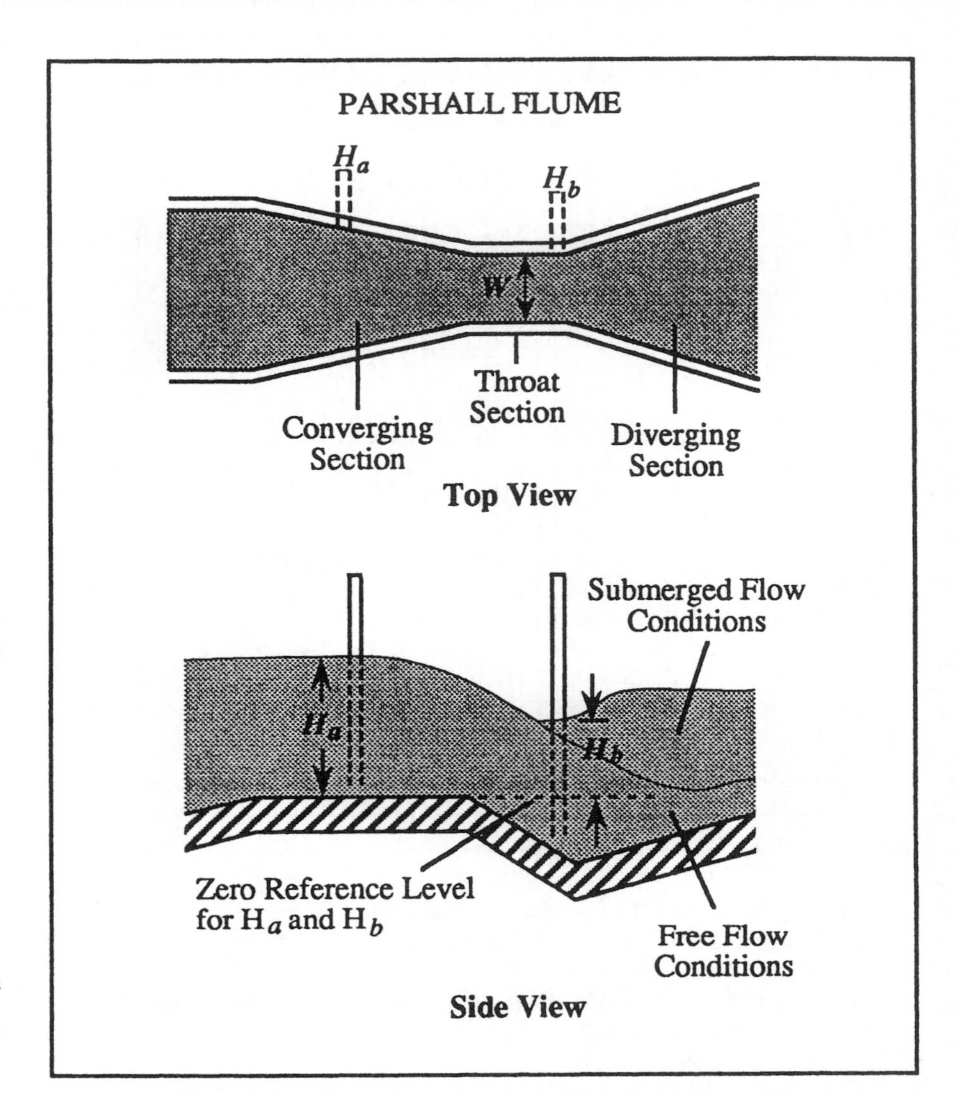

DISCHARGE OF 6-INCH PARSHALL FLUME							
Head ft	cfs	gps	MGD	Head ft	cfs	gps	MGD
0.31	.3238	2.422	.2093	0.41	.5036	3.767	.3255
0.32	.3404	2.546	.2200	0.42	.5231	3.913	.3381
0.33	.3574	2.673	.2309	0.43	.5429	4.061	.3509
0.34	.3746	2.802	.2421	0.44	.5630	4.211	.3638
0.35	.3922	2.934	.2535	0.45	.5834	4.364	.3770
0.36	.4100	3.067	.2650	0.46	.6040	4.518	.3904
0.37	.4282	3.203	.2767	0.47	.6249	4.675	.4039
0.38	.4466	3.341	.2887	0.48	.6460	4.832	.4175
0.39	.4653	3.480	.3007	0.49	.6674	4.992	.4313
0.40	.4843	3.623	.3130	0.50	.6890	5.154	.4453

$cfs = 2.06H^{1.58}$

Example 6: (Flow Measurement)

❑ The head measured at the upstream gage on a 6-inch Parshall flume is 0.38 ft. What is the MGD flow through the channel? (Assume there are no submergence conditions in the channel.)

First, find 0.38 ft in the head column. Then move to the right to the MGD flow rate column:

0.2887 MGD

Example 7: (Flow Measurement)

❑ What is the cfs flow through a 6-inch Parshall flume if the upstream gage indicates a depth of 0.35 ft? (Assume no submergence condition exists.)

Find 0.35 ft in the head column. Then move to the right to the cfs column. The flow rate indicated is:

0.3922 cfs

Example 8: (Flow Measurement)

❑ The head measured at the upstream gage of a 6-inch Parshall flume is 0.45 ft. What is the gpm flow through the flume? (Assume no submergence condition exists.)

First determine the gps flow rate indicated:

4.364 gps

Then convert gps flow to gpm:*

$$(4.364 \frac{\text{gal}}{\text{sec}})(60 \frac{\text{sec}}{\text{min}}) = \boxed{261.8 \text{ gpm}}$$

PARSHALL FLUME—FREE FLOW CONDITIONS

Free flow conditions means there are no downstream conditions causing the water to "back up" or restrict the water flowing out of the flume.

For free flow conditions, only the upstream depth measurement (H_a) is needed to determine flow through the flume. Examples 6-8 illustrate calculations when free flow conditions exist.

The table shown on the previous page is a partial table of heads and corresponding flow rates for a 6-inch Parshall flume. Reference handbooks contain tables for various throat widths and head conditions.

* For a review of flow conversions refer to Chapter 8 in *Basic Math Concepts*.

PARSHALL FLUME —SUBMERGED CONDITIONS

Submerged flow conditions indicate that there are downstream conditions which restrict the free flow of water and result in a false or inaccurate depth reading. Because the water is "backed up", the depth reading at H_a is higher than actual flow conditions would warrant.

To determine whether submerged flow conditions exist, calculate the percent submergence. Remember, percent equals "part over whole". Therefore since the downstream depth indicates the part submergence, percent submergence is calculated as:

$$\% \text{ Submerg.} = \frac{\text{Part Submerg.}}{\text{Total}} \times 100$$

This equation can be restated as:

$$\% \text{ Submergence} = \frac{H_b}{H_a} \times 100$$

Submerged conditions exist when the percent submergence exceeds:

- 50% for flumes 1 in., 2 in. and 3 in. wide,
- 60% for flumes 6 in. and 9 in. wide,
- 70% for flumes 1 to 8 ft wide, and
- 80% for flumes 8 to 50 ft wide.

PARSHALL FLUME NOMOGRAPH

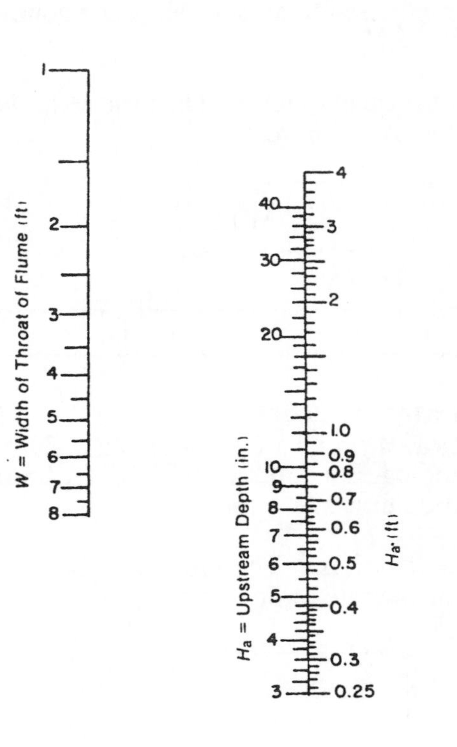

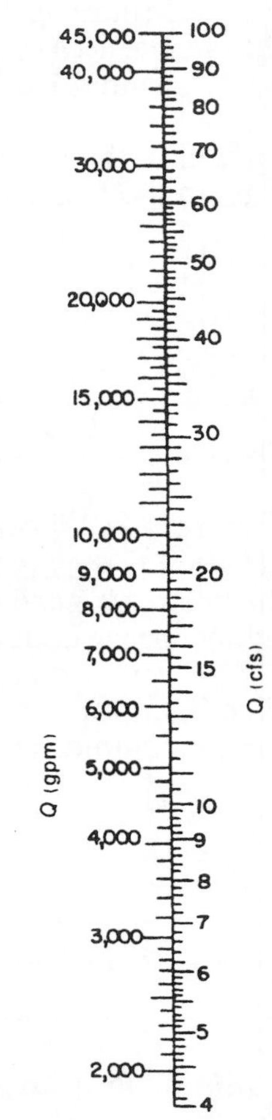

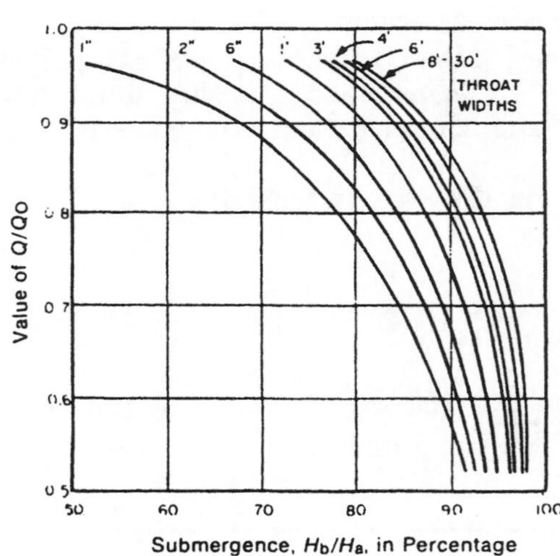

Reprinted with permission of Public Works Magazine, September, October, and November 1968 issues.

Example 9: (Flow Measurement)

❑ What is the cfs flow through a Parshall flume with a throat width of 3 ft if the water depth at the upstream gage is 14 inches? The downstream depth (H_b) is 11 inches.

First determine the flow rate indicated by the nomograph. Place a straight-edge at the mark on the width scale to the left. Then while holding the top of the straight-edge on the 3 mark, rotate the straight-edge until it lines up with the 14 mark on the left side of the middle scale. Now draw a line from the 3 mark, through the 14 and on to the scale on the right side.

The flow rate scale is marked off for gpm on the left side and cfs on the right side. The cfs flow rate indicated is 15 cfs.

Now that the flow rate has been determined, we must determine whether a correction factor is required. Therefore percent submergence must be determined:

$$\% \text{ Submergence} = \frac{H_b}{H_a} \times 100$$

$$= \frac{11 \text{ in}}{14 \text{ in}} \times 100$$

$$= 79\%$$

For a 3 ft Parshall flume, submergence exists when the percent exceeds 70%. Therefore the correction factor graph must be used to determine the appropriate correction factor.

First, find 79% on the bottom scale and move upward to the 3 ft throat width curve. Then move directly left to the scale. The correction factor indicated is 0.95. Multiply the flow rate determined from the nomograph (15 cfs) by the correction factor:

$$(15 \text{ cfs})\ (0.95) = \boxed{14.25 \text{ cfs}}$$

Where submerged conditions exist, a correction factor must be used with values shown in the table or on a nomograph. Using the percent submergence, you use a graph to determine the correction factor needed. Then **multiply the flow rate obtained in the table or on the nomograph by the correction factor.** Example 9 illustrates the use of a nomograph in determining Parshall flume flow rate. The correction factor graph is also used in this example.

NOTES:

9 Sedimentation

SUMMARY

1. Detention Time

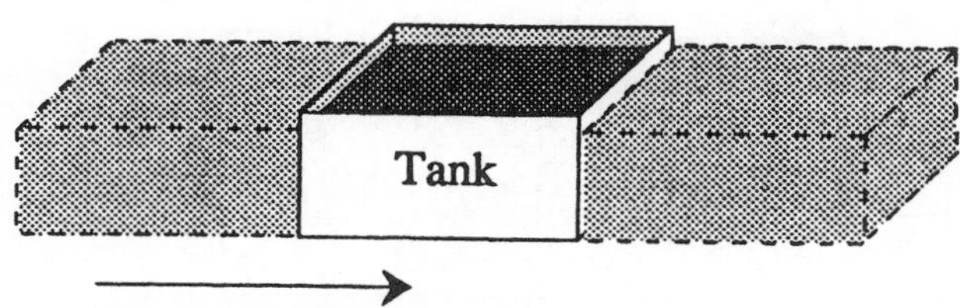

$$\text{Detention Time, hrs} = \frac{\text{Volume of Tank, gal}}{\text{Flow, gal/hr}}$$

2. Weir Overflow Rate

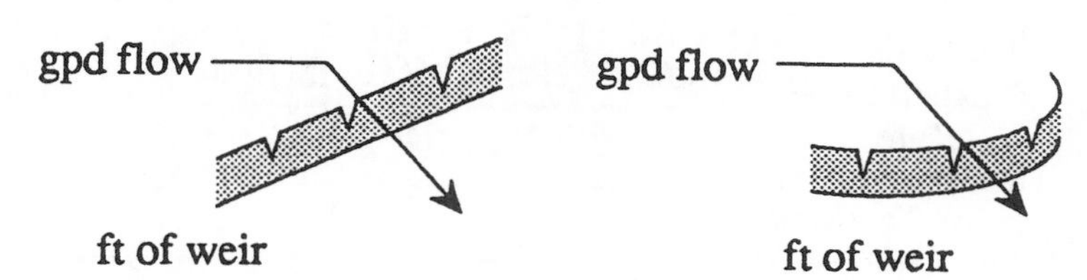

$$\text{Weir Overflow Rate} = \frac{\text{Flow, gpd}}{\text{Weir Length, ft}}$$

3. Surface Overflow Rate

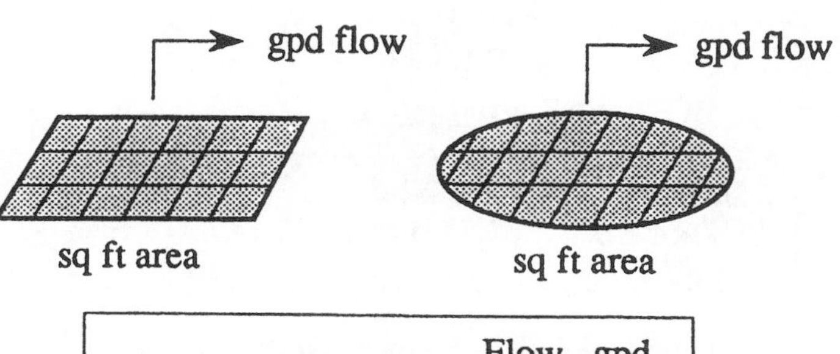

$$\text{Surface Overflow Rate} = \frac{\text{Flow, gpd}}{\text{Area, sq ft}}$$

(Surface overflow rate does not include recirculated flows.)

SUMMARY—Cont'd

4. Solids Loading Rate (Secondary Clarifier)

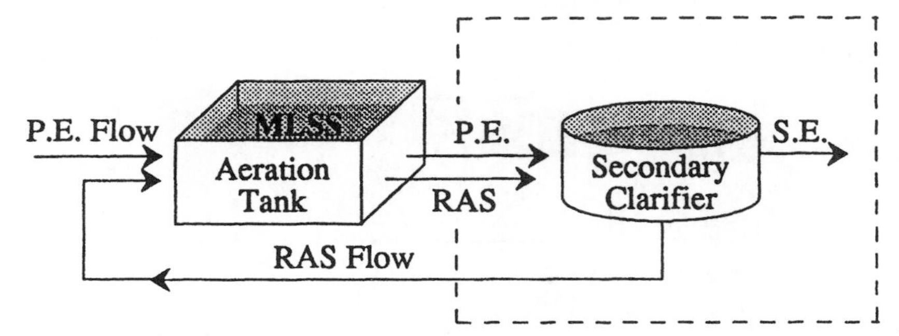

Simplified Equation:

$$\text{Solids Loading Rate} = \frac{\text{Solids Applied, lbs/day}}{\text{Surface Area, sq ft}}$$

Expanded Equation:

$$\text{Solids Loading Rate} = \frac{(\underset{\text{mg/}L}{\text{MLSS}})\ (\underset{\text{flow, MGD}}{\text{P.E. + RAS}})\ (\underset{\text{lbs/gal}}{8.34})}{(0.785)\ (D^2)}$$

5. BOD and Suspended Solids Removed, lbs/day

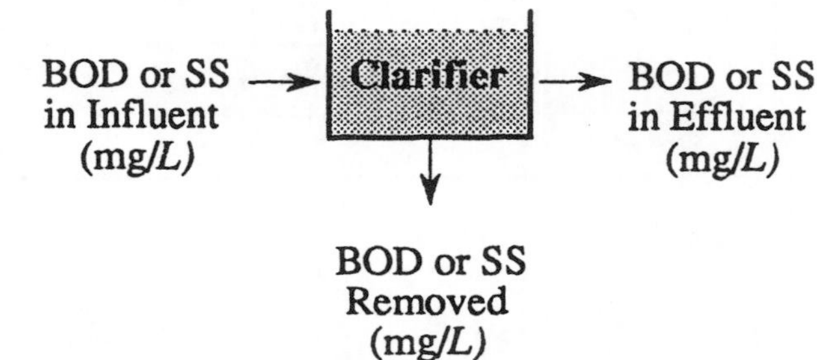

First calculate BOD or SS removed, mg/L:

$$\underset{\text{mg/}L}{\text{Influent BOD or SS}} - \underset{\text{mg/}L}{\text{Effluent BOD or SS}} = \underset{\text{mg/}L}{\text{Removed BOD or SS}}$$

Then calculate lbs/day BOD or SS removed:

$$(\underset{\text{BOD or SS Removed}}{\text{mg/}L})\ (\underset{\text{flow}}{\text{MGD}})\ (\underset{\text{lbs/gal}}{8.34}) = \underset{\text{BOD or SS Removed}}{\text{lbs/day}}$$

SUMMARY—Cont'd

6. Unit Process Efficiency

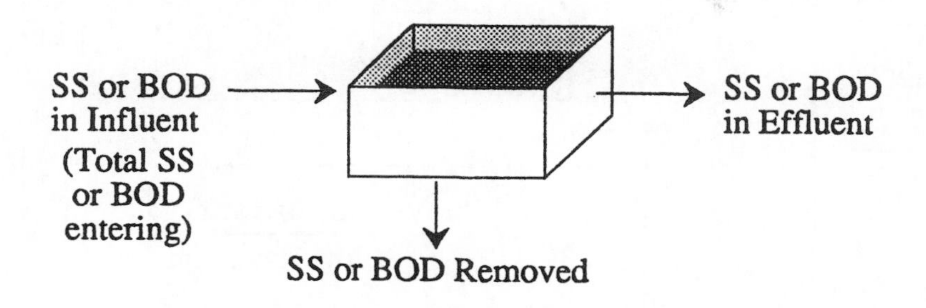

$$\%\ \text{SS Removed} = \frac{\text{SS Removed, mg}/L}{\text{SS Total, mg}/L} \times 100$$

$$\%\ \text{BOD Removed} = \frac{\text{BOD Removed, mg}/L}{\text{BOD Total, mg}/L} \times 100$$

NOTES:

9.1 DETENTION TIME

The detention time for clarifiers may vary from one to three hours. The equation used to calculate detention time is:

$$\text{Detention Time, hrs} = \frac{\text{Volume of Tank, gal}}{\text{Flow Rate, gph}}$$

MATCHING UNITS

There are many possible ways of calculating detention time, depending on the time unit desired (seconds, minutes, hours, days) and the expression of volume and flow rate.

When calculating detention time, therefore, it is essential that the time and volume units used in the equation are consistent on both sides of the equation, as illustrated to the right.

The flow rate to the clarifier is normally expressed as MGD or gpd. However, since the detention time is desired in hours, it is important to express the flow rate in the same time frame—gal/hr (or gph):*

$$\frac{\text{Flow, gpd}}{\text{24 hrs/day}} = \text{Flow, gph}$$

DETENTION TIME IS FLOW-THROUGH TIME

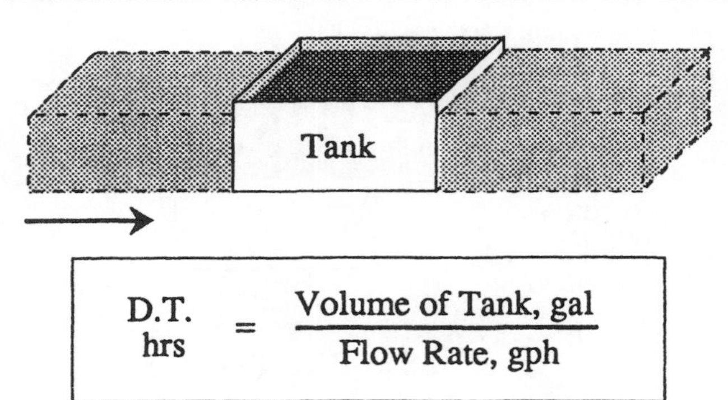

BE SURE YOUR TIME AND VOLUME UNITS MATCH

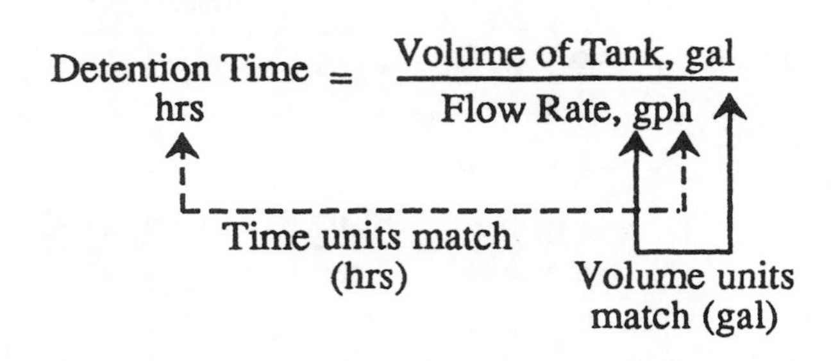

Example 1: (Detention Time)

❑ The flow to a sedimentation tank 70 ft long, 25 ft wide, and 10 ft deep is 2.78 MGD. What is the detention time in the tank, in hours?

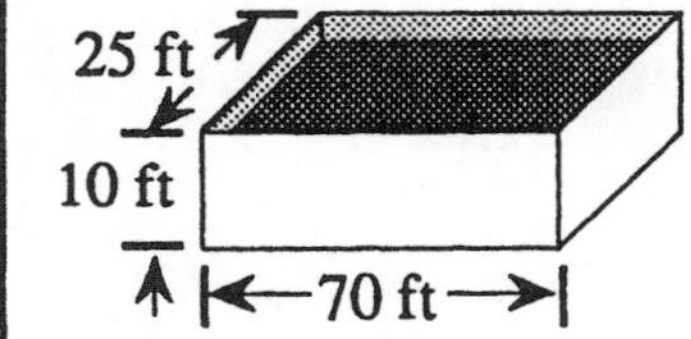

Tank Volume:

(70 ft) (25 ft) (10 ft) (7.48 gal/cu ft)
= 130,900 gal

First, MGD flow rate must be converted to gph so time units will match. (2,780,000 gpd ÷ 24 hrs/day = 115,833 gph.) Now fill in the equation and solve for the unknown.

$$\text{Detention Time, hrs} = \frac{\text{Volume of Tank, gal}}{\text{Flow Rate, gph}}$$

$$= \frac{130{,}900\ \text{gal}}{115{,}833\ \text{gph}}$$

$$= \boxed{1.1\ \text{hours}}$$

* For a review of flow rate conversions, refer to Chapter 8 in *Basic Math Concepts*.

Example 2: (Detention Time)

❑ A circular clarifier receives a flow of 4,752,000 gpd. If the clarifier is 65 ft in diameter and 12 ft deep, what is the clarifier detention time? (Assume the flow is steady and continuous.)

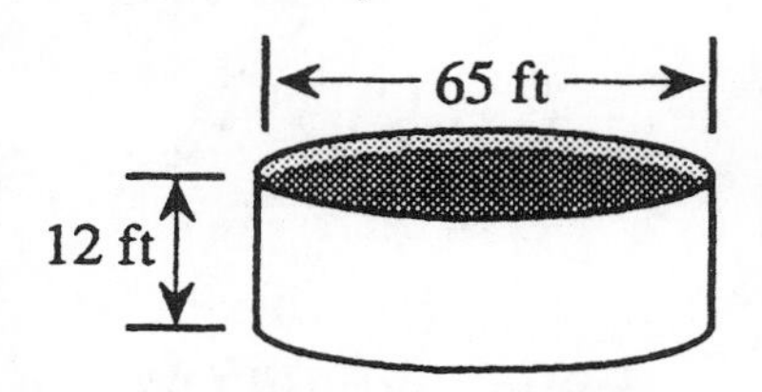

Tank Volume:

(0.785) (65) (65) (12) (7.48)
ft ft ft gal/cu ft

= 297,700 gal

First, convert the flow rate from gpd to gph so that time units will match. (4,752,000 gpd ÷ 24 hrs/day = 198,000 gph). Then calculate detention time:

$$\text{Detention Time, hrs} = \frac{\text{Volume of Tank, gal}}{\text{Flow Rate, gph}}$$

$$= \frac{297{,}700 \text{ gal}}{198{,}000 \text{ gph}}$$

$$= \boxed{1.5 \text{ hours}}$$

Example 3: (Detention Time)

❑ A circular clarifier has a capacity of 120,000 gallons. If the flow to the clarifier is 1,600,000 gpd, what is the clarifier detention time? (Assume the flow is steady and continuous.)

First, convert the flow rate from gpd to gph so that time units will match. (1,600,000 gpd ÷ 24 hrs/day = 66,667 gph). Then calculate detention time:

$$\text{Detention Time, hrs} = \frac{\text{Volume of Tank, gal}}{\text{Flow Rate, gph}}$$

$$= \frac{120{,}000 \text{ gal}}{66{,}667 \text{ gph}}$$

$$= \boxed{1.8 \text{ hours}}$$

9.2 WEIR OVERFLOW RATE

The calculation of weir overflow rate is important in detecting high velocities near the weir which affect the efficiency of the sedimentation process. With excessively high velocities near the weir, the settling solids are pulled over the weirs and into the effluent troughs, thus preventing desired settling.

Weir overflow rate is a measure of the **gallons** per day flowing **over each foot of weir**. (The weir overflow rate may be less for secondary clarifiers—5,000-15,000 gpd/ft).

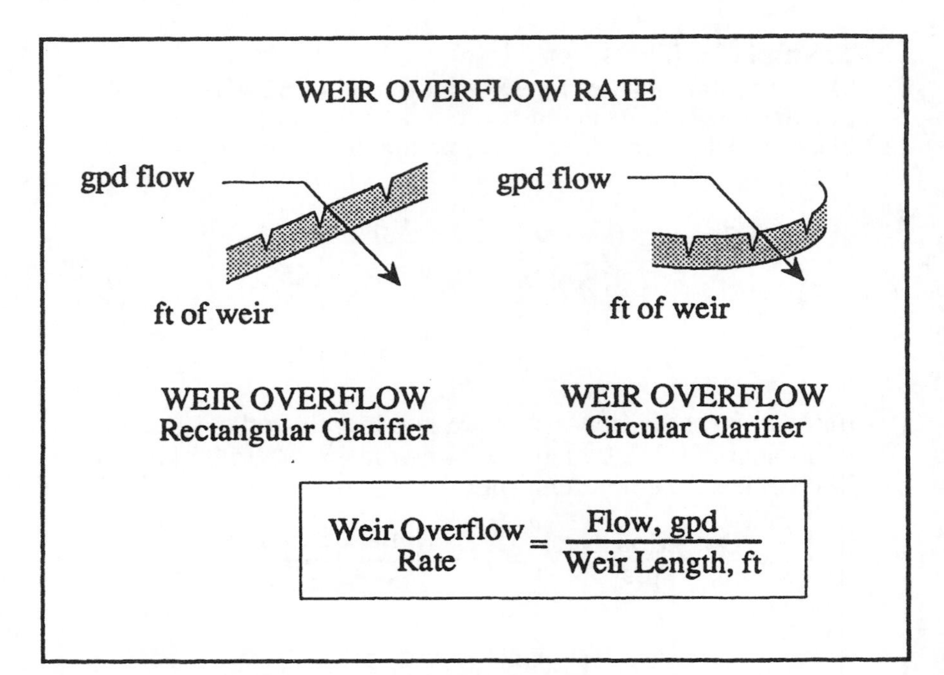

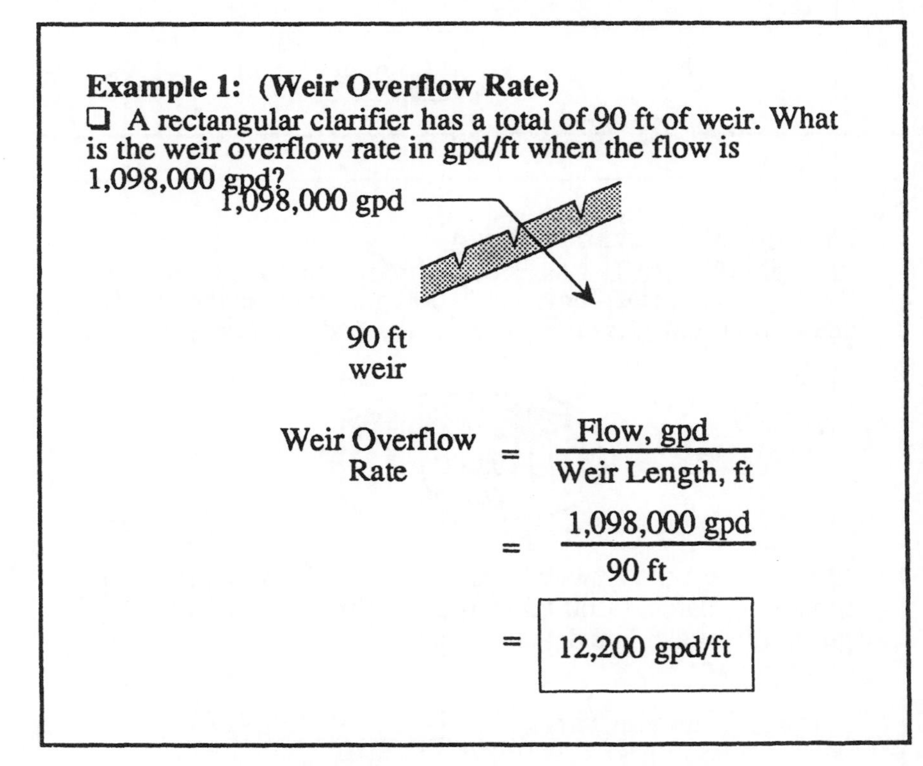

Example 1: (Weir Overflow Rate)

❑ A rectangular clarifier has a total of 90 ft of weir. What is the weir overflow rate in gpd/ft when the flow is 1,098,000 gpd?

$$\text{Weir Overflow Rate} = \frac{\text{Flow, gpd}}{\text{Weir Length, ft}}$$

$$= \frac{1{,}098{,}000 \text{ gpd}}{90 \text{ ft}}$$

$$= \boxed{12{,}200 \text{ gpd/ft}}$$

Example 2: (Weir Overflow Rate)

❑ A circular clarifier receives a flow of 3.6 MGD. If the diameter of the weir is 80 ft, what is the weir overflow rate in gpd/ft?

The total ft of weir is not given directly in this problem. However, weir diameter is given (80 ft) and from that information, the total ft of weir can be determined.

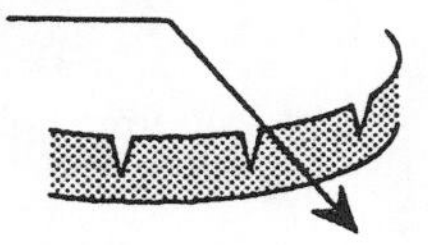

ft of weir:
= (3.14) (80 ft)
= 251 ft

$$\text{Weir Overflow Rate} = \frac{\text{Flow, gpd}}{\text{Weir Length, ft}}$$

$$= \frac{3{,}600{,}000 \text{ gpd}}{251 \text{ ft}}$$

$$= \boxed{14{,}343 \text{ gpd/ft}}$$

CALCULATING WEIR CIRCUMFERENCE

In some calculations of weir overflow rate, you will have to calculate the total weir length, given the weir diameter. To calculate the length of weir around a circular clarifier, you need to know the relationship between the diameter and circumference of a circle. **The distance around any circle (circumference) is about three times the distance across that circle (diameter).** Or more precisely, the circumference is 3.14 times the diameter.* Therefore, when you know the weir diameter, you can calculate the total feet of weir:

$$\text{Total Ft of Weir} = (3.14)\ (\text{Weir Diam. in ft})$$

Example 3: (Weir Overflow Rate)

❑ A clarifier receives a flow of 1.87 MGD. If the diameter of the weir is 60 ft, what is the weir over flow rate in gpd/ft?

First calculate the gpd flow, then express the answer in MGD.

$$\text{Weir Overflow Rate} = \frac{\text{Flow, gpd}}{\text{Weir Length, ft}}$$

$$= \frac{1{,}870{,}000 \text{ gpd}}{(3.14)\ (60 \text{ ft})}$$

$$= \boxed{9{,}926 \text{ gpd/ft}}$$

---

* For a review of circumference calculations, refer to Chapter 9, "Linear Measurement", in *Basic Math Concepts*.

9.3 SURFACE OVERFLOW RATE

Surface overflow rate is used to determine loading on clarifiers. It is similar to hydraulic loading rate—flow per unit area. However, hydraulic loading rate measures the total water entering the process (plant flow plus recirculation) whereas **surface overflow rate measures only the water overflowing the process (plant flow only).**

As indicated in the diagram to the right, **surface overflow rate calculations do not include recirculated flows.** This is because recirculated flows are taken from the bottom of the clarifier and hence do not flow up and out of the clarifier (overflow).

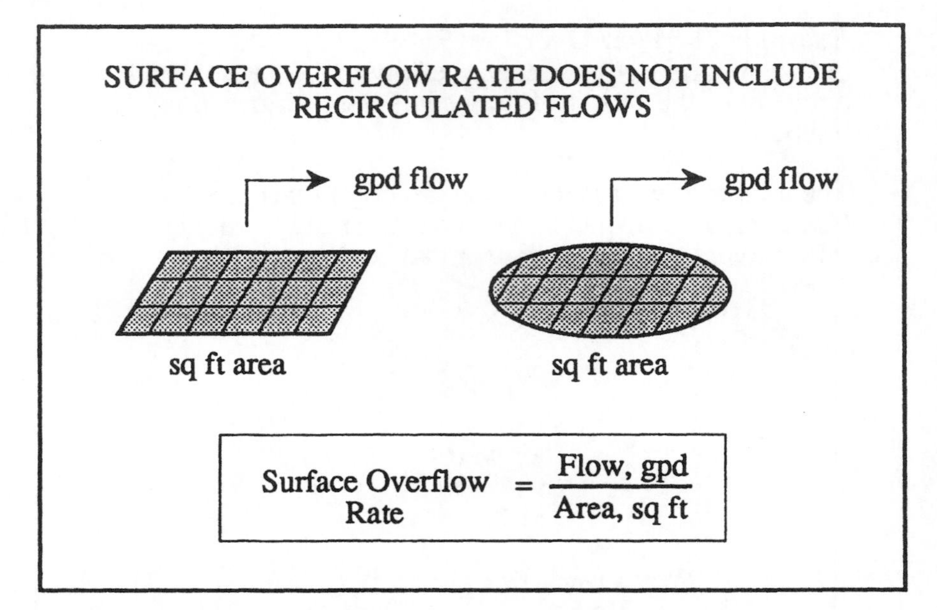

Since surface overflow rate is a measure of flow (Q) divided by area (A), surface overflow rate is an indirect measure of the **upward velocity** of water as it overflows the clarifier:*

$$V = \frac{Q}{A}$$

This calculation is important in maintaining proper clarifier operation since settling solids will be drawn upward and out of the clarifier if surface loading rates are too high.**

Other terms used synonymously with surface loading rate are:

- Surface Loading Rate, and
- Surface Settling Rate

Example 1: (Surface Overflow Rate)

❑ A circular clarifier has a diameter of 55 ft. If the primary effluent flow is 2,075,000 gpd, what is the surface overflow rate in gpd/sq ft?

2,075,000 gpd

Area = (0.785) (55 ft) (55 ft)

$$\text{Surface Overflow Rate} = \frac{\text{Flow, gpd}}{\text{Area, sq ft}}$$

$$= \frac{2{,}075{,}000 \text{ gpd}}{(0.785)(55 \text{ ft})(55 \text{ ft})}$$

$$= \boxed{874 \text{ gpd/sq ft}}$$

* Refer to Chapter 8 for a review of $Q = AV$ problems.

** Surface overflow rates should be calculated for both average and peak flow conditions.

Example 2: (Surface Overflow Rate)

❑ A sedimentation basin 70 ft by 15 ft receives a flow of 1.2 MGD. What is the surface overflow rate in gpd/sq ft?

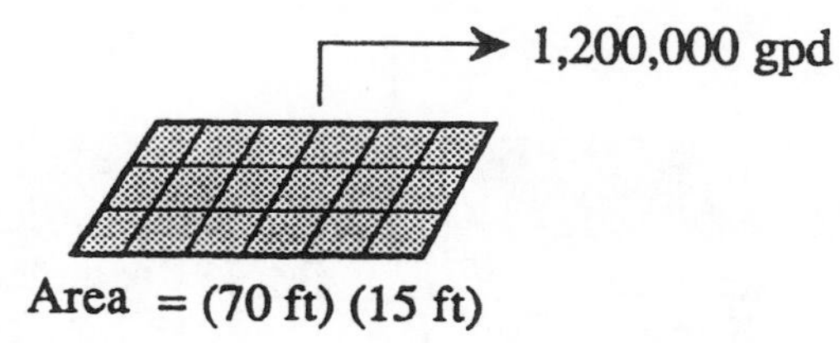

$$\text{Surface Overflow Rate} = \frac{\text{Flow, gpd}}{\text{Area, sq ft}}$$

$$= \frac{1{,}200{,}000 \text{ gpd}}{(70 \text{ ft})(15 \text{ ft})}$$

$$= \boxed{1143 \text{ gpd/sq ft}}$$

Example 3: (Surface Overflow Rate)

❑ A sedimentation basin 90 ft long and 25 ft wide receives a flow of 2,180,400 gpd. What is the surface overflow rate in gpd/sq ft?

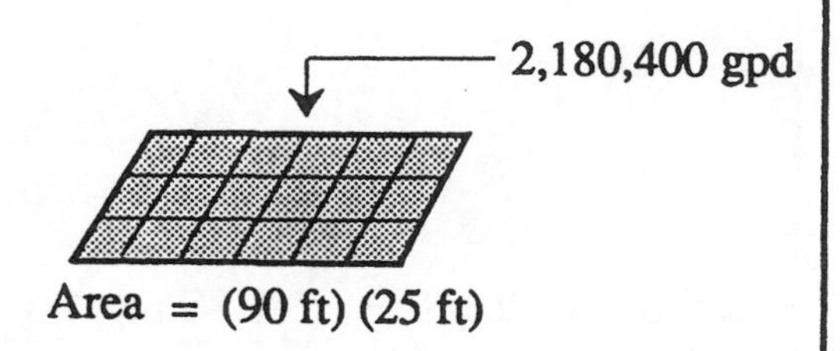

$$\text{Surface Overflow Rate} = \frac{\text{Flow, gpd}}{\text{Area, sq ft}}$$

$$689 \text{ gpd/sq ft} = \frac{2{,}180{,}400 \text{ gpd}}{(90 \text{ ft})(25 \text{ ft})}$$

$$= \boxed{969 \text{ gpd/sq ft}}$$

9.4 SOLIDS LOADING RATE (SECONDARY CLARIFIER)

The solids loading rate calculation is used to determine solids loading on activated sludge secondary clarifiers and gravity sludge thickeners.* It indicates the lbs/day solids loaded to each square foot of clarifier surface area. The general solids loading rate equation is:

$$\text{Solids Loading Rate} = \frac{\text{Solids Applied, lbs/day}}{\text{Surface Area, sq ft}}$$

In expanded form, the equation is:

$$\text{S.L.R.} = \frac{(\underset{\text{mg}/L}{\text{MLSS}})\ (\underset{\text{MGD flow}}{\text{P.E.} + \text{RAS}})\ (\underset{\text{lbs/gal}}{8.34})}{(0.785)\ (D^2)}$$

The vast majority of solids coming into the secondary clarifier come in as mixed liquor suspended solids (MLSS) from the aeration tank. (Remember, much of the suspended solids have already been removed by the primary system.) Therefore, the expanded equation substitutes **lbs/day MLSS applied for lbs/day solids applied** in the numerator.** The area equation used in the denominator depends on the shape of the tank.***

MOST OF THE SOLIDS LOADED TO THE SECONDARY CLARIFIER ENTER AS MLSS

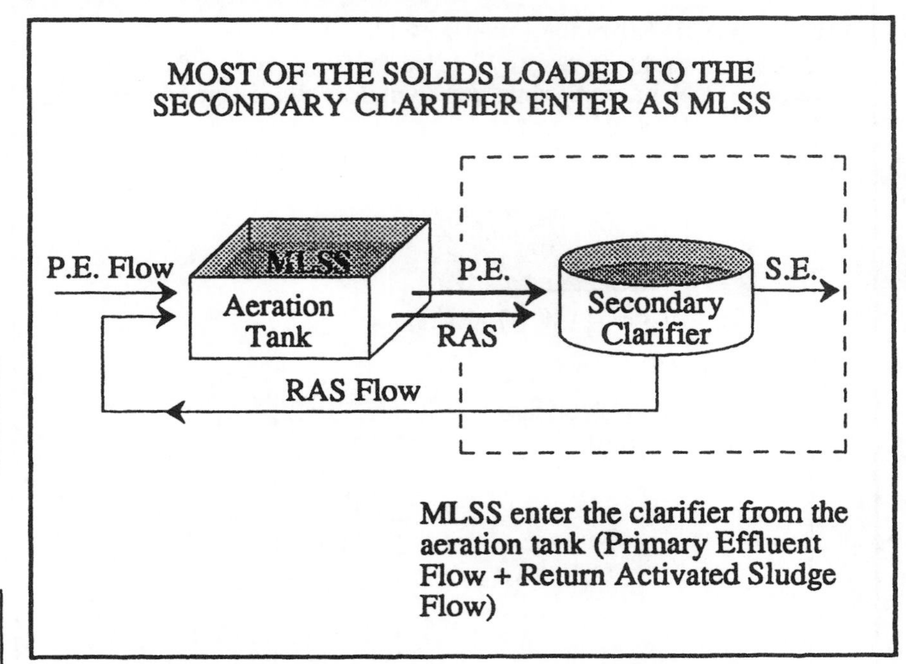

MLSS enter the clarifier from the aeration tank (Primary Effluent Flow + Return Activated Sludge Flow)

Example 1: (Solids Loading Rate)

❑ A secondary clarifier is 70 ft in diameter and receives a combined primary effluent (P.E.) and return activated sludge (RAS) flow of 3.6 MGD. If the MLSS concentration in the aerator is 3000 mg/L, what is the solids loading rate on the secondary clarifier in lbs/day/sq ft?

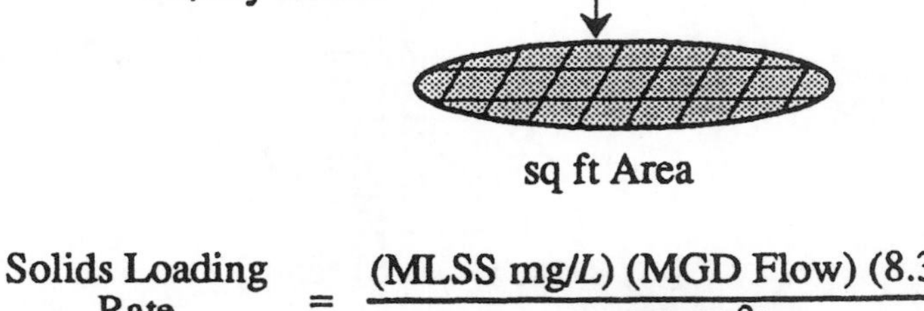

$$\text{Solids Loading Rate} = \frac{(\text{MLSS mg}/L)\ (\text{MGD Flow})\ (8.34\ \text{lbs/gal})}{(0.785)\ (D^2)}$$

$$= \frac{(3000\ \text{mg}/L)\ (3.6\ \text{MGD})\ (8.34\ \text{lbs/gal})}{(0.785)\ (70\ \text{ft})\ (70\ \text{ft})}$$

$$= \boxed{\frac{23.4\ \text{lbs solids/day}}{\text{sq ft}}}$$

* Surface overflow rates should be calculated for both average and peak flow conditions.

** For a review of mg/L to lbs/day calculations refer to Chapter 3.

*** Normally secondary clarifiers are circular. For a rectangular clarifier, the surface area would be $(l)\ (w)$. Area calculations are discussed in Chapter 10 in *Basic Math Concepts*.

Example 2: (Solids Loading Rate)

❑ A secondary clarifier 85 ft in diameter receives a primary effluent flow of 2.8 MGD and a return sludge flow of 0.75 MGD. If the MLSS concentration is 3510 mg/*L*, what is the solids loading rate on the clarifier?

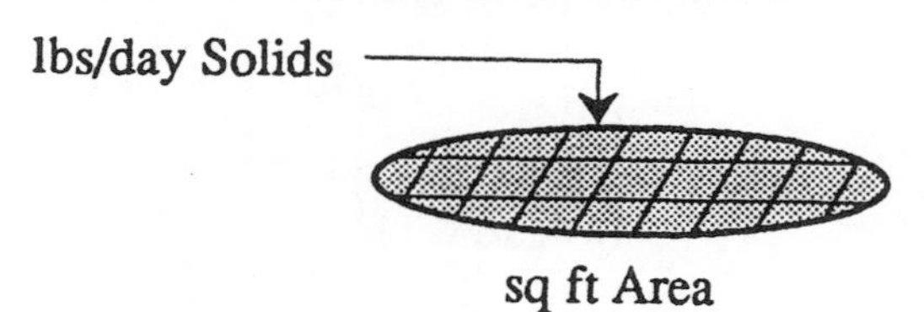

$$\text{Solids Loading Rate} = \frac{\text{Solids, lbs/day}}{\text{Area, sq ft}}$$

$$= \frac{(3510 \text{ mg}/L)\ (3.55 \text{ Tot. MGD})\ (8.34 \text{ lbs/gal})}{(0.785)\ (85 \text{ ft})\ (85 \text{ ft})}$$

$$= \boxed{\frac{18.3 \text{ lbs solids/day}}{\text{sq ft}}}$$

Example 3: (Solids Loading Rate)

❑ A secondary clarifier 75 ft in diameter receives a total flow of 4,270,000 gpd (P.E. + RAS flows). If the MLSS concentration is 3100 mg/*L*, what is the solids loading rate on the clarifier?

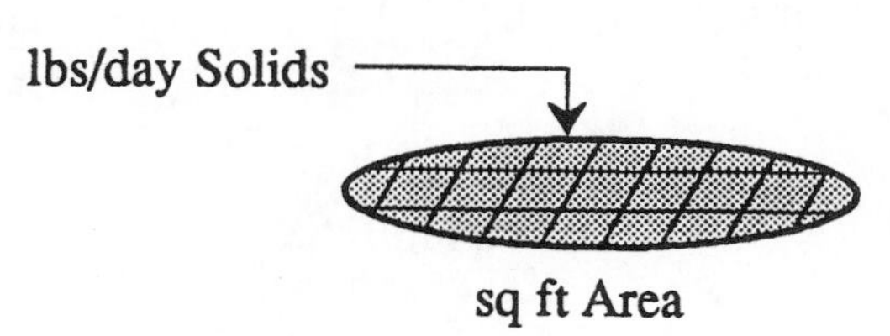

$$\text{Solids Loading Rate} = \frac{\text{Solids, lbs/day}}{\text{Area, sq ft}}$$

$$= \frac{(3100 \text{ mg}/L)\ (4.27 \text{ MGD})\ (8.34 \text{ lbs/gal})}{(0.785)\ (75 \text{ ft})\ (75 \text{ ft})}$$

$$= \boxed{\frac{25 \text{ lbs solids/day}}{\text{sq ft}}}$$

9.5 BOD AND SS REMOVED, lbs/day

To calculate the pounds of BOD or suspended solids removed each day, you will need to know the mg/*L* BOD or SS removed and the plant flow. Then you can use the mg/*L* to lbs/day equation:

(mg/*L*)	(MGD)	(8.34)	= lbs/day
Removed	flow	lbs/gal	

For most problems involving BOD or SS removal, you will not be given information stating how many mg/*L* BOD or SS have been removed. This is something you will calculate based on the mg/*L* concentrations entering (influent) and leaving (effluent) the system.

The influent BOD or SS concentration indicates how much BOD or SS is entering the system. The effluent concentration indicates how much is still in the wastewater (the part not removed). The mg/*L* SS or BOD removed, therefore, would be:

Influent BOD or SS mg/*L*	−	Effluent BOD or SS mg/*L*	=	Removed BOD or SS mg/*L*

Once you have determined the mg/*L* BOD or SS removed, continue with the usual mg/*L* to lbs/day equation to calculate lbs/day BOD or SS removed. Examples 2-4 illustrate this calculation.

Example 1: (BOD and SS Removal)

❑ If 110 mg/*L* suspended solids are removed by a primary clarifier, how many lbs/day suspended solids are removed when the flow is 6,150,000 gpd?

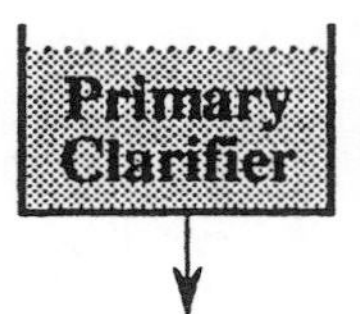

110 mg/*L* SS
Removed

(mg/*L*) (MGD flow) (8.34 lbs/gal) = lbs/day

(110 mg/*L*) (6.15 MGD) (8.34 lbs/gal) = **5642 lbs/day**

Example 2: (BOD and SS Removal)

❑ The flow to a secondary clarifier is 1.8 MGD. If the influent BOD concentration is 210 mg/*L* and the effluent BOD concentration is 74 mg/*L*, how many pounds of BOD are removed daily?

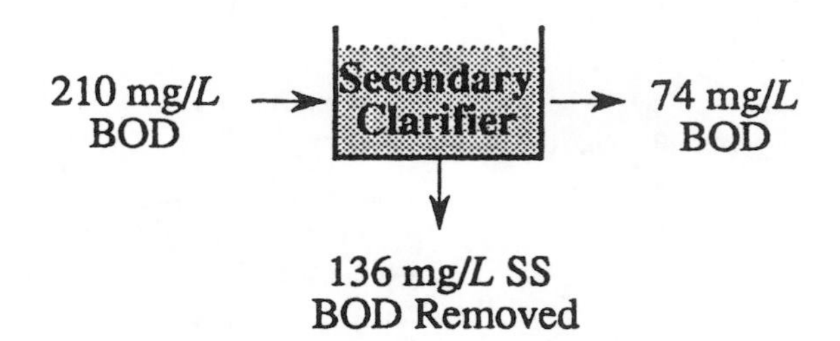

After calculating mg/*L* BOD removed, you can now calculate lbs/day BOD removed:

(mg/*L*) (MGD flow) (8.34 lbs/gal) = lbs/day
Removed   Removed

(136 mg/*L*) (1.8 MGD) (8.34 lbs/gal) = **2042 lbs/day Removed**

Example 3: (BOD and SS Removal)

❑ The flow to a primary clarifier is 5,310,000 gpd. If the influent to the clarifier has a suspended solids concentration of 190 mg/L and the primary effluent has 103 mg/L SS, how many lbs/day suspended solids are removed by the clarifier?

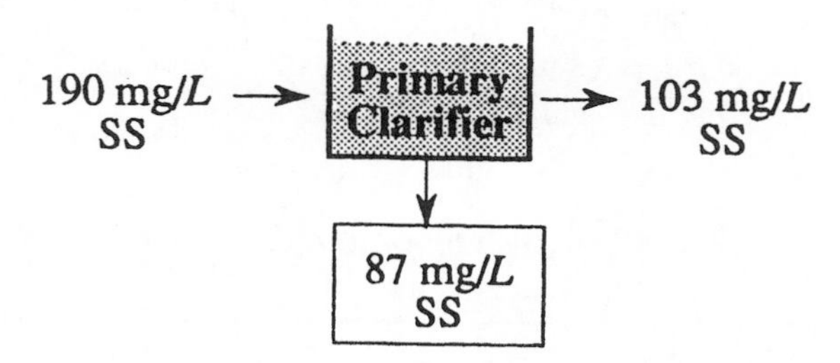

To calculate lbs/day SS removed:

(mg/L) (MGD flow) (8.34 lbs/gal) = lbs/day

(87 mg/L) (5.31 MGD) (8.34 lbs/gal) = **3853 lbs/day**

Example 4: (BOD and SS Removal)

❑ The flow to a primary clarifier is 3,040,000 gpd. If the influent to the clarifier has a suspended solids concentration of 215 mg/L and the primary effluent has 112 mg/L SS, how many lbs/day suspended solids are removed by the clarifier?

215 mg/L SS → Primary Clarifier → 112 mg/L SS

↓

103 mg/L SS

To calculate lbs/day SS removed:

(mg/L) (MGD flow) (8.34 lbs/gal) = lbs/day

(103 mg/L) (3.04 MGD) (8.34 lbs/gal) = **2611 lbs/day**

9.6 UNIT PROCESS EFFICIENCY

The efficiency of a treatment process is its effectiveness in removing various constituents from the water or wastewater. Suspended solids and BOD removal are therefore the most common calculations of unit process efficiency.

The efficiency of a clarifier may be affected by such factors as the types of solids in the wastewater, the temperature of the wastewater, and the age of the solids. Typical removal efficiencies for a primary clarifier are as follows:

Settleable Solids——90-99%

Suspended Solids——40-60%

Total Solids————10-15%

BOD——————20-50%

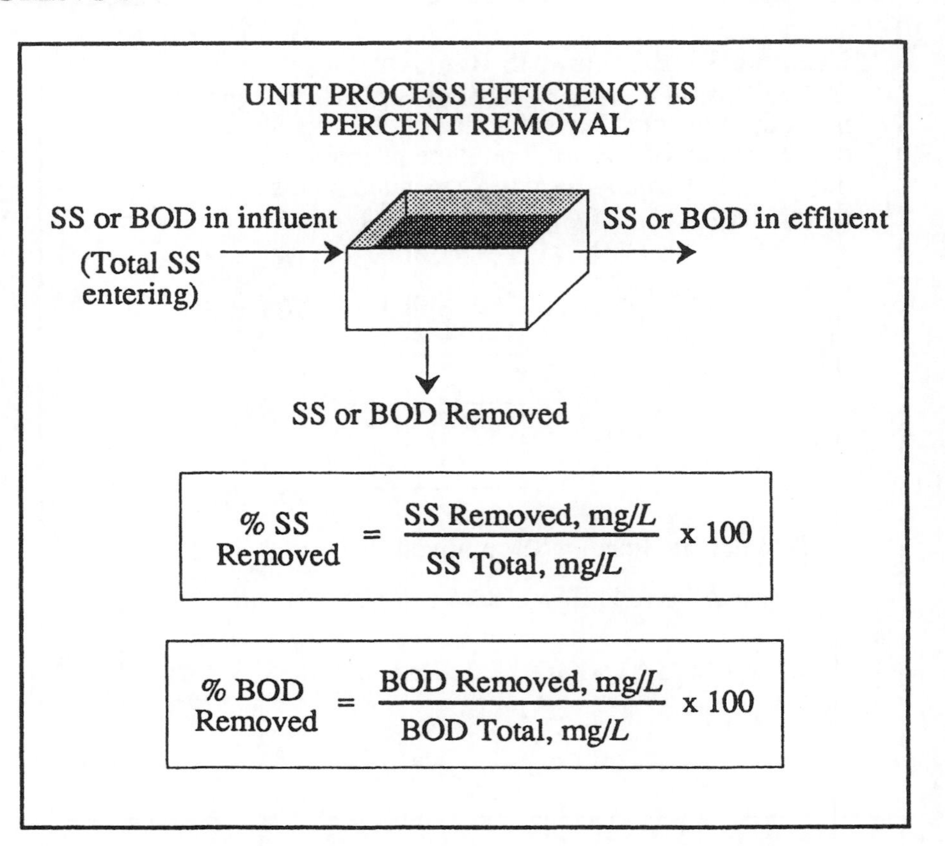

$$\text{\% SS Removed} = \frac{\text{SS Removed, mg}/L}{\text{SS Total, mg}/L} \times 100$$

$$\text{\% BOD Removed} = \frac{\text{BOD Removed, mg}/L}{\text{BOD Total, mg}/L} \times 100$$

Example 1: (Unit Process Efficiency)

❑ The suspended solids entering a primary clarifier is 182 mg/*L*. If the suspended solids in the primary clarifier effluent is 79 mg/*L*, what is suspended solids removal efficiency of the primary clarifier?

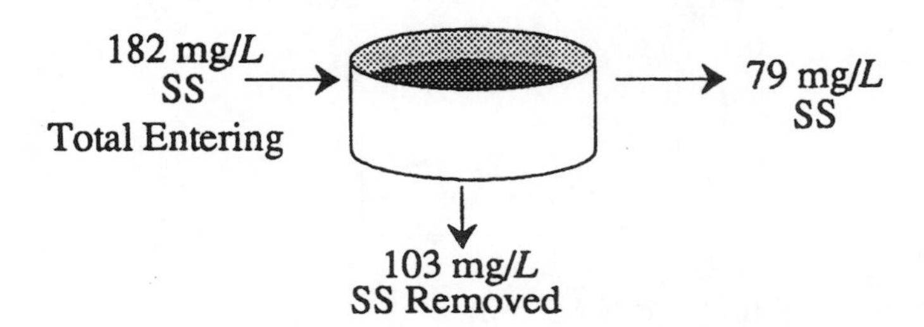

$$\text{\% SS Removed} = \frac{\text{SS Removed, mg}/L}{\text{SS Total, mg}/L} \times 100$$

$$= \frac{103 \text{ mg}/L}{182 \text{ mg}/L} \times 100$$

$$= \boxed{57\%}$$

Example 2: (Unit Process Efficiency)

❑ The influent to a primary clarifier has a BOD concentration of 245 mg/*L*. If the BOD content of the primary clarifier effluent is 140 mg/*L*, what is the BOD removal efficiency of the primary clarifier?

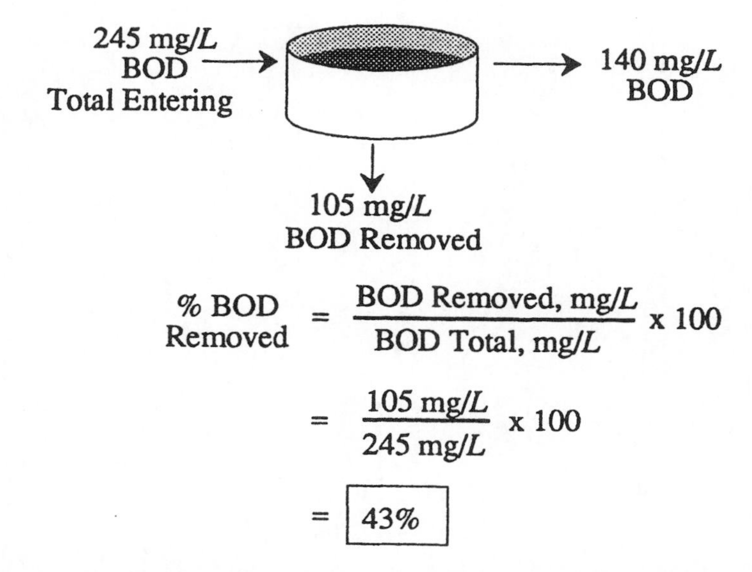

$$\%\ \text{BOD Removed} = \frac{\text{BOD Removed, mg}/L}{\text{BOD Total, mg}/L} \times 100$$

$$= \frac{105\ \text{mg}/L}{245\ \text{mg}/L} \times 100$$

$$= \boxed{43\%}$$

Example 3: (Unit Process Efficiency)

❑ The suspended solids entering a primary clarifier is 238 mg/*L*. If the suspended solids in the primary clarifier effluent is 124 mg/*L*, what is the suspended solids removal efficiency of the primary clarifier?

238 mg/*L*
SS
Total Entering

124 mg/*L*
SS

114 mg/*L*
SS Removed

$$\%\ \text{SS Removed} = \frac{\text{SS Removed, mg}/L}{\text{SS Total, mg}/L} \times 100$$

$$= \frac{114\ \text{mg}/L}{238\ \text{mg}/L} \times 100$$

$$= \boxed{48\%}$$

NOTES:

10 Trickling Filters

SUMMARY

1. Hydraulic Loading Rate

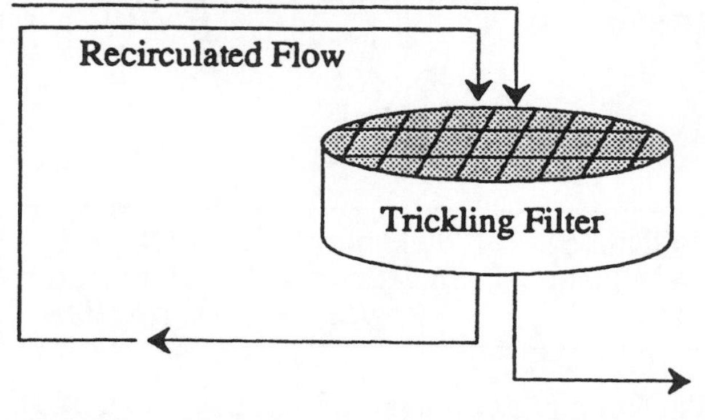

$$\text{Hydraulic Loading Rate} = \frac{\text{Total Flow Applied, gpd}}{\text{sq ft Area}}$$

(Hydraulic loading rate calculations include recirculated flows.)

2. Organic Loading Rate

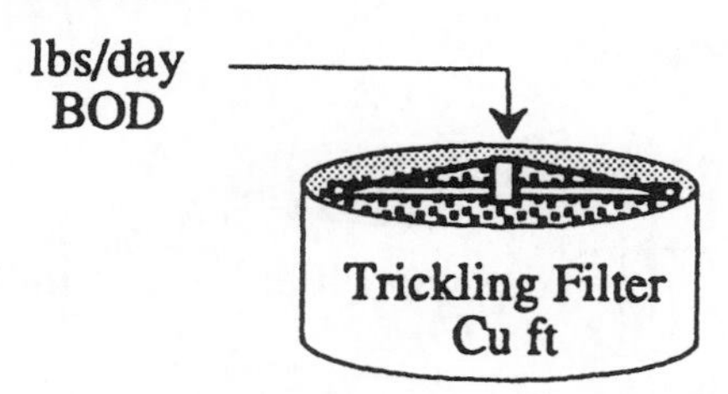

Simplified Equation:

$$\text{Organic Loading Rate, lbs/day/1000 cu ft} = \frac{\text{BOD, lbs/day}}{\text{1000 cu ft}}$$

Expanded Equation:

$$\text{Organic Loading Rate} = \frac{(\text{mg}/L\ \text{BOD})(\text{MGD})(8.34\ \text{lbs/gal})}{\text{1000 cu ft}}$$

SUMMARY—Cont'd

3. BOD and SS Removal, lbs/day

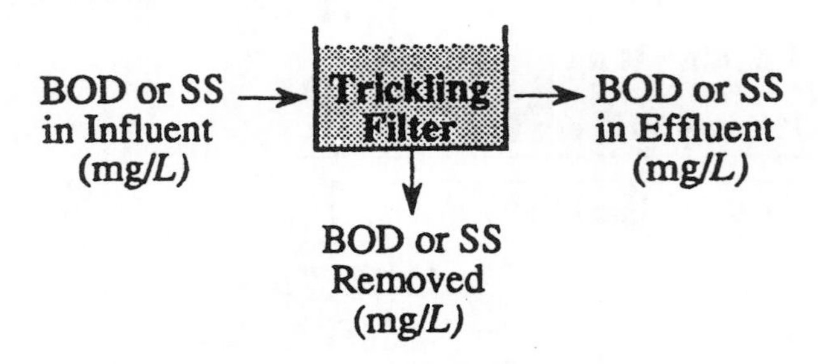

BOD or SS in Influent (mg/L)	−	BOD or SS in Effluent (mg/L)	=	BOD or SS Removed (mg/L)

After calculating mg/L BOD or SS removed, you can calculate lbs/day BOD or SS removed:

(mg/L) BOD or SS Removed	(MGD) flow	(8.34) lbs/gal	=	lbs/day BOD or SS Removed

4. Unit Process or Overall Efficiency

Unit Process Efficiency:

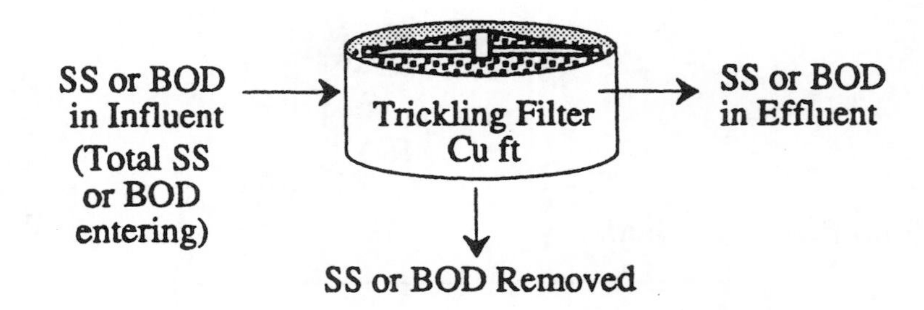

$$\%\ \text{SS Removed} = \frac{\text{mg/}L\ \text{SS Removed}}{\text{mg/}L\ \text{SS Total}} \times 100$$

$$\%\ \text{BOD Removed} = \frac{\text{mg/}L\ \text{BOD Removed}}{\text{mg/}L\ \text{BOD Total}} \times 100$$

SUMMARY—Cont'd

Overall Efficiency:

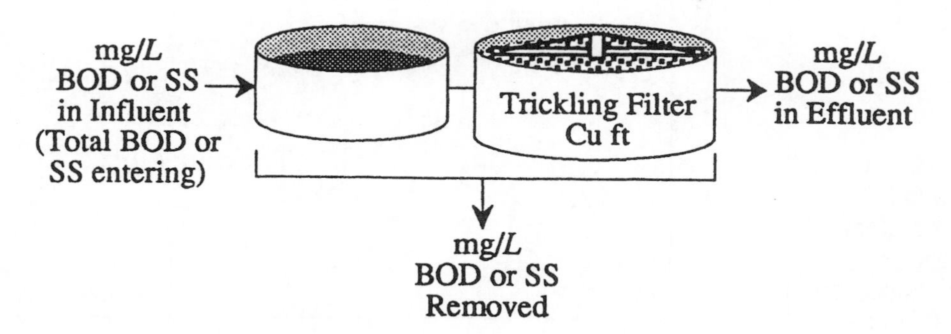

First calculate mg/*L* BOD or SS removed:

$$\text{BOD or SS in Influent (mg/}L) - \text{BOD or SS in Effluent (mg/}L) = \text{BOD or SS Removed (mg/}L)$$

Then calculate the overall efficiency:

$$\text{\% Overall Efficiency} = \frac{\text{BOD or SS Removed, mg/}L}{\text{BOD or SS Total, mg/}L} \times 100$$

5. Recirculation Ratio

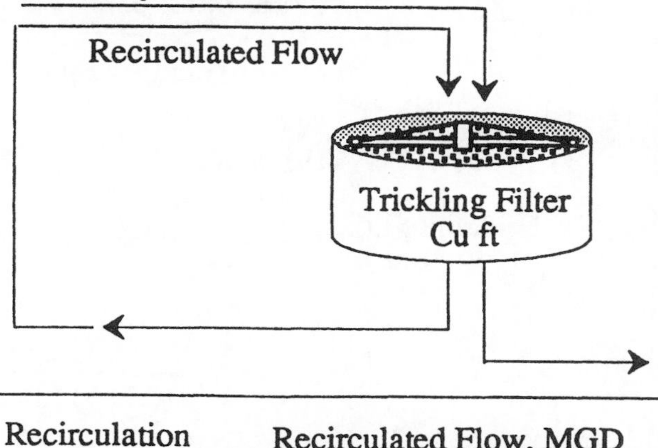

$$\text{Recirculation Ratio} = \frac{\text{Recirculated Flow, MGD}}{\text{Primary Effluent Flow, MGD}}$$

10.1 HYDRAULIC LOADING RATE

Hydraulic loading rate is an important trickling filter calculation, for it is associated with the contact time between the organisms of the zoogleal mass and the food which is entering the trickling filter with the influent flow.

Hydraulic loading rate is the total flow loaded or entering each square foot of water surface area. Mathematically, it is the total gpd flow to the process divided by the water surface area of the tank. As shown in the diagram to the right **recirculated flows must be included** as part of the total flow (total Q) to the process.

The normal hydraulic loading rate ranges for standard rate and high rate trickling filters are:

Standard Rate—25-100 gpd/sq ft
or 1-4 MGD/ac

High Rate——100-1000 gpd/sq ft
or 4-40 MGD/ac

If the hydraulic loading rate for a particular trickling filter is too low, septic conditions will begin to develop. **Recirculation of trickling filter or final clarifier effluent may be necessary when this occurs to increase the hydraulic loading rate.** When the loading is too high, other problems are created depending upon the recirculation flow pattern.

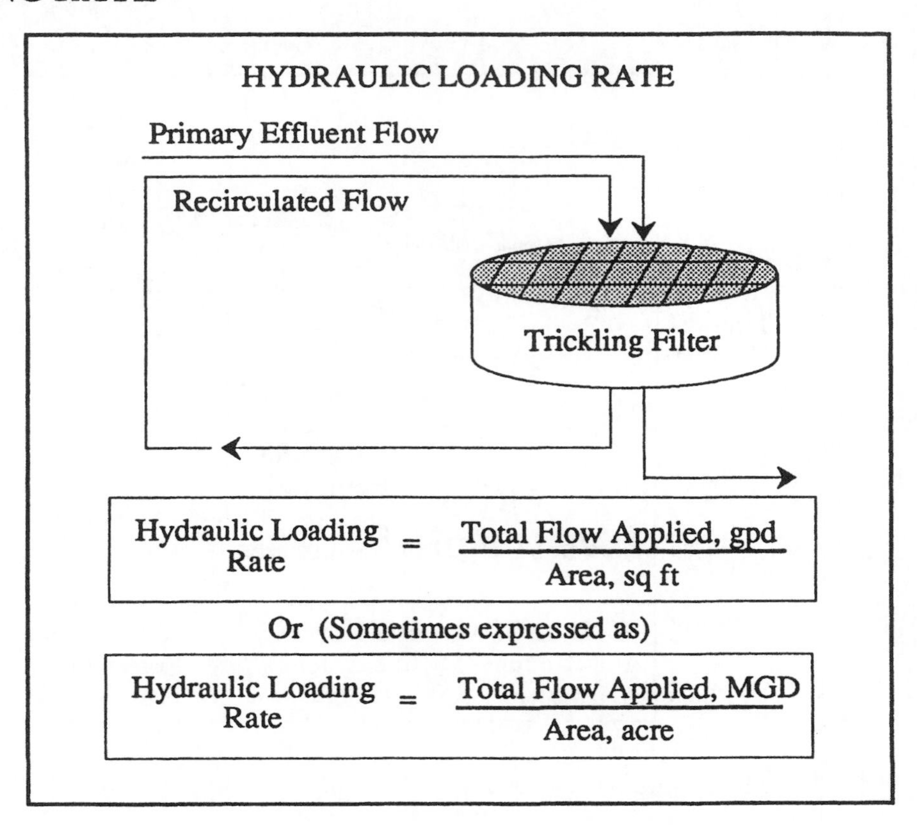

Example 1: (Hydraulic Loading)

❑ A trickling filter 80 ft in diameter treats a primary effluent flow of 450,000 gpd. If the recirculated flow to the clarifier is 0.1 MGD, what is the hydraulic loading on the trickling filter?

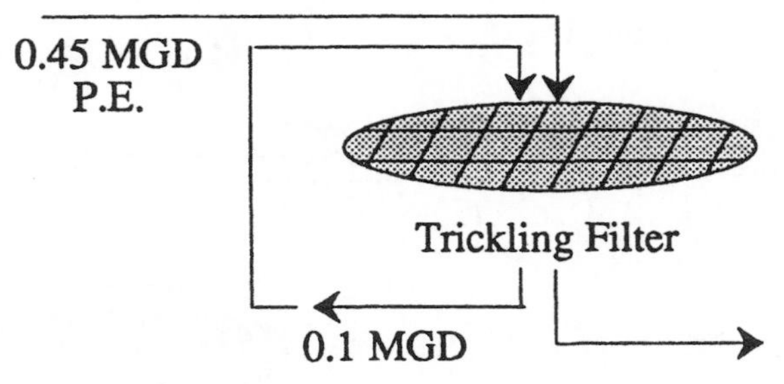

$$\text{Hydraulic Loading Rate} = \frac{\text{Total Flow, gpd}}{\text{Area, sq ft}}$$

$$= \frac{550{,}000 \text{ gpd total flow}}{(0.785)(80 \text{ ft})(80 \text{ ft})}$$

$$= \boxed{109 \text{ gpd/sq ft}}$$

Example 2: (Hydraulic Loading)

❑ A high rate trickling filter receives a flow of 2200 gpm. If the filter has a diameter of 95 ft, what is the hydraulic loading onto the filter?

2,200 gpm

(2,200 gpm) (1440 min/day) = 3,168,000 gpd

95 ft diam.

$$\text{Hydraulic Loading Rate} = \frac{\text{Total Flow, gpd}}{\text{Area, sq ft}}$$

$$= \frac{3{,}168{,}000 \text{ gpd}}{(0.785)\ (95 \text{ ft})\ (95 \text{ ft})}$$

$$= \boxed{447 \text{ gpd/sq ft}}$$

HYDRAULIC LOADING RATE AS MGD/ACRE

Occasionally, the hydraulic loading rate is expressed as MGD per acre. However, this is still an expression of gallons flow over surface area of trickling filter.

Example 3: (Hydraulic Loading)

❑ A high rate trickling filter receives a daily flow of 1.6 MGD. What is the hydraulic loading rate in MGD/acre if the filter is 90 ft in diameter and 4 ft deep?

1.6 MGD

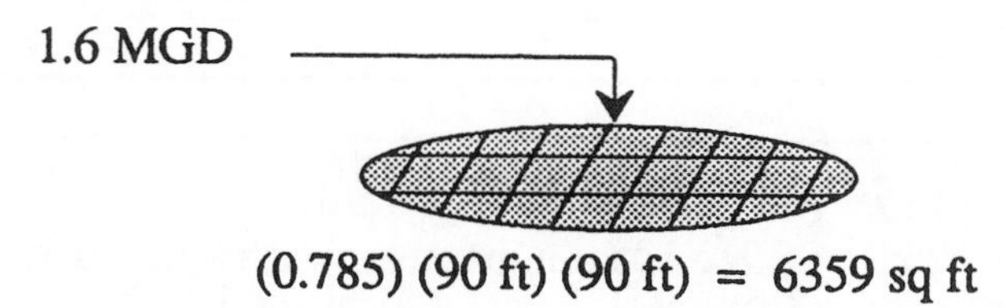

(0.785) (90 ft) (90 ft) = 6359 sq ft

First convert sq ft to acres:*

$$\frac{6{,}359 \text{ sq ft}}{43{,}560 \text{ sq ft/ac}} = 0.146 \text{ acres}$$

Then calculate hydraulic loading rate:

$$\text{Hydraulic Loading Rate} = \frac{1.6 \text{ MGD}}{0.146 \text{ acres}}$$

$$= \boxed{10.96 \text{ MGD/ac}}$$

* Refer to Chapter 8 in *Basic Math Concepts* for a discussion of sq ft to acres conversions.

10.2 ORGANIC LOADING RATE

When calculating the lbs/day BOD entering a wastewater process, you are calculating the organic load on the process—the food entering that process. Organic loading for trickling filters is most often calculated as lbs BOD/day per 1000 cu ft media. However, it is sometimes calculated as lbs/day BOD/per cu ft or lbs BOD/day per acre-foot.

In many cases, BOD will be expressed as a mg/*L* concentration and must be converted to lbs/day BOD,* as shown in the expanded equation.

Note that the "1000" in the denominator of organic loading equations is a unit of measure ("thousand cu ft") and is **not part of the numerical calculation.**

The normal organic loading rates for standard-rate and high-rate trickling filters are:

Standard Rate—

5-25 lbs BOD/day/1000 cu ft
or 200-1000 lbs BOD/day/ac-ft

High Rate—

25-300 lbs/day BOD/1000 cu ft
or 1000-13,000 lbs/day BOD/ac-ft

Increasing the BOD loading does not have the same effect as increasing the hydraulic loading (which lessens the contact time between the food and organisms and may decrease the efficiency of the trickling filter). To a degree, increasing the BOD loading appears to increase the number of organisms (more food, more organisms).

ORGANIC LOADING RATE

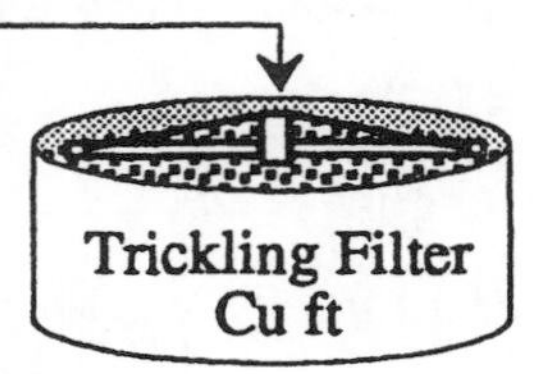

Simplified Equation:

$$\text{Organic Loading Rate} = \frac{\text{lbs BOD/day}}{\text{1000 cu ft}}$$

Expanded Equation:

$$\text{Organic Loading Rate} = \frac{(\text{mg}/L\ \text{BOD})\ (\text{MGD})\ (8.34)}{\frac{(0.785)\ (D^2)\ (\text{depth, ft})}{1000}}$$

Organic loading rate is sometimes calculated as:

$$\text{Organic Loading Rate} = \frac{\text{lbs BOD/day}}{\text{cu ft}}$$

OR

$$\text{Organic Loading Rate} = \frac{\text{lbs BOD/day}}{\text{acre-feet}}$$

Example 1: (Organic Loading Rate)

❑ A trickling filter 80 ft in diameter with a media depth of 8 ft receives a flow of 950,000 gpd. If the BOD concentration of the primary effluent is 210 mg/*L*, what is the organic loading on the trickling filter in lbs/day/1000 cu ft?

$$\text{Organic Loading Rate} = \frac{\text{lbs BOD/day}}{\text{1000 cu ft}}$$

$$= \frac{(210\ \text{mg}/L)\ (0.95\ \text{MGD})\ (8.34\ \text{lbs/gal})}{\frac{(0.785)\ (80\ \text{ft})\ (80\ \text{ft})\ (8\ \text{ft})}{1000}}$$

$$= \frac{1663.83\ \text{lbs BOD/day}}{40.192\quad \text{1000-cu ft}}$$

$$= \boxed{\frac{41\ \text{lbs BOD/day}}{\text{1000-cu ft}}}$$

---

* For a review of mg/*L* to lbs/day BOD, refer to Chapter 3.

Example 2: (Organic Loading Rate)

❑ A 75-ft diameter trickling filter with a media depth of 5 ft receives a primary effluent flow of 3.17 MGD with a BOD of 107 mg/*L*. What is the organic loading on the trickling filter in lbs/day/1000 cu ft?

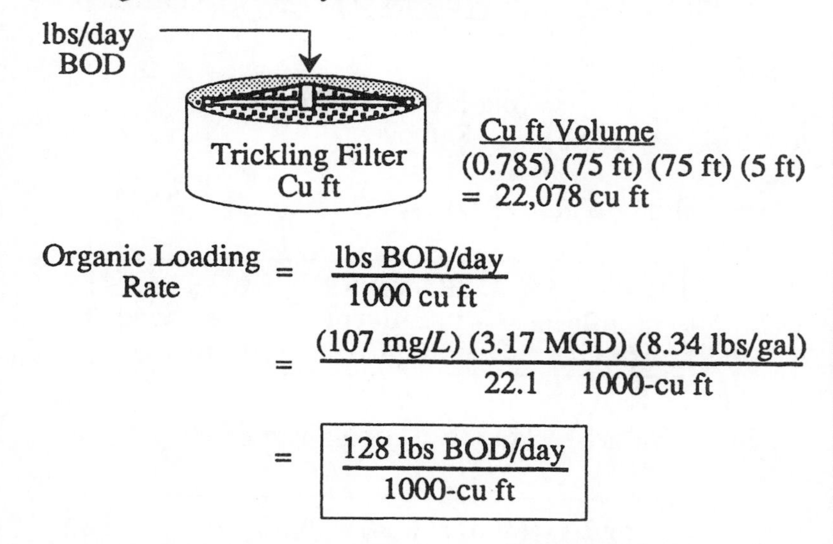

$$\text{Organic Loading Rate} = \frac{\text{lbs BOD/day}}{\text{1000 cu ft}}$$

$$= \frac{(107\ \text{mg/}L)\ (3.17\ \text{MGD})\ (8.34\ \text{lbs/gal})}{22.1 \quad \text{1000-cu ft}}$$

$$= \boxed{\frac{\text{128 lbs BOD/day}}{\text{1000-cu ft}}}$$

Example 3: (Organic Loading Rate)

❑ A standard rate trickling filter has a diameter of 100 feet and an average media depth of 7 feet. If the 1.2 MGD primary effluent has a BOD concentration of 120 mg/*L*, what is the organic loading in lbs per acre-feet?

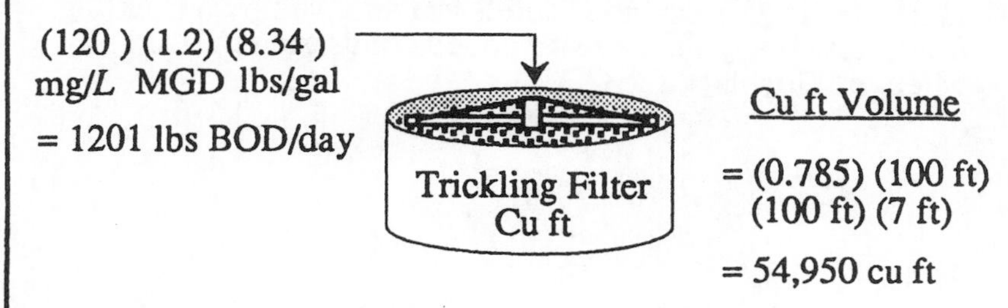

Since organic loading is desired in lbs/day BOD per acre foot, convert cu ft to acre feet:* (54,950 cu ft ÷ 43,560 cu ft/ac-ft = 1.26 ac-ft)

Now calculate organic loading rate:

$$\text{Organic Loading Rate} = \frac{\text{1201 lbs BOD/day}}{\text{1.26 ac-ft}}$$

$$= \boxed{\frac{\text{953 lbs BOD/day}}{\text{ac-ft}}}$$

ORGANIC LOADING RATE AS lbs BOD/day/ac-ft

Although organic loading is most often expressed as lbs BOD/day per cu ft or lbs BOD/day per 1000 cu ft, it may also be expressed as lbs BOD/day per acre-foot. This is still an expression of lbs BOD/day per volume.

* For a review of cu ft to ac-ft conversions, refer to Chapter 8 in *Basic Math Concepts*.

10.3 BOD AND SS REMOVED

To calculate the pounds of BOD or suspended solids removed each day, you will need to know the mg/L BOD or SS removed and the plant flow. Then you can use the mg/L to lbs/day equation:

(mg/L) (MGD) (8.34) = lbs/day
Removed flow lbs/gal

BOD AND SS REMOVED

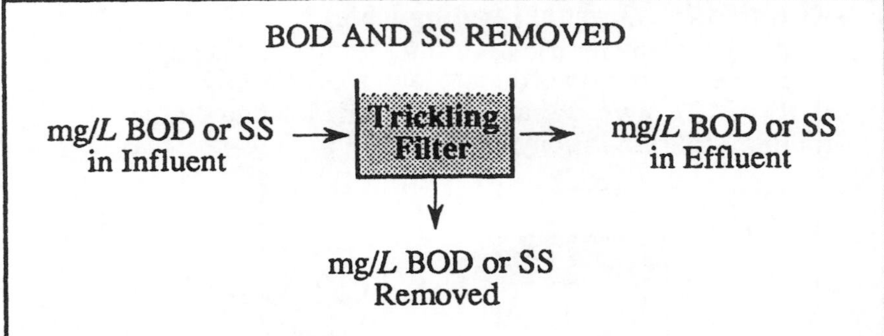

First calculate mg/L BOD or SS removed:

BOD or SS in Influent (mg/L)	–	BOD or SS in Effluent (mg/L)	=	BOD or SS Removed (mg/L)

Then calculate lbs/day BOD or SS removed:

(mg/L) (MGD) (8.34) = lbs/day BOD or SS
BOD or SS Removed flow lbs/gal Removed

Example 1: (BOD and SS Removal)

❑ If 110 mg/L suspended solids are removed by a trickling filter, how many lbs/day suspended solids are removed when the flow is 4.2 MGD?

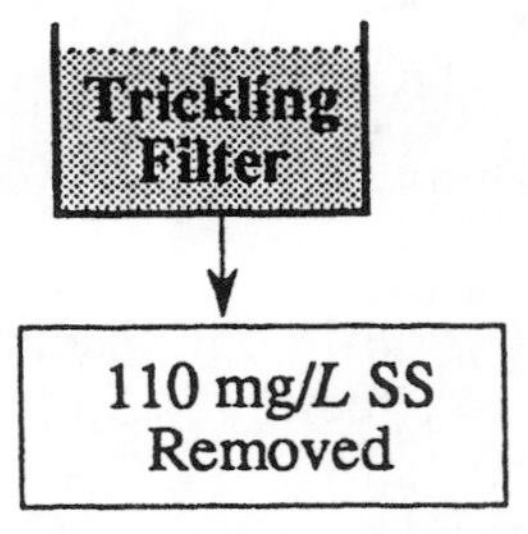

110 mg/L SS
Removed

(mg/L) (MGD flow) (8.34 lbs/gal) = lbs/day

(110 mg/L) (4.2 MGD) (8.34 lbs/gal) = **3853 lbs SS/day**

* To review mg/L to lbs/day conversions, refer to Chapter 3.

Example 2: (BOD and SS Removal)

❑ The flow to a trickling filter is 4.6 MGD. If the primary effluent has a BOD concentration of 150 mg/*L* and the trickling filter effluent has a BOD concentration of 25 mg/*L*, how many pounds of BOD are removed daily?

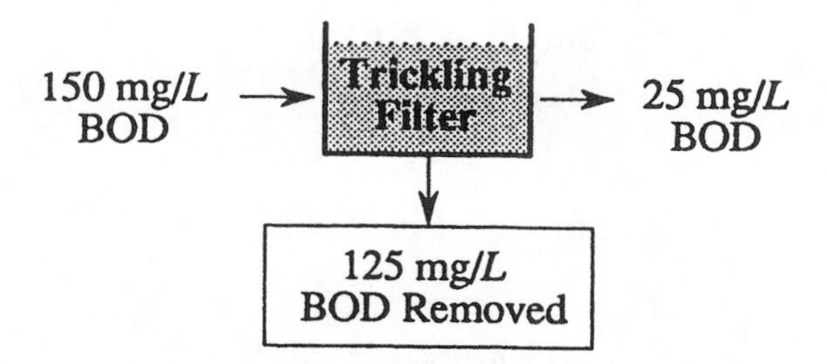

After calculating mg/*L* BOD removed, you can calculate lbs/day BOD removed:

(mg/*L*) (MGD flow) (8.34 lbs/gal) = lbs/day
Removed   Removed

(125 mg/*L*) (4.6 MGD) (8.34 lbs/gal) = **4796 lbs/day Removed**

Example 3: (BOD and SS Removal)

❑ The 3,700,000 gpd influent flow to a trickling filter has a BOD content of 185 mg/*L*. If the trickling filter effluent has a BOD content of 68 mg/*L*, how many pounds of BOD are removed daily?

185 mg/*L* BOD → Trickling Filter → 68 mg/*L* BOD

117 mg/*L* BOD Removed

(mg/*L*) (MGD flow) (8.34 lbs/gal) = lbs/day
Removed   Removed

(117 mg/*L*) (3.7 MGD) (8.34 lbs/gal) = **3610 lbs/day Removed**

10.4 UNIT PROCESS OR OVERALL EFFICIENCY

The efficiency of a treatment process is its effectiveness in removing various constituents from the water or wastewater. Unit process and overall efficiency are calculated using the same equation, as shown to the right.

When calculating **unit process efficiency** you will need to know the BOD or SS concentration of the trickling filter influent (sometimes referred to as "primary effluent") and the trickling filter effluent.

When calculating **overall efficiency**, you will need to know the BOD or SS concentration of the plant influent (or "primary influent") and that of the plant effluent.

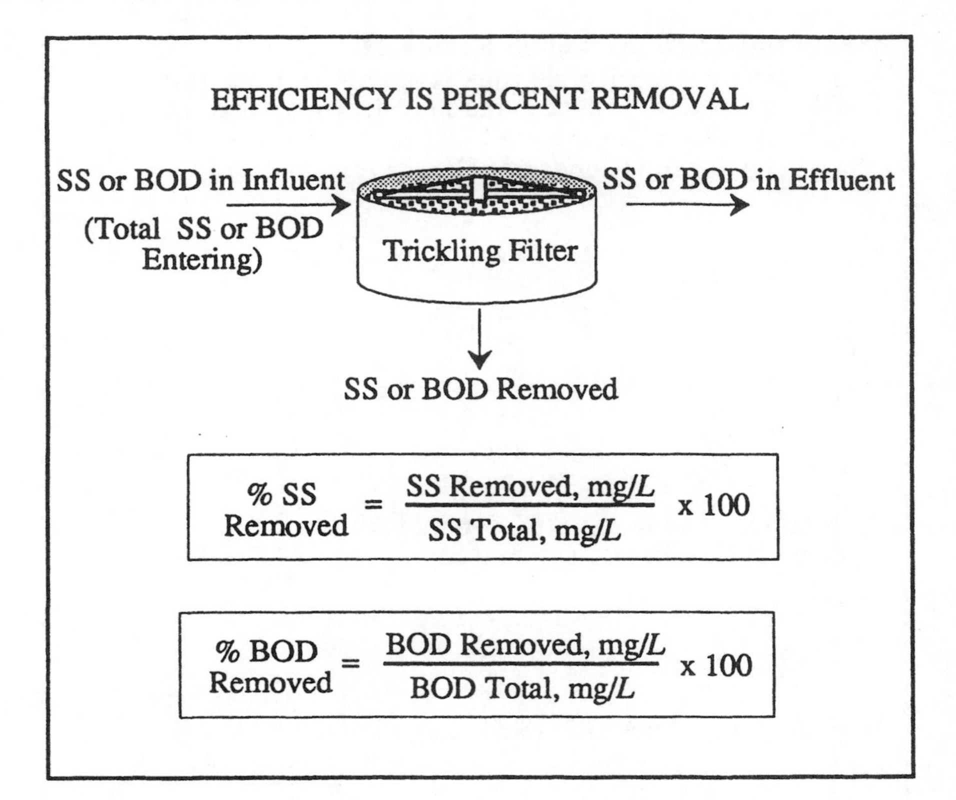

$$\frac{\%\ \text{SS}}{\text{Removed}} = \frac{\text{SS Removed, mg}/L}{\text{SS Total, mg}/L} \times 100$$

$$\frac{\%\ \text{BOD}}{\text{Removed}} = \frac{\text{BOD Removed, mg}/L}{\text{BOD Total, mg}/L} \times 100$$

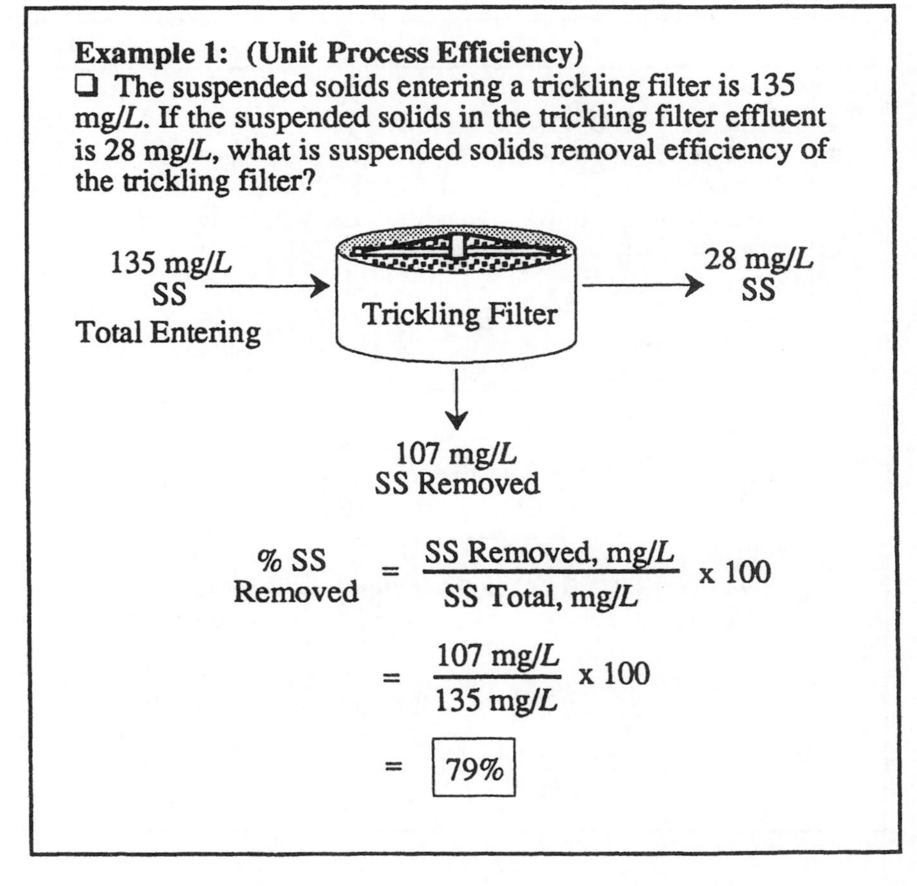

Example 1: (Unit Process Efficiency)

❑ The suspended solids entering a trickling filter is 135 mg/*L*. If the suspended solids in the trickling filter effluent is 28 mg/*L*, what is suspended solids removal efficiency of the trickling filter?

$$\frac{\%\ \text{SS}}{\text{Removed}} = \frac{\text{SS Removed, mg}/L}{\text{SS Total, mg}/L} \times 100$$

$$= \frac{107\ \text{mg}/L}{135\ \text{mg}/L} \times 100$$

$$= \boxed{79\%}$$

Example 2: (Unit Process Efficiency)

❑ The influent of a primary clarifier has a BOD content of 215 mg/*L*. The trickling filter effluent BOD is 30 mg/*L*. What is the BOD removal efficiency of the treatment plant?

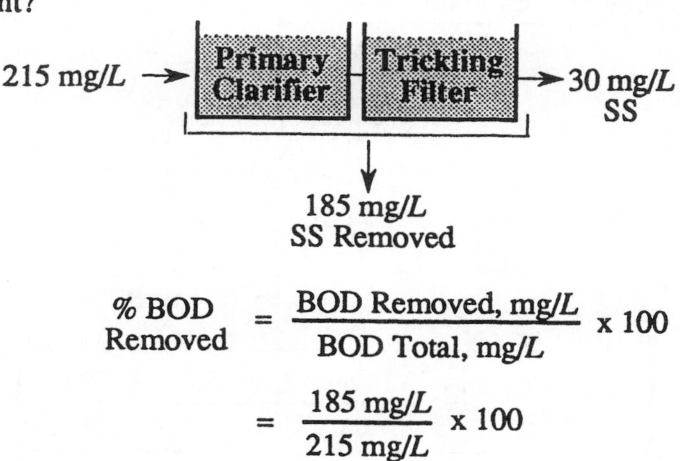

$$\%\ \text{BOD Removed} = \frac{\text{BOD Removed, mg}/L}{\text{BOD Total, mg}/L} \times 100$$

$$= \frac{185\ \text{mg}/L}{215\ \text{mg}/L} \times 100$$

$$= \boxed{86\%}$$

Example 3: (Unit Process Efficiency)

❑ The BOD concentration of a trickling filter influent is 210 mg/*L*. If the trickling filter effluent flow has a BOD content of 38 mg/*L*, what is the BOD removal efficiency of the trickling filter?

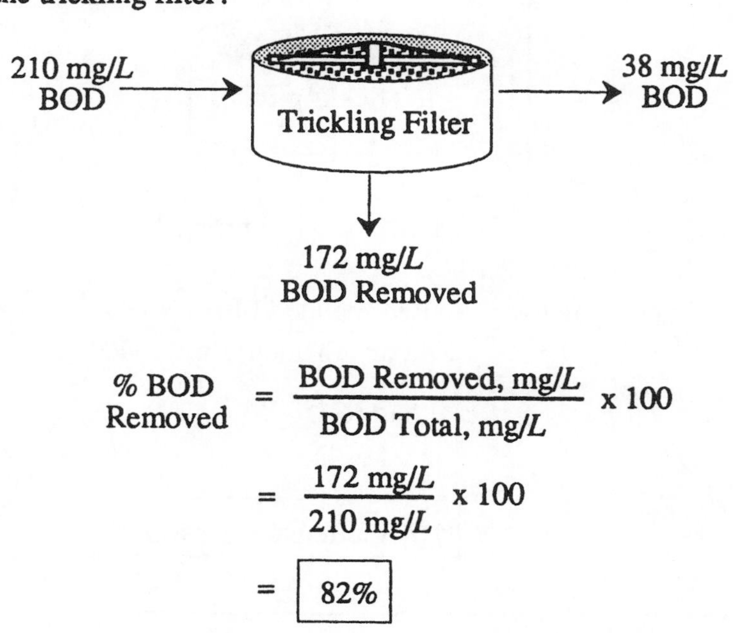

$$\%\ \text{BOD Removed} = \frac{\text{BOD Removed, mg}/L}{\text{BOD Total, mg}/L} \times 100$$

$$= \frac{172\ \text{mg}/L}{210\ \text{mg}/L} \times 100$$

$$= \boxed{82\%}$$

10.5 RECIRCULATION RATIO

The trickling filter recirculation ratio is the ratio* of the recirculated trickling filter flow to the primary effluent flow.

The trickling filter recirculation ratio may range from 0.5:1 (a ratio of 0.5/1 = .5) to 5:1 (a ratio of 5/1 = 5). However, the ratio is often found to be 1:1 or 2:1.

In the treatment plant process, recirculation of trickling filter or final clarifier effluent may be used for various reasons such as reducing the trickling filter detention time, increasing the hydraulic loading rate, or decreasing the trickling filter influent strength, thereby improving the ability to receive shock loads. It is also used to keep the filter wet during periods of low flow.

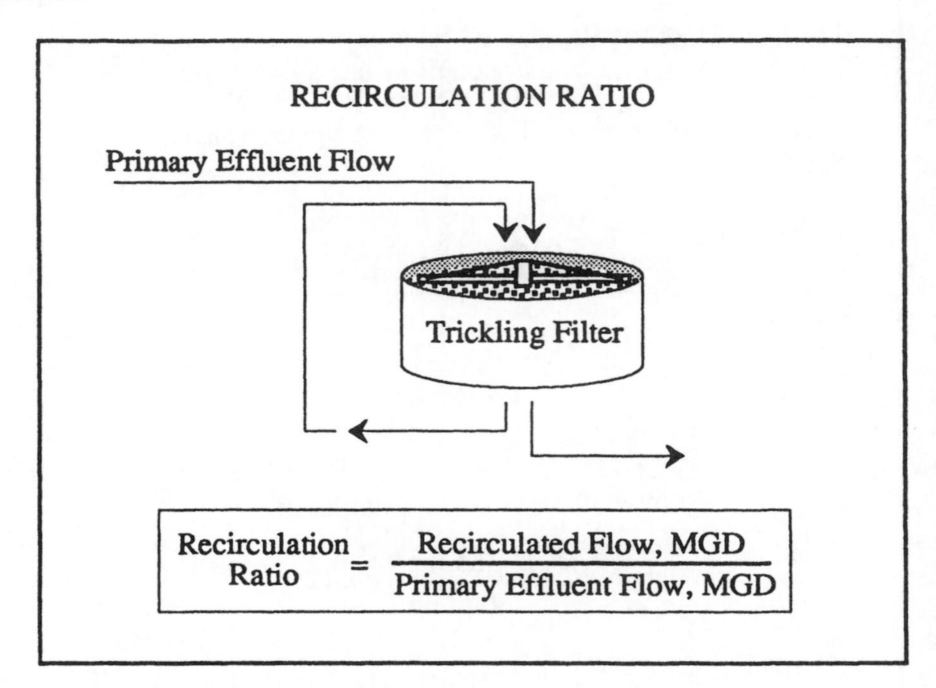

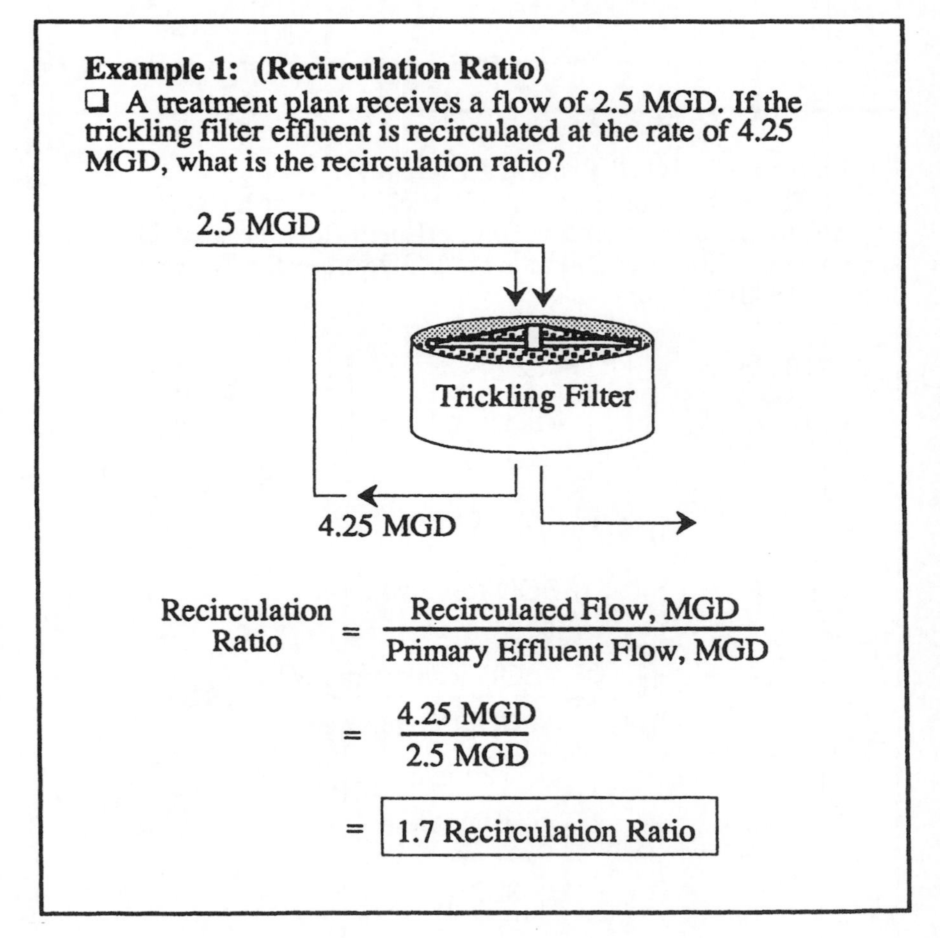

* For a review of ratios, refer to Chapter 7 in *Basic Math Concepts*.

Example 2: (Recirculation Ratio)

❑ The influent to the trickling filter is 3 MGD. If the recirculated flow is 5.4 MGD, what is the recirculation ratio?

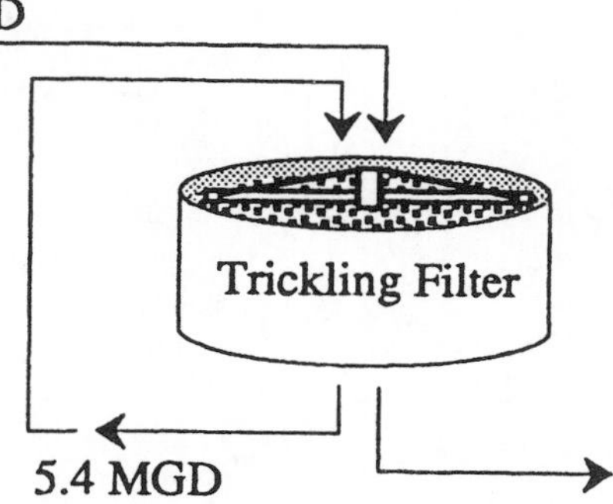

$$\text{Recirculation Ratio} = \frac{\text{Recirculated Flow, MGD}}{\text{Primary Effluent Flow, MGD}}$$

$$= \frac{5.4 \text{ MGD}}{3 \text{ MGD}}$$

$$= \boxed{1.8 \text{ Recirculation Ratio}}$$

Example 3: (Recirculation Ratio)

❑ A trickling filter receives a primary effluent flow of 4 MGD. If the recirculated flow is 4.8 MGD, what is the recirculation ratio?

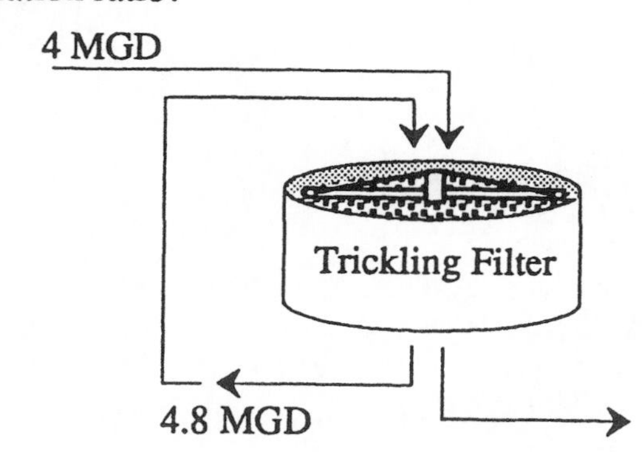

$$\text{Recirculation Ratio} = \frac{\text{Recirculated Flow, MGD}}{\text{Primary Effluent Flow, MGD}}$$

$$= \frac{4.8 \text{ MGD}}{4 \text{ MGD}}$$

$$= \boxed{1.2 \text{ Recirculation Ratio}}$$

NOTES:

11 *Rotating Biological Contactors*

SUMMARY

1. Hydraulic Loading Rate

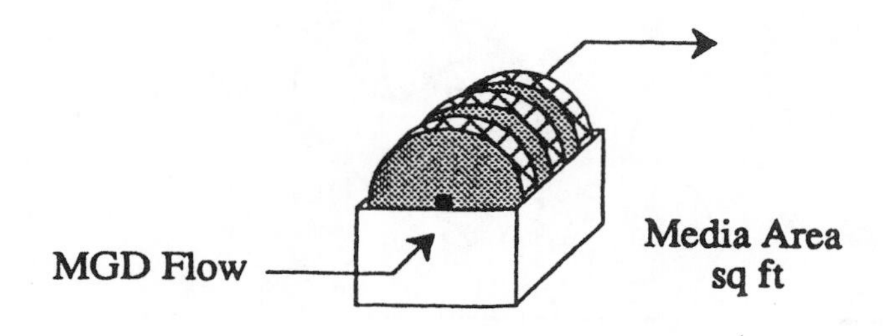

$$\text{Hydraulic Loading Rate} = \frac{\text{Total Flow Applied, gpd}}{\text{sq ft Area}}$$

(Hydraulic loading rate calculations include recirculated flows.)

2. Soluble BOD, mg/L

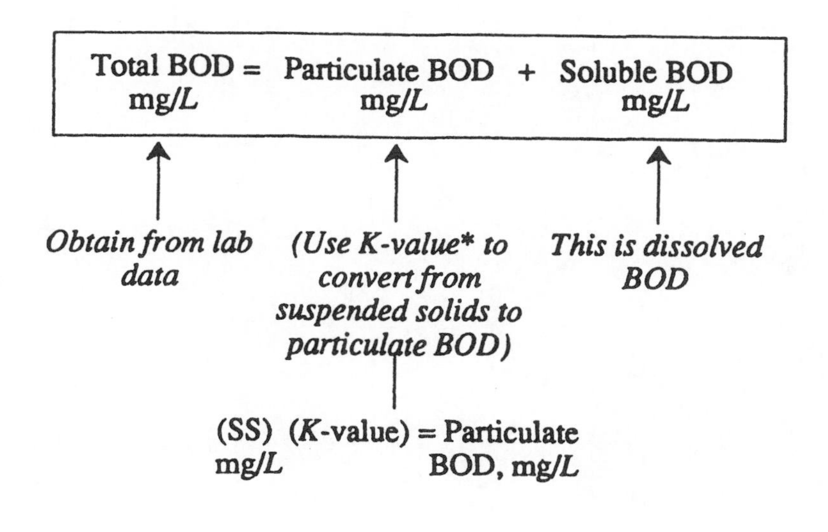

* The K-value for most domestic wastewaters is 0.5-0.7. Although this method allows estimation of soluble BOD, soluble BOD test results are preferred to using the *K*-value approach.

SUMMARY—Cont'd

3. Organic Loading Rate

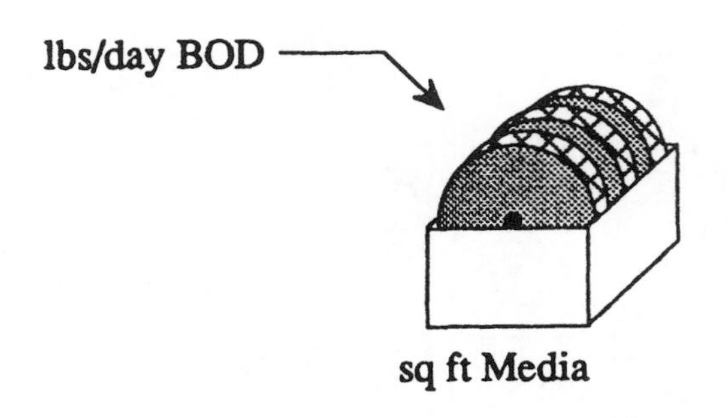

$$\text{System Organic Loading Rate, lbs/day/1000 sq ft} = \frac{\text{lbs/day Soluble BOD}}{\text{1000 sq ft Total Stages}}$$

$$\text{First Stage Organic Loading Rate, lbs/day/1000 sq ft} = \frac{\text{lbs/day Soluble BOD}}{\text{1000 sq ft First Stage}}$$

NOTES:

11.1 HYDRAULIC LOADING RATE

When calculating the hydraulic loading rate on a rotating biological contactor (RBC), use the **sq ft area of the media** rather than the sq ft area of the water surface, as in other hydraulic loading calculations. The RBC manufacturer provides media area information.

HYDRAULIC LOADING RATE INCLUDES RECIRCULATED FLOWS

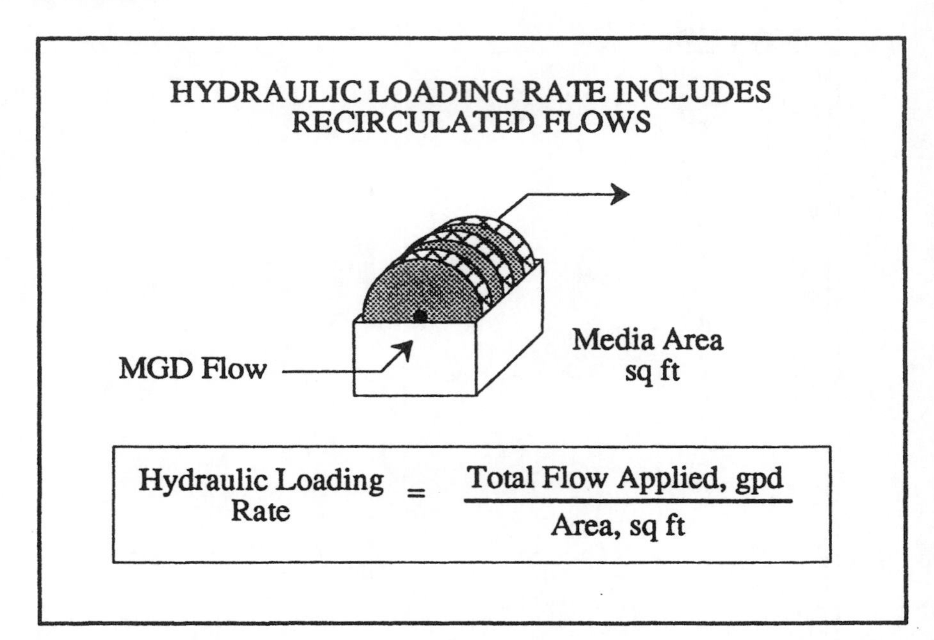

$$\text{Hydraulic Loading Rate} = \frac{\text{Total Flow Applied, gpd}}{\text{Area, sq ft}}$$

Example 1: (Hydraulic Loading)

❑ A rotating biological contactor (RBC) treats a primary effluent flow of 1.85 MGD. If the media surface area is 600,000 sq ft, what is the hydraulic loading on the RBC?

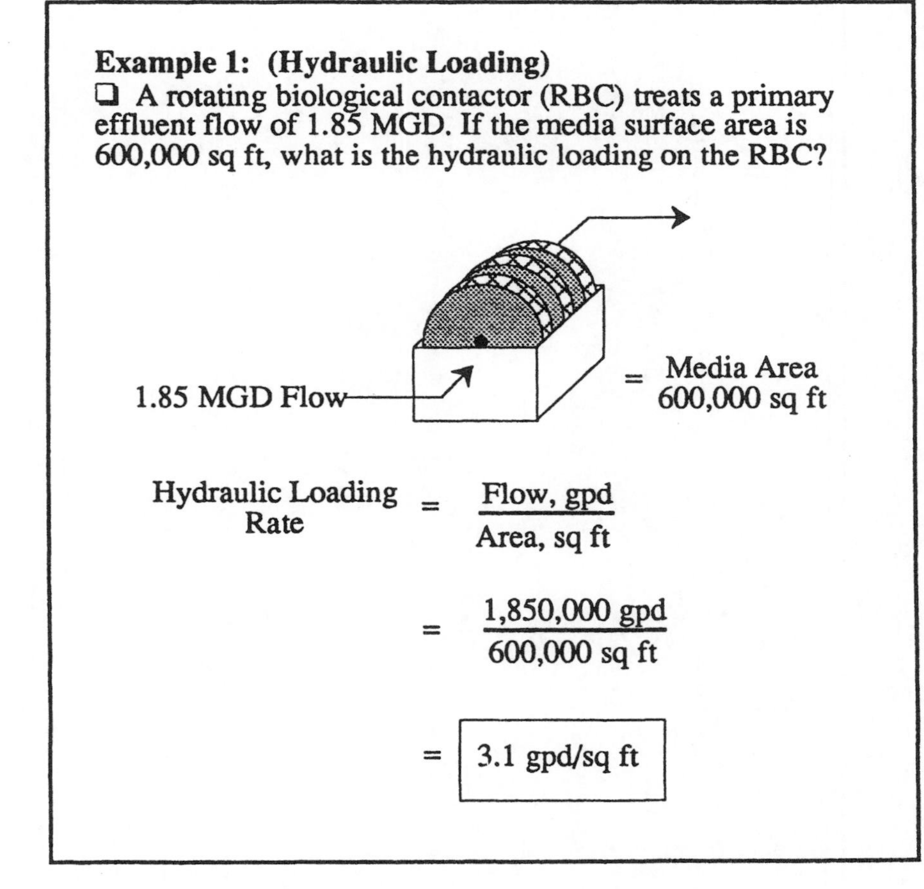

$$\text{Hydraulic Loading Rate} = \frac{\text{Flow, gpd}}{\text{Area, sq ft}}$$

$$= \frac{1{,}850{,}000 \text{ gpd}}{600{,}000 \text{ sq ft}}$$

$$= \boxed{3.1 \text{ gpd/sq ft}}$$

Example 2: (Hydraulic Loading)

❑ A rotating biological contactor treats a flow of 3.6 MGD. The manufacturer's data indicates a media surface area of 700,000 sq ft. What is the hydraulic loading rate on the RBC?

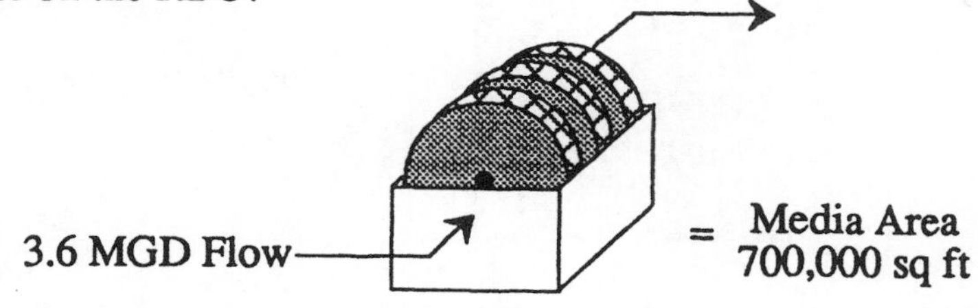

$$\text{Hydraulic Loading Rate} = \frac{\text{Flow, gpd}}{\text{Area, sq ft}}$$

$$= \frac{3{,}600{,}000\ \text{gpd}}{700{,}000\ \text{sq ft}}$$

$$= \boxed{5.1\ \text{gpd/sq ft}}$$

Example 3: (Hydraulic Loading)

❑ A rotating biological contactor treats a primary effluent flow of 1,270,000 gpd. The manufacturer's data indicates that the media surface area is 500,000 sq ft. What is the hydraulic loading rate on the filter?

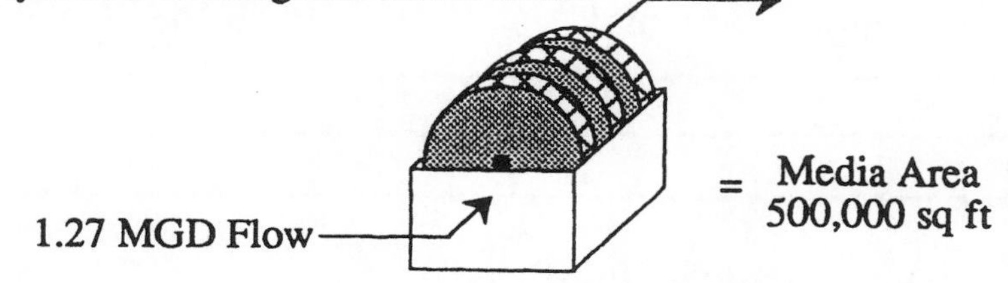

$$\text{Hydraulic Loading Rate} = \frac{\text{Flow, gpd}}{\text{Area, sq ft}}$$

$$= \frac{1{,}270{,}000\ \text{gpd}}{500{,}000\ \text{sq ft}}$$

$$= \boxed{2.5\ \text{gpd/sq ft}}$$

11.2 SOLUBLE BOD, mg/*L*

Organic loading on rotating biological contactors is based on lbs/day soluble BOD rather than lbs/day total BOD as in other calculations of organic loading rate. Therefore, in order to calculate the organic loading rate for RBCs, you must first understand how to determine soluble BOD.

The BOD in wastewater can be categorized into two types:

- BOD exerted by organic suspended particles in the wastewater (particulate BOD), and
- BOD exerted by dissolved substances in the wastewater (soluble BOD).

The diagram to the right illustrates these two components of total BOD.

Many labs do not test for soluble BOD. The BOD test normally conducted is for **total BOD.** However, there is a method to estimate soluble BOD using the equation shown to the right.*

If you know both total and particulate BOD, you can calculate soluble BOD using the equation shown to the right. Total BOD data is available as lab data. Particulate BOD must be estimated using a K-value.

The K-value indicates how much of the suspended solids is **organic suspended solids** (i.e., particulate BOD). For most domestic wastewater, about 50-70% of the suspended solids are organic suspended solids. (In other words, about 50-70% of the suspended solids are suspended BOD or particulate BOD.) The K-value is normally given as a decimal number rather than a percent (0.5-0.7).

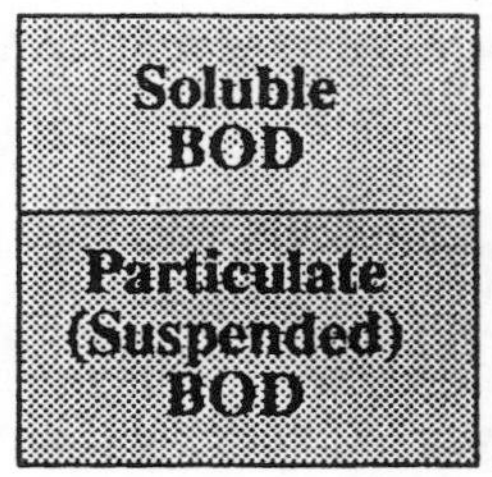

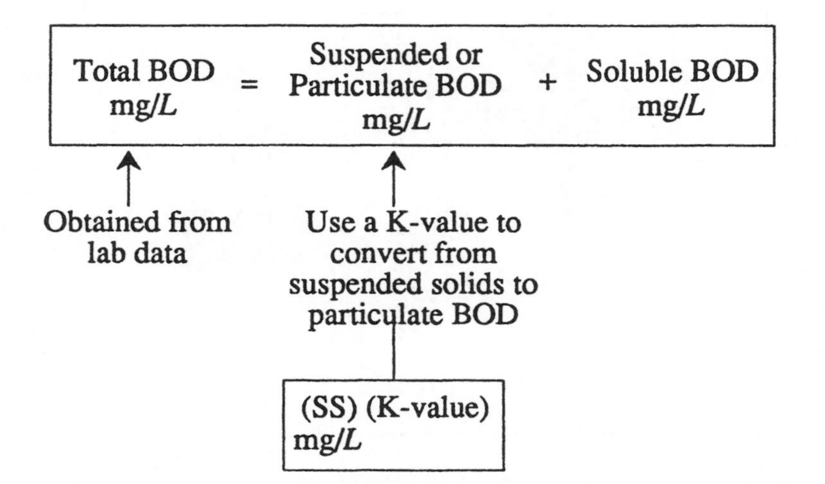

Substituting the suspended solids and K-value calculation into the equation above:

$$\underset{\text{mg/}L}{\text{Total BOD}} = \underset{\text{mg/}L}{\text{(SS) (K-value)}} + \underset{\text{mg/}L}{\text{Soluble BOD}}$$

Example 1: (Soluble BOD)

❑ The suspended solids concentration of a wastewater is 260 mg/*L*. If the normal K-value at the plant is 0.5, what is the estimated particulate BOD concentration of the wastewater?

The K-value of 0.5 indicates that about 50% of the suspended solids are organic suspended solids (or "particulate BOD"):

$$\underset{\text{SS}}{(260\ \text{mg/}L)}\ \underset{\text{K-value}}{(0.5)} = \boxed{\begin{array}{l}130\ \text{mg/}L\\ \text{Particulate BOD}\end{array}}$$

---

* Although the *K*-value method allows estimation of soluble BOD, soluble BOD test results are preferred to using the *K*-value approach.

Example 2: (Soluble BOD)

❑ The wastewater entering a rotating biological contactor has a BOD content of 205 mg/*L*. The suspended solids content is 245 mg/*L*. If the K-value is 0.5, what is the estimated soluble BOD (mg/*L*) of the wastewater?

$$\underset{\text{mg/}L}{\text{Total BOD}} = \underset{\text{mg/}L}{\text{Particulate BOD}} + \underset{\text{mg/}L}{\text{Soluble BOD}}$$

$$\underset{\text{BOD}}{205 \text{ mg/}L} = \underset{\text{SS}}{(245 \text{ mg/}L)}(0.5) + \underset{\text{Sol. BOD}}{x \text{ mg/}L}$$

$$205 \text{ mg/}L = 122.5 \text{ mg/}L + \underset{\text{Sol. BOD}}{x \text{ mg/}L}$$

$$205 - 122.5 = x$$

$$\boxed{\underset{\text{Soluble BOD}}{82.5 \text{ mg/}L}} = x$$

Example 3: (Soluble BOD)

❑ A rotating biological contactor receives a flow of 2.3 MGD with a BOD content of 180 mg/*L* and SS concentration of 150 mg/*L*. If the K-value is 0.7, how many pounds soluble BOD enter the RBC daily?

$$\underset{\text{mg/}L}{\text{Total BOD}} = \underset{\text{mg/}L}{\text{Particulate BOD}} + \underset{\text{mg/}L}{\text{Soluble BOD}}$$

$$\underset{\text{BOD}}{180 \text{ mg/}L} = \underset{\text{SS}}{(150 \text{ mg/}L)}(0.7) + \underset{\text{Sol. BOD}}{x \text{ mg/}L}$$

$$180 \text{ mg/}L = 105 \text{ mg/}L + \underset{\text{Sol. BOD}}{x \text{ mg/}L}$$

$$180 - 105 = x$$

$$\boxed{\underset{\text{Soluble BOD}}{75 \text{ mg/}L}} = x$$

Now lbs/day soluble BOD may be determined:

$$(\text{mg/}L \text{ Sol. BOD})(\text{MGD flow})(8.34 \text{ lbs/gal}) = \text{lbs/day}$$

$$(75 \text{ mg/}L)(2.3 \text{ MGD})(8.34 \text{ lbs/gal}) = \boxed{\underset{\text{Soluble BOD}}{1439 \text{ lbs/day}}}$$

CALCULATING LBS/DAY SOLUBLE BOD

Once you have determined the soluble BOD in mg/*L*, you can then determine lbs/day soluble BOD using the mg/*L* to lbs/day equation.*

$$\boxed{\underset{\text{Sol. BOD}}{(\text{mg/}L)}\ \underset{\text{flow}}{(\text{MGD})}\ \underset{\text{lbs/gal}}{(8.34)} = \text{lbs/day}}$$

* Refer to Chapter 3 for a discussion of mg/*L* to lbs/day calculations.

11.3 ORGANIC LOADING RATE

When calculating the lbs/day BOD entering a wastewater process, you are calculating the organic load on the process—the food entering that process. Organic loading for most processes is calculated as lbs BOD/day per 1000 cu ft media:

$$\text{Organic Loading Rate} = \frac{\text{BOD, lbs/day}}{\text{1000 cu ft}}$$

There are two different aspects to calculating organic loading on rotating biological contactors:

1. **Soluble BOD** is used to measure organic content rather than total BOD, and

2. The calculation of organic loading is **per 1000-sq ft media** rather than 1000-cu ft media as with other organic loading calculations. To find how many 1000 sq ft, find the thousands comma and place a decimal at that position. In Example 1, 500,000 sq ft is 500 units of 1000 sq ft. So 500 is placed in the denominator.

In most instances BOD (total or soluble) will be expressed as a mg/*L* concentration and must be converted to lbs/day BOD.* Therefore, the equation commonly used for organic loading is:

$$\text{O.L. Rate} = \frac{(\text{mg/}L\text{ BOD})\ (\text{MGD})\ (8.34)}{\text{1000 sq ft}}$$

Note that the "1000" in the denominator of both equations is a unit of measure ("thousand sq ft") and is **not part of the numerical calculation.**

ORGANIC LOADING RATE

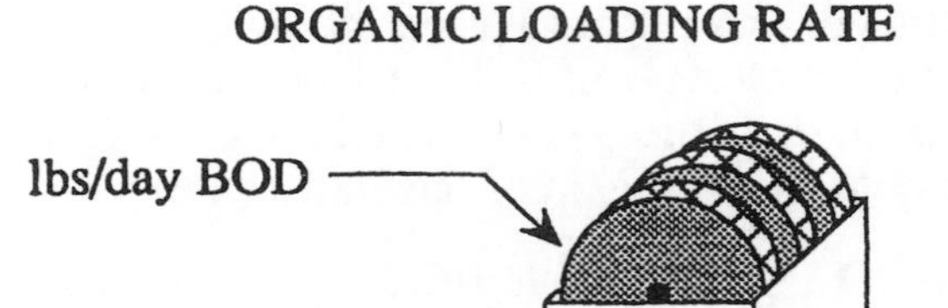

Media
1000-sq ft

Simplified Equation:

$$\text{Organic Loading Rate} = \frac{\text{Soluble BOD, lbs/day}}{\text{Media Area, 1000 sq ft}}$$

Expanded Equation:

$$\text{Organic Loading Rate} = \frac{\underset{\text{Sol. BOD}}{(\text{mg/L})}\ \underset{\text{Flow}}{(\text{MGD})}\ \underset{\text{lbs/gal}}{(8.34)}}{\text{Media Area, 1000 sq ft}}$$

Example 1: (Organic Loading Rate)

❑ A rotating biological contactor (RBC) has a media surface area of 500,000 sq ft and receives a flow of 1,200,000 gpd. If the Sol. BOD concentration of the primary effluent is 170 mg/*L*, what is the organic loading on the RBC in lbs/day/1000 sq ft?

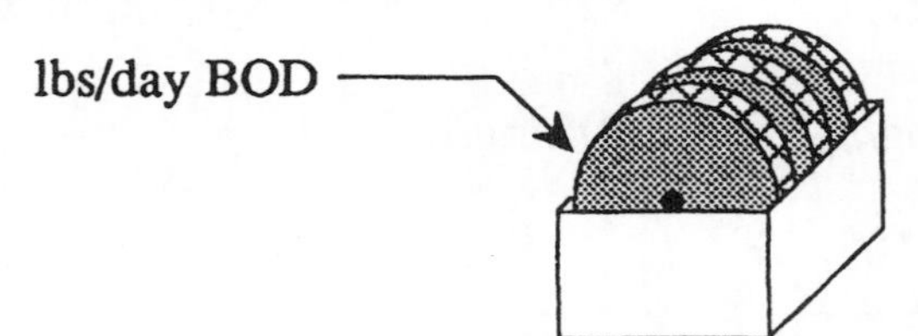

$$\text{Organic Loading Rate} = \frac{\text{Sol. BOD, lbs/day}}{\text{Media Area, 1000-sq ft}}$$

$$= \frac{(170\text{ mg/}L)\ (1.2\text{ MGD})\ (8.34\text{ lbs/gal})}{500\quad\text{1000-sq ft}}$$

$$= \boxed{\frac{\text{3.4 lbs/day Sol. BOD}}{\text{1000 sq ft}}}$$

* For a review of milligrams per liter (mg/*L*) to pounds per day (lbs/day) BOD, refer to Chapter 3.

Example 2: (Organic Loading Rate)

❑ The wastewater flow to an RBC is 3,110,000 gpd. The wastewater has a soluble BOD concentration of 122 mg/*L*. The RBC consists of six shafts (each 100,000 sq ft), with two shafts comprising the first stage of the system. What is the organic loading rate lbs/day/1000 sq ft on the first stage of the system?

$$\text{Organic Loading Rate} = \frac{\text{Sol. BOD, lbs/day}}{\text{Media Area, 1000 sq ft}}$$

$$= \frac{(122 \text{ mg}/L)\ (3.11 \text{ MGD})\ (8.34 \text{ lbs/gal})}{200 \quad 1000\text{-sq ft}}$$

$$= \boxed{15.8 \text{ lbs Sol. BOD/day/1000 sq ft}}$$

SYSTEM VS. SINGLE STAGE ORGANIC LOADING RATE

Typically, an RBC process includes several shafts of rotating media. These shafts are often grouped as "stages". For example, a system comprised of six shafts may have a first stage which includes two shafts. The organic loading rate may be calculated for the entire RBC system or for a single stage.

In Example 1, the organic loading rate was calculated for the entire system. To calculate the organic loading rate for a single stage, include only the sq ft media for that stage. Example 2 illustrates this calculation.

Example 3: (Organic Loading Rate)

❑ A rotating biological contactor (RBC) receives a flow of 2.4 MGD. The BOD of the influent wastewater to the RBC is 170 mg/*L*, and the surface area of the media is 700,000 sq ft. If the suspended solids concentration of the wastewater is 130 mg/*L* and the K-value is 0.5, what is the organic loading rate lbs/day/1000 sq ft?

First calculate mg/*L* soluble BOD:

$$\underset{\text{mg}/L}{\text{Total BOD}} = \underset{\text{mg}/L}{\text{Particulate BOD}} + \underset{\text{mg}/L}{\text{Soluble BOD}}$$

$$\underset{\text{BOD}}{170 \text{ mg}/L} = \underset{\text{SS}}{(130 \text{ mg}/L)\ (0.5)} + \underset{\text{Sol. BOD}}{x \text{ mg}/L}$$

$$170 \text{ mg}/L = 65 \text{ mg}/L + \underset{\text{Sol. BOD}}{x \text{ mg}/L}$$

$$170 - 65 = x$$

$$\boxed{\underset{\text{Sol. BOD}}{105 \text{ mg}/L}} = x$$

Then calculate organic loading rate:

$$\text{Organic Loading Rate} = \frac{\text{Sol. BOD, lbs/day}}{\text{Media Area, 1000 sq ft}}$$

$$= \frac{(105 \text{ mg}/L)\ (2.4 \text{ MGD})\ (8.34 \text{ lbs/gal})}{700 \quad 1000\text{-sq ft}}$$

$$= \boxed{\frac{3 \text{ lbs Sol. BOD/day}}{1000 \text{ sq ft}}}$$

NOTES:

12 Activated Sludge

SUMMARY

1. Volume—Aeration Tanks

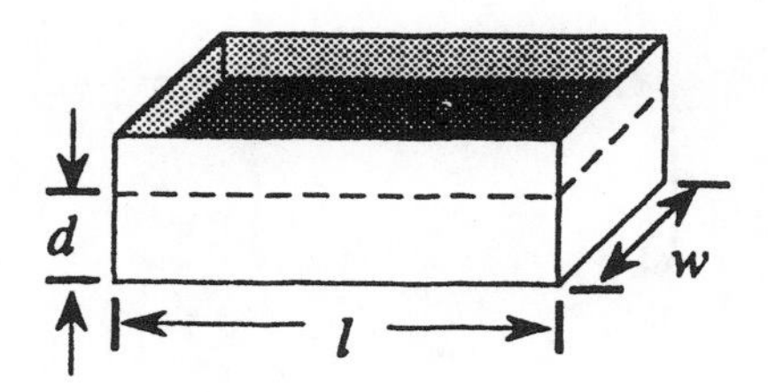

Volume, cu ft = (length, ft) (width, ft) (depth, ft)

$$V = lwd$$

Volume—Circular Clarifiers

Circular clarifier volume is normally calculated using the side wall depth, SWD. (The cone-shaped bottom of the tank is not included.)

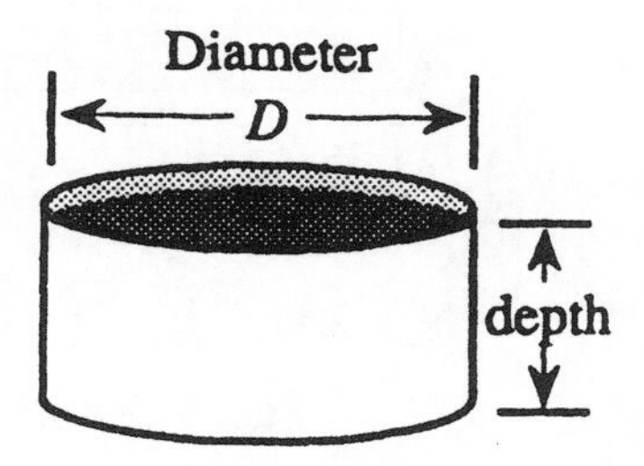

Volume, cu ft = (Area of Circle, sq ft) (depth, ft)

$$V = (0.785)(D^2)(\text{depth})$$

SUMMARY—Cont'd

Volume—Circular Clarifiers (Cont'd)

To include the volume of the cone, the following equation must be used:

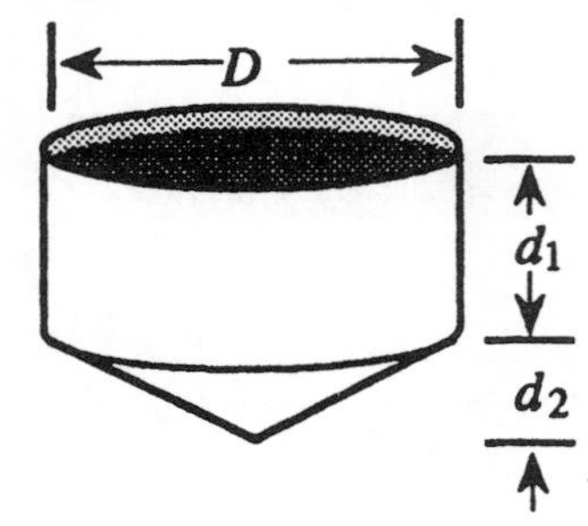

$$\text{Total Volume, cu ft} = \text{Volume of Cylinder, cu ft} + \text{Volume of Cone, cu ft}$$

$$V = (0.785)(D^2)(\text{depth}_1) + \frac{(0.785)(D^2)(\text{depth}_2)}{3}$$

Volume—Oxidation Ditch

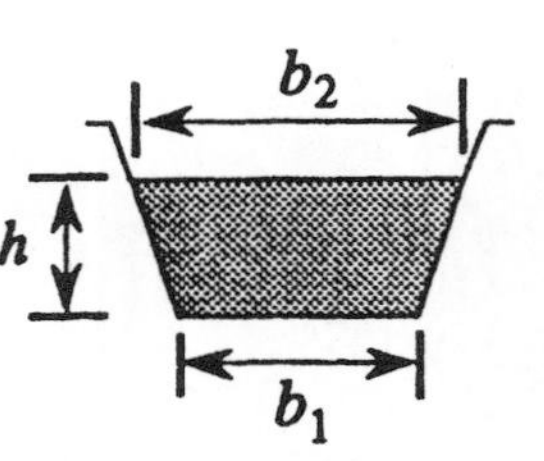

Cross Section of Ditch

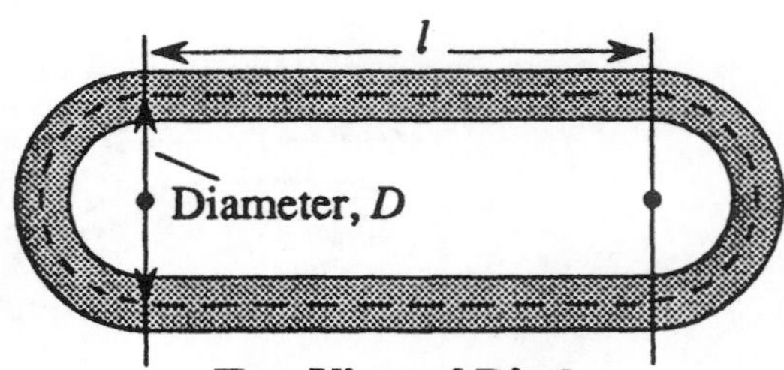

Top View of Ditch

Dashed line represents total ditch length (L). This is equal to 2 half circumferences + 2 lengths.

$$\text{Volume cu ft} = (\text{Trapezoidal Area})(\text{Total Length})$$

$$= \left[\frac{(b_1+b_2)(h)}{2}\right]\left[\begin{pmatrix}\text{Length of}\\ \text{2 Sides}\end{pmatrix} + \begin{pmatrix}\text{Length Around}\\ \text{2 Half Circles}\end{pmatrix}\right]$$

$$V = \frac{(b_1+b_2)(h)(2l + \pi D)}{2}$$

SUMMARY—Cont'd

2. BOD or COD Loading, lbs/day

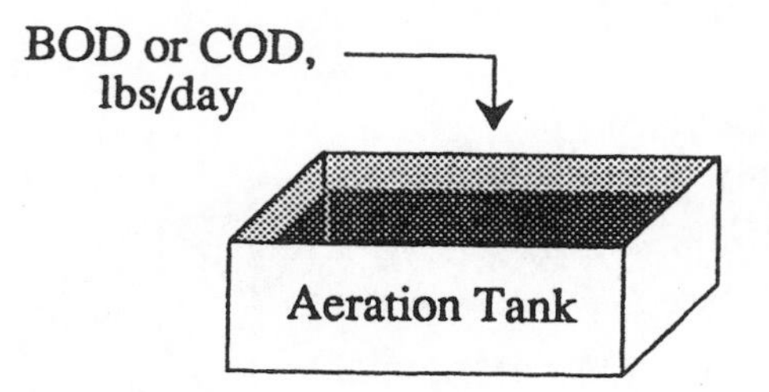

$$\underset{\text{lbs/day}}{\text{BOD Loading}} = \underset{\text{BOD}}{(\text{mg/}L)}\ \underset{\text{flow}}{(\text{MGD})}\ \underset{\text{lbs/gal}}{(8.34)}$$

$$\underset{\text{lbs/day}}{\text{COD Loading}} = \underset{\text{COD}}{(\text{mg/}L)}\ \underset{\text{flow}}{(\text{MGD})}\ \underset{\text{lbs/gal}}{(8.34)}$$

3. Solids Inventory in the Aeration Tank

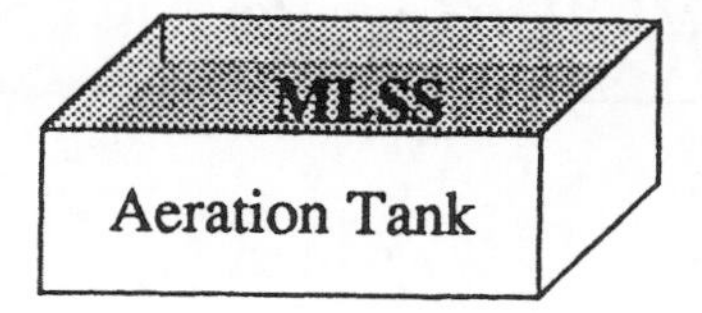

To determine the lbs MLSS in the aeration tank:

$$\underset{\text{MLSS}}{(\text{mg/}L)}\ \underset{\text{MG}}{(\text{Aer. Vol.})}\ \underset{\text{lbs/gal}}{(8.34)} = \text{lbs MLSS}$$

To determine the lbs MLVSS in the aeration tank:

$$\underset{\text{MLSS}}{(\text{mg/}L)}\ \underset{\text{MG}}{(\text{Aer. Vol.})}\ \underset{\text{lbs/gal}}{(8.34)}\ \frac{(\%\ \text{Volatile Sol.})}{100} = \text{lbs MLVSS}$$

SUMMARY—Cont'd

4. Food/Microorganism Ratio

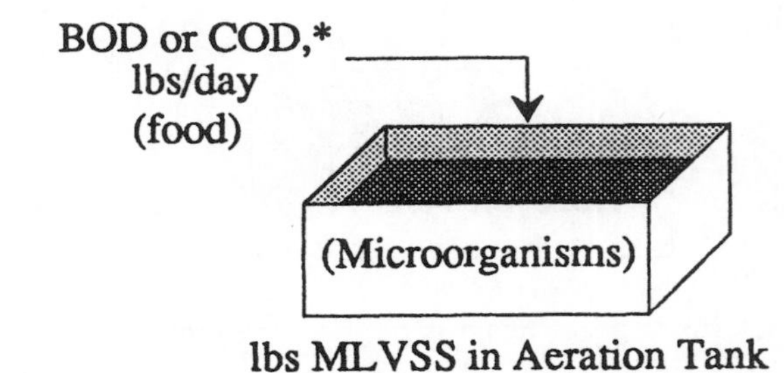

lbs MLVSS in Aeration Tank

Simplified Equation:

$$F/M = \frac{\text{BOD, lbs/day}}{\text{MLVSS, lbs}}$$

Expanded Equation:

$$F/M = \frac{(\text{mg}/L\ \text{BOD})\ (\text{MGD Flow})\ (8.34\ \text{lbs/gal})}{(\text{mg}/L\ \text{MLVSS})\ (\text{Aer Vol, MG})\ (8.34\ \text{lbs/gal})}$$

* COD may be used if there is generally a good correlation in BOD and COD characteristics of the wastewater.

SUMMARY—Cont'd

5. Sludge Age (Gould)

SLUDGE AGE IS BASED ON SUSPENDED SOLIDS
ENTERING THE AERATION TANK

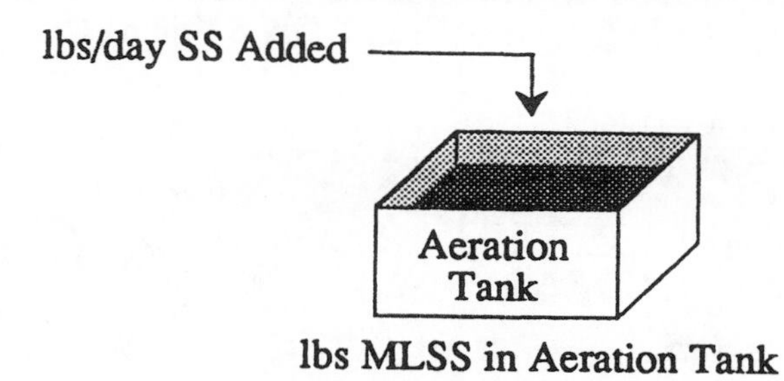

Simplified Equation:

$$\text{Sludge Age, days} = \frac{\text{MLSS, lbs}}{\text{SS Added, lbs/day}}$$

Expanded Equation:

$$\text{Sludge Age, days} = \frac{(\text{MLSS mg}/L)\ (\text{Aer. Vol., MG})\ (8.34\ \text{lbs/gal})}{(\text{Prim. Eff. SS, mg}/L)\ (\text{MGD Flow})\ (8.34\ \text{lbs/gal})}$$

SUMMARY—Cont'd

6. Solids Retention Time (SRT)
(also called Mean Cell Residence Time, MCRT)

SOLIDS RETENTION TIME IS BASED ON SUSPENDED SOLIDS LEAVING THE AERATION SYSTEM

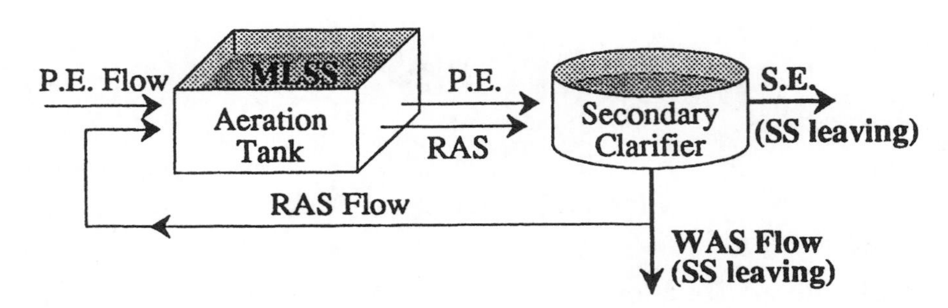

Simplified Equation:

$$\text{Solids Retention Time, days} = \frac{\text{Suspended Solids in System, lbs}}{\text{Suspended Solids Leaving System, lbs/day}}$$

Or

$$\text{Solids Retention Time, days} = \frac{\text{Suspended Solids in System, lbs}}{\text{WAS SS, lbs/day} + \text{S.E. SS, lbs/day}}$$

Expanded Equation:*

$$\text{SRT, days} = \frac{\underset{\text{mg/L}}{(\text{MLSS})}\ \underset{\text{MG}}{(\text{Aer. Vol.})}\ \underset{\text{lbs/gal}}{(8.34)} + \underset{\text{mg/L}}{(\text{CCSS})}\ \underset{\text{MG}}{(\text{Fin. Clar. Vol.})}\ \underset{\text{lbs/gal}}{(8.34)}}{\underset{\text{mg/L}}{(\text{WAS SS})}\ \underset{\text{MGD}}{(\text{WAS Flow})}\ \underset{\text{lbs/gal}}{(8.34)} + \underset{\text{mg/L}}{(\text{S.E. SS})}\ \underset{\text{MGD}}{(\text{Plant Flow})}\ \underset{\text{lbs/gal}}{(8.34)}}$$

Note: There are four ways to account for system solids (represented by the numerator of the SRT equation). The most accurate calculation of system solids is given in the SRT equation above. The other three methods are described in Section 12.6 of this chapter.

* CCSS is the average clarifier core SS concentration of the entire water column sampled by a core sampler.

SUMMARY—Cont'd

7. Return Sludge Rate —Using Settleability

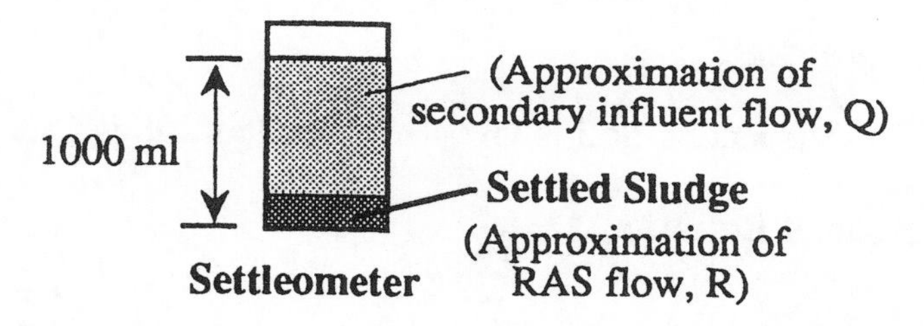

The ratio of RAS flow to secondary influent flow can be estimated using the following equation:

$$\frac{R}{Q} = \frac{\text{Settled Sludge Volume, ml/L}}{1000\text{ ml/L} - \text{Settled Sludge Volume, ml/L}}$$

The RAS ratio can then be used to calculate RAS flow rate in MGD:

$$\frac{R}{Q} = \frac{\text{RAS Flow Rate, MGD}}{\text{Secondary Influent Flow Rate, MGD}}$$

Return Sludge Rate — Using Secondary Clarifier Solids Balance

Simplified Equation:

Suspended Solids In = Suspended Solids Out

Expanded Equation:

Suspended Solids In, lbs/day = Suspended Solids Out, lbs/day*

$$\left[(\text{MLSS})(Q + R)\right]\ (8.34\text{ lbs/gal}) = \left[(\text{RAS SS})(R) + (\text{WAS SS})(W)\right]\ (8.34\text{ lbs/gal})$$

- $(\text{MLSS})(Q+R)$: *Solids entering from the aeration tank*
- $(\text{RAS SS})(R)$: *Solids leaving via the RAS flow*
- $(\text{WAS SS})(W)$: *Solids leaving via the WAS flow*

Where:

MLSS = mixed liquor suspended solids, mg/L
Q = secondary influent flow, MGD
R = return sludge flow, MGD
RAS SS = return activated sludge SS, mg/L
WAS SS = waste activated sludge SS, mg/L
W = waste sludge flow, MGD
R = return sludge flow, MGD

SUMMARY—Cont'd

Return Sludge Rate — Using Aeration Tank Solids Balance

Simplified Equation:

Suspended Solids In = Suspended Solids Out *

Expanded Equation:**

$$\underbrace{\left[(\text{RAS SS})\,(\text{R})\right](8.34)}_{\text{Suspended Solids In, lbs/day}} = \underbrace{\left[(\text{MLSS})\,(\text{Q} + \text{R})\right](8.34)}_{\text{Suspended Solids Out, lbs/day}}$$

(8.34 = lbs/gal)

The equation may be rearranged so that both R terms are on the same side of the equation. (*Note that since the 8.34 factor is on both sides of the equation, it can be divided out.*)

8. Wasting Rate—Using Constant F/M Ratio

Use the **desired F/M ratio** and BOD or COD applied (food) to calculate the **desired lbs MLVSS**:

$$\text{Desired F/M} = \frac{(\text{BOD, mg/L})\,(\text{MGD flow})\,(8.34\text{ lbs/gal})}{\text{Desired lbs MLVSS}}$$

Then determine the **desired lbs MLSS** using % volatile solids:

$$\text{Desired MLSS, lbs} = \frac{\text{Desired lbs MLVSS}}{\frac{\%\text{ VS}}{100}}$$

Compare the actual and desired MLSS to determine lbs SS to be wasted:

$$\text{Actual lbs MLSS} - \text{Desired lbs MLSS} = \text{lbs SS to be Wasted}$$

* For the aeration tank, this is true only when new cell growth in the tank is considered negligible.

** Abbreviation of terms is the same as that given for the secondary clarifier solids balance equation.

SUMMARY—Cont'd

Wasting Rate—Using Constant Sludge Age

Use the **desired sludge age** and the suspended solids entering to calculate the **desired lbs MLSS**:

$$\text{Desired Sludge Age} = \frac{\text{Desired lbs MLSS}}{(\text{Prim. Eff. SS, mg/}L)(\text{Flow, MGD})(8.34\text{ lbs/gal})}$$

Then **calculate the actual lbs MLSS** :

$$(\text{MLSS, mg/}L)(\text{Aer. Vol., MG})(8.34\text{ lbs/gal}) = \text{lbs MLSS}$$

Compare the desired and actual lbs MLSS to determine lbs SS to be wasted:

$$\text{Actual lbs MLSS} - \text{Desired lbs MLSS} = \text{lbs SS to be Wasted}$$

Wasting Rate (lbs/day)—Using Constant SRT

Use the SRT equation to determine wasting rate. The WAS SS lbs/day are the lbs/day SS to be wasted:

Simplified Equation:

$$\text{SRT, days} = \frac{\text{Suspended Solids in System, lbs}}{\text{WAS SS, lbs/day} + \text{S.E. SS, lbs/day}}$$

Expanded Equation:*

$$\text{SRT, days} = \frac{(\text{MLSS, mg/L})(\text{Aer. Vol., MG})(8.34\text{ lbs/gal}) + (\text{CCSS, mg/L})(\text{Fin. Clar. Vol., MG})(8.34\text{ lbs/gal})}{\text{WAS SS, lbs/day} + (\text{S.E. SS, mg/L})(\text{MGD plant flow})(8.34\text{ lbs/gal})}$$

* CCSS is the average clarifier core SS concentration of the entire water column sampled by a core sampler.

SUMMARY—Cont'd

9. WAS Pumping Rate — Using the mg/L to lb/day equation

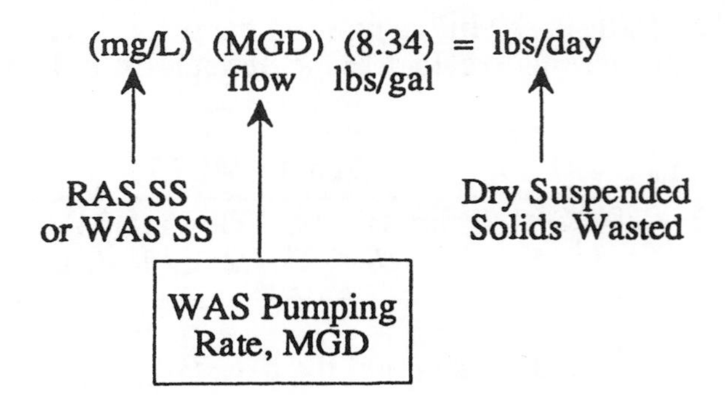

When the WAS pumping rate has been calculated in MGD, it can then be easily converted to gpm pumping rate:

$$\underset{\text{rate, MGD}}{(\text{Pumping})}\ (694\ \text{gpm/MGD}) = \underset{\text{rate, gpm}}{\text{Pumping}}$$

If the WAS pumping rate in MGD is first written as gpd, it may be converted to gpm as follows:

$$\frac{\text{Pumping rate, gpd}}{1440\ \text{min/day}} = \text{Pumping rate, gpm}$$

WAS Pumping Rate — Using the SRT Equation*

$$\text{SRT} = \frac{\underset{\text{mg/L}}{(\text{MLSS})}\ \underset{\text{MG}}{(\text{Aer. Vol.})}\ \underset{\text{lbs/gal}}{(8.34)} + \underset{\text{mg/L}}{(\text{CCSS})}\ \underset{\text{MG}}{(\text{Fin. Clar. Vol.})}\ \underset{\text{lbs/gal}}{(8.34)}}{\underset{\text{mg/L}}{(\text{WAS SS})}\ \underset{\text{MGD}}{(\text{WAS Flow})}\ \underset{\text{lbs/gal}}{(8.34)} + \underset{\text{mg/L}}{(\text{S.E. SS})}\ \underset{\text{MGD}}{(\text{Plant Flow})}\ \underset{\text{lbs/gal}}{(8.34)}}$$

↑ WAS Pumping Rate

* CCSS is the average clarifier core SS concentration of the entire water column sampled by a core sampler.

SUMMARY—Cont'd

10. Oxidation Ditch Detention Time

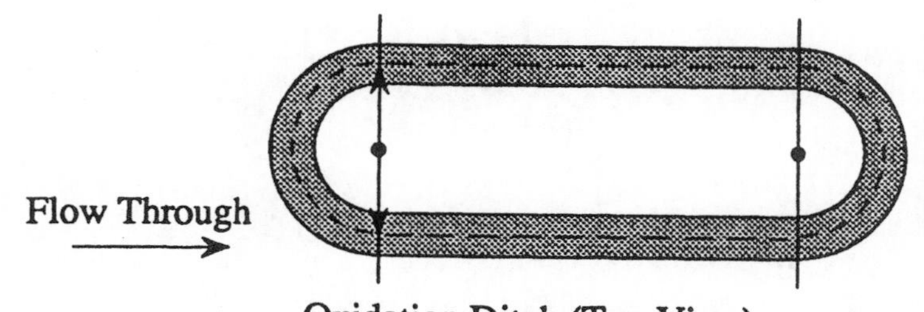

Oxidation Ditch (Top View)

$$\text{Detention Time, hrs} = \frac{\text{Volume, gal}}{\text{Flow, gal/hr}}$$

NOTES:

12.1 TANK VOLUME

The two common tank shapes in water and wastewater treatment are rectangular and cylindrical.

Rectangular Tank:

$$\underset{\text{cu ft}}{\text{Vol.}} = \underset{\text{ft}}{(\text{length})}\ \underset{\text{ft}}{(\text{width})}\ \underset{\text{ft}}{(\text{depth})}$$

Cylindrical Tank:

$$\underset{\text{cu ft}}{\text{Vol.}} = (0.785)\ \underset{\text{ft}^2}{(D^2)}\ \underset{\text{ft}}{(\text{depth})}$$

The volume of these tanks can be expressed as cubic feet or gallons. The equations shown above are for cubic feet volume.

The volume may also be expressed as gallons by multiplying cubic feet volume by 7.48 gal/cu ft, as illustrated in Example 2. You may wish to include the 7.48 gal/cu ft factor in the volume equation, as shown in Example 3.

Example 1: (Tank Volume)

❑ The dimensions of a tank are given below. Calculate the cubic feet volume of the tank.

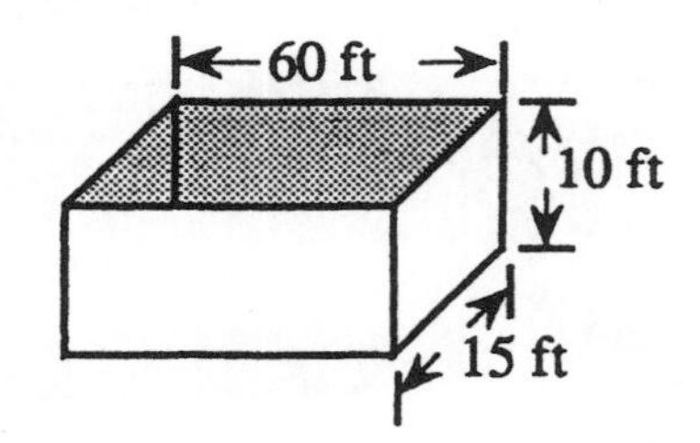

$$\underset{\text{cu ft}}{\text{Volume}} = (lw)\ (\text{depth})$$

$$= (60\text{ ft})\ (15\text{ ft})\ (10\text{ ft})$$

$$= \boxed{9000\text{ cu ft}}$$

Example 2: (Tank Volume)

❑ A tank is 25 ft wide, 75 ft long, and can hold water to a depth of 10 ft. What is the volume of the tank, in gallons?

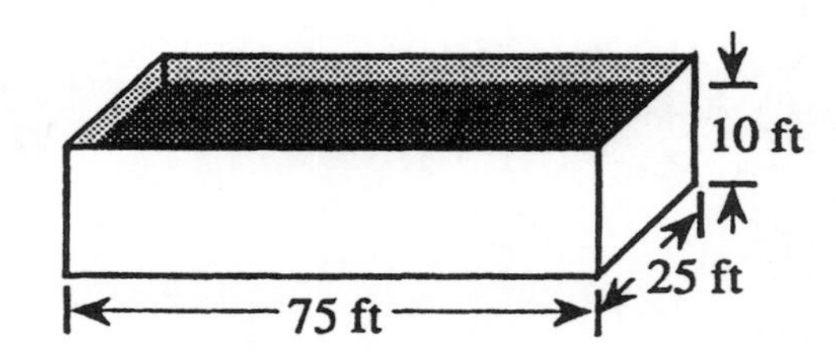

$$\text{Volume} = (lw)\ (\text{depth})$$

$$= (75\text{ ft})\ (25\text{ ft})\ (10\text{ ft})$$

$$= 18{,}750\text{ cu ft}$$

Now convert cu ft capacity to gal capacity:

$$(18{,}750\text{ cu ft})\ (7.48\text{ gal/cu ft}) = \boxed{140{,}250\text{ gal}}$$

Example 3: (Tank Volume)

❑ The diameter of a tank is 80 ft and the maximum water depth is 12 ft. What is the gallon volume of that tank?

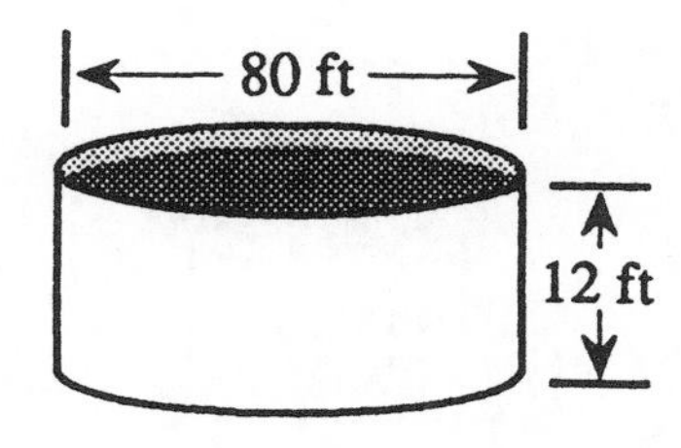

$$\text{Volume, gal} = (0.785)\,(D^2)\,(\text{depth})\,(7.48 \text{ gal/cu ft})$$

$$= (0.785)\,(80 \text{ ft})\,(80 \text{ ft})\,(12 \text{ ft})\,(7.48 \text{ gal/cu ft})$$

$$= \boxed{450{,}954 \text{ gal}}$$

The volume of a circular clarifier is normally calculated using average depth. (The cone-shaped bottom of the tank is not included.)

To include the cone-shaped bottom in the volume calculation, the following equation would be used:*

$$V = \text{Volume of Cylinder} + \text{Volume of Cone}$$

$$V = (0.785)\,(D^2)(\text{depth}_1) + \frac{(0.785)\,(D^2)(\text{depth}_2)}{3}$$

Example 3: (Volume Calculations)

❑ Calculate the cu ft volume of the oxidation ditch shown below. The cross section of the ditch is trapezoidal.

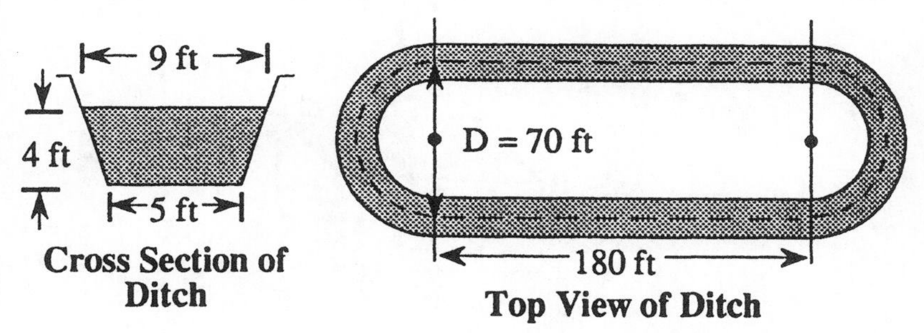

Cross Section of Ditch **Top View of Ditch**

$$\text{Total Volume} = (\text{Trapezoidal Area})\,(\text{Total Length})$$

$$= \left[\frac{(b_1 + b_2)\,(h)}{2}\right]\left[\begin{matrix}\text{(Length of)}\\ \text{2 Sides}\end{matrix} + \begin{matrix}\text{(Length Around)}\\ \text{2 Half Circles}\end{matrix}\right]$$

$$= \left[\frac{(5 \text{ ft} + 9 \text{ ft})\,(4 \text{ ft})}{2}\right]\left[360 \text{ ft} + (3.14)\,(70 \text{ ft})\right]$$

$$= \left[(7 \text{ ft})\,(4 \text{ ft})\right]\left[579.8 \text{ ft}\right]$$

$$= \boxed{16{,}234 \text{ cu ft}}$$

OXIDATION DITCH CAPACITIES

Normally oxidation ditches and ponds are trapezoidal in cross section. Example 3 illustrates one such calculation.

In Example 3 the oxidation ditch has sloping sides (trapezoidal cross section). The total volume of the oxidation ditch is the trapezoidal area times the total length, as shown in the equation to the left. (The total length is measured at the center of the ditch. Note that the length around two half circles is the circumference of one circle.)

* For a review of this type volume calculation, refer to Chapter 11 in *Basic Math Concepts*.

12.2 BOD OR COD LOADING

When calculating BOD), COD, or SS loading on a treatment system, the following equation is used:

(mg/L)	(MGD)	(8.34) = lbs/day
Conc.	flow	lbs/gal

Loading on a system is usually calculated as lbs/day. Given the BOD, COD, or SS mg/L concentration and flow information, the lbs/day loading may be calculated as demonstrated in Examples 1-4.

BOD, COD AND SS LOADING

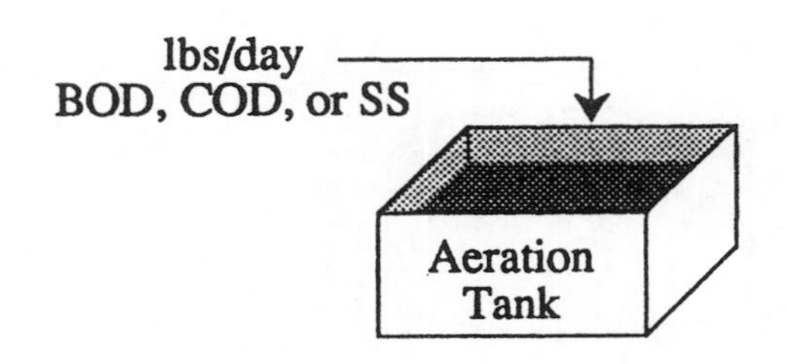

BOD, COD, or SS Loading, lbs/day = (mg/L) BOD COD or SS (MGD) Flow (8.34) lbs/gal

Example 1: (Loading Calculations)

❑ The BOD concentration of the wastewater entering an aerator is 215 mg/L. If the flow to the aerator is 1,440,000 gpd, what is the lbs/day BOD loading?

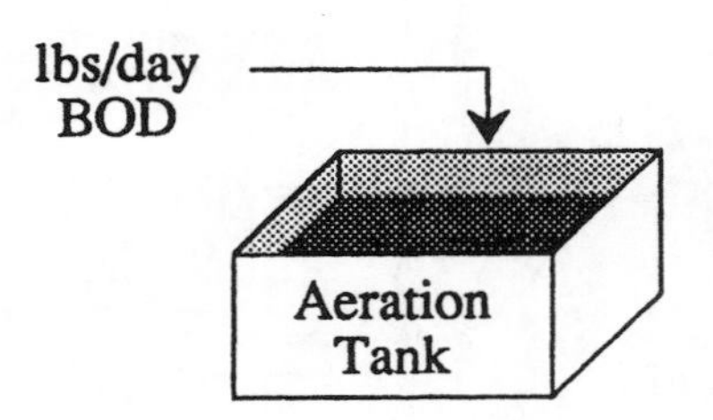

lbs/day BOD = (mg/L) BOD (MGD) Flow (8.34) lbs/gal

= (215 mg/L) (1.44 MGD) (8.34 lbs/gal)

= **2582 lbs/day BOD**

Example 2: (Loading Calculations)

❑ The flow to an aeration tank is 2850 gpm. If the BOD concentration of the wastewater is 130 mg/*L*, how many pounds of BOD are applied to the aeration tank daily?

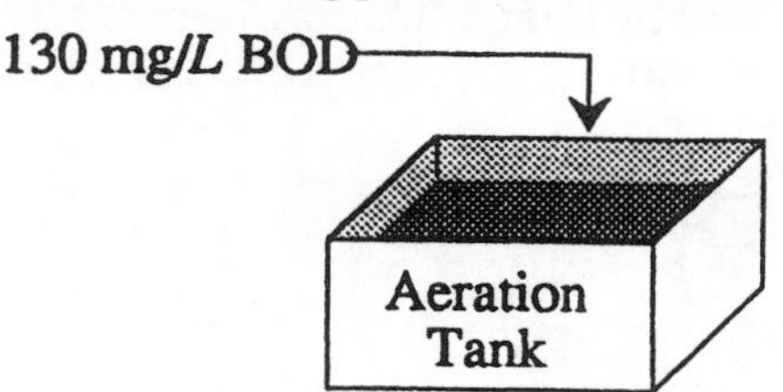

First convert the gpm flow to gpd flow:

(2850 gal/min) (1440 min/day) = 4,104,000 gpd

Then calculate lbs/day BOD:

(mg/*L* /BOD) (MGD flow) (8.34 lbs/gal) = lbs/day

(130 mg/*L* /BOD) (4.104 MGD) (8.34 lbs/gal) = **4450 lbs/day BOD**

Sometimes flow information is not given in the desired terms (MGD flow). When this is the case, convert the given flow rate (such as gpd, gpm, or cfs) to MGD flow.*

Example 3: (Loading Calculations)

❑ The flow to an aeration tank is 3400 gpm. If the COD concentration of the wastewater is 120 mg/*L*, how many pounds of COD are applied to the aeration tank daily?

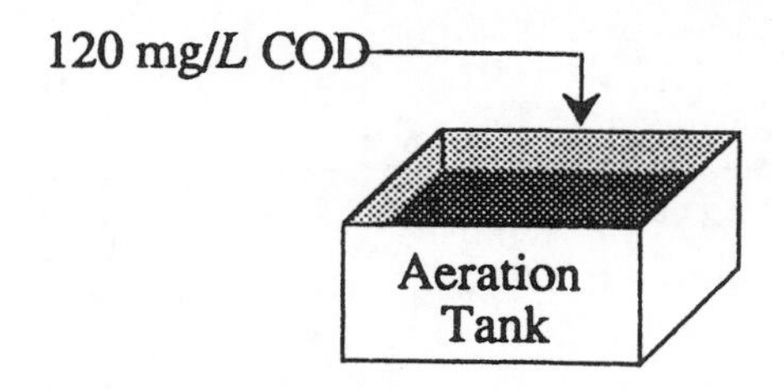

Before the mg/*L* to lbs/day equation can be used the gpm flow must be expressed as MGD flow:

(3400 gal/min) (1440 min/day) = 4,896,000 gpd

or = 4.9 MGD

Now use the mg/*L* equation to calculate lbs/day COD:

(mg/*L* COD) (MGD flow) (8.34 lbs/gal) = lbs/day

(120 mg/*L* COD) (4.9 MGD) (8.34 lbs/gal) = **4904 lbs/day COD**

* Refer to Chapter 8 in *Basic Math Concepts* for a review of flow conversions.

12.3 SOLIDS INVENTORY IN THE AERATION TANK

In any activated sludge system it is important to control the amount of solids under aeration. The suspended solids in an aeration tank are called Mixed Liquor Suspended Solids (MLSS). To calculate the pounds of solids in the aeration tank, you will need to know the mg/*L* MLSS concentration and the aeration tank volume. Then lbs MLSS can be calculated as follows:*

(mg/*L*) MLSS	(MG) Vol	(8.34) lbs/gal	= lbs MLSS

Example 1: (Solids Inventory in Aeration Tank)

❑ If the mixed liquor suspended solids concentration is 1100 mg/*L*, and the aeration tank has a volume of 525,000 gallons, how many pounds of suspended solids are in the aeration tank?

Aeration Tank
1100 mg/*L*
MLSS

Vol = 0.525 MG

(mg/*L*) (MG Vol.) (8.34 lbs/gal) = lbs

(1100 mg/*L*) (0.525 MG) (8.34 lbs/gal) = **4816 lbs MLSS**

Another important measure of solids in the aeration tank is the amount of volatile suspended solids.** The volatile solids content of the aeration tank is used as an estimate of the microorganism population in the aeration tank . The Mixed Liquor Volatile Suspended Solids (MLVSS) usually comprises about 70% of the MLSS. The other 30% of the MLSS are fixed (inorganic) solids. To calculate the lbs MLVSS, use the following equation:

(mg/*L*) MLVSS	(MG) Vol	(8.34) lbs/gal	= lbs MLVSS

Example 2: (Solids Inventory in Aeration Tank)

❑ The volume of an aeration tank is 175,000 gallons. If the MLVSS concentration is 3220 mg/*L*, how many pounds of volatile solids are under aeration?

Aeration Tank
3220 mg/*L*
MLVSS

Vol = 175,000 gal
or = 0.175 MG

(3220 mg/*L*) (0.175 MG Vol.) (8.34 lbs/gal) = **4700 lbs MLVSS**

---

* To review mg/*L* to lbs conversions, refer to Chapter 3.

** For a discussion of volatile suspended solids calculations, refer to Chapter 6, "Efficiency and Other Percent Calculations".

Example 3: (Solids Inventory in Aeration Tank)

❑ The aeration tank of a conventional activated sludge plant has a mixed liquor volatile suspended solids concentration of 2980 mg/*L*. If the aeration tank is 100 ft long, 40 wide and has wastewater to a depth of 12 ft, how many pounds of MLVSS are under aeration?

Aeration Tank 2300 mg/*L* MLVSS

Vol = (100 ft) (40 ft) (12 ft) (7.48 gal/cu ft)

= 359, 040 gal

or = 0.36 MG

Now calculate lbs MLVSS using the usual equation and fill in the given information:

(2980 mg/*L*) (0.36 MG) (8.34 lbs/gal) = **8947 lbs MLVSS**

Example 4: (Solids Inventory in Aeration Tank)

❑ The aeration tank of a conventional activated sludge plant has a mixed liquor suspended solids concentration of 2833 mg/*L*. with a volatile solids content of 72%. The aeration tank is 85 ft long, 35 wide and has wastewater to a depth of 15 ft. How many pounds of MLVSS are under aeration?

Aeration Tank 2040 mg/*L* MLVSS

Vol = (85 ft) (35 ft) (15 ft) (7.48 gal/cu ft)

= 333,795 gal

or = 0.33 MG

The lbs MLVSS can be calculated as follows:

(2833 mg/*L*) (0.33 MG) (8.34 lbs/gal) (0.72) = **5614 lbs MLVSS**

12.4 FOOD/MICROORGANISM RATIO

In order for the activated sludge process to operate properly, there must be a balance between food entering the system (as measured by BOD or COD) and micro-organisms in the aeration tank (as measured by the MLVSS). The best F/M ratio for a particular system depends on several factors including the type of activated sludge process and the characteristics of the wastewater entering the system.

COD is sometimes used as the measure of food entering the system. COD may be used if there is generally a good correlation in BOD and COD characteristics of the wastewater. The COD test can be completed in only a few hours compared with 5 days for a BOD test.

Note that the F/M equation is given in two forms—the simplified equation and the expanded equation. If BOD or COD data is given in lbs/day and MLVSS data is given in lbs, then the simplified equation should be used. In most instances, however, BOD, COD, and MLVSS data will be given as mg/*L* and the expanded equation will be required.

FOOD SUPPLY (BOD OR COD)
AND MICROORGANISMS (MLVSS)
MUST BE IN BALANCE

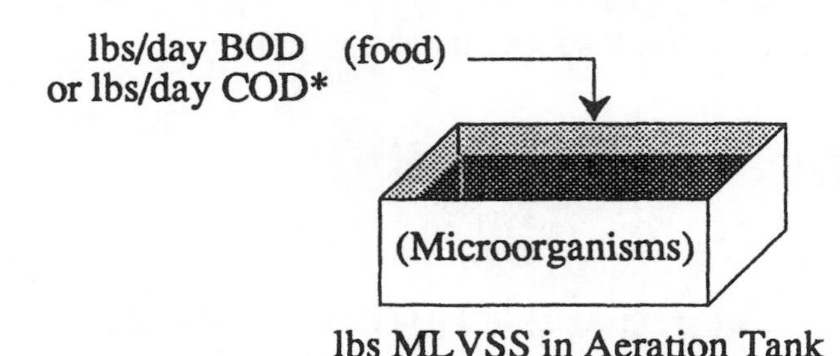

lbs MLVSS in Aeration Tank

Simplified Equation:

$$F/M = \frac{\text{BOD, lbs/day}}{\text{MLVSS, lbs}}$$

Expanded Equation:

$$F/M = \frac{(\text{mg}/L \text{ BOD})(\text{MGD Flow})(8.34 \text{ lbs/gal})}{(\text{mg}/L \text{ MLVSS})(\text{Aer Vol, MG})(8.34 \text{ lbs/gal})}$$

Example 1: (F/M Ratio)

❑ An activated sludge aeration tank receives a primary effluent flow of 2.42 MGD with a BOD of 170 mg/*L*. The mixed liquor volatile suspended solids is 1980 mg/*L* and the aeration tank volume is 350,000 gallons. What is the current F/M ratio?

Since BOD and MLVSS data are given as mg/*L*, the expanded equation will be used. Note that the 8.34 factor can be divided out of the numerator and denominator:

$$F/M = \frac{(\text{mg}/L \text{ BOD})(\text{MGD})(8.34 \text{ lbs/gal})}{(\text{mg}/L \text{ MLVSS})(\text{Aer Vol, MG})(8.34 \text{ lbs/gal})}$$

$$F/M = \frac{(170 \text{ mg}/L)(2.42 \text{ MGD})(\cancel{8.34} \text{ lbs/gal})}{(1980 \text{ mg}/L)(0.35 \text{ MG})(\cancel{8.34} \text{ lbs/gal})}$$

$$F/M = \boxed{0.59}$$

Example 2: (F/M Ratio)

❑ The flow to a 0.18-MG oxidation ditch is 300,000 gpd. The BOD concentration of the wastewater is 210 mg/*L*. If the mixed liquor suspended solids is 3800 mg/*L* with a volatile solids content of 68%, what is the F/M ratio?

First calculate the mg/*L* MLVSS:

$$\underset{\text{MLSS}}{(3800 \text{ mg/}L)}\ \underset{\text{Vol. Sol.}}{(0.68)} = \underset{\text{MLVSS}}{2584 \text{ mg/}L}$$

Then calculate the F/M ratio:

$$\text{F/M} = \frac{(\text{mg/}L \text{ BOD})(\text{MGD flow})(8.34 \text{ lbs/gal})}{(\text{mg/}L \text{ MLVSS})(\text{Ditch Vol, MG})(8.34 \text{ lbs/gal})}$$

$$= \frac{(210 \text{ mg/}L)(0.3 \text{ MGD})(\cancel{8.34} \text{ lbs/gal})}{(2584 \text{ mg/}L)(0.18 \text{ MGD})(\cancel{8.34} \text{ lbs/gal})}$$

$$= \boxed{0.14}$$

WHEN MLVSS IS NOT KNOWN

The denominator of the F/M calculation is the mixed liquor volatile suspended solids (MLVSS) concentration. Sometimes, however, you will not have MLVSS information.

The MLVSS concentration can be calculated if you know the MLSS concentration and, percent volatile solids content:

$$\frac{\underset{\text{mg/}L}{(\text{MLSS})}(\%\text{ Vol. Sol.})}{100} = \underset{\text{mg/}L}{\text{MLVSS}}$$

Example 3: (F/M Ratio)

❑ The desired F/M ratio at a particular activated sludge plant is 0.5 lbs COD/1 lb mixed liquor volatile suspended solids. If the 3.6 MGD primary effluent flow has a COD of 165 mg/*L* how many lbs of MLVSS should be maintained in the aeration tank?

$$\text{F/M} = \frac{\text{COD, lbs/day}}{\text{MLVSS, lbs}}$$

Fill in the equation with the known information. Since COD is given as mg/*L*, the lbs/day COD must be calculated using the expanded equation.*

$$0.5 = \frac{(165 \text{ mg/}L)(3.6 \text{ MGD})(8.34 \text{ lbs/gal})}{x \text{ lbs MLVSS}}$$

Then solve for the unknown value:**

$$x = \frac{(165)(3.6)(8.34)}{0.5}$$

$$= \boxed{9908 \text{ lbs MLVSS}}$$

CALCULATING DESIRED LBS MLVSS

The F/M ratio equation can be used to calculate desired pounds of MLVSS, given a desired F/M ratio. Use the same F/M equation, fill in the given information, then solve for the unknown value (desired lbs MLVSS), as illustrated in Example 3.

* For a review of mg/*L* to lbs/day calculations, refer to Chapter 3.

** To review solving for the unknown value, refer to Chapter 2 in *Basic Math Concepts*.

12.5 SLUDGE AGE (GOULD)

Sludge age refers to the average number of days a particle of suspended solids remains under aeration. It is a calculation used to maintain the proper amount of activated sludge in the aeration tank. This calculation is sometimes referred to as Gould Sludge Age so that it is not confused with similar calculations such as Solids Retention Time (or Mean Cell Residence Time).

When considering sludge age, in effect you are asking, "how many days of suspended solids are in the aeration tank?" For example, if 2000 lbs SS enter the aeration tank daily and the aeration tank contains 10,000 lbs of suspended solids, then 5 days of solids are in the aeration tank—a sludge age of 5 days.

Notice the similarity of this calculation to that of detention time—sludge age is **solids retained** calculated using units of lbs and lbs/day; detention time is **water retained**, using units of gal and gal/time or cu ft and cu ft/time.

$$\text{Sludge Age, days} = \frac{\text{SS in Tank, lbs}}{\text{SS Added, lbs/day}}$$

$$\text{Detention Time, min} = \frac{\text{Volume of Tank, gal}}{\text{Flow Rate, gpm}}$$

SLUDGE AGE IS BASED ON SUSPENDED SOLIDS ENTERING THE AERATION TANK

lbs/day SS Added

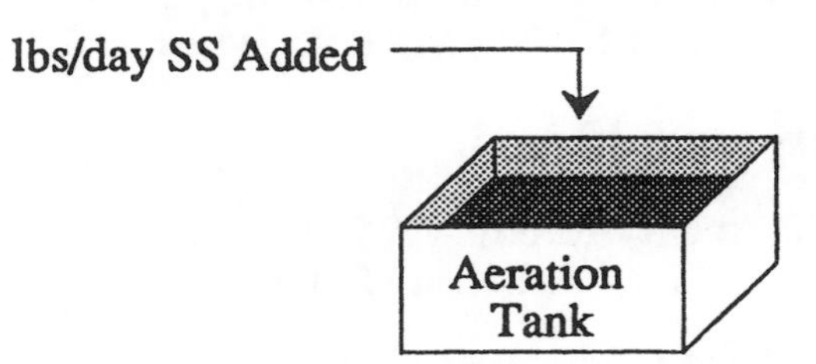

lbs MLSS in Aeration Tank

Simplified Equation:

$$\text{Sludge Age, days} = \frac{\text{MLSS, lbs}}{\text{SS Added, lbs/day}}$$

Expanded Equation:

$$\text{Sludge Age, days} = \frac{(\text{MLSS mg/}L)\ (\text{Aer. Vol., MG})\ (8.34\ \text{lbs/gal})}{(\text{P.E.* SS, mg/}L)\ (\text{Flow MGD})\ (8.34\ \text{lbs/gal})}$$

Example 1: (Sludge Age)

❑ A total of 2640 lbs/day suspended solids enter an aeration tank in the primary effluent flow. If the aeration tank has a total of 13,700 lbs of mixed liquor suspended solids, what is the sludge age in the aeration tank?

2640 lbs/day SS

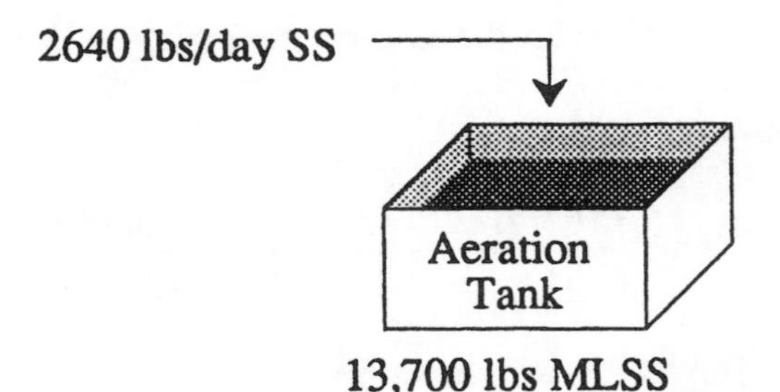

13,700 lbs MLSS

$$\text{Sludge Age, days} = \frac{\text{MLSS, lbs}}{\text{SS Added, lbs/day}}$$

$$= \frac{13{,}700\ \text{lbs}}{2640\ \text{lbs/day}}$$

$$= \boxed{5.2\ \text{days}}$$

* P.E. refers to primary effluent.

Example 2: (Sludge Age)

❑ An aeration tank contains 450,000 gallons of wastewater with a MLSS concentration of 2430 mg/*L*. If the primary effluent flow is 2.3 MGD with a suspended solids concentration of 105 mg/*L*, what is the sludge age?

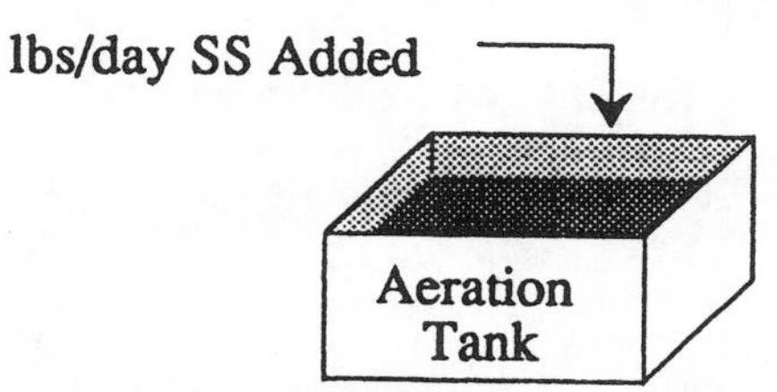

$$\text{Sludge Age, days} = \frac{\text{MLSS, lbs}}{\text{SS Added, lbs/day}}$$

$$= \frac{(2430\ \text{mg}/L)\ (0.45\ \text{MG})\ (8.34\ \text{lbs/gal})}{(105\ \text{mg}/L)\ (2.3\ \text{MGD})\ (8.34\ \text{lbs/gal})}$$

$$= \frac{9120\ \text{lbs MLSS}}{2014\ \text{lbs/day SS}}$$

$$= \boxed{4.5\ \text{days}}$$

Example 3: (Sludge Age)

❑ An aeration tank is 80 ft long, 25 ft wide with wastewater to a depth of 15 ft. The mixed liquor suspended solids concentration is 2640 mg/*L*. If the primary effluent flow is 1.73 MGD with a suspended solids concentration of 85 mg/*L*, what is the sludge age in the aeration tank?

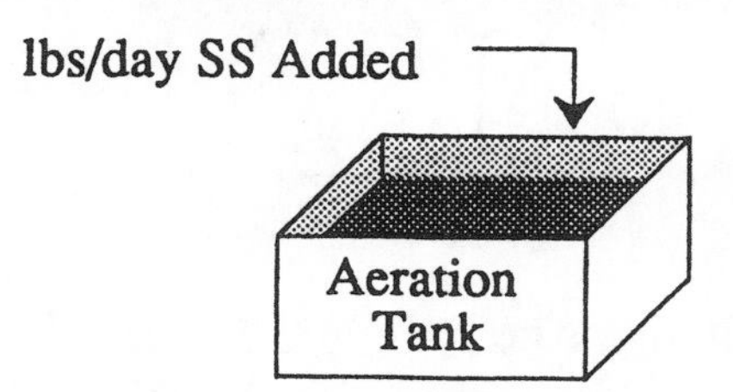

Aerator Volume

(80 ft) (25 ft) (15 ft) (7.48 gal/cu ft) = 224,400 gal

$$\text{Sludge Age, days} = \frac{(2640\ \text{mg}/L)\ (0.224\ \text{MG})\ (\cancel{8.34}\ \text{lbs/gal})}{(85\ \text{mg}/L)\ (1.73\ \text{MGD})\ (\cancel{8.34}\ \text{lbs/gal})}$$

$$= \boxed{4.0\ \text{days}}$$

Example 4: (Sludge Age)

❑ An oxidation ditch has a volume of 190,000 gallons. The 220,000-gpd flow to the oxidation ditch has a suspended solids concentration of 210 mg/L. If the MLSS concentration is 3900 mg/L, what is the sludge age in the oxidation ditch?

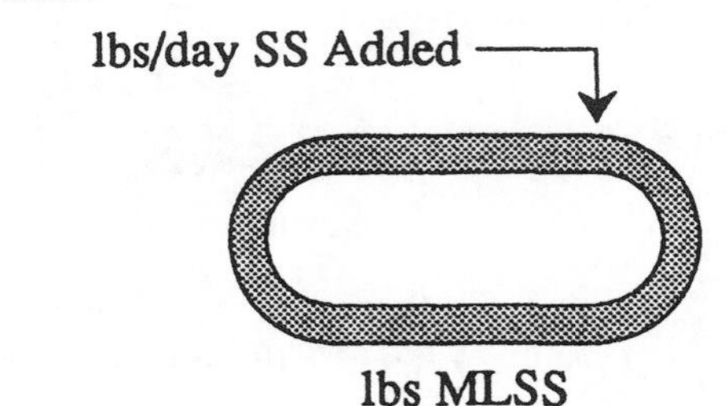

$$\text{Sludge Age, days} = \frac{\text{MLSS, lbs}}{\text{SS Added, lbs/day}}$$

$$= \frac{(3900 \text{ mg/L})(0.19 \text{ MG})(\cancel{8.34} \text{ lbs/gal})}{(210 \text{ mg/L})(0.22 \text{ MGD})(\cancel{8.34} \text{ lbs/gal})}$$

$$= \boxed{16 \text{ days}}$$

USING SLUDGE AGE TO CALCULATE DESIRED MLSS

The sludge age equation may be used to calculate the desired mg/L or pounds of MLSS to be maintained in the aeration tank.

In Examples 1-3, sludge age was the unknown variable. However, pounds MLSS (the denominator of the simplified sludge age equation) or mg/L MLSS (in the denominator of the expanded sludge age equation) can also be the unknown variable. Examples 4 - 7 illustrate this type of calculation.

Example 5: (Sludge Age)

❑ A sludge age of 5.5 days is desired. Assume 1500 lbs/day suspended solids enter the aeration tank in the primary effluent. To maintain the desired sludge age, how many lbs of MLSS must be maintained in the aeration tank?

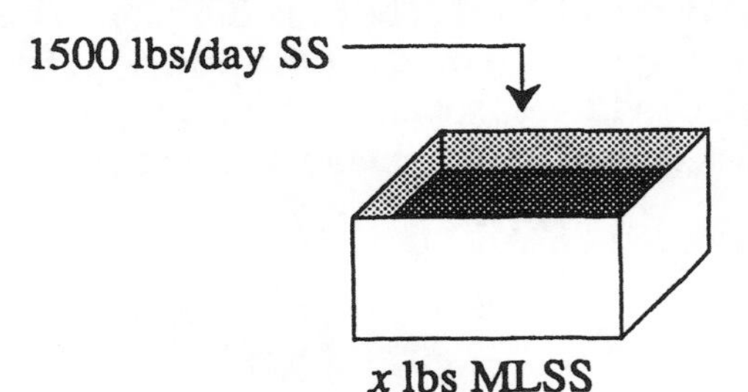

$$\text{Sludge Age, days} = \frac{\text{MLSS, lbs}}{\text{SS Added, lbs/day}}$$

$$5.5 \text{ days} = \frac{x \text{ lbs MLSS}}{1500 \text{ lbs/day SS Added}}$$

$$(5.5)(1500) = x$$

$$\boxed{8250 \text{ lbs}} = x$$

Example 6: (Sludge Age)

❑ The 1.6-MGD influent flow to an aeration tank has a suspended solids concentration of 70 mg/*L*. The aeration tank volume is 235,000 gallons. If a sludge age of 6 days is desired, what is the desired mg/*L* MLSS concentration?

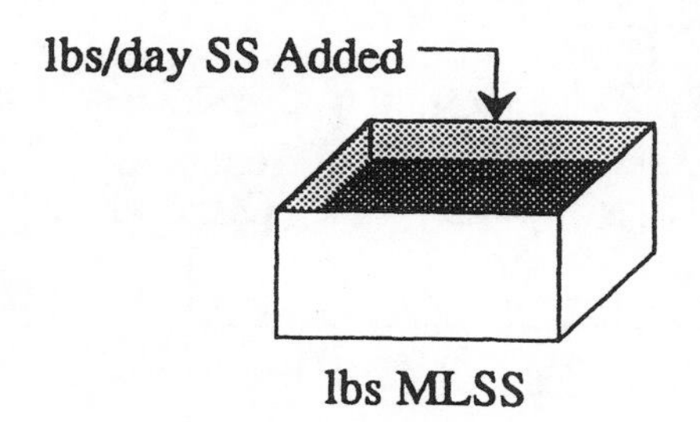

$$6 \text{ days} = \frac{(x \text{ mg/}L \text{ MLSS})(0.235 \text{ MG})(\cancel{8.34} \text{ lbs/gal})}{(70 \text{ mg/}L)(1.6 \text{ MGD})(\cancel{8.34} \text{ lbs/gal})}$$

After dividing out the 8.34 factor from the numerator and denominator*, solve for x:

$$\frac{(70)(1.6)(6)}{0.235} = x$$

$$\boxed{2860 \text{ mg/}L \text{ MLSS}} = x$$

Example 7: (Sludge Age)

❑ A 330,000-gallon aeration tank receives a flow of 2,100,000 gpd with a suspended solids concentration of 75 mg/*L*. If a sludge age of 5.5 days is desired, what is the desired mg/*L* MLSS concentration?

$$5.5 \text{ days} = \frac{(x \text{ mg/}L \text{ MLSS})(0.33 \text{ MG})(\cancel{8.34} \text{ lbs/gal})}{(75 \text{ mg/}L)(2.1 \text{ MGD})(\cancel{8.34} \text{ lbs/gal})}$$

After dividing out the 8.34 factor from the numerator and denominator, solve for x:

$$\frac{(75)(2.1)(5.5)}{0.33} = x$$

$$\boxed{2625 \text{ mg/}L \text{ MLSS}} = x$$

* Sometimes it is advisable not to divide out the 8.34 lbs/gal factor. If the lbs MLSS and lbs SS are used in other calculations, it is best to leave the 8.34 factor in the equation.

12.6 SOLIDS RETENTION TIME

Solids Retention Time (SRT), also called Mean Cell Residence Time (MCRT), is a calculation very similar to the sludge age calculation. There are two principal differences in calculating SRT:

1. **The SRT calculation is based on suspended solids leaving the system.** (Sludge age is based on suspended solids entering the aeration tank.)

2. **When calculating lbs MLSS, both the aeration tank and final clarifier volumes are normally used.** (In sludge age calculations, only the aeration tank volume is used in calculating lbs MLSS.)

In making SRT calculations, you must determine the pounds suspended solids in the system (solids in the aeration tank and final clarifier). There are actually four different approaches to calculating lbs SS. Each approach results in a slightly different numerator for the SRT equation.

The **most accurate method** of determining system solids includes a measure of aeration tank solids (mg/*L* MLSS) and a measure of final clarifier solids (mg/*L* SS determined by a core sampler). This **first method** is given as the "expanded equation", shown to the right.

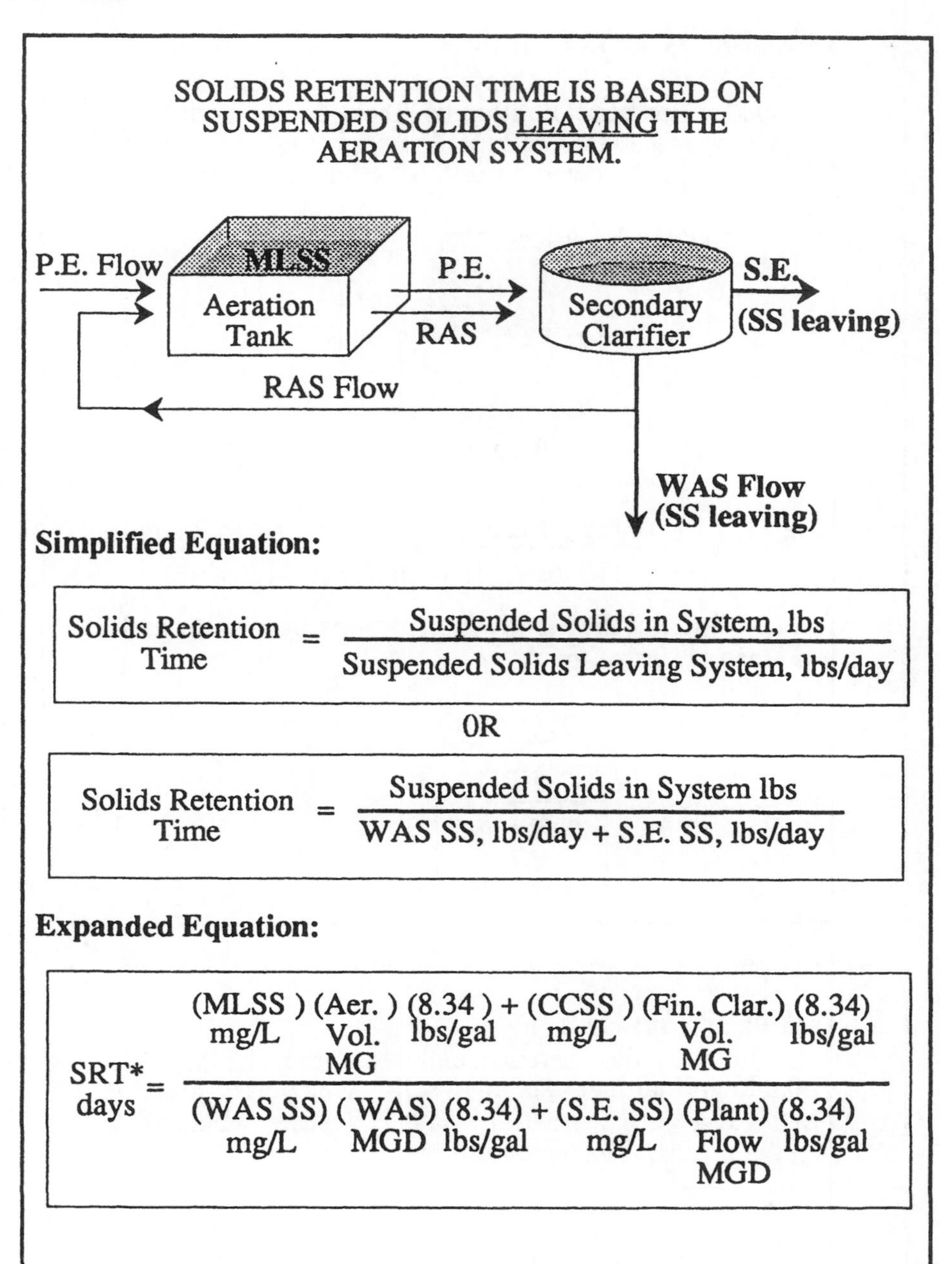

OTHER METHODS OF DETERMINING SYSTEM SOLIDS

2. To measure aeration tank solids and estimate clarifier solids:

$$(\text{MLSS mg/L})(\text{Aer. Vol. MG})(8.34 \text{ lbs/day}) + \frac{(\text{MLSS mg/L} + \text{RAS SS mg/L})}{2}(\text{Sludge Blanket Vol. MG})(8.34 \text{ lbs/gal}) = \text{SS lbs}$$

3. To measure aeration tank solids and estimate clarifier solids:

(MLSS mg/L) (Aer. Vol., MG + Fin. Clar., MG) (8.34 lbs/gal) = lbs MLSS

4. To measure aeration tank solids only:

(MLSS mg/L) (Aer. Vol., MG) (8.34 lbs/gal) = lbs MLSS

* CCSS is the average clarifier core SS concentration of the entire water column sampled by a core sampler.
S.E. is Secondary Effluent.

Example 1: (Solids Retention Time)

❑ An activated sludge system has a total of 30,800 lbs of mixed liquor suspended solids. The suspended solids leaving the final clarifier in the effluent is calculated to be 425 lbs/day. The lbs suspended solids wasted from the final clarifier is 3077 lbs/day. What is the solids retention time, in days?

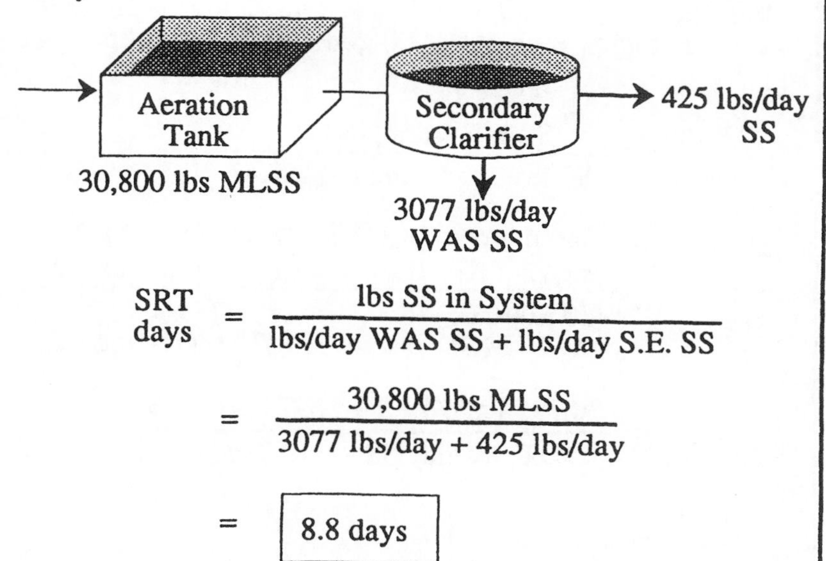

$$\text{SRT days} = \frac{\text{lbs SS in System}}{\text{lbs/day WAS SS} + \text{lbs/day S.E. SS}}$$

$$= \frac{30{,}800 \text{ lbs MLSS}}{3077 \text{ lbs/day} + 425 \text{ lbs/day}}$$

$$= \boxed{8.8 \text{ days}}$$

Example 2: (Solids Retention Time)

❑ An aeration tank has a volume of 425,000 gal. The final clarifier has a volume of 120,000 gal. The MLSS concentration in the aerator is 2780 mg/*L*. If a total of 1640 lbs/day SS are wasted and 340 lbs/day SS are in the secondary effluent, what is the solids retention time for the activated sludge system? (To determine system solids use the method that combines aeration tank and clarifier volumes, numerator 3 shown to the left.)

$$\text{SRT, days} = \frac{\text{lbs SS in System}}{\text{lbs/day WAS SS} + \text{lbs/day S.E. SS}}$$

$$= \frac{(2780 \text{ mg/}L)\,(0.545 \text{ MGD})\,(8.34)}{1640 \text{ lbs/day} + 340 \text{ lbs/day}}$$

$$= \frac{12{,}636 \text{ lbs MLSS}}{1980 \text{ lbs/day SS leaving}}$$

$$= \boxed{6.4 \text{ days}}$$

Note that the solids in the aeration tank and final clarifier are determined separately and are then added to determine total solids in the system.

The **second method** of calculating system solids includes a measure of aeration tank solids (using mg/*L* MLSS) and an estimate of final clarifier solids (using an average SS value of the sludge blanket). This average SS value of the sludge blanket is obtained by determining the SS concentration at the top of the sludge blanket (represented by the MLSS concentration) and the SS concentration at the bottom of the sludge blanket (represented by the RAS SS concentration).* Again, note that the solids in the aeration tank are added to the solids in the sludge blanket to determine total solids in the system.

The **third method** of calculating system solids includes a measure of the aeration tank solids (using mg/*L* MLSS) and an estimate of the final clarifier solids (using the same mg/*L* MLSS). With this method, the MLSS concentration is multiplied by the combined volumes of both tanks. This equation for SRT is used frequently.

The **fourth method** of calculating system solids includes a measure of aeration tank solids only. This method is often used when most of the solids are in the aeration tank and there are not many solids in the final clarifier.

Use the equation for SRT that works best for your plant and stay with it.

* Refer to Chapter 6 in *Basic Math Concepts* for a review of average calculations.

Example 3: (Solids Retention Time)

❑ Determine the solids retention time (SRT) given the following data: (Use the equation that includes core sampler suspended solids data, Equation 1 given on the previous page for the SRT numerator.)

Aer. Vol. — 1.5 MG	MLSS — 2460 mg/*L*
Fin. Clar. Vol. — 0.11 MG	WAS SS — 8040 mg/*L*
P.E. Flow — 3.4 MGD	S.E. SS — 18 mg/*L*
WAS Pumping Rate —60,000 gpd	CCSS* — 1850 mg/*L*

$$\text{SRT days} = \frac{\text{SS in System, lbs}}{\text{SS leaving System, lbs/day}}$$

$$= \frac{\underset{\text{mg/}L}{(2460)}\ \underset{\text{MG}}{(1.5)}\ \underset{\text{lbs/gal}}{(8.34)} + \underset{\text{mg/}L}{(1850)}\ \underset{\text{MG}}{(0.11)}\ \underset{\text{lbs/gal}}{(8.34)}}{\underset{\text{mg/}L}{(8040)}\ \underset{\text{MGD}}{(0.06)}\ \underset{\text{lbs/gal}}{(8.34)} + \underset{\text{mg/}L}{(18)}\ \underset{\text{MGD}}{(3.4)}\ \underset{\text{lbs/gal}}{(8.34)}}$$

$$= \frac{30{,}775 \text{ lbs MLSS} + 1697 \text{ lbs CCSS}}{4023 \text{ lbs/day SS} + 510 \text{ lbs/day SS}}$$

$$= \boxed{7.2 \text{ days}}$$

Example 4: (Solids Retention Time)

❑ Calculate the solids retention time (SRT) given the following data: (Use the SRT equation that includes CCSS.)

Aer. Vol.— 525,000 gal	MLSS —2810 mg/*L*
Fin. Clar. Vol. —120,000 gal	WAS SS —6900 mg/*L*
P.E. Flow —1.8 MGD	S.E. SS —15 mg/*L*
WAS Pumping Rate —24,000 gpd	CCSS* —1920 mg/L

$$\text{SRT, days} = \frac{\text{SS in System, lbs}}{\text{SS Leaving System, lbs/day}}$$

$$= \frac{\underset{\text{mg/}L}{(2810)}\ \underset{\text{MG}}{(0.525)}\ \underset{\text{lbs/gal}}{(8.34)} + \underset{\text{mg/}L}{(1920)}\ \underset{\text{MG}}{(0.12)}\ \underset{\text{lbs/gal}}{(8.34)}}{\underset{\text{mg/}L}{(6900)}\ \underset{\text{MGD}}{(0.024)}\ \underset{\text{lbs/gal}}{(8.34)} + \underset{\text{mg/}L}{(15)}\ \underset{\text{MGD}}{(1.8)}\ \underset{\text{lbs/gal}}{(8.34)}}$$

$$= \frac{12{,}304 \text{ lbs MLSS} + 1922 \text{ lbs CCSS}}{1381 \text{ lbs/day SS} + 225 \text{ lbs/day SS}}$$

$$= \boxed{8.9 \text{ days}}$$

---

* CCSS is the average clarifier core SS concentration of the entire water column sampled by a core sampler.

Example 5: (Solids Retention Time)

❑ What WAS pumping rate, in MGD, will be required given the information listed below? (Use the SRT equation that includes CCSS.)

Aer. Vol.— 355,000 gal
Fin. Clar. Vol. —100,000 gal
P.E. Flow —1.7 MGD
WAS Pumping Rate —x MGD

MLSS —2340 mg/L
WAS SS —6600 mg/L
S.E. SS —16 mg/L
CCSS —1850 mg/L
Desired SRT — 7 days

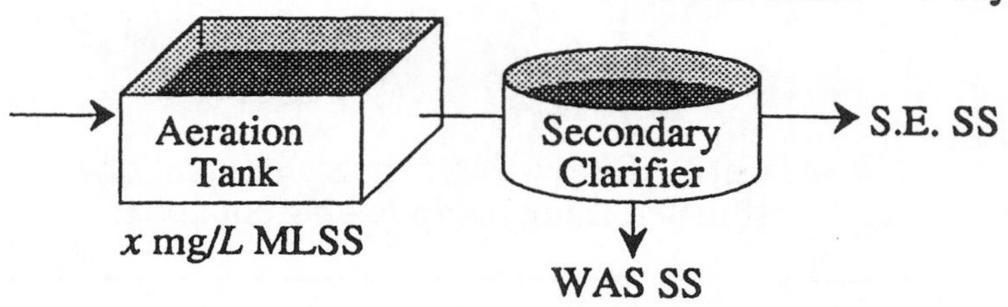

$$\underset{\text{days}}{\text{SRT}} = \frac{\text{SS in System, lbs}}{\text{SS Leaving System, lbs/day}}$$

First fill in the SRT equation with the known information:

$$7 \text{ days} = \frac{\underset{\text{mg/}L}{(2340)}\ \underset{\text{MG}}{(0.355)}\ \underset{\text{lbs/gal}}{(8.34)} + \underset{\text{mg/}L}{(1850)}\ \underset{\text{MG}}{(0.1)}\ \underset{\text{lbs/gal}}{(8.34)}}{\underset{\text{mg/}L}{(6600)}\ \underset{\text{MGD}}{(x)}\ \underset{\text{lbs/gal}}{(8.34)} + \underset{\text{mg/}L}{(16)}\ \underset{\text{MGD}}{(1.7)}\ \underset{\text{lbs/gal}}{(8.34)}}$$

Next, simplify as many terms as possible before continuing with the calculation:

$$7 \text{ days} = \frac{6928 \text{ lbs MLSS} + 1543 \text{ lbs CCSS}}{\underset{\text{mg/}L}{(6600)}\ \underset{\text{MGD}}{(x)}\ \underset{\text{lbs/gal}}{(8.34)} + 227 \text{ lbs/day SS}}$$

$$7 \text{ days} = \frac{8471 \text{ lbs SS}}{\underset{\text{mg/}L}{(6600)}\ \underset{\text{MGD}}{(x)}\ \underset{\text{lbs/gal}}{(8.34)} + 227 \text{ lbs/day SS}}$$

Then solve for the unknown value:*

$$(6600)(x)(8.34) + 227 = \frac{8471}{7}$$

$$(6600)(x)(8.34) = 1210 - 227$$

$$x = \frac{983}{(6600)(8.34)}$$

$$x = \boxed{0.0179 \text{ MGD}}$$

CALCULATING OTHER UNKNOWN FACTORS

The SRT equation incorporates several variables. In Examples 1-4, SRT was the unknown variable. Other variables can also be the unknown value, as shown below:

- Desired MLSS, mg/L
- Desired WAS SS, mg/L
- WAS Pumping Rate, MGD

Regardless of which variable is unknown, use the same SRT equation, fill in the given data, then solve for the unknown value.*

* For a review of solving for the unknown value, refer to Chapter 2 in *Basic Math Concepts*.

12.7 RETURN SLUDGE RATE

RETURN SLUDGE RATE —USING SETTLEABILITY

A key aspect in the proper operation of an activated sludge system is maintaining a balance between the food entering the system (measured by BOD or COD) and the microorganisms within that system (measured by the MLSS). Since there is not much control in the amount of food entering the system, most of the control in an activated sludge system is focused on maintaining an adequate solids inventory. One calculation important in this consideration is determination of the return activated sludge (RAS) flow rate.

The most direct method of determining an appropriate return sludge rate is observation of the sludge blanket depth. Depending on whether the sludge blanket depth is rising or falling, the RAS rate is increased or decreased accordingly.

Another method of determining the return sludge rate (RAS pumping rate) is based on the settleability test after 30 minutes.

THE RATIO OF RAS RATE TO INFLUENT FLOW

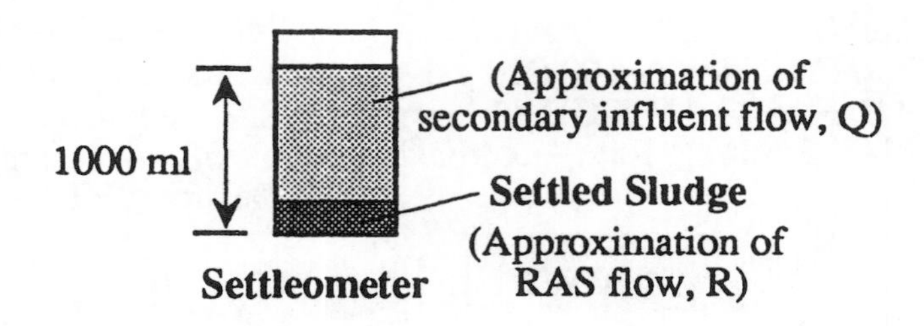

The ratio of RAS flow to secondary influent flow can be estimated using the following equation:

$$\frac{R}{Q} = \frac{\text{Settled Sludge Volume, ml/L}}{\text{1000 ml/L} - \text{Settled Sludge Volume, ml/L}}$$

The RAS ratio can then be used to calculate RAS flow rate in MGD:

$$\frac{R}{Q} = \frac{\text{RAS Flow Rate, MGD}}{\text{Secondary Influent Flow Rate, MGD}}$$

Example 1: (Return Sludge Rate)

❑ The settleability test after 30 minutes indicates a sludge settling volume of 210 ml/L. Calculate the ratio of RAS flow, R, to the secondary influent flow, Q.

$$\frac{R}{Q} = \frac{\text{Settled Sludge Volume, ml/L}}{\text{1000 ml/L} - \text{Settled Sludge Volume, ml/L}}$$

$$= \frac{\text{210 ml/L Settled Sludge}}{\text{1000 ml/L} - \text{210 ml/L Set. Sludge}}$$

$$= \frac{\text{210 ml/L}}{\text{790 ml/L}}$$

$$= \boxed{0.27}$$

Example 2: (Return Sludge Rate)

❑ A total of 280 ml/*L* sludge settled during a settleability test after 30 minutes. Calculate the ratio of the RAS flow to the secondary influent flow.

$$\frac{R}{Q} = \frac{\text{ml}/L \text{ Settled Sludge}}{1000\text{-ml}/L - \text{ml}/L \text{ Settled Sludge}}$$

$$= \frac{280 \text{ ml}/L \text{ Settled Sludge}}{1000\text{-ml}/L - 280 \text{ ml}/L \text{ Settled Sludge}}$$

$$= \frac{280 \text{ ml}/L}{720 \text{ ml}/L}$$

$$= \boxed{0.39}$$

Example 3: (Return Sludge Rate)

❑ The secondary influent flow to an aeration tank is 2.95 MGD. If the results of the settleability test after 30 minutes indicate that 318 ml/*L* sludge settled, what is the ratio of the RAS flow to the secondary influent flow? (b) What is the RAS flow expressed in MGD?

(a)

$$\frac{R}{Q} = \frac{\text{ml}/L \text{ Settled Sludge}}{1000\text{-ml}/L - \text{ml}/L \text{ Settled Sludge}}$$

$$= \frac{318 \text{ ml}/L \text{ Settled Sludge}}{1000\text{-ml}/L - 318 \text{ ml}/L \text{ Settled Sludge}}$$

$$= \frac{318 \text{ ml}/L}{682 \text{ ml}/L}$$

$$= \boxed{0.47}$$

(b)

$$\frac{R}{Q} = \frac{\text{RAS, MGD}}{\text{Secondary Influent Flow, MGD}}$$

$$0.47 = \frac{x \text{ MGD}}{2.95 \text{ MGD}}$$

$$(2.95 \text{ MGD})(0.47) = x$$

$$\boxed{1.39 \text{ MGD}} = x$$

RETURN SLUDGE RATE —USING SECONDARY CLARIFIER SOLIDS BALANCE

SOLIDS ENTERING THE CLARIFIER ARE EQUAL TO THE SOLIDS LEAVING THE CLARIFIER

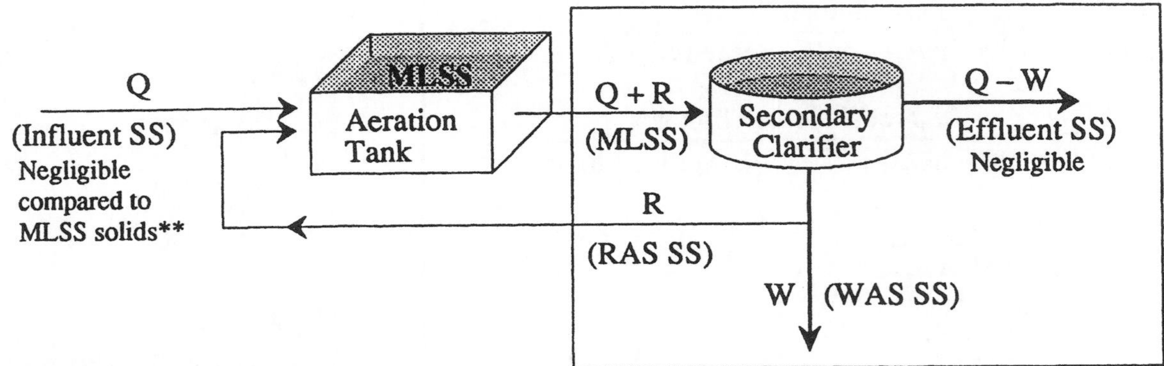

Simplified Equation:

$$\text{Suspended Solids In} = \text{Suspended Solids Out}$$

Expanded Equation:

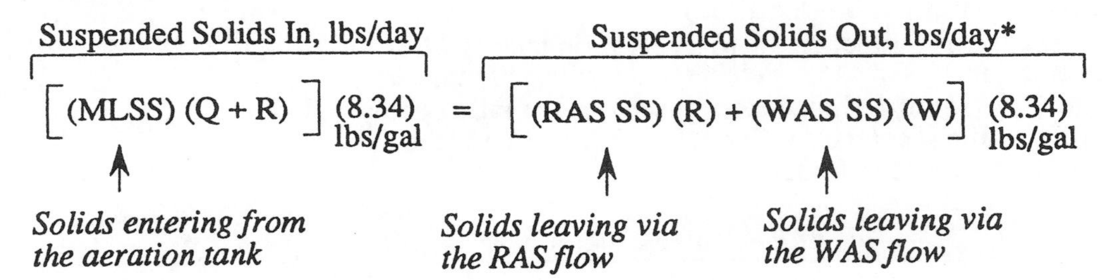

$$\left[(\text{MLSS})(Q + R)\right]\ (8.34\ \text{lbs/gal}) = \left[(\text{RAS SS})(R) + (\text{WAS SS})(W)\right]\ (8.34\ \text{lbs/gal})$$

Where:

MLSS = mixed liquor suspended solids, mg/L
Q = secondary influent flow, MGD
R = return sludge flow, MGD
RAS SS = return activated sludge SS, mg/L
WAS SS = waste activated sludge SS, mg/L
W = waste sludge flow, MGD
R = return sludge flow, MGD

The equation may be rearranged so that both R terms are on the same side of the equation: *(Note that since the 8.34 factor is on both sides of the equation, it can be divided out. However, if either of the sludges has a density different than 8.34 lb/gal,* ***both*** *density factors must be included as part of the equation.)*

$$(\text{MLSS})(Q) + (\text{MLSS})(R) = (\text{RAS SS})(R) + (\text{WAS SS})(W)$$

$$(\text{MLSS})(Q) - (\text{WAS SS})(W) = (\text{RAS SS})(R) - (\text{MLSS})(R)$$

$$(\text{MLSS})(Q) - (\text{WAS SS})(W) = \left[(\text{RAS SS}) - (\text{MLSS})\right](R)$$

Then solve for R:

$$\frac{(\text{MLSS})(Q) - (\text{WAS SS})(W)}{(\text{RAS SS}) - (\text{MLSS})} = R$$

* This equation assumes negligible loss of solids in the effluent.
**Except for modified aeration processes which may have very low MLSS concentrations.

Example 4: (Return Sludge Rate)

❑ Given the following data, calculate the RAS return rate, R, using the secondary clarifier solids balance equation:

MLSS = 2460 mg/L   W = 60,000 gpd
RAS SS = 7850 mg/L   Q = 3.4 MGD
WAS SS = 7850 mg/L

$$\frac{(\text{MLSS})(\text{Q}) - (\text{WAS SS})(\text{W})}{(\text{RAS SS}) - (\text{MLSS})} = \text{R}$$

$$\frac{\underset{\text{mg/L}}{(2460)}\ \underset{\text{MGD}}{(3.4)} - \underset{\text{mg/L}}{(7850)}\ \underset{\text{MGD}}{(0.06)}}{7850 \text{ mg/L} - 2460 \text{ mg/L}} = \text{R}$$

$$\boxed{1.46 \text{ MGD}} = \text{R}$$

Example 5: (Return Sludge Rate)

❑ Given the following data, calculate the RAS return rate, R, using the secondary clarifier solids balance equation:

MLSS = 2460 mg/L   W = 60,000 gpd
RAS SS = 7850 mg/L   Q = 3.4 MGD
WAS SS = 7850 mg/L

Suspended Solids In* = Suspended Solids Out*

$$\left[(\text{MLSS})(\text{Q} + \text{R})\right] = \left[(\text{RAS SS})(\text{R}) + (\text{WAS SS})(\text{W})\right]$$

First, fill in the equation with the given information:

$$\underset{\text{mg/L}}{(2460)}\ (\underset{\text{MGD}}{3.4} + \underset{\text{MGD}}{\text{R}}) = \underset{\text{mg/L}}{(7850)}\ \underset{\text{MGD}}{(\text{R})} + \underset{\text{mg/L}}{(7850)}\ \underset{\text{MGD}}{(0.06)}$$

Next, simplify terms where possible:

$$(2460)(3.4 + \text{R}) = (7850)(\text{R}) + (471)$$

$$8364 + 2460\text{ R} = 7850\text{ R} + 471$$

Then group R terms on the right side of the equation and numerical terms on the left side of the equation:

$$7893 = 5390\text{ R}$$

$$\frac{7893}{5390} = \text{R}$$

$$\boxed{1.46 \text{ MGD}} = \text{R}$$

A solids balance (or mass balance) calculation is a comparison of the solids going into a process with those coming out. In other words:

$$\boxed{\text{Solids In} = \text{Solids Out}}$$

The secondary clarifier solids balance can be used to determine the return sludge rate (RAS rate) because return sludge rate is one of the variables of the solids balance calculation, as shown in the expanded equation on the opposite page. Note this equation assumes that negligible solids leave the system in the clarifier effluent flow.

The expanded equation may be rearranged and simplified, as shown in the last equation on the opposite page. Example 4 illustrates the calculation of RAS using this simplified equation.

Sometimes a simplified equation is difficult to remember, since it is no longer an obvious expression of "solids in = solids out". You may, therefore, wish to use the first expanded equation since it remains closer in concept to the basic "solids in = solids out" equation. Example 5 illustrates the use of the first expanded equation.* Example 5 uses the same problem as presented in Example 4 so that you can compare the use of the two forms of the equation.

* Note that the 8.34 lbs/gal factor has been divided out on both sides of the equation.

RETURN SLUDGE RATE —USING AERATION TANK SOLIDS BALANCE

SOLIDS ENTERING THE AERATION TANK ARE EQUAL TO THE SOLIDS LEAVING THE AERATION TANK

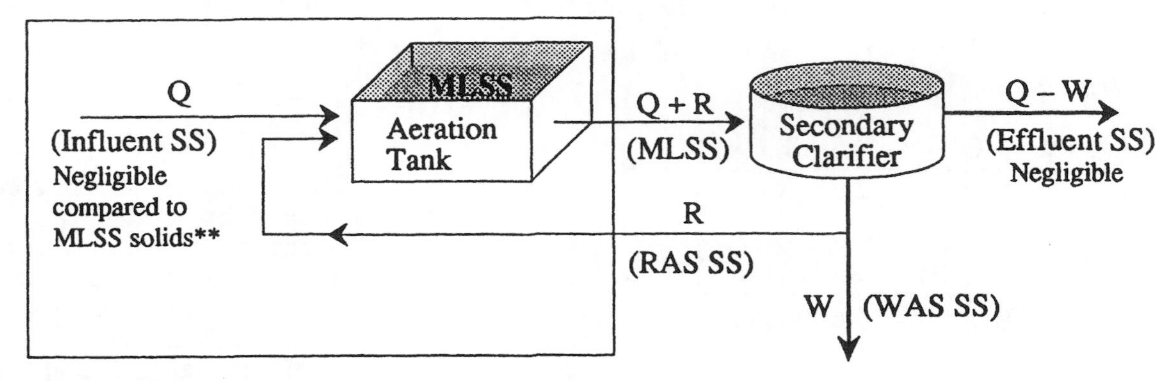

Simplified Equation:

Suspended Solids In = Suspended Solids Out*

Expanded Equation:**

$$\underbrace{\left[(\text{RAS SS})(R)\right](8.34\ \text{lbs/gal})}_{\text{Suspended Solids In, lbs/day}} = \underbrace{\left[(\text{MLSS})(Q+R)\right](8.34\ \text{lbs/gal})}_{\text{Suspended Solids Out, lbs/day}}$$

Solids entering the aeration tank — *Solids leaving the aeration tank*

Where:

MLSS = mixed liquor suspended solids, mg/L
Q = secondary influent flow, MGD
R = return sludge flow, MGD
RAS SS = return activated sludge SS, mg/L

The equation may be rearranged so that both R terms are on the same side of the equation: *(Note that since the 8.34 factor is on both sides of the equation, it can be dropped.)*

$$(\text{RAS SS})(R) = (\text{MLSS})(Q) + (\text{MLSS})(R)$$

$$(\text{RAS SS})(R) - (\text{MLSS})(R) = (\text{MLSS})(Q)$$

$$\left[(\text{RAS SS}) - (\text{MLSS})\right](R) = (\text{MLSS})(Q)$$

Then solve for R:

$$R = \frac{(\text{MLSS})(Q)}{(\text{RAS SS}) - (\text{MLSS})}$$

Note that this is the **same equation** as for the secondary clarifier solids balance **except** that the aeration tank equation has no WAS term in the numerator (because there is no wasting from the aeration tank), as illustrated in the diagram above.

* For the aeration tank, this is true only when new cell growth in the tank is considered negligible.
** Except for modified aeration processes which may have very low MLSS concentrations.

Example 6: (Return Sludge Rate)

❑ Given the following data, calculate the RAS return rate, R, using the aeration tank solids balance equation:

MLSS = 2100 mg/*L* Q = 6.3 MGD
RAS SS = 7490 mg/*L*

$$\frac{\text{(MLSS) (Q)}}{\text{(RAS SS)} - \text{(MLSS)}} = \text{R}$$

$$\frac{(2100\ \text{mg}/L)\ (6.3\ \text{MGD})}{7490\ \text{mg}/L - 2100\ \text{mg}/L} = \text{R}$$

$$\boxed{2.45\ \text{MGD}} = \text{R}$$

The aeration tank solids balance can also be used to determine the return sludge rate (RAS rate) since return sludge rate is one of the variables of the solids balance calculation.

The expanded equation may be rearranged and simplified, as shown in the last equation on the opposite page. Example 6 illustrates the calculation of RAS using this simplified equation. Example 7 illustrates the use of the first expanded equation.*

Both examples use the same problem so that you can compare the use of both forms of the equation.

Example 7: (Return Sludge Rate)

❑ Given the following data, calculate the RAS return rate, R, using the aeration tank solids balance equation:

MLSS = 2100 mg/*L* Q = 6.3 MGD
RAS SS = 7490 mg/*L*

$$\text{Suspended Solids In*} = \text{Suspended Solids Out*}$$

$$\text{(RAS SS) (R)} = \text{(MLSS) (Q + R)}$$

First, fill in the equation with the given information:

$$\underset{\text{mg}/L}{(7490)}\ \underset{\text{MGD}}{(\text{R})} = \underset{\text{mg}/L}{(2100)}\ (\underset{\text{MGD}}{6.3} + \underset{\text{MGD}}{\text{R}})$$

Multiply terms as indicated:

$$7490\ \text{R} = 13{,}230 + 2100\ \text{R}$$

Then group R terms on the left side of the equation and numerical terms on the right side of the equation:

$$5390\ \text{R} = 13{,}230$$

$$\frac{13{,}230}{5390} = \text{R}$$

$$\boxed{2.45\ \text{MGD}} = \text{R}$$

* Note that the 8.34 lbs/gal factor has been divided out on both sides of the equation.

12.8 WASTING RATE

One of the most critical aspects in the operation of an activated sludge treatment system is maintaining a proper balance between the food entering the system (measured by BOD or COD entering) and the microorganisms in the system (measured in a general way by the the mixed liquor suspended solids, MLSS, or more precisely by the mixed liquor volatile suspended solids, MLVSS).

Since there is not much control possible in the amount of food entering the treatment system (the wastewater coming in must be treated), much of the control of an activated sludge system is focused on controlling (or adjusting) the amount of microorganisms in the system (MLVSS).

The size of the microorganism population naturally increases as food is consumed (measured by BOD or COD removed). Therefore, to maintain the same food-to-microorganism balance, a portion of the microorganism population must be removed or wasted periodically from the system. **The question is—how much should be wasted?**

Although the microorganism growth rate represents most of the increase in solids within an activated sludge system, another source of solids must also be considered—the suspended solids entering with the primary effluent flow. Wasting rate calculations should consider both sources of solids.

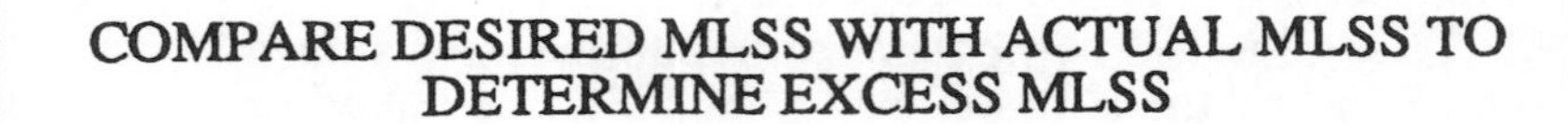
COMPARE DESIRED MLSS WITH ACTUAL MLSS TO DETERMINE EXCESS MLSS

Desired MLSS

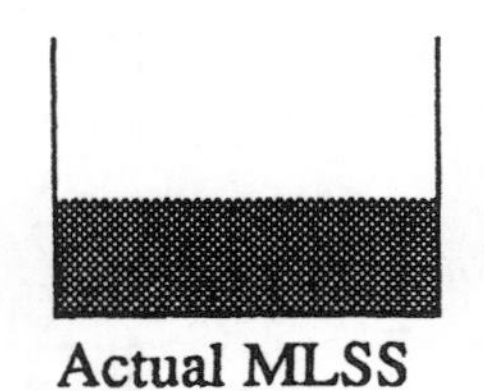
Actual MLSS

First calculate ***desired lbs MLVSS*** *(microorganisms) using the* ***desired F/M ratio*** *and BOD or COD applied (food):*

$$\text{Desired F/M} = \frac{\underset{\text{mg/L}}{(\text{BOD})}\ \underset{\text{flow}}{(\text{MGD})}\ \underset{\text{lbs/gal}}{(8.34)}}{\text{Desired lbs MLVSS}}$$

Next, determine ***desired lbs MLSS*** *using % volatile solids data:*

$$\text{Desired MLSS, lbs} = \frac{\text{Desired lbs MLVSS}}{\frac{\%\ \text{VS}}{100}}$$

$$\underset{\text{MLSS}}{(\text{mg/L})}\ \underset{\text{Aer. Vol.}}{(\text{MG})}\ \underset{\text{lbs/gal}}{(8.34)} = \text{lbs MLSS}$$

Then compare the actual and desired MLSS to determine lbs SS to be wasted:

$$\text{Actual lbs MLSS} - \text{Desired lbs MLSS} = \text{lbs SS to be Wasted}$$

Example 1: (Wasting Rate)

❑ The desired F/M ratio for an activated sludge plant is 0.3 lbs BOD/lb MLVSS. It has been calculated that 5900 lbs/day BOD enter the aeration tank. The MLSS volatile solids content is 70%. Based on the desired F/M ratio, what is the desired lbs MLSS in the aeration tank?

First calculate the desired lbs MLVSS using the desired F/M ratio:

$$\frac{0.3 \text{ lbs/day BOD}}{1 \text{lb MLVSS}} = \frac{5900 \text{ lbs/day BOD}}{x \text{ lbs MLVSS}}$$

$$x = \frac{5900}{0.3} = \boxed{\text{19,667 lbs MLVSS desired}}$$

Then calculate the desired lbs MLSS, using % VS content:

$$\frac{\text{19,667 lbs MLVSS}}{\frac{70}{100} \text{ VS Content}} = \boxed{\text{28,096 lbs MLSS desired}}$$

Example 2: (Wasting Rate)

❑ Given the following data, use the desired F/M ratio to determine the lbs SS to be wasted:

Aer. Vol.—1,300,000 gal
Influent Flow—3,190,000 gpd
COD— 115 mg/*L*
Desired F/M—0.15 lbs COD/day/lb MLVSS
MLSS—2980 mg/*L*
% VS—70%

First calculate the **desired lbs MLVSS:**

$$\frac{0.15 \text{ lbs/day COD}}{1 \text{ lb MLVSS}} = \frac{(115 \text{ mg/}L)(3.19 \text{ MGD})(8.34 \text{ lbs/gal})}{x \text{ lbs MLVSS}}$$

$$x = \frac{(115)(3.19)(8.34)}{0.15}$$

$$= \boxed{\text{20,397 lbs MLVSS desired}}$$

Then calculate the **desired MLSS:**

$$\frac{\text{20,397 lbs MLVSS}}{\frac{70}{100} \text{ VS Content}} = \boxed{\text{29,139 lbs MLSS desired}}$$

Now calculate the **actual lbs MLSS:**

$$\underset{\text{MLSS}}{(2980 \text{ mg/}L)}\ \underset{\text{Aer. Vol.}}{(1.3 \text{ MG})}\ \underset{\text{lbs/gal}}{(8.34)} = \boxed{\text{32,309 lbs MLSS in Aer. Tank}}$$

And compare desired and actual MLSS to determine lbs SS to be wasted:

$$\underset{\text{Actual MLSS}}{\text{32,309 lbs}} - \underset{\text{Desired MLSS}}{\text{29,139 lbs}} = \boxed{\text{3170 lbs SS to be Wasted}}$$

One or more of the following calculations may be used to determine the desired wasting rate for the activated sludge system:

- F/M Ratio
- Sludge Age
- Solids Retention Time, SRT (also called Mean Cell Residence Time, MCRT)

CALCULATING WASTING RATE USING F/M RATIO

The Food/Microorganism (F/M) ratio can be used to calculate wasting rates. However, because this calculation focuses solely on the organic components of the wastewater, it is a good idea to use this calculation in conjunction with another wasting rate calculation that includes both organic and inorganic solids, such as Solids Retention Time, SRT (also called Mean Cell Residence Time, MCRT).

The F/M ratio calculated on one day may vary significantly from an F/M ratio calculated the next day. This is because the F/M ratio reflects changes in flow rate and organic content of the wastewater. For this reason, when using the F/M ratio to estimate wasting rates, you should use a 7-day moving average* for the flow rate, organic content (food—BOD or COD), and microorganism concentration (MLVSS). The data given in Examples 1 and 2 reflect 7-day moving averages.

CALCULATING WASTING RATE USING SLUDGE AGE

Sludge age can also be used to determine pounds of suspended solids to be wasted. The approach is similar to that described on the previous two pages.

To calculate wasting rate based on sludge age:

1. Calculate the desired lbs of MLSS, using the desired sludge age and lbs/day suspended solids added.
2. Calculate the actual lbs of MLSS.
3. Determine the lbs of MLSS to be wasted ("excess" MLSS):

Actual MLSS	−	Desired MLSS	=	Excess MLSS

As with data used in the F/M ratio to determine wasting rates, the data used in these calculations (mg/*L* SS, MGD flow, and mg/*L* MLSS) should be 7-day moving averages so that significant shifts in data will not result in wasting rate calculations that vary significantly from day to day.

COMPARE DESIRED MLSS WITH ACTUAL MLSS TO DETERMINE EXCESS MLSS

Desired MLSS

Actual MLSS

*First calculate **desired MLSS** using **desired sludge age** and suspended solids added daily:*

(SS Added) lbs/day × (Desired Sludge Age) days = desired lbs MLSS

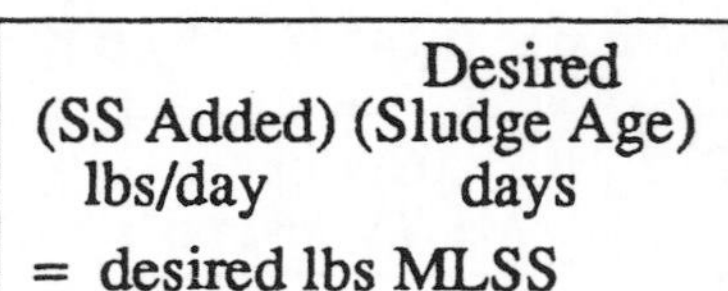

Or

(mg/*L*) SS × (MGD) flow × (8.34) lbs/gal × (Desired Sludge Age) days = desired lbs MLSS

(mg/*L*) MLSS × (MG) Aer. Vol. × (8.34) lbs/gal = lbs MLSS

Example 3: (Wasting Rate)

❑ Using the desired sludge age, it was calculated that 15,000 lbs MLSS are desired in the aeration tank. If the aeration tank volume is 800,000 gallons and the MLSS concentration is 2720 mg/*L*, how many lbs MLSS should be wasted?

Since desired lbs MLSS has already been determined, the actual lbs MLSS should be calculated:

(mg/*L*) MLSS × (MG) Aer. Vol. × (8.34) lbs/gal = lbs MLSS

(2720 mg/*L*) MLSS × (0.8 MG) Aer. Vol. × (8.34) lbs/gal = **18,148 lbs MLSS in Aeration Tank**

Now compare actual versus desired lbs MLSS:

18,148 lbs Actual MLSS − 15,000 lbs Desired MLSS = **3148 lbs MLSS to be Wasted**

* Refer to Chapter 6, "Averages", in *Basic Math Concepts* for a discussion of moving averages.

Example 4: (Wasting Rate)

❑ The desired sludge age for a plant is 4 days. The aeration tank volume is 600,000 gpd. If 3000 lbs/day suspended solids enter the aeration tank and the MLSS concentration is 2800 mg/*L*, how many lbs MLSS (suspended solids) should be wasted?

First calculate the desired lbs MLSS, using desired sludge age and lbs/day SS added:

(3000 lbs/day) (4 days) = **12,000 lbs MLSS Desired**
SS Added / Desired Sludge Age

Then calculate the actual lbs MLSS:

(mg/*L*) (MG) (8.34) = lbs MLSS
MLSS / Aer. Vol. / lbs/gal

(2800 mg/*L*) (0.6 MG) (8.34) = **14,011 lbs MLSS in Aeration Tank**
MLSS / Aer. Vol. / lbs/gal

And compare desired with actual lbs MLSS:

14,011 lbs Actual MLSS − 12,000 lbs Desired MLSS = **2011 lbs MLSS to be Wasted**

Example 5: (Wasting Rate)

❑ The desired sludge age for a plant is 5 days. The aeration tank has a volume of 750,000 gal, with a MLSS concentration of 2980 mg/*L*. The flow to the aeration tank is 3.9 MGD with a suspended solids concentration of 105 mg/*L*. Calculate the lbs MLSS to be wasted.

First determine the lbs MLSS desired ,using desired sludge age and lbs/day SS added:

(105 mg/L) (3.9) (8.34) (5 days) = **17,076 lbs MLSS Desired**
SS / MGD / lbs/gal (SS Added, lbs/day) / Desired Sludge Age

Then calculate the actual lbs MLSS:

(2980 mg/*L*) (0.75 MG) (8.34) = **18,640 lbs MLSS in Aeration Tank**
MLSS / Aer. Vol. / lbs/gal

Now compare actual and desired lbs MLSS:

18,640 lbs Actual MLSS − 17,076 lbs Desired MLSS = **1564 lbs MLSS to be Wasted**

CALCULATING WASTING RATE USING SRT

The calculation of lbs/day suspended solids to be wasted using the SRT equation is a little different than the previous two methods. Since wasting rate is part of the SRT equation (WAS SS, lbs/day), you can leave that factor as the unknown variable* in the SRT calculation and fill in all other terms. Examples 6-8 illustrate this calculation.**

As with the three other methods of wasting rate calculation, the data used in this calculation (MLSS mg/*L*, S.E. SS mg/*L*, and plant flow) should be 7-day moving averages.

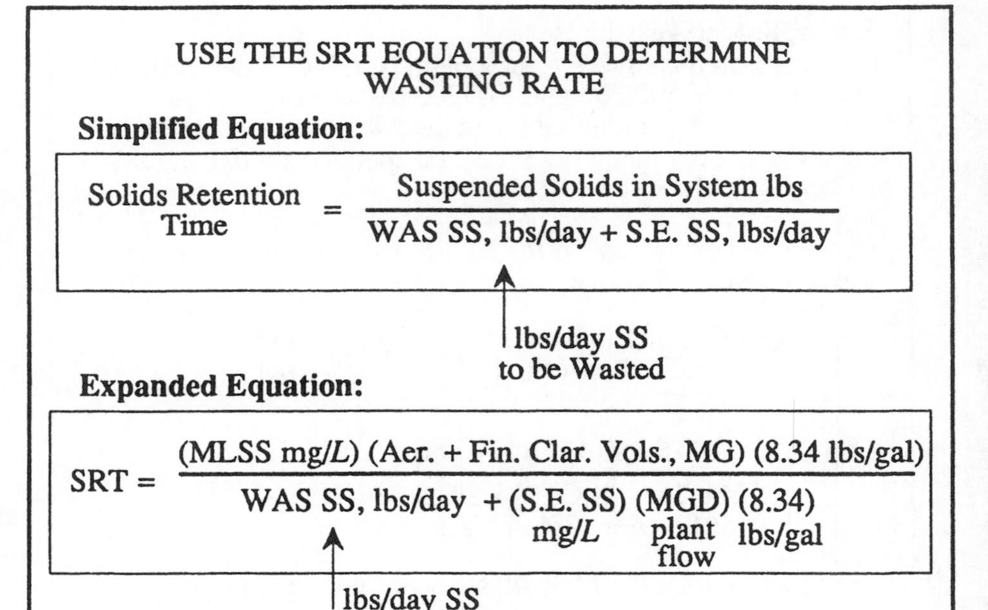

Example 6: (Wasting Rate)

❑ The desired SRT for an activated sludge system is 8.5 days. There are a total of 30,000 lbs SS in the system. The secondary effluent flow has 690 lbs/day suspended solids. To maintain the desired SRT, how many lbs/day suspended solids should be removed?

Write the SRT equation, filling in all information except WAS SS, lbs/day. The simplified equation can be used in this calculation:

$$\text{SRT} = \frac{\text{Suspended Solids in System, lbs}}{\text{WAS SS, lbs/day} + \text{S.E. SS, lbs/day}}$$

$$8.5 \text{ days} = \frac{30{,}000 \text{ lbs SS in System}}{x \text{ lbs/day WAS SS} + 690 \text{ SS, lbs/day}}$$

$$x + 690 = \frac{30{,}000}{8.5}$$

$$x + 690 = 3529$$

$$x = 3529 - 690$$

$$x = \boxed{2839 \text{ lbs/day WAS SS}}$$

---

* For a review of solving for the unknown variable in calculations such as these, refer to Chapter 2 in *Basic Math Concepts*.

** As described in Section 12.6, the numerator of the SRT equation can be determined using any of four methods. In Section 12.6, the method of determining system solids using core sampler data was used primarily. In this discussion the combined volume method of determining system solids is utilized.

Example 7: (Wasting Rate)

❑ The desired SRT for an activated sludge plant is 8 days. There are a total of 28,700 lbs SS in the system. The secondary effluent flow is 3,420,000 gpd, with a suspended solids content of 18 mg/*L*. How many lbs/day WAS SS must be wasted to maintain the desired SRT?

Use a modified version of the simplified equation:

$$8 \text{ days} = \frac{28{,}700 \text{ lbs SS in System}}{x \text{ lbs/day WAS SS} + \underset{\text{S.E. SS}}{(18 \text{ mg}/L)}\ \underset{\text{flow}}{(3.42 \text{ MGD})}\ \underset{\text{lbs/gal}}{(8.34)}}$$

$$8 = \frac{28{,}700}{x + 513}$$

$$x + 513 = \frac{28{,}700}{8}$$

$$x + 513 = 3588$$

$$x = 3588 - 513$$

$$x = \boxed{3075 \text{ lbs/day WAS SS}}$$

Example 8: (Wasting Rate)

❑ Given the following data, calculate the lbs/day WAS SS to be wasted.

Desired SRT—9 days
Clarifier + Aerator Vol.—1.2 MG
MLSS—3150 mg/*L*
S.E. SS—16 mg/*L*
Inf. Flow— 6.9 MGD

$$9 \text{ days} = \frac{(3150 \text{ mg}/L)\ (1.2 \text{ MG})\ (8.34 \text{ lbs/gal})}{x \text{ lbs/day WAS SS} + \underset{\text{mg}/L}{(16)}\ \underset{\text{MGD}}{(6.9)}\ \underset{\text{lbs/gal}}{(8.34)}}$$

$$9 = \frac{31{,}525 \text{ lbs SS in System}}{x \text{ lbs/day WAS SS} + 921 \text{ lbs S.E. SS}}$$

$$x + 921 = \frac{31{,}525}{9}$$

$$x + 921 = 3503$$

$$x = \boxed{2582 \text{ lbs/day WAS SS}}$$

12.9 WAS PUMPING RATE

WAS PUMPING RATE USING THE mg/L to lbs/day EQUATION

Waste activated sludge (WAS) pumping rate calculations are calculations that involve mg/*L* and flow. Therefore, the equation used in these calculations is:

(mg/*L*) (MGD) (8.34) = lbs/day flow lbs/gal

In WAS pumping rate calculations, the "mg/*L* SS" refers to the suspended solids content of the waste activated sludge being pumped away, and "MGD flow" refers to the WAS Pumping Rate of the sludge being wasted.

Sometimes waste activated sludge SS is not known but return activated sludge (RAS) SS is known. Remember that **RAS SS and WAS SS are the same measurement**. It is a measurement taken of secondary clarifier sludge. This sludge is either pumped back to the aeration tank (RAS) or wasted (WAS).

Since a biological system does not generally respond well to rapid changes in environment, any changes in WAS pumping rates should not be greater than 10-15% of the WAS pumping rate on the previous day. For the same reason, pumping continuously is preferred to wasting intermittently (batch wasting).

WAS PUMPING RATE CALCULATIONS ARE mg/*L* TO lbs/day PROBLEMS

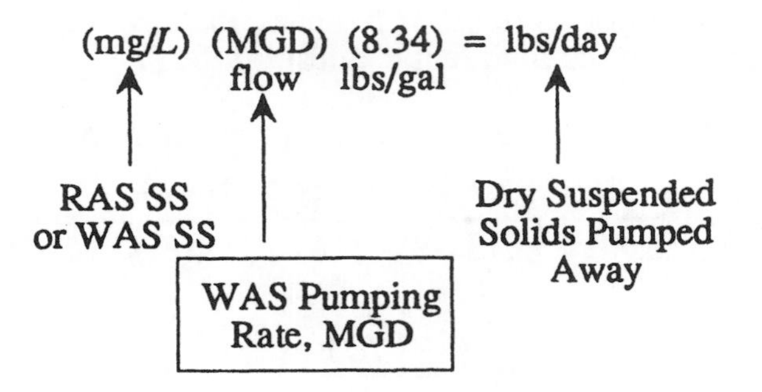

Example 1: (WAS Pumping Rate Calculations)

❑ The WAS suspended solids concentration is 5900 mg/*L*. If 4100 lbs/day solids are to be wasted, (a) What must the WAS pumping rate be, in MGD? (b) What is this rate expressed in gpm?

(a) First calculate the MGD pumping rate required, using the mg/*L* to lbs/day equation:

$$(\text{mg/}L)\ (\text{MGD flow})\ (8.34\ \text{lbs/gal}) = \text{lbs/day}$$

$$\underset{\text{mg/}L}{(5900)}\ \underset{\text{flow}}{(x\ \text{MGD})}\ \underset{\text{lbs/gal}}{(8.34)} = 4100\ \text{lbs/day}$$

$$x = \frac{4100\ \text{lbs/day}}{\underset{\text{mg/}L}{(5900)}\ \underset{\text{lbs/gal}}{(8.34)}}$$

$$x = 0.083\ \text{MGD}$$

(b) Then convert the MGD flow to gpm flow: *

$$0.083\ \text{MGD} = 83{,}000\ \text{gpd}$$

$$= \frac{83{,}000\ \text{gpd}}{1440\ \text{min/day}}$$

$$= \boxed{57.6\ \text{gpm}}$$

* Refer to Chapter 8 in *Basic Math Concepts* for a review of flow conversions.

Example 2: (WAS Pumping Rate Calculations)

❑ It has been determined that 5260 lbs/day of solids must be removed from the secondary system. If the RAS SS concentration is 6940 mg/L, what must be the WAS pumping rate, in MGD?

Calculate the MGD pumping rate required:

$$\underset{\text{mg/L}}{(6940)}\ \underset{\text{flow}}{(x\text{ MGD})}\ \underset{\text{lbs/gal}}{(8.34)} = 5260\text{ lbs/day}$$

$$x = \frac{5260\text{ lbs/day}}{\underset{\text{mg/L}}{(6940)}\ \underset{\text{lbs/gal}}{(8.34)}}$$

$$x = \boxed{0.0909\text{ MGD}}$$

Example 3: (WAS Pumping Rate Calculations)

❑ Given the following data, calculate the WAS pumping rate required (in MGD): (Use the combined volume method of determining system solids.)

Desired SRT—10 days	RAS SS—6290 mg/L
Clarifier + Aerator Vol.—1.8 MG	S.E. SS—14 mg/L
MLSS—2710 mg/L	Inf. Flow— 4.7 MGD

Use the expanded SRT equation:*

$$10\text{ days} = \frac{(2710\text{ mg/L})\ (1.8\text{ MG})\ (8.34\text{ lbs/gal})}{\underset{\text{WAS SS}}{(6290\text{ mg/L})}\ \underset{\text{WAS Flow}}{\underset{\text{MGD}}{(x)}}\ (8.34) + \underset{\text{S.E. SS}}{(14\text{ mg/L})}\ \underset{\text{Plant Flow}}{\underset{\text{MGD}}{(4.7)}}\ (8.34)}$$

$$10\text{ days} = \frac{40{,}683\text{ lbs}}{(6290)\ (x\text{ MGD})\ (8.34) + 549\text{ lbs}}$$

$$(6290)\ (x)\ (8.34) + 549 = \frac{40{,}683}{10}$$

$$(6290)\ (x)\ (8.34) + 549 = 4068.3$$

$$(6290)\ (x)\ (8.34) = 4068.3 - 549$$

$$(6290)\ (x)\ (8.34) = 3519.3$$

$$x = \frac{3519.3}{(6290)\ (8.34)}$$

$$x = \boxed{\begin{array}{l}0.067\text{ MGD}\\ \text{WAS Pumping Rate}\end{array}}$$

WAS PUMPING RATE USING THE SRT EQUATION

When the **lbs/day suspended solids to be wasted** is calculated using the F/M ratio or sludge age methods (described in Section 12.8), the WAS pumping rate would be calculated as shown in Examples 1 and 2.

However, when the SRT method is used to determing solids wasting, the WAS pumping rate can be calculated directly, using the SRT equation. (WAS flow, MGD, is the unknown value.) Example 3 illustrates this calculation. To review this method of determining wasting rate, refer to Section 12.8.

* To review solving for the unknown value for calculations such as these, refer to Chapter 2 in *Basic Math Concepts*.

12.10 OXIDATION DITCH DETENTION TIME

Detention time is the length of time required for a given flow rate to pass through a tank. Although detention time is not normally calculated for aeration basins, it is calculated for oxidation ditches.

When calculating detention time it is essential that the time and volume units used in the equation are consistent with each other, as illustrated to the right.

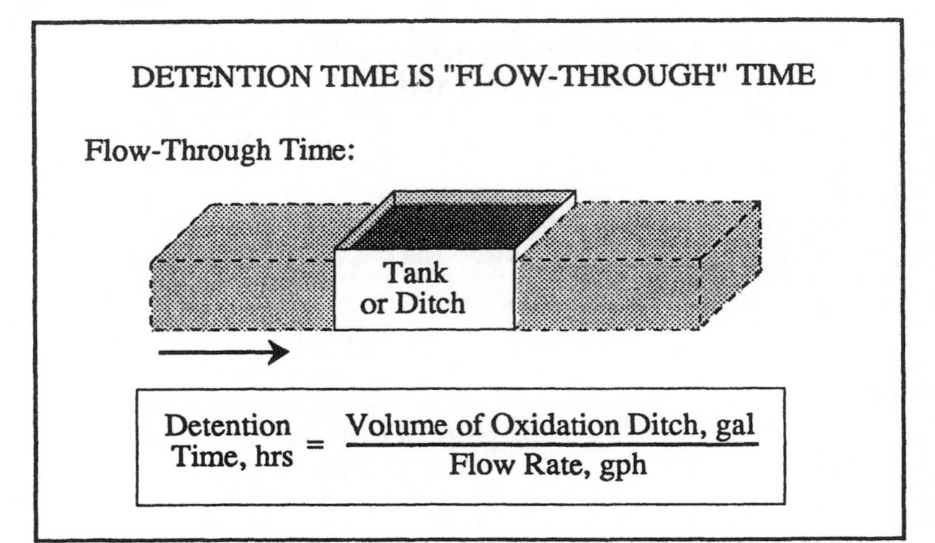

$$\text{Detention Time, hrs} = \frac{\text{Volume of Oxidation Ditch, gal}}{\text{Flow Rate, gph}}$$

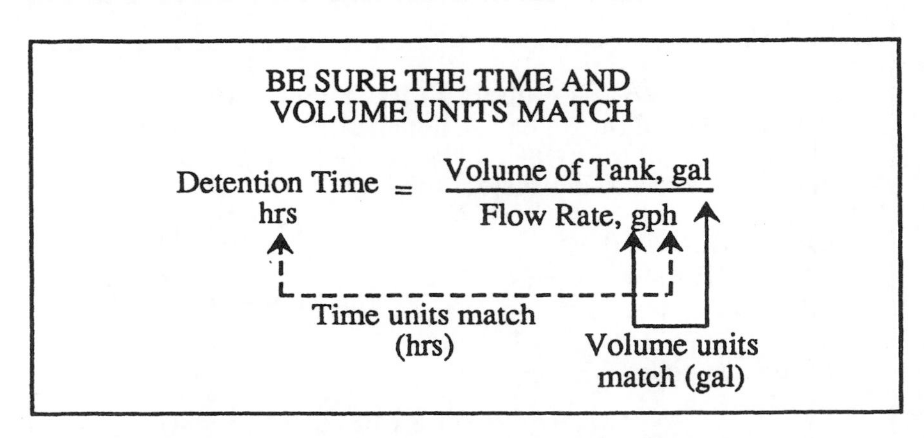

Example 1: (Detention Time)

❑ An oxidation ditch has a volume of 150,000 gallons. If the flow to the oxidation ditch is 195,000 gpd, what is the detention time in hours?

Since detention time is desired in hours, the flow must be expressed as gph:

$$\frac{195{,}000 \text{ gpd}}{24 \text{ hrs/day}} = 8125 \text{ gph}$$

Now calculate detention time:

$$\text{Detention Time, hrs} = \frac{\text{Volume of Oxidation Ditch, gal}}{\text{Flow Rate, gph}}$$

$$= \frac{150{,}000 \text{ gal}}{8125 \text{ gph}}$$

$$= \boxed{18.5 \text{ hrs}}$$

Example 2: (Detention Time)

❑ An oxidation ditch receives a flow of 150,000 gpd. If the volume of the oxidation ditch is 138,000 gallons, what is the detention time in hours?

$$\frac{150{,}000 \text{ gpd}^*}{24 \text{ hrs/day}}$$

$= 6250$ gph

Volume = 138,000 gal

$$\text{Detention Time, hrs} = \frac{\text{Volume of Oxidation Ditch, gal}}{\text{Flow Rate, gph}}$$

$$= \frac{138{,}000 \text{ gal Volume}}{6250 \text{ gph}}$$

$$= \boxed{22 \text{ hrs}}$$

Example 3: (Detention Time)

❑ An oxidation ditch receives a flow of 250,000 gpd. If the volume of the oxidation ditch is 187,000 gallons, what is the detention time in hours?

$$\frac{250{,}000 \text{ gpd}}{24 \text{ hrs/day}}$$

$= 10{,}417$ gph

Volume = x gal

$$\text{Detention Time, hrs} = \frac{\text{Volume of Ditch, gal}}{\text{Flow Rate, gph}}$$

$$= \frac{187{,}000 \text{ gal Volume}}{10{,}417 \text{ gph}}$$

$$= \boxed{18 \text{ hrs}}$$

* Refer to Chapter 8 in *Basic Math Concepts* for a review of flow rate conversions.

NOTES:

13 Waste Treatment Ponds

SUMMARY

1. BOD Loading

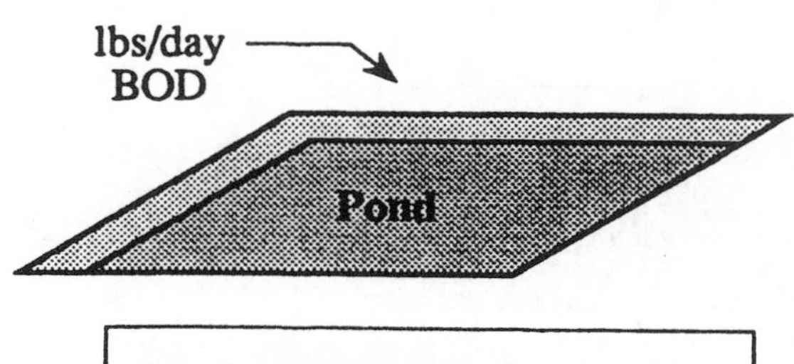

$$\underset{\text{BOD}}{(\text{mg}/L)}\ \underset{\text{flow}}{(\text{MGD})}\ \underset{\text{lbs/gal}}{(8.34)} = \underset{\text{BOD}}{\text{lbs/day}}$$

2. Organic Loading Rate

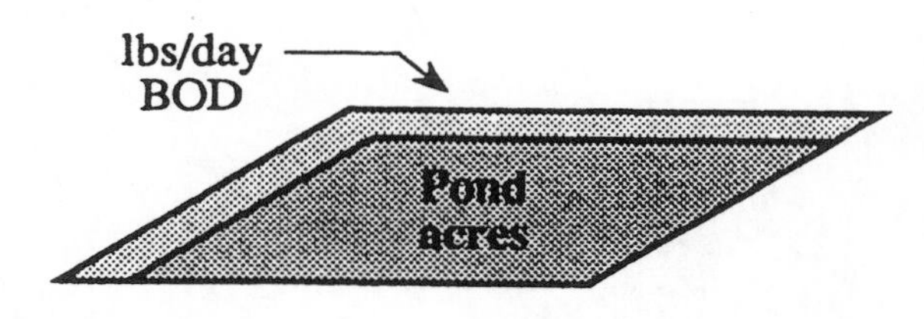

$$\text{Organic Loading} = \frac{(\text{mg}/L\ \text{BOD})\ (\text{MGD Flow})\ (8.34\ \text{lbs/gal})}{\text{acres}}$$

3. BOD Removal Efficiency

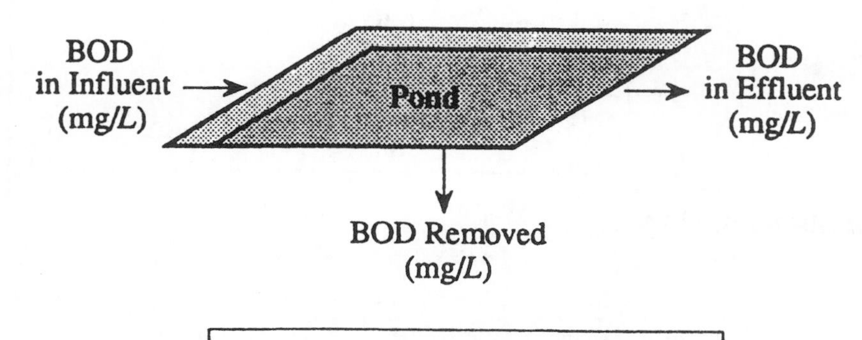

$$\%\ \text{BOD Removed} = \frac{\text{mg}/L\ \text{BOD Removed}}{\text{mg}/L\ \text{BOD Total}}$$

SUMMARY—Cont'd

4. Hydraulic Loading Rate

Hydraulic loading rate is often calculated as gpd/sq ft. However the common expression of hydraulic loading rate for ponds is ac-ft/day/acre or in./day.

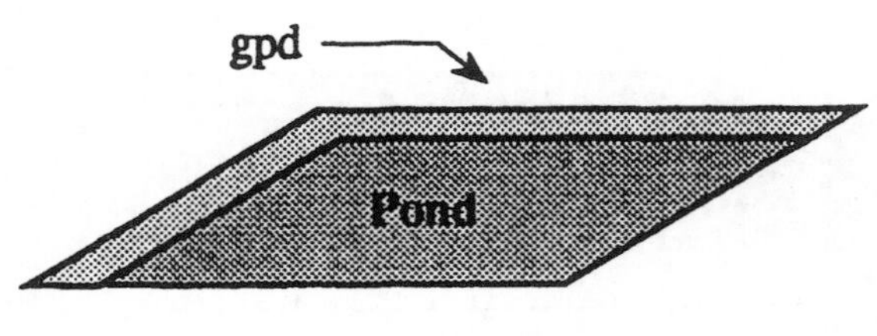

$$\text{Hydraulic Loading Rate} = \frac{\text{gpd flow}}{\text{sq ft Area}}$$

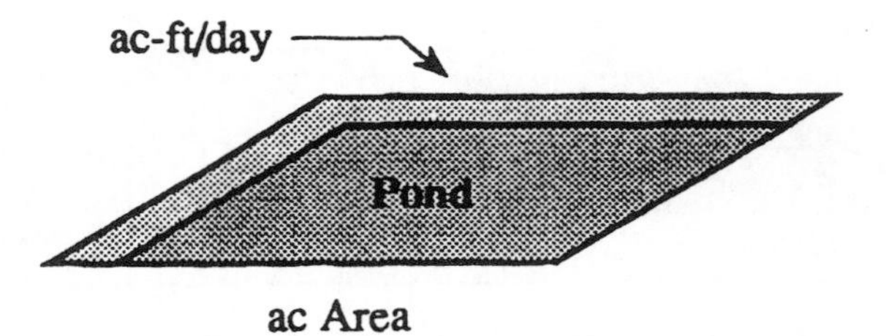

$$\text{Hydraulic Loading Rate} = \frac{\text{ac-ft/day}}{\text{ac}}$$

OR

$$\text{Hydraulic Loading Rate} = \frac{\text{in.}}{\text{day}}$$

5. Population Loading

$$\text{Population Loading} = \frac{\text{persons}}{\text{acre}}$$

SUMMARY—Cont'd

6. Detention Time

Depending on flow data, the detention time can be calculated using either equation shown below.

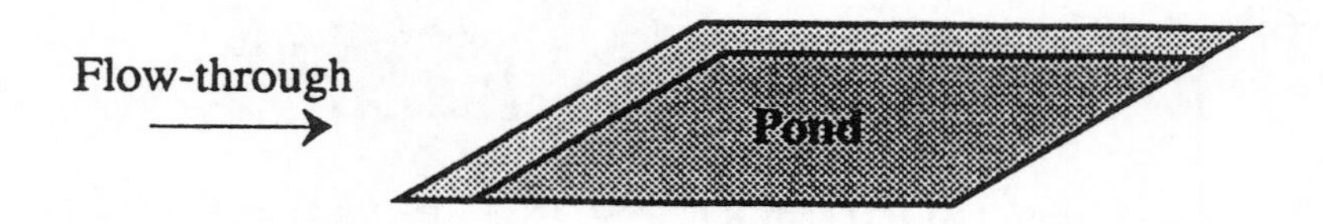

$$\text{Detention Time, days} = \frac{\text{Volume of Pond, gal}}{\text{Flow Rate, gpd}}$$

Or

$$\text{Detention Time, days} = \frac{\text{Volume of Pond, ac-ft}}{\text{Flow Rate, ac-ft/day}}$$

NOTES:

13.1 BOD LOADING

When calculating BOD loading on a waste treatment pond, the following equation is used:

(mg/L) (MGD) (8.34) = lbs/day
BOD flow lbs/gal

Loading on a system is usually calculated as lbs/day. Given the BOD concentration and flow information, the lbs/day loading may be calculated as demonstrated in Examples 1-4.

Example 1: (BOD Loading)

❑ Calculate the BOD loading (lbs/day) on a pond if the influent flow is 0.2 MGD with a BOD of 210 mg/L.

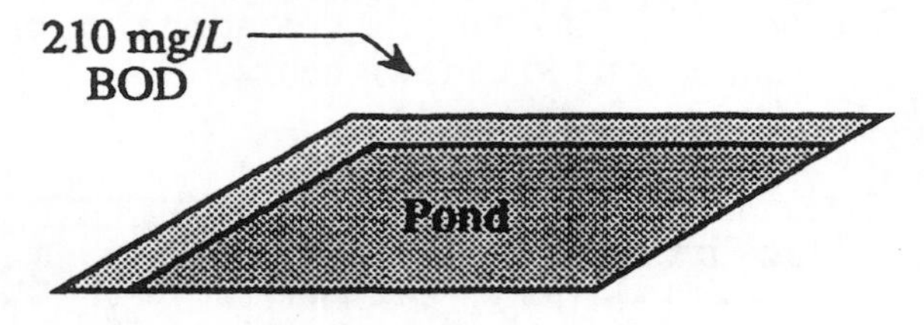

(mg/L BOD) (MGD flow) (8.34 lbs/gal) = lbs/day

(210 mg/L) (0.2 MGD) (8.34 lbs/gal) = **350 lbs/day BOD**

Example 2: (BOD Loading)

❑ The BOD concentration of the wastewater entering a pond is 195 mg/L. If the flow to the pond is 180,000 gpd, how many lbs/day BOD enter the pond?

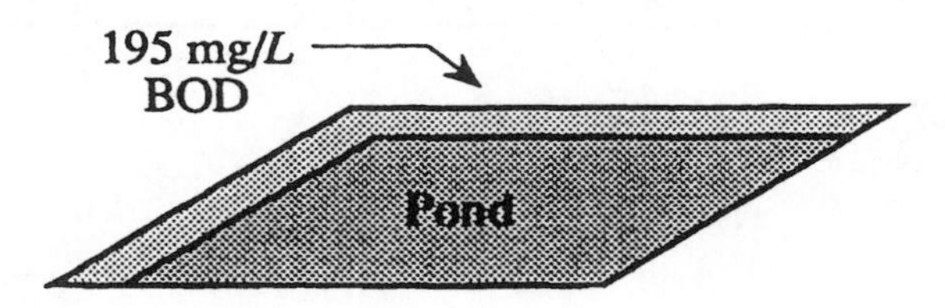

(mg/L) (MGD flow) (8.34 lbs/gal) = lbs/day
BOD

(195 mg/L) (0.18 MGD) (8.34 lbs/gal) = **293 lbs/day BOD**

Sometimes flow information will not be given in the desired terms (MGD flow). When this is the case, convert the flow rate given (such as gpd, gpm, or cfs) to MGD flow. *

Example 3: (BOD Loading)

❑ The flow to a waste treatment pond is 155 gpm. If the BOD concentration of the water is 270 mg/L, how many pounds of BOD are applied to the pond daily?

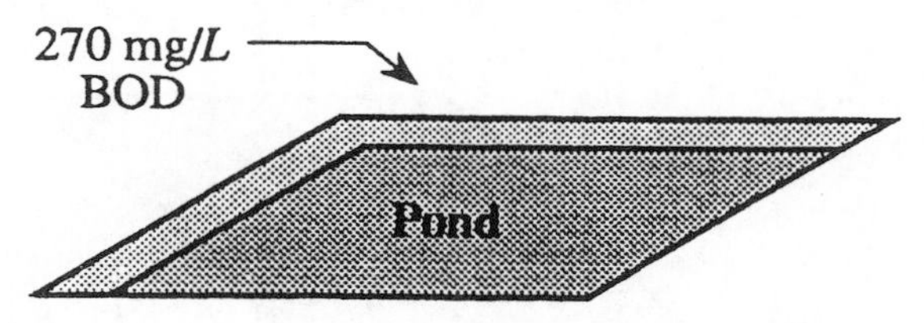

Before the mg/*L* to lbs/day equation can be used the gpm flow must be expressed in terms of MGD flow:*

$$\left(155\ \frac{\text{gal}}{\text{min}}\right)\left(1440\ \frac{\text{min}}{\text{day}}\right) = 223{,}200\ \text{gpd}$$

$$\text{or} = 0.223\ \text{MGD}$$

Now use the mg/*L* equation to solve the problem:

$$(\text{mg/}L)\ (\text{MGD flow})\ (8.34\ \text{lbs/gal}) = \text{lbs/day}$$

$$\underset{\text{BOD}}{(270\ \text{mg/}L)}\ (0.223\ \text{MGD})\ (8.34\ \text{lbs/gal}) = \boxed{\underset{\text{BOD}}{502\ \text{lbs/day}}}$$

Example 4: (BOD Loading)

❑ The daily flow to a pond is 310,000 gpd. If the BOD concentration of the wastewater is 410 mg/L, how many pounds of BOD are applied to the pond daily?

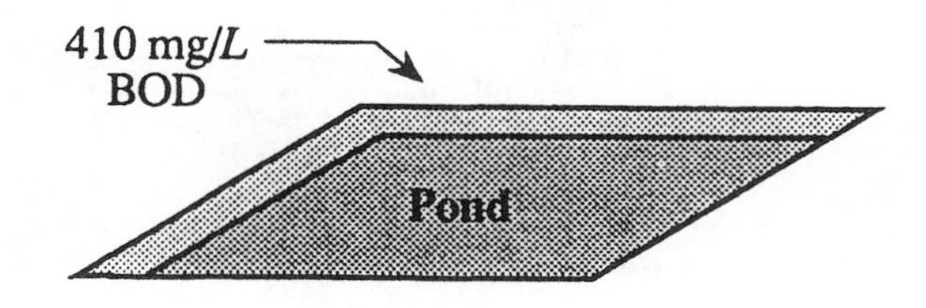

$$(\text{mg/}L)\ (\text{MGD flow})\ (8.34\ \text{lbs/gal}) = \text{lbs/day}$$

$$\underset{\text{BOD}}{(410\ \text{mg/}L)}\ (0.31\ \text{MGD})\ (8.34\ \text{lbs/gal}) = \boxed{\underset{\text{BOD}}{1060\ \text{lbs/day}}}$$

---

* Refer to Chapter 8 in *Basic Math Concepts* for a review of flow conversions.

13.2 ORGANIC LOADING RATE

Organic loading rate is a calculation which expresses BOD loading in lbs BOD/day per acre of pond area. Examples 1-3 illustrate this calculation.

ORGANIC LOADING RATE IS
BOD LOADING PER ACRE OF POND

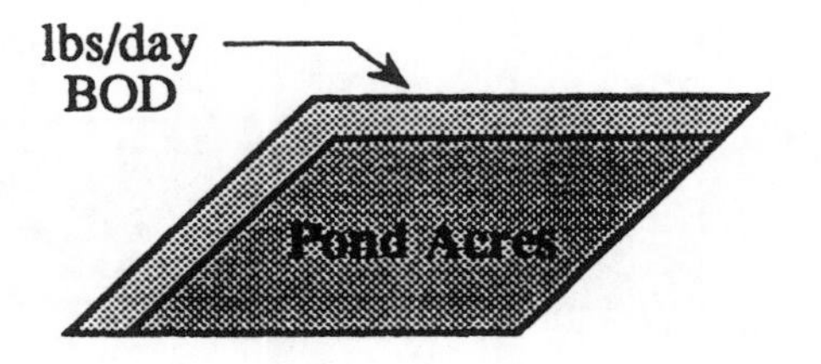

Simplified Equation:

$$\text{Organic Loading Rate} = \frac{\text{BOD, lbs/day}}{\text{Area, acres}}$$

Expanded Equation:

$$\text{Organic Loading Rate} = \frac{(\text{mg/}L\text{ BOD})\ (\text{MGD Flow})\ (8.34\text{ lbs/gal})}{\text{Area, acres}}$$

Example 1: (Organic Loading Rate)

❑ A 5.5-acre pond receives a flow of 170,000 gpd. If the influent flow has a BOD content of 160 mg/*L*, what is the organic loading rate on the pond in lbs BOD/day/ac?

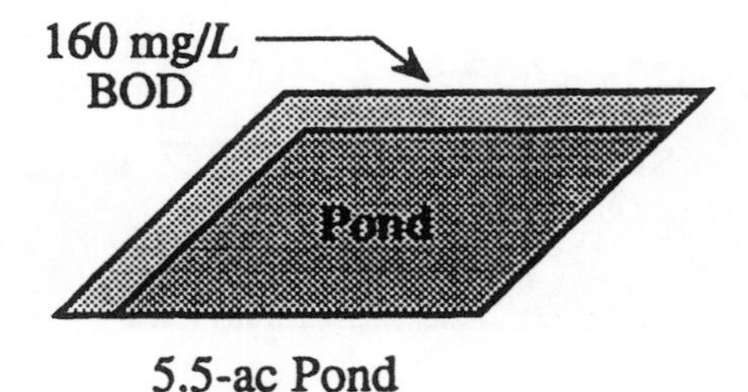

$$\text{Organic Loading Rate} = \frac{(160\text{ mg/}L\text{ BOD})\ (0.17\text{ MGD})\ (8.34\text{ lbs/gal})}{5.5\text{ ac}}$$

$$= \boxed{\frac{41\text{ lbs BOD/day}}{\text{ac}}}$$

Example 2: (Organic Loading Rate)

❑ A 22-acre pond receives a flow of 600,000 gpd. If the influent flow to the pond has a BOD concentration of 220 mg/*L*, what is the organic loading on the pond in lbs BOD/day/ac?

$$\text{Organic Loading Rate} = \frac{(\text{mg/}L\ \text{BOD})\ (\text{MGD Flow})\ (8.34\ \text{lbs/gal})}{\text{ac}}$$

$$= \frac{(220\ \text{mg/}L\ \text{BOD})\ (0.6\ \text{MGD})\ (8.34\ \text{lbs/gal})}{22\ \text{ac}}$$

$$= \boxed{\frac{50\ \text{lbs BOD/day}}{\text{ac}}}$$

Example 3: (Organic Loading Rate)

❑ A pond has an average width of 380 ft and an average length of 620 ft. The flow to the pond is 121,000 gpd, with a BOD content of 185 mg/*L*. What is the organic loading rate on the pond in lbs BOD/day/ac?

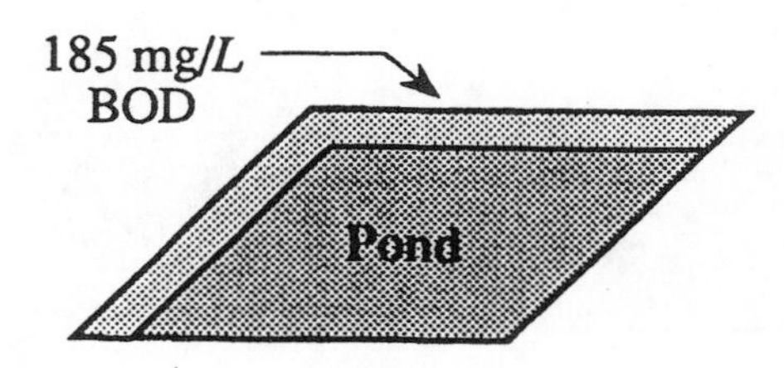

First calculate the pond area, in acres*:

$$(380\ \text{ft})\ (620\ \text{ft}) = 235{,}600\ \text{sq ft}$$

$$\frac{235{,}600\ \text{sq ft}}{43{,}560\ \text{sq ft/ac}} = 5.4\ \text{ac}$$

Now continue with the organic loading rate calculation:

$$\text{Organic Loading Rate} = \frac{(185\ \text{mg/}L\ \text{BOD})\ (0.121\ \text{MGD})\ (8.34\ \text{lbs/gal})}{5.4\ \text{ac}}$$

$$= \boxed{\frac{35\ \text{lbs BOD/day}}{\text{ac}}}$$

WHEN POND DIMENSIONS ARE GIVEN

The denominator of the organic loading rate calculation is acres. Therefore, if pond dimensions are given instead of acres, you will need to calculate the sq ft area for the pond and convert sq ft area to acres area.*

* To review the calculation of acres area, refer to Chapter 10 in *Basic Math Concepts*.

13.3 BOD REMOVAL EFFICIENCY

The efficiency of a treatment process is its effectiveness in removing various constituents from the water or wastewater. BOD removal efficiency is therefore a measure of the effectiveness of the waste treatment pond in removing BOD from the wastewater.

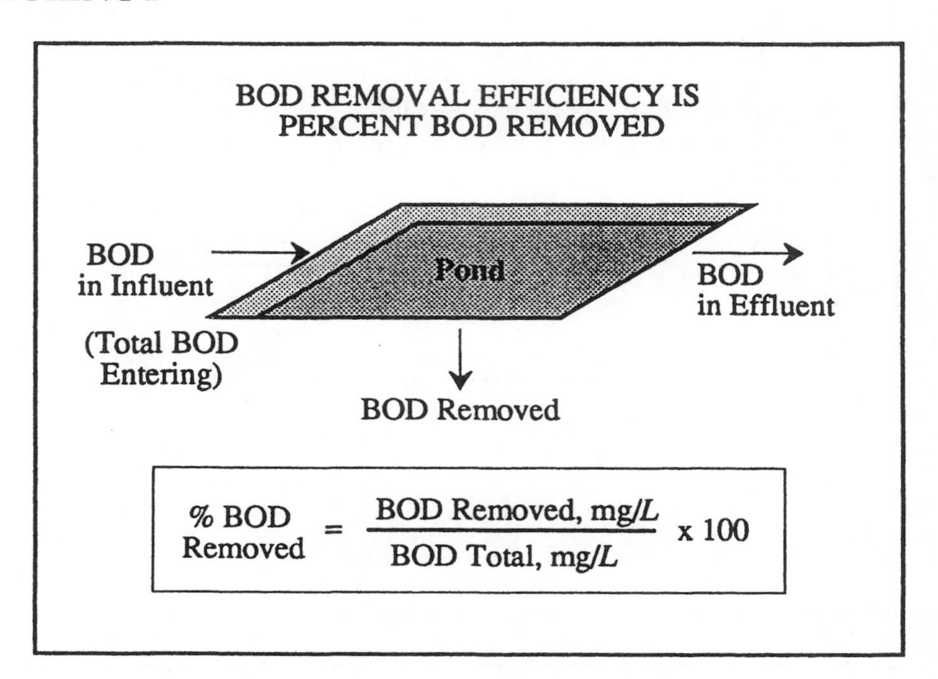

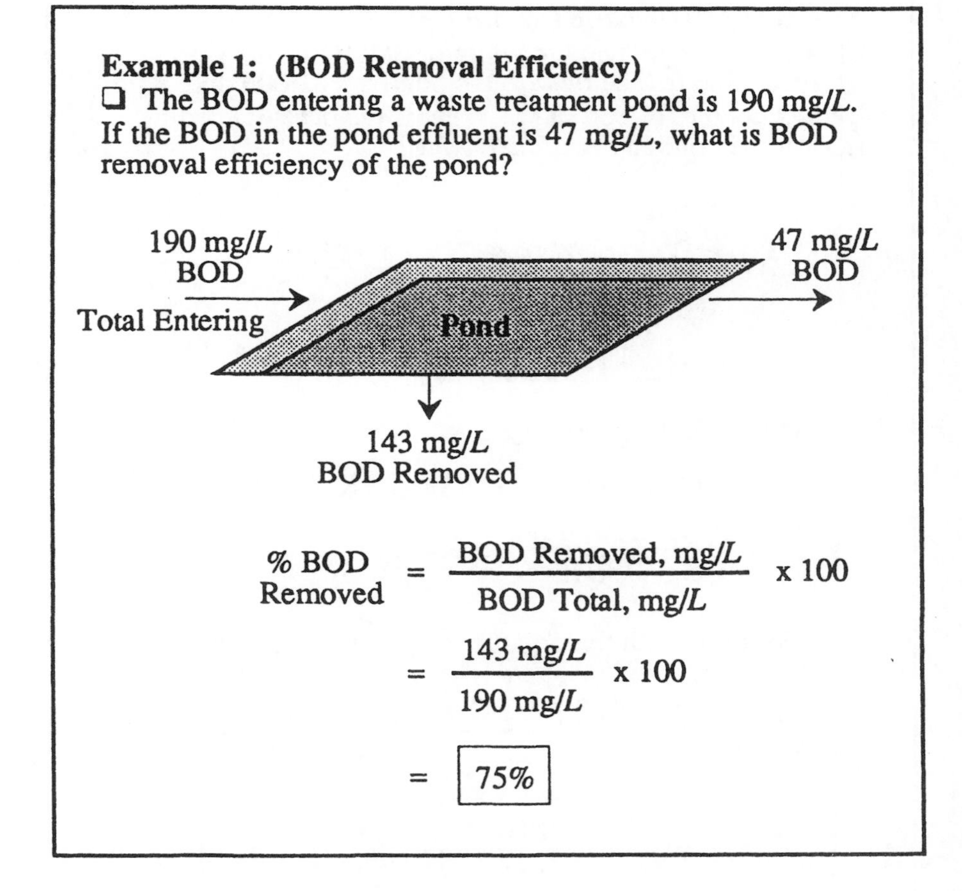

Example 2: (BOD Removal Efficiency)

❑ The influent of a waste treatment pond has a BOD content of 250 mg/*L*. If the BOD content of the pond effluent is 70 mg/*L*, what is the BOD removal efficiency of the pond?

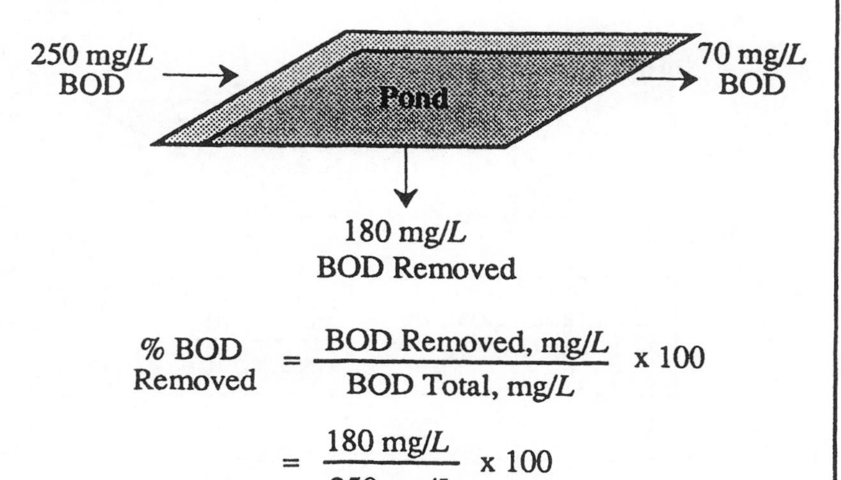

$$\%\ \text{BOD Removed} = \frac{\text{BOD Removed, mg}/L}{\text{BOD Total, mg}/L} \times 100$$

$$= \frac{180\ \text{mg}/L}{250\ \text{mg}/L} \times 100$$

$$= \boxed{72\%}$$

Example 3: (BOD Removal Efficiency)

❑ The BOD entering a waste treatment pond is 220 mg/*L*. The BOD concentration of the pond effluent is 44 mg/*L*. What is the BOD removal efficiency of the pond?

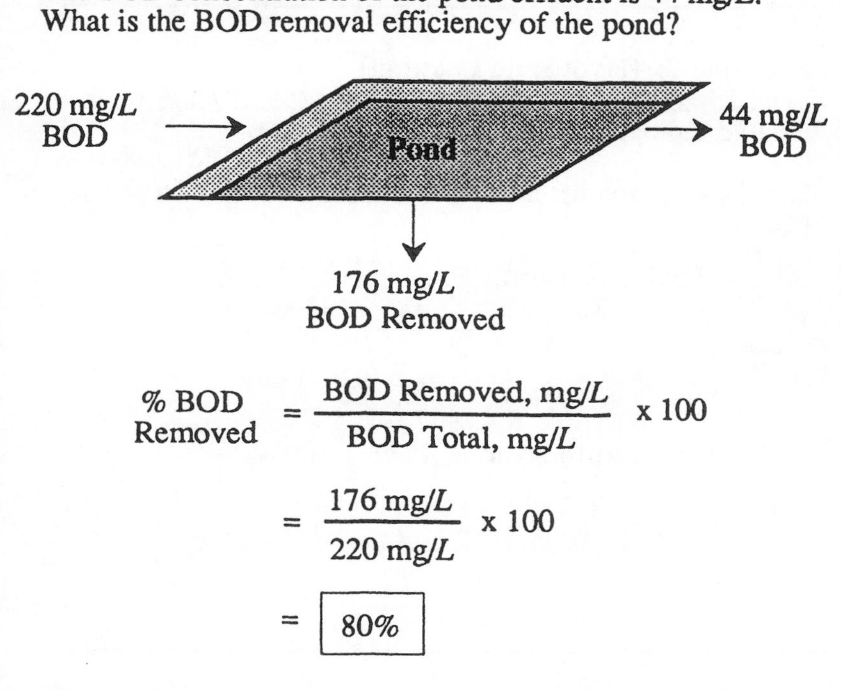

$$\%\ \text{BOD Removed} = \frac{\text{BOD Removed, mg}/L}{\text{BOD Total, mg}/L} \times 100$$

$$= \frac{176\ \text{mg}/L}{220\ \text{mg}/L} \times 100$$

$$= \boxed{80\%}$$

13.4 HYDRAULIC LOADING RATE

Hydraulic loading rate is a term used to indicate the total flow, in gpd, loaded or entering each square foot of water surface area. It is the total gpd flow to the process divided by the water surface area of the pond. **Recirculated flows must be included** as part of the total flow (total Q) to the pond.

HYDRAULIC LOADING RATE CAN BE CALCULATED USING THREE DIFFERENT EQUATIONS, DEPENDING ON WHICH DATA IS GIVEN

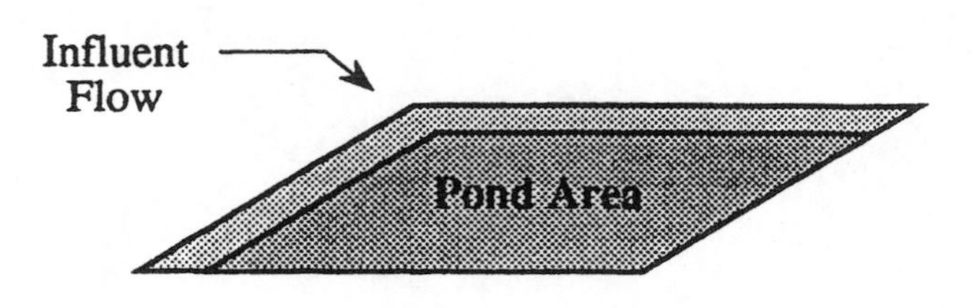

$$\text{Hydraulic Loading Rate} = \frac{\text{Flow, gpd}}{\text{Area, sq ft}}$$

OR

$$\text{Hydraulic Loading Rate} = \frac{\text{Flow, ac-ft/day}}{\text{Acre, ac}}$$

OR

$$\text{Hydraulic Loading Rate} = \frac{\text{in.}}{\text{day}}$$

When calculating hydraulic loading for wastewater ponds, the answer is often expressed as in./day rather than gpd/sq ft.

There are two ways to calculate in./day hydraulic loading, depending on how the flow to the pond is expressed.

If the flow to the pond is expressed in acre-feet/day:

1. Set up the hydraulic loading equation using acre-ft/day flow per acres area.

2. Canceling terms results in ft/day hydraulic loading.

$$\frac{\text{ac-ft/day}}{\text{ac}} = \text{ft/day}$$

3. Then convert ft/day to in./day:

$$\left(\frac{\text{ft}}{\text{day}}\right)\left(\frac{\text{in.}}{\text{ft}}\right) = \frac{\text{in.}}{\text{day}}$$

Example 1: (Hydraulic Loading)

❑ A 25-acre pond receives a flow of 6.2 ac-ft/day. What is the hydraulic loading on the pond in in./day?

Use the equation for hydraulic loading that uses acre-ft/day flow:

$$\text{Hydraulic Loading Rate} = \frac{6.2 \text{ ac-ft/day}}{25 \text{ ac}}$$

$$= 0.25 \text{ ft/day}$$

Then convert ft/day to in./day:

$$(0.25 \text{ ft/day})(12 \text{ in./ft}) = \boxed{3 \text{ in./day}}$$

Example 2: (Hydraulic Loading)

❑ A 25-acre pond receives a flow of 5.8 ac-ft/day. What is the hydraulic loading on the pond in ac-ft/day/ac?

$$\text{Hydraulic Loading Rate} = \frac{\text{Flow, gpd}}{\text{Area, sq ft}}$$

$$= \frac{5.8 \text{ ac-ft/day}}{25 \text{ ac}}$$

$$= \boxed{0.23 \text{ ac-ft/day/ac}}$$

Example 3: (Hydraulic Loading)

❑ A waste treatment pond receives a flow of 2,400,000 gpd. If the surface area of the pond is 15 acres, what is the hydraulic loading in in./day?

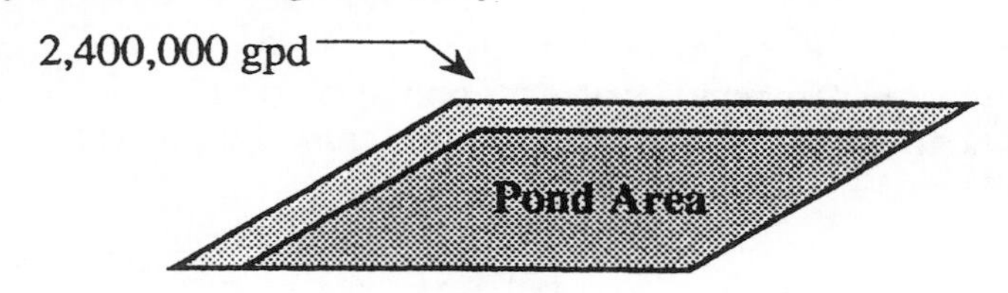

$$\text{Area} = 15 \text{ ac}$$

or $(15 \text{ ac})(43{,}560 \text{ sq ft/ac})$

$$= 653{,}400 \text{ sq ft}$$

$$\text{Hydraulic Loading Rate} = \frac{2{,}400{,}000 \text{ gpd}}{653{,}400 \text{ sq ft}}$$

Convert gpd flow to ft^3/day flow (2,400,000 gpd ÷ 7.48 gal/cu ft = 320,856 ft^3/day):

$$= \frac{320{,}856 \text{ ft}^3\text{/day}}{653{,}400 \text{ ft}^2}$$

$$= 0.5 \text{ ft/day}$$

Then convert to in./day:

$$(0.5 \text{ ft/day})(12 \text{ in./ft}) = \boxed{6 \text{ in./day}}$$

If the flow to the pond is expressed in gpd:

1. Set up the hydraulic loading equation as gpd/sq ft.
2. Convert gpd flow to cubic feet per day flow (cfd or ft^3/day). This is done by dividing gpd by 7.48 gal/cu ft.*
3. Cancel terms to obtain ft/day hydraulic loading.**

$$\frac{\text{cu ft/day}}{\text{sq ft}} = \frac{\text{ft}^3\text{/day}}{\text{ft}^2} = \text{ft/day}$$

4. Then convert ft/day to in./day (by multiplying by 12 in./ft).

(Note: There is a shortcut method to determine in./day hydraulic loading using the conversion factor of 1 gpd/sq ft = 1.6 in./day. To use the shortcut method, first calculate the hydraulic loading rate in gpd/sq ft. Then multiply by the conversion factor of 1.6 to obtain hydraulic loading, in./day.)

* For a review of flow conversions, refer to Chapter 8 in *Basic Math Concepts*.

** To review cancellation of terms, refer to Chapter 15, "Dimensional Analysis", in *Basic Math Concepts*.

13.5 POPULATION LOADING AND POPULATION EQUIVALENT

POPULATION LOADING

Population loading is a calculation associated with wastewater treatment by ponds. Population loading is an indirect measure of both water and solids loading to a system. It is calculated as the number of persons served per acre of pond:

$$\text{Population Loading} = \frac{\text{Persons}}{\text{Acre}}$$

Example 1: (Population Loading)

❑ A 3.5-acre wastewater pond serves a population of 1500. What is the population loading on the pond?

$$\text{Population Loading} = \frac{\text{Persons}}{\text{Acre}}$$

$$= \frac{1500 \text{ persons}}{3.5 \text{ acres}}$$

$$= \boxed{429 \frac{\text{persons}}{\text{acre}}}$$

Example 2: (Population Loading)

❑ A wastewater pond serves a population of 4000. If the pond is 16 acres, what is the population loading on the pond?

$$\text{Population Loading} = \frac{\text{persons}}{\text{acre}}$$

$$= \frac{4000 \text{ persons}}{16 \text{ acres}}$$

$$= \boxed{250 \frac{\text{persons}}{\text{acre}}}$$

POPULATION EQUIVALENT

Industrial or commercial wastewater generally has a higher organic content than domestic wastewater. Population equivalent calculations equate these concentrated flows with the number of people that would produce a domestic wastewater with that organic load. For a domestic wastewater system, each person served by the system contributes about 0.17 to 0.2 lbs BOD/day. To determine the population equivalent of a wastewater flow, therefore, divide the lbs BOD/day content by the lbs BOD/day contributed per person (e.g., 0.2 lbs BOD/day).

Example 3: (Population Equivalent)

❏ A 0.4-MGD wastewater flow has a BOD concentration of 1800 mg/*L* BOD. Using an average of 0.2 lbs BOD/day/person, what is the population equivalent of this wastewater flow?

$$\text{Population Equivalent} = \frac{\text{BOD, lbs/day}}{\text{lbs BOD/day/person}}$$

Convert mg/*L* BOD to lbs/day BOD* then divide by 0.2 lbs BOD/day/person:

$$\text{Population Equivalent} = \frac{(1800\ \text{mg}/L)\ (0.4\ \text{MGD})\ (8.34\ \text{lbs/gal})}{0.2\ \text{lbs BOD/day/person}}$$

$$= \boxed{30{,}024\ \text{people}}$$

Example 4: (Population Equivalent)

❏ A 100,000-gpd wastewater flow has a BOD content of 2800 mg/*L*. Using an average of 0.2 lbs/day BOD/person, what is the population equivalent of this flow?

$$\text{Population Equivalent} = \frac{\text{BOD, lbs/day}}{\text{lbs BOD/day/person}}$$

$$= \frac{(2800\ \text{mg}/L)\ (0.1\ \text{MGD})\ (8.34\ \text{lbs/gal})}{0.2\ \text{lbs BOD/day/person}}$$

$$= \boxed{11{,}676\ \text{people}}$$

* For a review of mg/*L* to lbs/day calculations, refer to Chapter 3.

13.6 DETENTION TIME

There are many possible ways of writing the detention time equation, depending on the time unit desired (seconds, minutes, hours, days) and the expression of volume and flow rate.

When calculating detention time, it is essential that the time and volume units used in the equation are consistent with each other, as illustrated to the right.

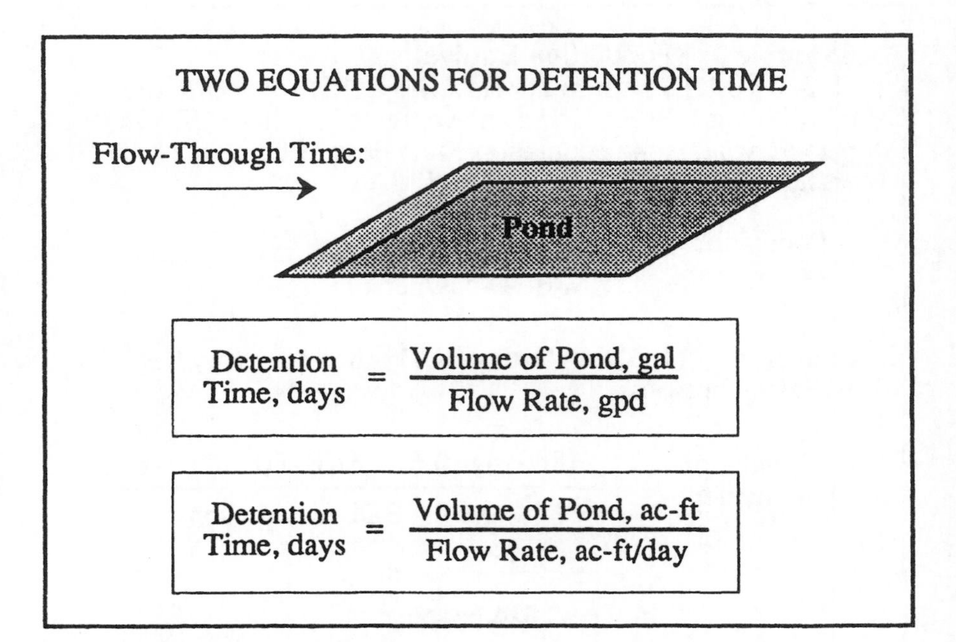

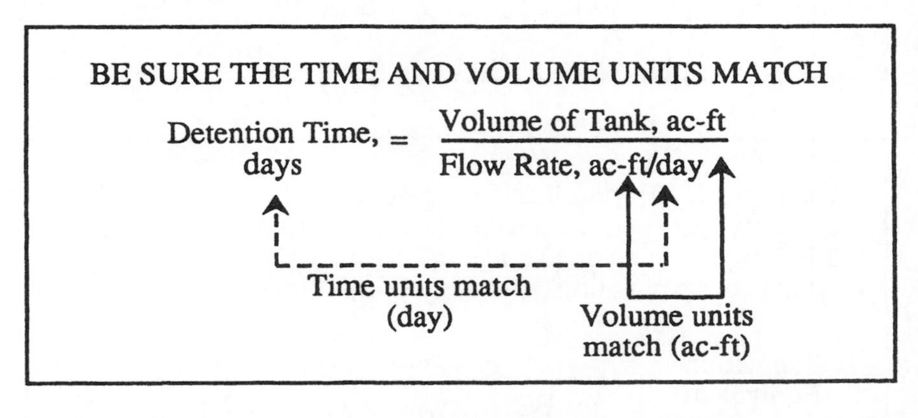

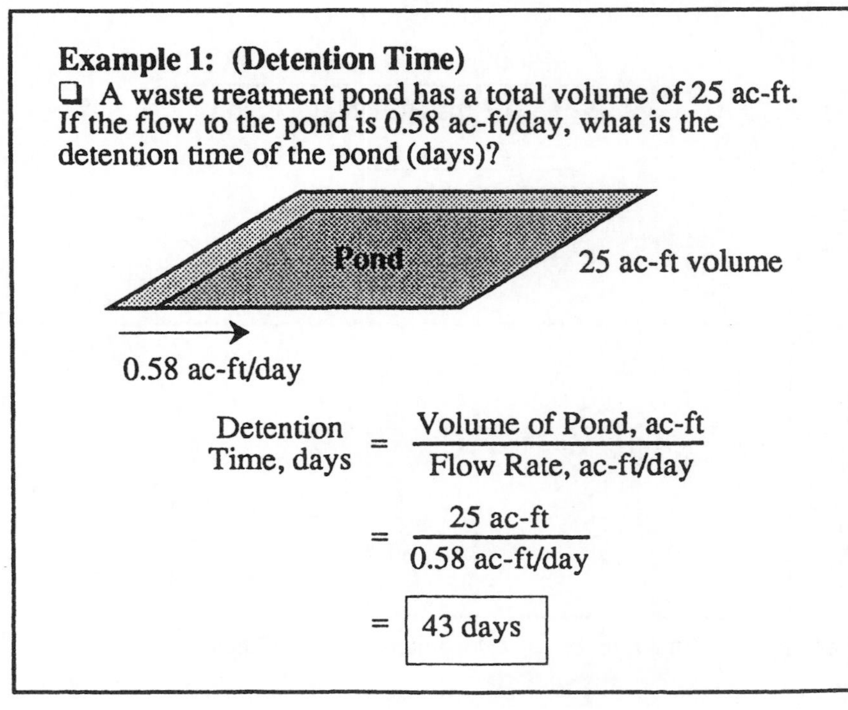

Example 2: (Detention Time)

❑ A waste treatment pond is operated at a depth of 5 feet. The average width of the pond is 390 ft and the average length is 640 ft. If the flow to the pond is 540,000 gpd, what is the detention time, in days?

$$(640 \text{ ft})(390 \text{ ft})(5 \text{ ft})(7.48 \text{ gal/cu ft}) = \underset{\text{Volume}}{9{,}335{,}040 \text{ gal}}$$

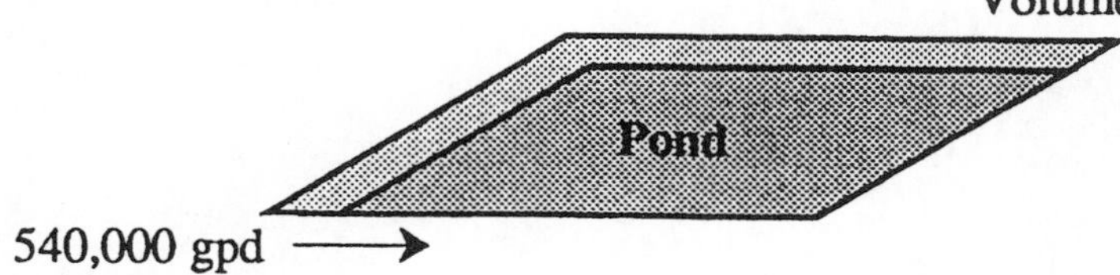

$$\text{Detention Time, days} = \frac{\text{Volume of Pond, gal}}{\text{Flow Rate, gpd}}$$

$$= \frac{9{,}335{,}040 \text{ gal Volume}}{540{,}000 \text{ gpd}}$$

$$= \boxed{17 \text{ days}}$$

Example 3: (Detention Time)

❑ A waste treatment pond has an average length of 650 ft, an average width of 450 ft, and a water depth of 4 ft. If the flow to the pond is 0.5 ac-ft/day what is the detention time for the pond in days?

First calculate cu ft volume then ac-ft volume:

$$(650 \text{ ft})(450 \text{ ft})(4 \text{ ft}) = 1{,}170{,}000 \text{ cu ft}$$

$$\frac{1{,}170{,}000 \text{ cu ft}}{43{,}560 \text{ cu ft/ac-ft}} = \boxed{26.9 \text{ ac-ft}}$$

Then calculate detention time:

$$\text{Detention Time, days} = \frac{26.9 \text{ ac-ft}}{0.5 \text{ ac-ft/day}}$$

$$= \boxed{53.8 \text{ days}}$$

CALCULATING ACRE-FEET VOLUME

Occasionally it will be necessary to calculate ac-ft volume of a pond in order to complete a detention time calculation. Once cu ft volume of the pond has been calculated,* ac-ft volume can be determined:

$$\text{Volume, ac-ft} = \frac{\text{Volume, cu ft}}{43{,}560 \text{ cu ft/ac ft}}$$

* Refer to Chapter 11 in *Basic Math Concepts* for a discussion of volume.

NOTES:

14 *Chemical Dosage*

SUMMARY

1. Chemical Feed Rate, lbs/day—Full-Strength Chemicals

$$(\text{mg}/L\ \text{Chem.})\ (\text{MGD flow})\ (8.34\ \text{lbs/gal}) = \begin{array}{c}\text{lbs/day}\\ \text{Chemical}\end{array}$$

2. Chlorine Dose, Demand and Residual

$$Cl_2\ \text{Dose} = Cl_2\ \text{Demand} + Cl_2\ \text{Residual}$$

3. Chemical Feed Rate, lbs/day—Less Than Full-Strength Chemicals

$$\frac{(\text{mg}/L\ \text{Chem.})\ (\text{MGD flow})\ (8.34\ \text{lbs/gal})}{\frac{\%\ \text{Strength of Chemical}}{100}} = \begin{array}{c}\text{lbs/day}\\ \text{Chemical}\end{array}$$

OR

$$\frac{(\text{mg}/L\ \text{Chem.})\ (\text{MG Tank Vol.})\ (8.34\ \text{lbs/gal})}{\frac{\%\ \text{Strength of Chemical}}{100}} = \begin{array}{c}\text{lbs/day}\\ \text{Chemical}\end{array}$$

4. Percent Strength of Solutions

Percent strength using dry chemicals:

$$\%\ \text{Strength} = \frac{\text{Chemical, lbs}}{\text{Solution, lbs}} \times 100$$

Percent strength using liquid chemicals:

$$\begin{array}{c}(\text{Liquid Polymer})\\ \text{lbs}\end{array} \left(\frac{\%\ \text{Strength of Liq. Poly.}}{100}\right) = \begin{array}{c}(\text{Polymer Solution})\\ \text{lbs}\end{array} \left(\frac{\%\text{Strength of Poly. Soln.}}{100}\right)$$

SUMMARY—Cont'd

5. Mixing Solutions of Different Strength

$$\text{\% Strength of Mixture} = \frac{\text{lbs Chemical in Mixture}}{\text{lbs Solution Mixture}} \times 100$$

OR, if target strength is desired:

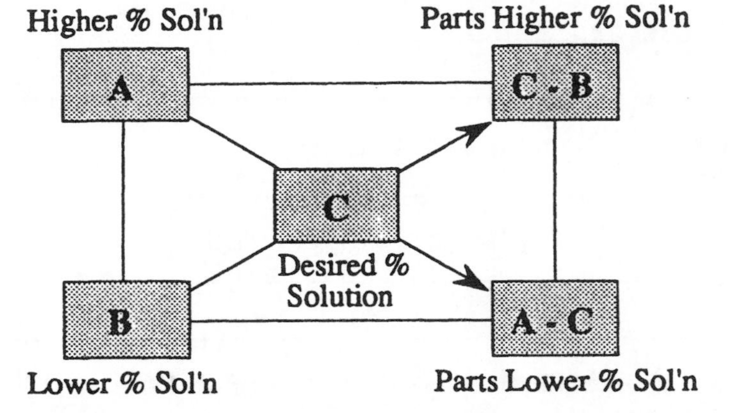

6. Solution Chemical Feeder Setting, gpd

Simplified Equation:

$$\underset{\text{lbs/day}}{\text{Desired Dose}} = \underset{\text{lbs/day}}{\text{Actual Dose}}$$

Expanded Equation:

$$\underset{\text{Dose}}{(\text{mg/}L)}\ \underset{\text{Treated}}{(\text{MGD Flow})}\ \underset{\text{lbs/day}}{(8.34)} = \underset{\text{Sol'n}}{(\text{mg/}L)}\ \underset{\text{Sol'n}}{(\text{MGD})}\ \underset{\text{lbs/gal}}{(8.34)}$$

7. Chemical Feed Pump—% Stroke Setting

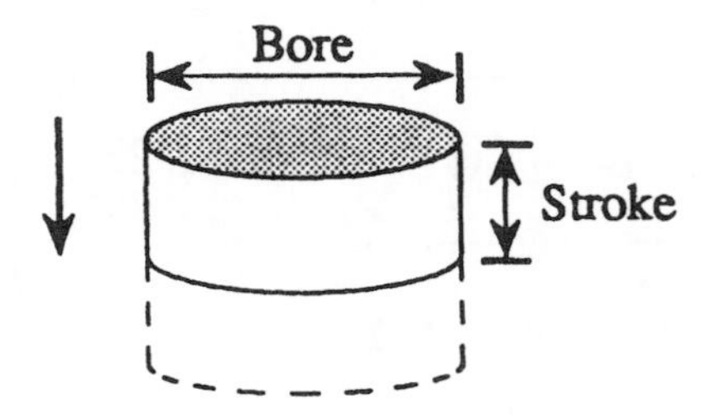

$$\text{\% Setting} = \frac{\text{Required Feed Pump, gpd}}{\text{Maximum Feed Pump, gpd}} \times 100$$

SUMMARY—Cont'd

8. Solution Chemical Feeder Setting, mL/min

First calculate gpd solution flow required:

Simplified Equation:

$$\text{Desired Dose, lbs/day} = \text{Actual Dose, lbs/day}$$

Expanded Equation:

$$\underset{\text{Dose}}{(\text{mg}/L)}\ \underset{\text{Flow Treated}}{(\text{MGD})}\ \underset{\text{lbs/gal}}{(8.34)} = \underset{\text{Sol'n}}{(\text{mg}/L)}\ \underset{\text{Sol'n Flow}}{(\text{MGD})}\ \underset{\text{lbs/gal}}{(8.34)}$$

Then convert mL/min solution flow required:

$$\frac{(\text{gpd flow})\ (3785\ \text{m}L/\text{gal})}{1440\ \text{min/day}} = \text{Chemical, m}L/\text{min}$$

9. Dry Chemical Feeder Calibration

$$\text{Chemical Feed Rate, lbs/day} = \frac{\text{Chemical Used, lbs}}{\text{Application Time, days}}$$

10. Solution Chemical Feeder Calibration
(Given mL/min Flow)

First convert m^/min flow to gpd flow:

$$\frac{(\text{m}L/\text{min})\ (1440\ \text{min/day})}{3875\ \text{m}L/\text{gal}} = \text{gpd}$$

Then calculate chemical dosage, lbs/day:

$$(\text{mg}/L\ \text{Chem.})\ (\text{MGD flow})\ (8.34\ \text{lbs/gal}) = \text{Chemical, lbs/day}$$

SUMMARY—Cont'd

11. Solution Chemical Feeder Calibration
(Given Drop in Solution Tank Level)

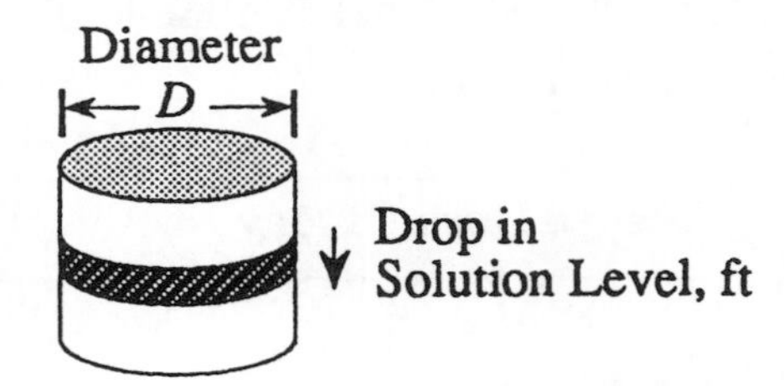

$$\text{Flow, gpm} = \frac{\text{Volume Pumped, gal}}{\text{Duration of Test, min}}$$

Or

$$\text{Flow, gpm} = \frac{(0.785)\,(D^2)\,(\text{Drop, ft})\,(7.48\text{ gal/cu ft})}{\text{Duration of Test, min}}$$

12. Average Use Calculations

First determine the average chemical use:

$$\text{Average Use, lbs/day} = \frac{\text{Total Chem. Used, lbs}}{\text{Number of Days}}$$

Or

$$\text{Average Use, gpd} = \frac{\text{Total Chem. Used, gal}}{\text{Number of Days}}$$

Then calculate day's supply in inventory

$$\text{Day's Supply in Inventory} = \frac{\text{Total Chem. in Inventory, lbs}}{\text{Average Use, lbs/day}}$$

Or

$$\text{Day's Supply in Inventory} = \frac{\text{Total Chem. in Inventory, gal}}{\text{Average Use, gpd}}$$

NOTES:

14.1 CHEMICAL FEED RATE—(Dosing Full-Strength Chemicals)

In chemical dosing, a measured amount of chemical is added to the water or wastewater. The amount of chemical required depends on such factors as the type of chemical used, the reason for dosing, and the flow rate being treated.

The two expressions most often used to describe the amount of chlorine added or required are:

- milligrams per liter (mg/*L*)*, and
- pounds per day (lbs/day)

CHLORINE DOSAGE

Wastewater may be chlorinated during various stages of treatment. For example, chlorination in the early stages of treatment may be practiced for odor control. In other cases, chlorination may be used to aid in grease removal or BOD reduction. In the chlorination of secondary effluent, disinfection is the principal objective. Chlorine is added to kill the disease-causing organisms which are a potential health hazard if discharged into receiving waters used for human consumption or water contact sports such as swimming.

MILLIGRAMS PER LITER IS A MEASURE OF CONCENTRATION

Assume each liter below is divided into 1 million parts:

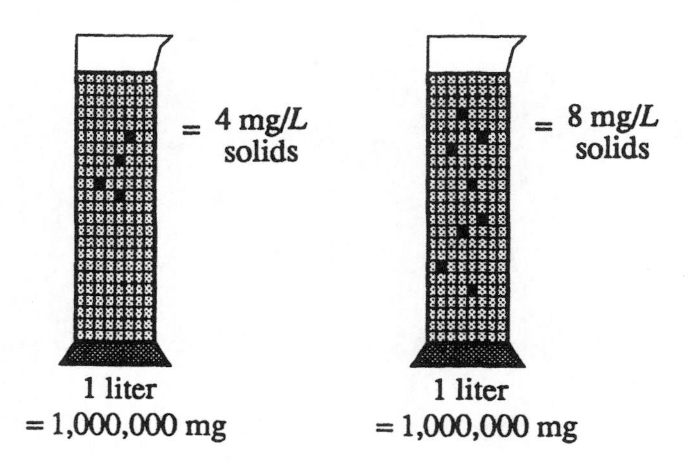

The mg/*L* concentration expresses a ratio of the milligrams chemical in each liter of water. For example, if a concentration of 4 mg/*L* is desired, then a total of 12 mg chemical would be required to treat 3 liters:

$$\frac{4 \text{ mg} \times 3}{L \times 3} = \frac{12 \text{ mg}}{3\,L}$$

The amount of chlorine required therefore depends on two factors:

- The desired concentration (mg/*L*), and
- The amount of water to be treated (normally expressed as MGD).

To convert from mg/*L* to lbs/day, the following equation is used:

$$\underset{\text{Chem.}}{(\text{mg/}L)}\ \underset{\text{flow}}{(\text{MGD})}\ \underset{\text{lbs/gal}}{(8.34)} = \text{lbs/day}$$

* For most water and wastewater calculations, mg/*L* concentration and ppm concentration may be used interchangeably. That is, 1 mg/*L* = 1 ppm. Of the two expressions, mg/*L* is preferred.

Example 1: (Chemical Feed Rate)
❑ Determine the chlorinator setting (lbs/day) needed to treat a flow of 3 MGD with a chlorine dose of 4 mg/*L*.

First write the equation. Then fill in the information given:

(mg/*L* Cl_2) (MGD flow) (8.34 lbs/gal) = lbs/day Cl_2

(4 mg/*L* Cl_2) (3 MGD) (8.34 lbs/gal) = **100 lbs/day Cl_2**

Example 2: (Chemical Feed Rate)
❑ The desired dosage for a dry polymer is 12 mg/*L*. If the flow to be treated is 2,160,000 gpd, how many lbs/day polymer will be required?

(mg/*L* Chem.) (MGD flow) (8.34 lbs/day) = lbs/day Polymer

(12 mg/*L* Polymer) (2.16 MGD) (8.34 lbs/day) = **216 lbs/day Polymer**

DOSAGE OF OTHER CHEMICALS

When calculating the dosage rate for other chemicals such as alum, polymer, or lime, the same equation is used as for chlorine dosage.

Example 3: (Chemical Feet Rate)
❑ To neutralize a sour digester, one pound of lime is to be added for every pound of volatile acids in the digester sludge. If the digester contains 250,000 gal of sludge with a volatile acid (VA) level of 2,300 mg/*L*, how many pounds of lime should be added?

Since the VA concentration is 2300 mg/*L*, the lime concentration should also be 2300 mg/*L*:

(mg/*L*) Lime (MG) Dig.Vol (8.34) lbs/gal = lbs Lime Required

(2300 mg/*L*) (0.25 MG) (8.34 lbs/gal) = **4,796 lbs lime**

DOSAGE IN A TANK

To calculate chemical dose for tanks or pipelines, a slightly modified equation must be used. Instead of MGD flow, MG volume is used:

(mg/*L*) Chem. (MG) Tank Vol. (8.34) lbs/gal = lbs Chem.

14.2 CHLORINE DOSE, DEMAND AND RESIDUAL

CHLORINE DOSAGE, DEMAND, AND RESIDUAL

In some chlorination calculations, the mg/*L* chlorine dose is not given directly but indirectly as chlorine demand and residual information.

Chlorine dose depends on two considerations—the chlorine demand and the desired chlorine residual:

Dose mg/*L*	=	Demand mg/*L*	+	Resid. mg/*L*

The **chlorine demand** is the amount of chlorine used in reacting with various components of the water such as harmful organisms and other organic and inorganic substances. When the chlorine demand has been satisfied, these reactions stop.

In some cases, such as perhaps during pretreatment, chlorinating just to meet the chlorine demand is sufficient. In other cases, however, it is desirable to have an additional amount of chlorine in the water available for disinfection.

Example 1: (Chlorine Dose, Demand, Residual)

❑ The secondary effluent is tested and found to have a chlorine demand of 6 mg/*L*. If the desired chlorine residual is 0.8 mg/*L*, what is the desired chlorine dose in mg/*L*?

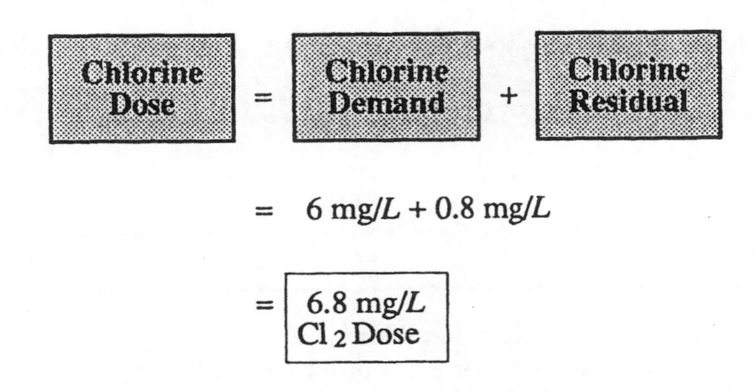

Example 2: (Chlorine Dose, Demand, Residual)

❑ The chlorine demand of a secondary effluent is 9.5 mg/*L*. If a chlorine residual of 0.6 mg/*L* is desired, what is the desired chlorine dosage in mg/*L*?

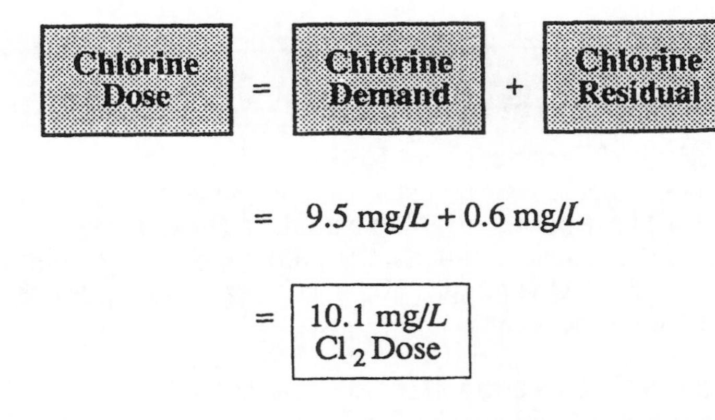

CALCULATING OTHER UNKNOWN VARIABLES

In Examples 1 and 2, the unknown variable was chlorine dosage, mg/*L*. However, the same equation may be used when chlorine demand or chlorine residual are unknown. Example 3 illustrates this calculation.

Example 3: (Chlorine Dose, Demand, Residual)

❑ The chlorine dosage for a secondary effluent is 7 mg/*L*. If the chlorine residual after 30 minutes contact time is found to be 0.6 mg/*L*, what is the chlorine demand expressed in mg/*L*?

Chlorine Dose = **Chlorine Demand** + **Chlorine Residual**

$$7 \text{ mg}/L = x \text{ mg}/L + 0.6 \text{ mg}/L$$

$$7 \text{ mg}/L - 0.6 \text{ mg}/L = x \text{ mg}/L$$

$$6.4 \text{ mg}/L = x$$

Cl_2 Demand

COMBINING WITH FEED RATE CALCULATIONS

Once the chlorine dosage (mg/*L*) has been calculated using the dose/demand/residual equation, the chlorine dosage in lbs/day can be calculated. Example 4 illustrates this type of problem.

Example 4: (Chlorine Dose, Demand, Residual)

❑ What should the chlorinator setting be (lbs/day) to treat a flow of 3.7 MGD if the chlorine demand is 9 mg/*L* and a chlorine residual of 2 mg/*L* is desired?

First calculate the chlorine dosage in mg/*L*:

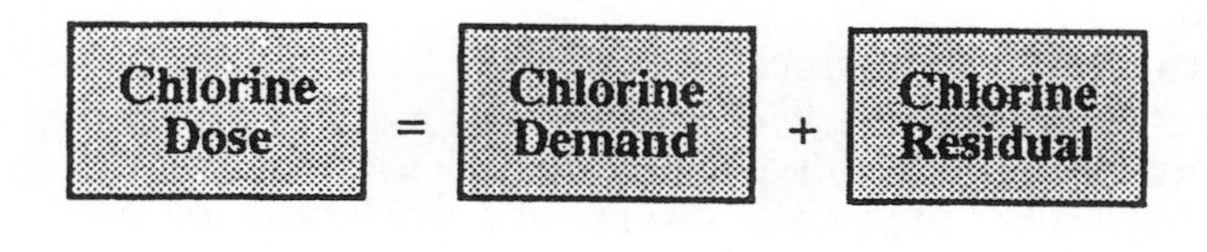

$$= 9 \text{ mg}/L + 2 \text{ mg}/L$$

$$= 11 \text{ mg}/L$$

Then calculate the chlorine dosage (feed rate) in lbs/day:

$$(\text{mg}/L\ Cl_2)\ (\text{MGD flow})\ (8.34 \text{ lbs/gal}) = \text{lbs/day } Cl_2$$

$$(11 \text{ mg}/L)\ (3.7 \text{ MGD})\ (8.34 \text{ lbs/gal}) = \boxed{339 \text{ lbs/day Chlorine}}$$

14.3 CHEMICAL FEED RATE—(Dosing Chemicals Less Than Full Strength)

HYPOCHLORITE COMPOUNDS

When chlorinating water or wastewater with chlorine gas, there is 100% available chlorine. Therefore, if the chlorine demand and residual require 50 lbs/day chlorine, the chlorinator setting would be just that—50 lbs/24 hrs.

Many times, however, a chlorine compound called hypochlorite is used to chlorinate wastewater. Hypochlorite compounds contain chlorine and are similar to a strong bleach.

Because hypochlorites are not 100% pure chlorine **more lbs/day must be fed into the system to obtain the same amount of chlorine for disinfection.**

To calculate the lbs/day hypochlorite required:

1. First calculate the lbs/day chlorine required.
2. Then calculate the lbs/day hypochlorite needed by dividing the lbs/day chlorine by the percent available chlorine.

$$\text{Hypochlorite lbs/day} = \frac{\text{lbs/day } Cl_2}{\dfrac{\%\text{ Available } Cl_2}{100}}$$

Example 1: (Chemical Feed Rate)

❑ A total chlorine dosage of 12 mg/*L* is required to treat a particular water. If the flow is 1.2 MGD and the hypochlorite has 65% available chlorine, how many lbs/day of hypochlorite will be required?

First, calculate the lbs/day chlorine required using the mg/*L* to lbs/day equation:

$$(\text{mg/}L\ Cl_2)\ (\text{MGD flow})\ (8.34\ \text{lbs/gal}) = \text{lbs/day } Cl_2$$

$$(12\ \text{mg/}L)\ (1.2\ \text{MGD})\ (8.34\ \text{lbs/gal}) = \boxed{120\ \text{lbs/day}}$$

Then calculate the lbs/day hypochlorite required. Since only 65% of the hypochlorite is chlorine, <u>more than 120 lbs/day</u> will be required:

$$\text{Hypochlorite lbs/day} = \frac{\text{lbs/day } Cl_2}{\dfrac{\%\text{ Available } Cl_2}{100}}$$

$$= \frac{120\ \text{lbs/day } Cl_2}{0.65\ \text{Avail. } Cl_2}$$

$$= \boxed{185\ \text{lbs/day Hypochlorite}}$$

Example 2: (Chemical Feed Rate)

❑ A wastewater flow of 850,000 gpd requires a chlorine dose of 25 mg/*L*. If hypochlorite (65% available chlorine) is to be used, how many lbs/day of hypochlorite are required?

First, calculate the lbs/day chlorine required:

$$(\text{mg/}L\ Cl_2)\ (\text{MGD flow})\ (8.34\ \text{lbs/gal}) = \text{lbs/day}$$

$$(25\ \text{mg/}L)\ (0.85\ \text{MGD})\ (8.34\ \text{lbs/gal}) = \boxed{177\ \text{lbs/day Chlorine}}$$

Then calculate the lbs/day hypochlorite:

$$\text{Hypochlorite lbs/day} = \frac{\text{lbs/day } Cl_2}{\dfrac{\%\text{ Available } Cl_2}{100}}$$

$$= \frac{177\ \text{lbs/day } Cl_2}{0.65\ \text{Avail. } Cl_2}$$

$$= \boxed{272\ \text{lbs/day Hypochlorite}}$$

Example 3: (Chemical Feed Rate)

❑ The desired dose of a polymer is 8 mg/L. The polymer literature indicates that the polymer compound provided is 60% active polymer. If a flow of 4.4 MGD is to be treated, how many lbs/day of the polymer compound will be required?

First calculate the lbs/day polymer required:

$$(\text{mg}/L \text{ Polymer})(\text{MGD flow})(8.34 \text{ lbs/gal}) = \text{lbs/day Polymer}$$

$$(8 \text{ mg}/L)(4.4 \text{ MGD})(8.34 \text{ lbs/gal}) = \boxed{294 \text{ lbs/day Polymer}}$$

Then calculate the lbs/day polymer compound required:

$$\text{Polymer Compound lbs/day} = \frac{\text{lbs/day Polymer}}{\frac{\%\text{ Active Polymer}}{100}}$$

$$= \frac{294 \text{ lbs/day}}{0.60}$$

$$= \boxed{490 \text{ lbs/day}}$$

OTHER CHEMICALS LESS THAN FULL STRENGTH

Other chemicals used in wastewater may be less than full strength, or less than 100% active. For example some polymers are less than 100% active. Be sure to check the chemical literature to determine whether or not it is 100% active. If a chemical is less than 100% "available chemical", calculate the lbs/day dosing requirement using the same equation as a hypochlorite problem. Example 3 illustrates this calculation.

Example 4: (Chemical Feed Rate)

❑ A total of 695 lbs of 65% hypochlorite are used in a day. If the flow rate treated is 4,780,000 gpd, what is the chlorine dosage in mg/L?

First calculate the lbs/day chlorine dosage:

$$\text{Hypochlorite lbs/day} = \frac{\text{lbs/day } Cl_2}{\frac{\%\text{ Available } Cl_2}{100}}$$

$$695 \text{ lbs/day Hypochlorite} = \frac{x \text{ lbs/day } Cl_2}{0.65}$$

$$(0.65)(695) = x$$

$$\boxed{452 \text{ lbs/day Chlorine}} = x$$

Then calculate mg/L Cl_2, using the mg/L to lbs/day equation and filling in the known information:*

$$(x \text{ mg}/L\ Cl_2)(4.78 \text{ MGD})(8.34 \text{ lbs/gal}) = 452 \text{ lbs/day } Cl_2$$

$$x = \frac{452 \text{ lbs/day}}{(4.78 \text{ MGD})(8.34 \text{ lbs/gal})}$$

$$x = \boxed{11.3 \text{ mg}/L \text{ Chlorine}}$$

CALCULATING mg/L CHLORINE GIVEN HYPOCHLORITE

Occasionally you will know the lbs/day hypochlorite and will want to determine either the lbs/day chlorine or the mg/L chlorine. To calculate either of these unknown, begin with the hypochlorite equation and then, if needed, use the mg/L to lbs/day equation. In effect, it is working the problem "backwards" from the problem shown in Example 2. Example 4 illustrates this type of calculation.

* Refer to Chapter 2 in *Basic Math Concepts* for a review of solving for the unknown value.

14.4 PERCENT STRENGTH OF SOLUTIONS

PERCENT STRENGTH USING DRY CHEMICALS

The strength of a solution is a measure of the amount of chemical (solute) dissolved in the solution. Since percent is calculated as "part over whole,"

$$\% = \frac{\text{Part}}{\text{Whole}} \times 100$$

percent strength is calculated as **part chemical,** in lbs, divided by the **whole solution,** in lbs:

$$\%\ \text{Strength} = \frac{\text{Chemical, lbs}}{\text{Solution, lbs}} \times 100$$

The denominator of the equation (lbs solution) includes both chemical (lbs) and water (lbs). Therefore the equation can be written in expanded form as:

$$\%\ \text{Strength} = \frac{\text{Chemical, lbs}}{\text{Water lbs} + \text{Chemical lbs}} \times 100$$

As the two equations above illustrate, **the chemical added must be expressed in pounds.** If the chemical weight is expressed in ounces (as in Example 1) or grams (as in Example 2), it must first be converted to pounds (to correspond with the other units in the problem) before percent strength is calculated.

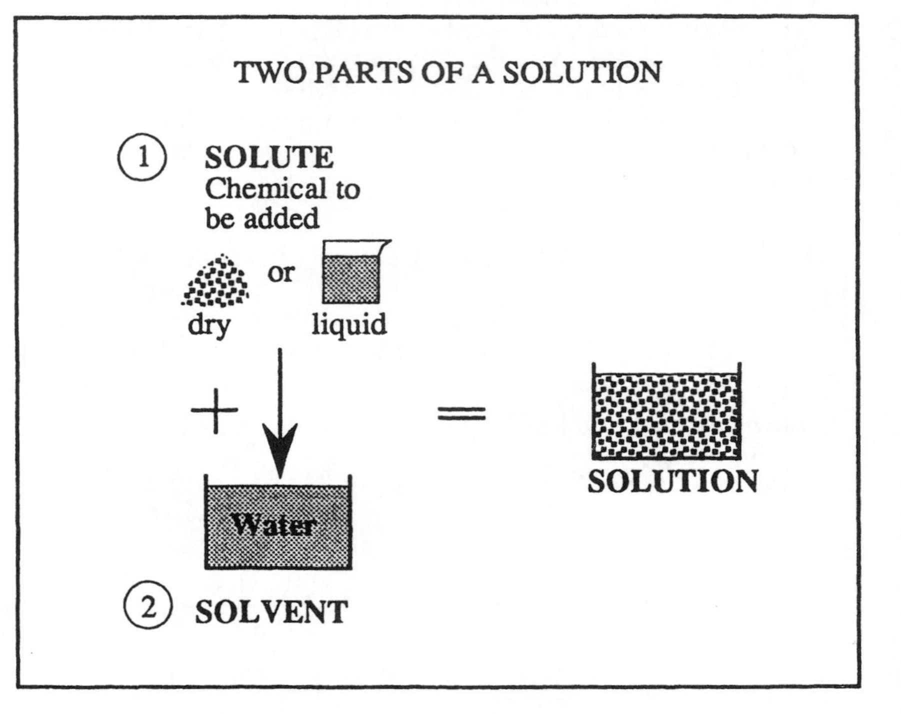

Example 1: (Percent Strength)

❑ If a total of 8 ounces of dry polymer are added to 10 gallons of water, what is the percent strength (by weight) of the polymer solution?

Before calculating percent strength, the ounces chemical must be converted to lbs chemical:*

$$\frac{8 \text{ ounces}}{16 \text{ ounces/pound}} = 0.5 \text{ lbs chemical}$$

Now calculate percent strength:

$$\%\ \text{Strength} = \frac{\text{Chemical, lbs}}{\text{Water, lbs} + \text{Chemical, lbs}} \times 100$$

$$= \frac{0.5 \text{ lbs Chemical}}{(10 \text{ gal})(8.34 \text{ lbs/gal}) + 0.5 \text{ lbs}} \times 100$$

$$= \frac{0.5 \text{ lbs Chemical}}{84 \text{ lbs Solution}} \times 100$$

$$= \boxed{0.6\%}$$

* To review ounces to pounds conversions refer to Chapter 8 in *Basic Math Concepts*.

Example 2: (Percent Strength)

❑ If 100 grams of dry polymer are dissolved in 5 gallons of water, what percent strength is the solution? (1 gram = 0.0022 lbs)

First, convert grams chemical to pounds chemical. Since 1 gram equals 0.0022 lbs, 100 grams is 100 times 0.0022 lbs:

$$(100 \text{ grams Chemical})(0.0022 \text{ lbs/gram}) = 0.22 \text{ lbs Chemical}$$

Now calculate percent strength of the solution:

$$\% \text{ Strength} = \frac{\text{lbs Chemical}}{\text{lbs Water} + \text{lbs Chemical}} \times 100$$

$$= \frac{0.22 \text{ lbs Chemical}}{(5 \text{ gal})(8.34 \text{ lbs/gal}) + 0.22 \text{ lbs}} \times 100$$

$$= \frac{0.22 \text{ lbs}}{41.92 \text{ lbs}} \times 100$$

$$= \boxed{0.5\%}$$

Example 3: (Percent Strength)

❑ How many pounds of dry polymer must be added to 25 gallons of water to make a 1% polymer solution:

First, write the equation as usual and fill in the known information. Then solve for the unknown value.**

$$\% \text{ Strength} = \frac{\text{lbs Chemical}}{\text{lbs Water} + \text{lbs Chemical}} \times 100$$

$$1 = \frac{x \text{ lbs Chemical}}{(25 \text{ gal})(8.34 \text{ lbs/gal}) + x \text{ lbs Chem.}} \times 100$$

$$1 = \frac{100\,x}{208.5 + x}$$

$$1\,(208.5 + x) = 100\,x$$

$$208.5 + x = 100\,x$$

$$208.5 = 99\,x$$

$$\boxed{2.1 \text{ lbs Chemical}} = x$$

WHEN GRAMS CHEMICAL ARE USED

The chemical (solute) to be used in making a solution may be measured in grams rather than pounds or ounces. When this is the case, convert grams of chemical to pounds of chemical before calculating percent strength. The following relationship is used for the conversion:

$$\boxed{1 \text{ gram} = 0.0022 \text{ lbs}}$$

SOLVING FOR OTHER UNKNOWN VARIABLES

In the percent strength equation there are three variables:

- % Strength
- lbs Chemical
- lbs Water

In Examples 1 and 2, the unknown value was percent strength. However, the same equation can be used to determine either one of the other two variables. Example 3 illustrates this type of calculation.

Note that gallons water can also be the unknown variable in percent strength calculations. First set pounds water as the unknown variable in the equation. Then after the pounds water has been calculated convert pounds water to gallons water, using the 8.34 lbs/gal factor.*

* For a review of lbs to gallons conversions refer to Chapter 8 in *Basic Math Concepts.*

** To review solving for the unknown value, refer to Chapter 2 in *Basic Math Concepts.*

PERCENT STRENGTH USING LIQUID CHEMICALS

When using a liquid chemical to make up a solution, such as liquid polymer, a different calculation is required.

The liquid chemical is shipped from the supplier at a certain percent strength—perhaps 10 or 12%. This chemical is then added to water to obtain a desired solution of lower percent polymer—such as 1 or 0.5%.

Percent strength calculations using liquid chemicals are very similar to sludge thickening calculations.* In sludge thickening problems, lbs solids are set equal to lbs solids. In these percent strength problems, lbs chemical are set equal to lbs chemical, as illustrated by the diagram to the right.

THE KEY TO THESE CALCULATIONS—POUNDS CHEMICAL REMAINS CONSTANT
(Liquid polymer is used to illustrate the concept.)

LIQUID POLYMER
(10% Polymer)

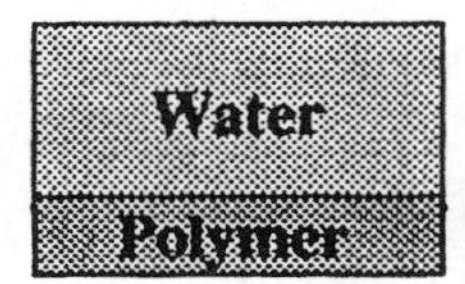

POLYMER SOLUTION
(0.5% Polymer)

Simplified Equation:

$$\text{lbs Polymer in Liquid Polymer} = \text{lbs Polymer in Polymer Solution}$$

Expanded Equation:

$$(\text{Liquid Polymer, lbs})\frac{(\%\text{ Strength of Liq. Poly.})}{100} = (\text{Polymer Solution, lbs})\frac{(\%\text{Strength of Poly. Soln.})}{100}$$

Or

$$(\text{Liq. Poly. gal})(10.2\text{ lbs/gal})\frac{(\%\text{ Strength of Liq. Poly.})}{100} = (\text{Poly. Soln gal})(8.34\text{ lbs/gal})\frac{(\%\text{Strength of Poly. Soln.})}{100}$$

Use actual density factor here. Liquid polymer generally weighs more than 8.34 lbs/gal. Other liquid chemicals may have the same density as water.

Use actual density factor here. Liquid polymer may have a density closer to or equal to 8.34 lbs/gal since the heavy polymer has been diluted.

Example 4: (Percent Strength)

❑ A 10% liquid polymer is to be used in making up a polymer solution. How many lbs of liquid polymer should be mixed with water to produce 150 lbs of a 0.8% polymer solution?

$$(\text{Liquid Polymer lbs})\frac{(\%\text{ Strength of Liq. Poly.})}{100} = (\text{Poly. Soln. lbs})\frac{(\%\text{Strength of Poly. Soln.})}{100}$$

$$(x\text{ lbs})\left(\frac{10}{100}\right) = (150\text{ lbs})\frac{(0.8)}{100}$$

$$x = \frac{(150)(0.008)}{0.1}$$

$$x = \boxed{12\text{ lbs}}$$

* Refer to Chapter 15, Section 15.3.

Example 5: (Percent Strength)

❑ How many gallons of 8% liquid polymer should be mixed with water to produce 75 gallons of a 0.5% polymer solution? The density of the polymer liquid is 10.2 lbs/gal. Assume the density of the polymer solution is 8.34 lbs/gal.

Use the expanded form of the equation, filling in known information:

$$\frac{(\text{Liq. Poly.})_{gal}\,(10.2)_{lbs/gal}\,(\%\text{ Strength of Liq. Poly.})}{100} = \frac{(\text{Poly. Soln})_{gal}\,(8.34)_{lbs/gal}\,(\%\text{Strength of Poly. Soln.})}{100}$$

$$(x \text{ gal})\,(10.2)\,\frac{(8)}{100} = (75 \text{ gal})\,(8.34)\,\frac{(0.5)}{100}$$

$$x = \frac{(75)\,(8.34)\,(0.005)}{(10.2)\,(0.08)}$$

$$x = \boxed{3.8 \text{ gallons}}$$

Example 6: (Percent Strength)

❑ A 10% liquid polymer will be used in making up a solution. How many gallons of liquid polymer should be added to the water to make up 60 gallons of 0.4% polymer solution? The liquid polymer has a specific gravity of 1.1 Assume the polymer solution has a specific gravity of 1.0.

First, convert specific gravity information to density information. The density of the liquid polymer is (8.34 lbs/gal) (1.1) = 9.2 lbs/gal. The density of the polymer solution is 8.34 lbs/gal, the same as water.

$$\frac{(\text{Liq. Poly.})_{gal}\,(9.2)_{lbs/gal}\,(\%\text{ Strength of Liq. Poly.})}{100} = \frac{(\text{Poly. Soln})_{gal}\,(8.34)_{lbs/gal}\,(\%\text{Strength of Poly. Soln.})}{100}$$

$$(x \text{ gal})\,(9.2)\,\frac{(10)}{100} = (60 \text{ gal})\,(8.34)\,\frac{(0.4)}{100}$$

$$x = \frac{(60)\,(8.34)\,(0.004)}{(9.2)\,(0.1)}$$

$$x = \boxed{2.2 \text{ gallons}}$$

DENSITY AND SPECIFIC GRAVITY CONSIDERATIONS

As shown in the second expanded equation on the opposite page, **the density of the solution must be included.** Density is the mass per unit volume.* In water and wastewater calculations, 8.34 lbs/gal is used as the density of water. However, the weight of a polymer solution can be as much as 10 or 11 lbs/gal. To obtain accurate results using the percent strength equation, it is important to use the appropriate density factor—one for the solute (such as liquid polymer) and another for the solution. When the solution strength is very low, such as 0.5% or 0.1%, the density of the solution is normally much closer to that of water—8.34 lbs/gal.

Occasionally **specific gravity** data may be given for a liquid chemical rather than density information. In fact, density and specific gravity are closely related terms. Density is a measure of the mass per unit volume, and is measured in such terms as lbs/gal. Specific gravity is a comparison of the density of a substance to a standard density. (For liquids, the standard is water. All other densities are compared to the density of water.) So, **a specific gravity of 1.0 means the liquid has the same density as water** (8.34 lbs/gal). A specific gravity of 0.5 means the liquid has a density half that of water, or 4.17 lbs/gal. A specific gravity of 1.5 means the liquid has a density 1.5 times that of water, or 12.51 lbs/gal. Example 6 illustrates a calculation including specific gravity data.

* Refer to Chapter 7, Section 7.1, "Density and Specific Gravity".

14.5 MIXING SOLUTIONS OF DIFFERENT STRENGTH

There are two types of solution mixture calculations. In one type of calculation, two solutions of different strengths are mixed with no particular target solution strength. The calculation involves determining the percent strength of the solution mixture.

The second type of solution mixture calculation includes a desired or target strength. This type of problem is described in the next section.

WHEN DIFFERENT PERCENT STRENGTH SOLUTIONS ARE MIXED

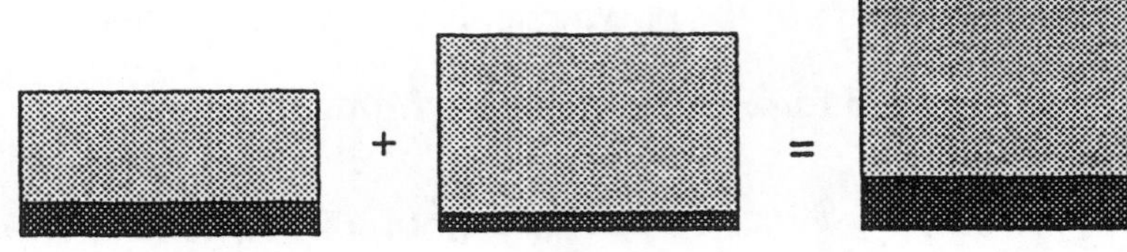

10% Strength Solution | 1% Strength Solution | Solution Mixture (% Strength somewhere between 10% and 1% depending on the quantity contributed by each.)

Simplified Equation:

$$\text{\% Strength of Mixture} = \frac{\text{Chemical in Mixture, lbs}}{\text{Solution Mixture, lbs}} \times 100$$

Expanded Equations:

$$\text{\% Strength of Mixture} = \frac{\text{lbs Chem. from Solution 1} + \text{lbs Chem. from Solution 2}}{\text{lbs Solution 1} + \text{lbs Solution 2}} \times 100$$

$$\text{\% Strength of Mixture} = \frac{(\text{Sol'n 1 lbs})\frac{(\text{\% Strength of Sol'n 1})}{100} + (\text{Sol'n 2 lbs})\frac{(\text{\% Strength of Sol'n 2})}{100}}{\text{lbs Solution 1} + \text{lbs Solution 2}} \times 100$$

Example 1: (Solution Mixtures)

❑ If 20 lbs of a 10% strength solution are mixed with 50 lbs of 1% strength solution, what is the percent strength of the solution mixture?

$$\text{\% Strength of Mixture} = \frac{(\text{Sol'n 1 lbs})\frac{(\text{\% Strength of Sol'n 1})}{100} + (\text{Sol'n 2 lbs})\frac{(\text{\% Strength of Sol'n 2})}{100}}{\text{lbs Solution 1} + \text{lbs Solution 2}} \times 100$$

$$= \frac{(20 \text{ lbs})(0.1) + (50 \text{ lbs})(0.01)}{20 \text{ lbs} + 50 \text{ lbs}} \times 100$$

$$= \frac{2 \text{ lbs} + 0.5 \text{ lbs}}{70 \text{ lbs}} \times 100$$

$$= \boxed{3.6\%}$$

Example 2: (Solution Mixtures)

❑ If 5 gallons of an 8% strength solution are mixed with 40 gallons of a 0.5% strength solution, what is the percent strength of the solution mixture? (Assume the 8% solution weighs 9.5 lbs/gal and the 0.5% solution weighs 8.34 lbs/gal.)

$$\text{\% Strength of Mixture} = \frac{\frac{(\text{Sol'n 1 lbs})(\text{\% Strength of Sol'n 1})}{100} + \frac{(\text{Sol'n 2 lbs})(\text{\% Strength of Sol'n 2})}{100}}{\text{lbs Solution 1} + \text{lbs Solution 2}} \times 100$$

$$= \frac{(5\text{ gal})(9.5\text{ lbs/gal})(0.08) + (40\text{ gal})(8.34\text{ lbs/gal})(0.005)}{(5\text{ gal})(9.5\text{ lbs/gal}) + (40\text{ gal})(8.34\text{ lbs/gal})} \times 100$$

$$= \frac{3.8\text{ lbs Chem.} + 1.7\text{ lbs Chem.}}{47.5\text{ lbs Soln 1} + 333.6\text{ lbs Soln 2}} \times 100$$

$$= \frac{5.5\text{ lbs Chemical}}{381.1\text{ lbs Solution}} \times 100$$

$$= \boxed{1.4\%\text{ Strength}}$$

Example 3: (Solution Mixtures)

❑ If 15 gallons of a 10% strength solution are added to 50 gallons of 0.8% strength solution, what is the percent strength of the solution mixture? (Assume the 10% strength solution weighs 10.2 lbs/gal and the 0.8% strength solution weighs 8.8 lbs/gal.)

$$\text{\% Strength of Mixture} = \frac{\frac{(\text{Sol'n 1 lbs})(\text{\% Strength of Sol'n 1})}{100} + \frac{(\text{Sol'n 2 lbs})(\text{\% Strength of Sol'n 2})}{100}}{\text{lbs Solution 1} + \text{lbs Solution 2}} \times 100$$

$$= \frac{(15\text{ gal})(10.2\text{ lbs/gal})(0.1) + (50\text{ gal})(8.8\text{ lbs/gal})(0.008)}{(15\text{ gal})(10.2\text{ lbs/gal}) + (50\text{ gal})(8.8\text{ lbs/gal})} \times 100$$

$$= \frac{15.3\text{ lbs Chem.} + 3.5\text{ lbs Chem.}}{153\text{ lbs Soln 1} + 440\text{ lbs Soln 2}} \times 100$$

$$= \frac{18.8\text{ lbs Chemical}}{593\text{ lbs Solution}} \times 100$$

$$= \boxed{3.2\%\text{ Strength}}$$

USE DIFFERENT DENSITY FACTORS WHEN APPROPRIATE

Percent strength should be expressed in terms of **pounds chemical per pounds solution**. Therefore, when solutions are expressed in terms of gallons, the gallons should be expressed as pounds before continuing with the percent strength calculation.

It is important to know what density factor should be used to convert from gallons to pounds. If the solution has a density the same as water, 8.34 lbs/gal is used. If, however, the solution has a higher density, such as some polymer solutions, then the higher density factor should be used. When the density is unknown, it is sometimes possible to weigh the chemical solution to determine the density.

SOLUTION MIXTURES TARGET PERCENT STRENGTH

In the previous section we examined the first type of solution mixture calculation—a calculation where there is no target percent strength. In this type calculation, two solutions are mixed and the percent strength of the mixture is determined.

In the second type of solution mixture calculation, **two different percent strength solutions are mixed in order to obtain a desired quantity of solution and a target percent strength**. These problems may be solved using the same equation shown in Examples 1-3. An illustration of this approach is given in Example 4.

Another and perhaps preferred approach in solving these problems is by using the dilution rectangle. Although the first use of the dilution rectangle can be confusing, the effort to master its use is rewarded— solution mixture problems are quickly calculated. Example 5 uses the dilution rectangle to solve the problem stated in Example 4. Compare the two methods of calculating this type of mixture problem.

Example 4: (Dilution Rectangle)

❑ What weights of a 2% solution and a 7% solution must be mixed to make 850 lbs of a 4% solution?

Use the same equation as shown for Examples 1-3 and fill in given information.* (Note that the lbs of Solution 1 is unknown, *x*. If lbs of Solution 1 is *x*, then the lbs of Solution 2 must be the balance of the 850 lbs, or 850-*x*.)

$$\text{\% Strength of Mixture} = \frac{\dfrac{(\text{Sol'n 1 lbs})(\text{\% Strength of Sol'n 1})}{100} + \dfrac{(\text{Sol'n 2 lbs})(\text{\% Strength of Sol'n 2})}{100}}{\text{lbs Solution 1} + \text{lbs Solution 2}} \times 100$$

$$4 = \frac{(x\text{ lbs})(0.02) + (850 - x\text{ lbs})(0.07)}{850\text{ lbs}} \times 100$$

$$\frac{(4)(850)}{100} = 0.02x + 59.5 - 0.07x$$

$$34 = -0.05x + 59.5$$

$$0.05x = 25.5$$

$$x = \boxed{510\text{ lbs of 2\% Solution}}$$

$$\text{Then } 850 - 510 = \boxed{340\text{ lbs of 7\% Solution}}$$

THE DILUTION RECTANGLE

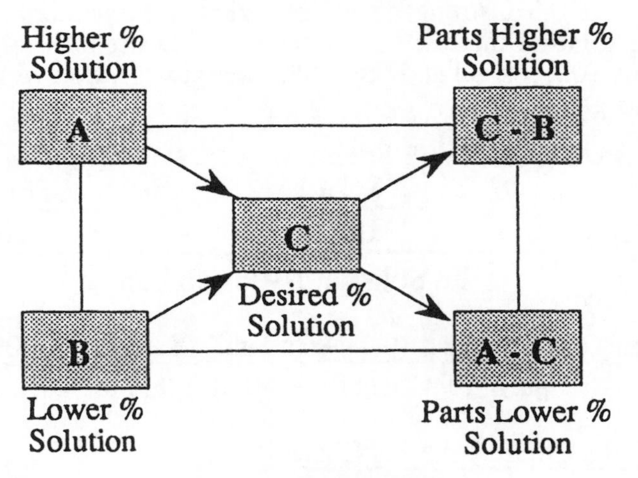

Steps in Using the Dilution Rectangle:

1. Place the % Strength numbers in positions A, B, and C.
2. Calculate parts higher % solution and parts lower % solution, subtracting as indicated.
3. Multiply fractional parts of each solution by the total lbs of solution desired.

* Refer to Chapter 2 in *Basic Math Concepts* for a review of solving for the unknown value.

Example 5: (Dilution Rectangle)

❑ What weights of a 2% solution and a 7% solution must be mixed to make 850 lbs of a 4% solution?

Use the Dilution Rectangle to solve this problem. First determine the parts required of each solution:

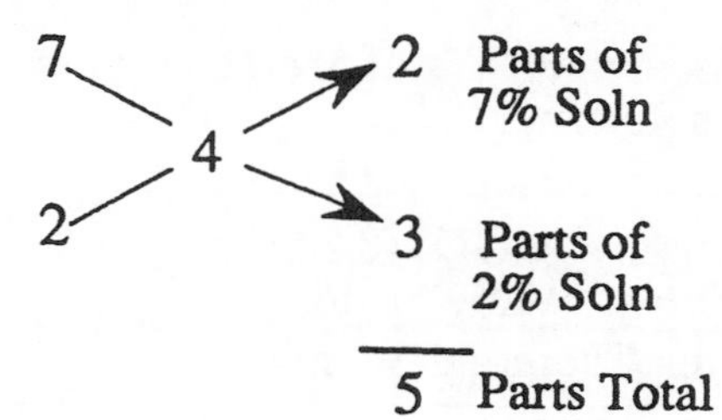

Thus, 2 parts of the total 5 parts (2/5) come from the 7% solution, and the other three parts (3/5) come from the 2% solution. Now calculate the lbs of 2% and 7% solution, using these fractions:

Amt. of 7% Soln: $\frac{2}{5}$ (850 lbs) = **340 lbs**

Amt. of 2% Soln: $\frac{3}{5}$ (850 lbs) = **510 lbs**

Example 6: (Dilution Rectangle)

❑ How many lbs of a 10% polymer solution and water should be mixed together to form 425 lbs of a 1% polymer solution?

First calculate the parts of each solution required:

10 → 1 → 1 Parts of 10% Soln

0 → 1 → 9 Parts of Water

10 Parts Total

Then calculate the actual lbs of each solution:

Amt. of 10% Soln: $\frac{(1)}{10}$ (425 lbs) = **42.5 lbs 10% Soln**

Amt. of Water: $\frac{(9)}{10}$ (425 lbs) = **382.5 lbs Water**

MIXING A SOLUTION AND WATER

In solution mixing Examples 1-5, two solutions of different strengths are blended. The solution mixing equation and dilution rectangle can also be used when only one solution and water are blended. (Water is considered a 0% strength solution.) Example 6 illustrates such a calculation using the dilution rectangle.

14.6 SOLUTION CHEMICAL FEEDER SETTING, GPD

WHEN SOLUTION CONCENTRATION IS EXPRESSED AS LBS CHEM/GAL SOL'N

When solution concentration is expressed as lbs chemical/gal solution calculate the gpd solution required:

1. Calculate the lbs/day dry chemical required.

$$\underset{\text{Chem.}}{(\text{mg}/L)}\ \underset{\text{flow}}{(\text{MGD})}\ \underset{\text{lbs/gal}}{(8.34)} = \text{Chem. lbs/day}$$

2. Convert the lbs/day dry chemical to gpd solution (using lbs chemical/gal solution information).

These two steps can be combined into one equation. Note the similarity between this equation and the hypochlorite type of calculation (see Section 14.3).

$$\frac{\underset{\text{Chem.}}{(\text{mg}/L)}\ \underset{\text{Flow}}{(\text{MGD})}\ \underset{\text{lbs/gal}}{(8.34)}}{\text{lbs Chem/gal Sol'n}} = \text{Sol'n gpd}$$

Example 1 illustrates this type of calculation.

CALCULATING GPD FEEDER SETTING DEPENDS ON HOW SOLUTION CONCENTRATION IS EXPRESSED: (LBS/GAL OR PERCENT)

If the solution strength is expressed as lbs/gal: (lbs chemical/gal solution)

$$\frac{\underset{\text{Chem.}}{(\text{mg}/L)}\ \underset{\text{Flow}}{(\text{MGD})}\ \underset{\text{lbs/gal}}{(8.34)}}{\text{lbs Chem/gal Sol'n}} = \text{gpd Sol'n}$$

If the solution strength is expressed as a percent:

Simplified Equation:

$$\text{Desired Dose, lbs/day} = \text{Actual Dose, lbs/day}$$

Expanded Equation:

$$\underset{\text{Chem.}}{(\text{mg}/L)}\ \underset{\text{Flow Treated}}{(\text{MGD})}\ \underset{\text{lbs/gal}}{(8.34)} = \underset{\text{Sol'n}}{(\text{mg}/L)}\ \underset{\text{Sol'n Flow}}{(\text{MGD})}\ \underset{\text{lbs/gal}}{(8.34)}$$

Example 1: (Feedeer Setting, gpd)

❑ Jar tests indicate that the best liquid alum dose for a water is 9 mg/*L*. The flow to be treated is 1.94 MGD. Determine the gpd setting for the liquid alum chemical feeder if the liquid alum contains 5.36 lbs of alum per gallon of solution.

First calculate the lbs/day of dry alum required, using the mg/*L* to lbs/day equation:

$$(\text{mg}/L)\ \underset{\text{flow}}{(\text{MGD})}\ \underset{\text{lbs/gal}}{(8.34)} = \text{lbs/day}$$

$$(9\ \text{mg}/L)\ (1.94\ \text{MGD})\ (8.34\ \text{lbs/gal}) = \boxed{146\ \text{lbs/day Dry Alum}}$$

Then calculate gpd solution required. (Each gallon of solution contains 5.36 lbs of dry chemical. To find how many gallons are required, therefore, you need to determine how many 5.36 lbs are needed.)

$$\frac{146\ \text{lbs/day alum}}{5.36\ \text{lbs alum/gal solution}} = \boxed{27\ \text{gpd Alum Solution}}$$

Example 2: (Feeder Setting, gpd)

❑ Jar tests indicate that the best liquid alum dose for a water is 9 mg/*L*. The flow to be treated is 1.94 MGD. Determine the gpd setting for the liquid alum chemical feeder if the liquid alum is a 64.3% solution.

First write the equation, then fill in given information. The solution concentration is 64.3%. This can be re-expressed as 643,000 mg/*L* for use in the equation.*

Desired Dose, lbs/day = Actual Dose, lbs/day

$$\underset{\text{Chem.}}{(\text{mg/L})}\ \underset{\text{Flow Treated}}{(\text{MGD})}\ \underset{\text{lbs/gal}}{(8.34)} = \underset{\text{Sol'n}}{(\text{mg/L})}\ \underset{\text{Sol'n Flow}}{(\text{MGD})}\ \underset{\text{lbs/gal}}{(8.34)}$$

$$(9\text{ mg/L})\ \underset{\text{MGD}}{(1.94)}\ \underset{\text{lbs/gal}}{(8.34)} = (643{,}000\text{ mg/L})\ (x\text{ MGD})\ \underset{\text{lbs/gal}}{(8.34)}$$

$$\frac{(9)\ (1.94)\ \cancel{(8.34)}}{(643{,}000)\ \cancel{(8.34)}} = x$$

$$\boxed{0.0000271\text{ MGD}} = x$$

Now convert MGD flow to gpd flow:**

$$0.0000271\text{ MGD} = \boxed{27.1\text{ gpd flow}}$$

Example 3: (Feeder Setting, gpd)

❑ The flow to a plant is 3.46 MGD. Jar testing indicates that the optimum alum dose is 12 mg/*L*. What should the gpd setting be for the solution feeder if the alum solution is a 55% solution?

A solution concentration of 55% is equivalent to 550,000 mg/*L*:

Desired Dose, lbs/day = Actual Dose, lbs/day

$$\underset{\text{Chem.}}{(\text{mg/L})}\ \underset{\text{Flow Treated}}{(\text{MGD})}\ \underset{\text{lbs/gal}}{(8.34)} = \underset{\text{Sol'n}}{(\text{mg/L})}\ \underset{\text{Sol'n Flow}}{(\text{MGD})}\ \underset{\text{lbs/gal}}{(8.34)}$$

$$(12\text{ mg/L})\ \underset{\text{MGD}}{(3.46)}\ \underset{\text{lbs/gal}}{(8.34)} = (550{,}000\text{ mg/L})\ (x\text{ MGD})\ \underset{\text{lbs/gal}}{(8.34)}$$

$$\frac{(12)\ (3.46)\ \cancel{(8.34)}}{(550{,}000)\ \cancel{(8.34)}} = x$$

$$\boxed{0.0000754\text{ MGD}} = x$$

This can be expressed as gpd flow:

$$0.0000754\text{ MGD} = \boxed{75.4\text{ gpd flow}}$$

WHEN SOLUTION CONCENTRATION IS EXPRESSED AS A PERCENT

When the solution concentration is expressed as a percent, it may be converted to mg/*L* and a different equation may be used, as shown on the facing page. The basis of this equation, stated in simple terms, is that **the desired dosage rate (lbs/day) must be equal to the actual dosage rate (lbs/day).** In expanded form, the desired dosage rate (lbs/day) is calculated using mg/*L* desired dosage, flow rate to be treated (in MGD) and 8.34 lbs/gal. The actual dosage rate is calculated using mg/*L* solution concentration, MGD solution flow, and 8.34 lbs/gal.

Note that Examples 1 and 2 have the same answers. In Example 1, the solution strength was expressed as lbs chemical/gal solution; whereas in Example 2, the same solution strength was expressed as a percent then converted to mg/*L*:

$$\boxed{1\% = 10{,}000\text{ mg/L}}$$

* To review the conversion from mg/*L* to %, and vice versa, refer to Chapter 8 in *Basic Math Concepts*.

** Refer to Chapter 8 in *Basic Math Concepts* for a discussion of flow conversions.

14.7 CHEMICAL FEED PUMP—PERCENT STROKE SETTING

Chemical feed pumps are generally piston pumps (also called "positive displacement" pumps). This type of pump operates on the principle of positive displacement. This means that it displaces, or pushes out, a volume of chemical equal to the volume of the piston. The length of the piston, called the stroke, can be lengthened or shortened to increase or decrease the amount of chemical delivered by the pump. Normally the piston pump is operated no faster than about 50 gpm.

EACH STROKE OF A PISTON PUMP "DISPLACES" OR PUSHES OUT CHEMICAL

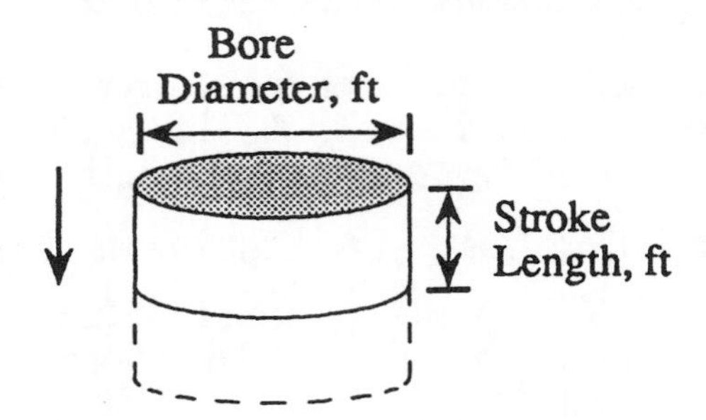

First calculate the gpd solution req'd, using either method described in the previous section (14.6).

Then compare the required gpd setting with the maximum gpd possible to determine the percent stroke setting:

$$\text{\% Stroke Setting} = \frac{\text{Desired Feed, gpd}}{\text{Maximum Feed, gpd}} \times 100$$

Example 1: (% Stroke Setting)

❑ The required chemical pumping rate has been calculated as 10 gpm. If the maximum pumping rate is 95 gpm, what should the percent stroke setting be?

The percent stroke setting is based on the ratio of the gpm required to the total possible gpm:

$$\text{\% Stroke Setting} = \frac{\text{Required Feed, gpd}}{\text{Maximum Feed, gpd}} \times 100$$

$$= \frac{10 \text{ gpm}}{95 \text{ gpm}} \times 100$$

$$= \boxed{10.5\%}$$

Example 2: (% Stroke Setting)

❑ The required chemical pumping rate has been calculated as 15 gpm. If the maximum pumping rate is 82 gpm, what should the percent stroke setting be?

$$\text{\% Stroke Setting} = \frac{\text{Required Feed, gpd}}{\text{Maximum Feed, gpd}} \times 100$$

$$= \frac{15 \text{ gpm}}{82 \text{ gpm}} \times 100$$

$$= \boxed{18.3\%}$$

Example 3: (% Stroke Setting)

❑ The required chemical pumping rate has been determined to be 75 gpm. If the maximum pumping rate is 85 gpm, what should the percent stroke setting be?

$$\text{\% Stroke Setting} = \frac{\text{Required Feed, gpd}}{\text{Maximum Feed, gpd}} \times 100$$

$$= \frac{75 \text{ gpm}}{85 \text{ gpm}} \times 100$$

$$= \boxed{88\%}$$

14.8 SOLUTION CHEMICAL FEEDER SETTING, mL/min

Some solution chemical feeders dispense chemical as milliliters per minute (mL/min). To calculate the mL/min solution required, first calculate the gpd feed rate, as described in Section 14.6. Then convert gpd flow rate to ml/min flow rate. The process, as shown in the equation to the right, involves the following conversions:**

$$\frac{\text{gal}}{\text{day}} \rightarrow \frac{\text{gal}}{\text{min}} \rightarrow \frac{\text{mL}}{\text{min}}$$

FIRST DETERMINE GPD FLOW THEN CALCULATE ML/MIN FLOW

Calculate gpd solution flow required:

Simplified Equation:

Desired Dose, lbs/day = Actual Dose, lbs/day

Expanded Equation:

$$\underset{\text{Dose}}{(\text{mg/L})}\ \underset{\text{Flow Treated}}{(\text{MGD})}\ \underset{\text{lbs/gal}}{(8.34)} = \underset{\text{Sol'n}}{(\text{mg/L})}\ \underset{\text{Sol'n Flow}}{(\text{MGD})}\ \underset{\text{lbs/gal}}{(8.34)}$$

Convert mL/min solution flow required:*

$$\frac{(\text{gal})}{\text{day}}\ \frac{(1\ \text{day})}{1440\ \text{min}}\ \frac{(3785\ \text{mL})}{1\ \text{gal}} = \frac{\text{mL}}{\text{min}}$$

Or, simplified as:

$$\frac{(\text{gpd})\ (3785\ \text{mL/gal})}{1440\ \text{min/day}} = \text{mL/min}$$

Example 1: (Feeder Setting, mL/min)

❑ The desired solution feed rate was calculated to be 8 gpd. What is this feed rate expressed as mL/min?

Since the gpd flow has already been determined, the mL/min flow rate can be calculated directly:

$$\frac{(\text{gpd})\ (3785\ \text{mL/gal})}{1440\ \text{min/day}} = \text{mL/min}$$

$$\frac{(8\ \text{gpd})\ (3785\ \text{mL/gal})}{1440\ \text{min/day}} = \boxed{21\ \text{mL/min Feed Rate}}$$

* This equation is written in a form so that dimensional analysis may be used to check the units of the answer. Refer to Chapter 15 in *Basic Math Concepts*.

** Refer to Chapter 8 in *Basic Math Concepts* for flow rate and metric conversions.

Example 2: (Feeder Setting, mL/min)

❑ The desired solution feed rate has been calculated to be 20 gpd. What is this feed rate expressed as mL/min?

Since the gpd solution feed rate has been determined, the ml/min may be calculated directly:

$$\frac{\text{(gpd) (3785 mL/gal)}}{\text{1440 min/day}} = \text{mL/min}$$

$$\frac{\text{(20 gpd) (3785 mL/gal)}}{\text{1440 min/day}} = \boxed{\text{52.6 mL/min Feed Rate}}$$

Example 3: (Feeder Setting, mL/min)

❑ The optimum polymer dose has been determined to be 12 mg/L. The flow to be treated is 980,000 gpd. If the solution to be used contains 60% active polymer, what should the solution chemical feeder setting be, in mL/min?

First calculate the gpd feed rate required:

$$\underset{\text{Chem Dose}}{\text{(mg/L)}}\ \underset{\text{flow treated}}{\text{(MGD)}}\ \underset{\text{lbs/gal}}{\text{(8.34)}} = \underset{\text{Sol'n}}{\text{(mg/L)}}\ \underset{\text{Sol'n flow}}{\text{(MGD)}}\ \underset{\text{lbs/gal}}{\text{(8.34)}}$$

$$\underset{\text{Polym.}}{\text{(12 mg/L)}}\ \text{(0.98 MGD)}\ \underset{\text{lbs/gal}}{\text{(8.34)}} = \underset{\text{mg/L}}{\text{(600,000)}}\ (x\ \text{MGD})\ \underset{\text{lbs/gal}}{\text{(8.34)}}$$

$$\frac{(12)\ (0.98)\ \cancel{(8.34)}}{(600{,}000)\ \cancel{(8.34)}} = x\ \text{MGD}$$

$$0.0000196\ \text{MGD} = x$$

$$20\ \text{gpd} = x$$

Then convert gpd flow rate to mL/min flow rate:

$$\frac{\text{(gpd) (3785 mL/gal)}}{\text{1440 min/day}} = \text{mL/min}$$

$$\frac{\text{(20 gpd) (3785 mL/gal)}}{\text{1440 min/day}} = \boxed{\text{53 mL/min Feed Rate}}$$

14.9 DRY CHEMICAL FEED CALIBRATION

Occasionally you will want to **compare the actual chemical feed rate with the feed rate indicated by the instrumentation**. This is called a calibration calculation.

To calculate the actual chemical feed rate for a dry chemical feeder, place a bucket under the feeder, weigh the bucket when empty, then weigh the bucket again after a specified length of time, such as 30 minutes.

The actual chemical feed rate can then be determined as:

$$\text{Chem. Feed Rate, lbs/min} = \frac{\text{Chem. Applied, lbs}}{\text{Length of Applic, min}}$$

The chemical feed rate can be converted to lbs/day, if desired:

$$(\text{Feed Rate, lbs/min})\left(1440\ \frac{\text{min}}{\text{day}}\right) = \text{Feed Rate, lbs/day}$$

Example 1: (Dry Chemical Feed Calibration)

❑ Calculate the actual chemical feed rate, lbs/day, if a bucket is placed under a chemical feeder and a total of 1.5 lbs is collected during a 30-minute period.

First calculate the lbs/min feed rate:

$$\text{Chem. Feed Rate, lbs/min} = \frac{\text{Chem. Applied, lbs}}{\text{Length of Application, min}}$$

$$= \frac{1.5\ \text{lbs}}{30\ \text{min}}$$

$$= \boxed{0.05\ \text{lbs/min Feed Rate}}$$

Then calculate the lbs/day feed rate:

$$\text{Chem. Feed Rate, lbs/day} = (0.05\ \text{lbs/min})(1440\ \text{min/day})$$

$$= \boxed{72\ \text{lbs/day Feed Rate}}$$

Example 2: (Dry Chemical Feed Calibration)

❑ Calculate the actual chemical feed rate, lbs/day, if a bucket is placed under a chemical feeder and a total of 1.3 lbs is collected during a 20-minute period.

First calculate the lbs/min feed rate:

$$\text{Chem. Feed Rate, lbs/min} = \frac{\text{Chem. Applied, lbs}}{\text{Length of Application, min}}$$

$$= \frac{1.3\ \text{lbs}}{20\ \text{min}}$$

$$= \boxed{0.065\ \text{lbs/min Feed Rate}}$$

Then calculate the lbs/day feed rate:

$$\text{Chem. Feed Rate, lbs/day} = (0.065\ \text{lbs/min})(1440\ \text{min/day})$$

$$= \boxed{94\ \text{lbs/day Feed Rate}}$$

Example 3: (Dry Chemical Feed Calibration)

❑ A chemical feeder is to be calibrated. The bucket to be used to collect chemical is placed under the chemical feeder and weighed (0.25 lbs). After 30 minutes, the weight of the bucket and chemical is found to be 2.6 lbs. Based on this test, what is the actual chemical feed rate, in lbs/day?

First calculate the lbs/min feed rate:
(Note that the chemical applied is the wt. of the bucket and chemical minus the wt. of the empty bucket.)

$$\text{Chem. Feed Rate, lbs/min} = \frac{\text{Chem. Applied, lbs}}{\text{Length of Application, min}}$$

$$= \frac{2.6 \text{ lbs} - 0.25 \text{ lbs}}{30 \text{ minutes}}$$

$$= \frac{2.35 \text{ lbs}}{30 \text{ min}}$$

$$= \boxed{\begin{array}{l}0.078 \text{ lbs/min} \\ \text{Feed Rate}\end{array}}$$

Then calculate the lbs/day feed rate:

$$(0.078 \text{ lbs/min})\ (1440 \text{ min/day}) = \boxed{\begin{array}{l}112 \text{ lbs/day} \\ \text{Feed Rate}\end{array}}$$

Example 4: (Dry Chemical Feed Calibration)

❑ To calibrate a chemical feeder, a bucket is first weighed (0.29 lbs) then placed under the chemical feeder. After 25 minutes the bucket is weighed again. If the weight of the bucket with chemical is 1.7 lbs, what is the actual chemical feed rate, in lbs/day?

First calculate the lbs/min feed rate:

$$\text{Chem. Feed Rate, lbs/min} = \frac{\text{Chem. Applied, lbs}}{\text{Length of Application, min}}$$

$$= \frac{1.7 \text{ lbs} - 0.29 \text{ lbs}}{25 \text{ minutes}}$$

$$= \frac{1.41 \text{ lbs}}{25 \text{ min}}$$

$$= \boxed{\begin{array}{l}0.056 \text{ lbs/min} \\ \text{Feed Rate}\end{array}}$$

Then calculate the lbs/day feed rate:

$$(0.056 \text{ lbs/min})\ (1440 \text{ min/day}) = \boxed{\begin{array}{l}81 \text{ lbs/day} \\ \text{Feed Rate}\end{array}}$$

14.10 SOLUTION FEED CALIBRATION (Given m*L*/min Flow)

The calibration calculation for a solution feeder is slightly more difficult than that for a dry chemical feeder.

As with other calibration calculations, the actual chemical feed rate is determined and then compared with the feed rate indicated by the instrumentation.

To calculate the actual chemical feed rate for a solution feeder, first express the solution feed rate in terms of MGD. (The equation for converting from m*L*/min to gpd is given to the right.) Once the MGD solution flow rate has been calculated, use the mg/*L* equation to determine chemical dosage in lbs/day.

SOLUTION FEED CALIBRATION

First convert m*L*/min flow rate to gpd flow rate

$$\frac{(\text{m}L/\text{min})\ (1440\ \text{min/day})}{3785\ \text{m}L/\text{gal}} = \text{gpd}$$

Then calculate chemical dosage, lbs/day

$$(\text{mg}/L\ \text{Chem.})\ (\text{MGD Flow})\ (8.34\ \text{lbs/day}) = \text{Chem., lbs/day}$$

Example 1: (Solution Chemical Feed Calibration)

❑ A calibration test is conducted for a solution chemical feeder. During 5 minutes, a total of 750 m*L* is delivered by the solution feeder. The polymer solution is a 1.2% solution. What is the lbs/day feed rate? (Assume the polymer solution weighs 8.34 lbs/gal.)

Normally the mg/*L* to lbs/day equation* is used to determine the lbs/day feed rate. And in making these calculations, the flow rate must be expressed as MGD. Therefore, the m*L*/min flow rate must first be converted to gpd and then MGD. The m*L*/min flow rate is calculated as:

$$\frac{750\ \text{m}L}{5\ \text{min}} = \boxed{150\ \text{m}L/\text{min}}$$

Then convert m*L*/min flow rate to gpd flow rate:

$$\frac{(150\ \text{m}L/\text{min})\ (1440\ \text{min/day})}{3785\ \text{m}L/\text{gal}} = \boxed{57\ \text{gpd flow rate}}$$

And calculate lbs/day feed rate:**

$$(\text{mg}/L\ \text{Chem.})\ (\text{MGD Flow})\ (8.34\ \text{lbs/day}) = \text{lbs/day Chem.}$$

$$(12{,}000\ \text{mg}/L)\ (0.000057\ \text{MGD})\ (8.34\ \text{lbs/day}) = \boxed{5.7\ \text{lbs/day Polymer}}$$

* A detailed discussion of the mg/*L* to lbs/day calculation is given in Chapter 3.

** A solution of 1.2% strength is equivalent to a solution of 12,000 mg/L concentration. Refer to Chapter 8 in *Basic Math Concepts* for a discussion of mg/L to % conversions.

Example 2: (Solution Chemical Feed Calibration)

❑ A calibration test is conducted for a solution chemical feeder. During the 5-minute test, the pump delivered 920 m*L* of the 1.25% polymer solution. What is the polymer dosage rate in lbs/day? (Assume the polymer solution weighs 8.34 lbs/gal.)

First determine the m*L*/min solution flow rate during the 5-minute test:

$$\frac{920 \text{ m}L}{5 \text{ min}} = \boxed{184 \text{ m}L\text{/min}}$$

Next convert the m*L*/min flow rate to gpd flow rate:

$$\frac{(184 \text{ m}L\text{/min})\ (1440 \text{ min/day})}{3785 \text{ m}L\text{/gal}} = \boxed{\begin{array}{l}70 \text{ gpd}\\ \text{flow rate}\end{array}}$$

Then calculate the lbs/day polymer feed rate:

$$(12{,}500 \text{ mg/}L)\ (0.000070 \text{ MGD})\ (8.34 \text{ lbs/day}) = \boxed{\begin{array}{c}7.3 \text{ lbs/day}\\ \text{Polymer}\end{array}}$$

Example 3: (Solution Chemical Feed Calibration)

❑ A calibration test is conducted for a solution chemical feeder. During the 5-minute test, the pump delivered 820 m*L* of the 1.1% polymer solution. The specific gravity of the polymer solution is 1.2. What is the polymer dosage rate in lbs/gal?

First calculate the m*L*/min flow rate during the 5-minute test:

$$\frac{820 \text{ m}L}{5 \text{ min}} = \boxed{164 \text{ m}L\text{/min}}$$

Then convert the m*L*/min flow rate to gpd flow rate:

$$\frac{(164 \text{ ml/min})\ (1440 \text{ min/day})}{3785 \text{ ml/gal}} = \boxed{\begin{array}{l}62.4 \text{ gpd}\\ \text{flow rate}\end{array}}$$

And calculate the lbs/day polymer feed rate:
(Remember, the specific gravity is 1.2, so the density of the solution is (8.34 lbs/gal) (1.2) = 10 lbs/gal)

$$(11{,}000 \text{ mg/}L)\ (0.0000624 \text{ MGD})\ (10 \text{ lbs/gal}) = \boxed{\begin{array}{c}6.9 \text{ lbs/day}\\ \text{Polymer}\end{array}}$$

TAKING DENSITY AND SPECIFIC GRAVITY INTO CONSIDERATION

In Examples 1 and 2, the polymer solution was assumed to have a density of 8.34 lbs/gal. In many instances, however, polymer solutions have densities different than water—sometimes higher and other times lower.

When the density is different than water, use a different factor in the equation other than 8.34 lbs/gal.

Density information is sometimes given as specific gravity. To determine the density when specific gravity information is given, simply multiply the density of water (8.34 lbs/gal) by the specific gravity number:

$$\boxed{\begin{array}{cc}(8.34) & (\text{Specific})\\ \text{lbs/gal} & \text{Gravity}\end{array} = \begin{array}{c}\text{New}\\ \text{Density}\\ \text{lbs/gal}\end{array}}$$

14.11 SOLUTION CHEMICAL FEED CALIBRATION (Given Drop in Solution Tank Level)

Actual pumping rates can be determined by calculating the volume pumped during a specified time frame. For example, if 50 gallons are pumped during a 10-minute test, the average pumping rate during the test is 5 gpm.

The gallons pumped can be determined by **measuring the drop in water level during the timed test.**

VOLUME PUMPED IS INDICATED BY DROP IN TANK LEVEL

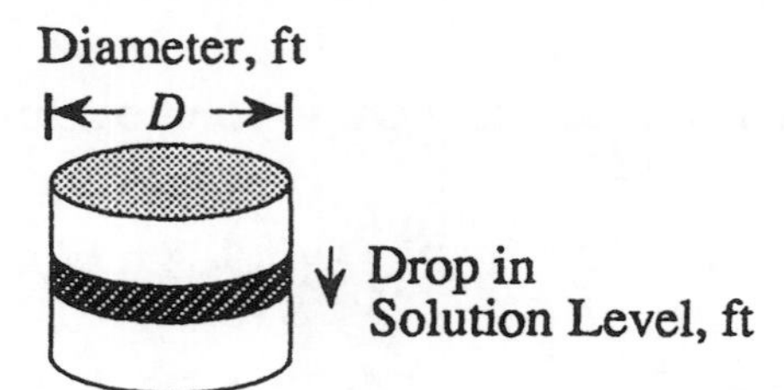

Simplified Equation:

$$\text{Flow, gpm} = \frac{\text{Volume Pumped, gal}}{\text{Duration of Test, min}}$$

Expanded Equation:

$$\text{Flow Rate, gpm} = \frac{(0.785)\,(D^2)\,(\text{Drop in Level, ft})\,(7.48\text{ gal/cu ft})}{\text{Duration of Test, min}}$$

Example 1: (Solution Feeder Calibration)

❑ A pumping rate calibration test is conducted for a 5-minute period. The liquid level in the 3-ft diameter solution tank is measured before and after the test. If the level drops 1.2 ft during the 5-min test, what is the pumping rate in gpm?

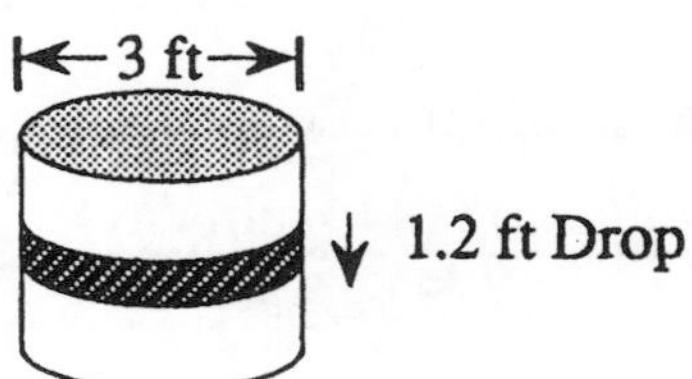

$$\text{Flow Rate, gpm} = \frac{(0.785)\,(D^2)\,(\text{Drop, ft})\,(7.48\text{ gal/cu ft})}{\text{Duration of Test, min}}$$

$$= \frac{(0.785)\,(3\text{ ft})\,(3\text{ ft})\,(1.2\text{ ft})\,(7.48\text{ gal/cu ft})}{5\text{ min}}$$

$$= \boxed{\text{13 gpm Pumping Rate}}$$

Example 2: (Solution Feeder Calibration)

❑ A pumping rate calibration test is conducted for a 4-minute period. The liquid level in the 4-ft diameter tank is measured before and after the pumping test. If the level drop is 10 inches during the test, what is the pumping rate in gpm?

4 ft

$$\frac{10 \text{ in}}{12 \text{ in/ft}} = 0.83 \text{ ft drop}$$

$$\text{Flow Rate, gpm} = \frac{(0.785)\ (D^2)\ (\text{Drop, ft})\ (7.48 \text{ gal/cu ft})}{\text{Duration of Test, min}}$$

$$= \frac{(0.785)\ (4 \text{ ft})\ (4 \text{ ft})\ (0.83 \text{ ft})\ (7.48 \text{ gal/cu ft})}{4 \text{ min}}$$

$$= \boxed{\begin{array}{c}19 \text{ gpm} \\ \text{Pumping Rate}\end{array}}$$

Example 3: (Solution Feeder Calibration)

❑ A pump test indicates that a pump delivers 45 gpm during a 5-minute pumping test. The diameter of the solution tank is 3 feet. What was the ft drop in solution level during the pumping test?

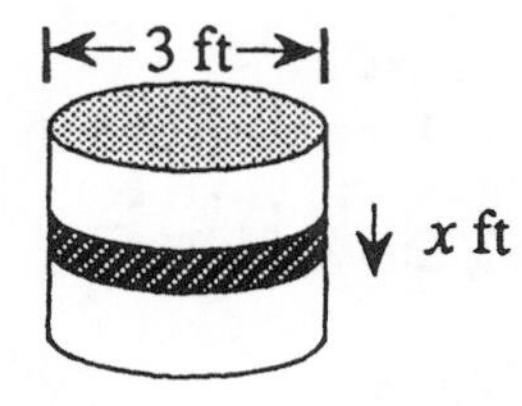

$$\text{Flow Rate, gpm} = \frac{(0.785)\ (D^2)\ (\text{Drop, ft})\ (7.48 \text{ gal/cu ft})}{\text{Duration of Test, min}}$$

$$45 = \frac{(0.785)\ (3 \text{ ft})\ (3 \text{ ft})\ (x \text{ ft})\ (7.48 \text{ gal/cu ft})}{5 \text{ min}}$$

Now solve for the unknown value*:

$$\frac{(45)\ (5)}{(0.785)\ (3)\ (3)\ (7.48)} = x \text{ ft Drop}$$

$$\boxed{\begin{array}{c}4.3 \text{ ft} \\ \text{Drop}\end{array}} = x$$

CALCULATING OTHER UNKNOWN VARIABLES

In Examples 1 and 2, the unknown variable was gpm pumping rate. The same equation can be used to solve for any one of the other variables: gpm pumping rate, tank diameter, level drop, or duration of test. In Example 3, the level drop is the unknown variable.

* Refer to Chapter 2 in *Basic Math Concepts* for a review of solving for the unknown value.

14.12 AVERAGE USE CALCULATIONS

The lbs/day or gpd chemical use should be recorded each day. From this data, you can calculate the average daily use of the chemical or solution.

From this information you can forecast expected chemical use, compare it with chemical in inventory, and determine when additional chemical supplies will be required.

AVERAGE CHEMICAL USE

First determine the average chemical use:

$$\text{Average Use lbs/day} = \frac{\text{Total Chem. Used, lbs}}{\text{Number of Days}}$$

Or

$$\text{Average Use gpd} = \frac{\text{Total Chem. Used, gal}}{\text{Number of Days}}$$

Then calculate day's supply in inventory:*

$$\text{Day's Supply in Inventory} = \frac{\text{Total Chem. in Inventory, lbs}}{\text{Average Use, lbs/day}}$$

Or

$$\text{Day's Supply in Inventory} = \frac{\text{Total Chem. in Inventory, gal}}{\text{Average Use, gpd}}$$

Example 1: (Average Use)

❑ The chemical used for each day during a week is given below. Based on this data, what was the average lbs/day chemical use during the week?

Monday—90 lbs/day
Tuesday—96 lbs/day
Wednesday—92 lbs/day
Thursday—89 lbs/day
Friday—98 lbs/day
Saturday—91 lbs/day
Sunday—87 lbs/day

$$\text{Average Use lbs/day} = \frac{\text{Total Chem. Used, lbs}}{\text{Number of Days}}$$

$$= \frac{643 \text{ lbs}}{7 \text{ days}}$$

$$= \boxed{91.9 \text{ lbs/day Aver. Use}}$$

* Note how similar these equations are to detention time equations.

Example 2: (Average Use)

❑ The average chemical use at a plant is 78 lbs/day. If the chemical inventory in stock is 2400 lbs, how many days' supply is this?

$$\text{Days' Supply in Inventory} = \frac{\text{Total Chem. in Inventory, lbs}}{\text{Average Use, lbs/day}}$$

$$= \frac{\text{2400 lbs in Inventory}}{\text{78 lbs/day Aver. Use}}$$

$$= \boxed{\text{30.8 days' Supply in Inventory}}$$

Example 3: (Average Use)

❑ The average gallons polymer solution used each day at a treatment plant is 86 gpd. A chemical feed tank has a diameter of 3 ft and contains solution to a depth of 4.1 ft. How many days' supply are represented by the solution in the tank?

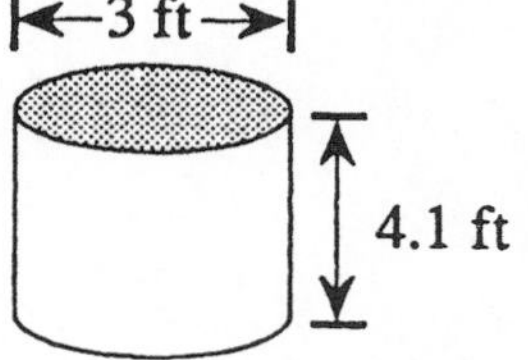

$$\text{Days' Supply in Tank} = \frac{\text{Total Solution in Tank, gal}}{\text{Average Use, gpd}}$$

$$x \text{ days} = \frac{(0.785)\text{ (3 ft) (3 ft) (4.1 ft) (7.48 gal/cu ft)}}{\text{86 gpd}}$$

$$= \boxed{\text{2.5 days' Supply in Tank}}$$

NOTES:

15 Sludge Production and Thickening

SUMMARY

1. Primary and Secondary Clarifier Solids Production

For Primary Clarifier:

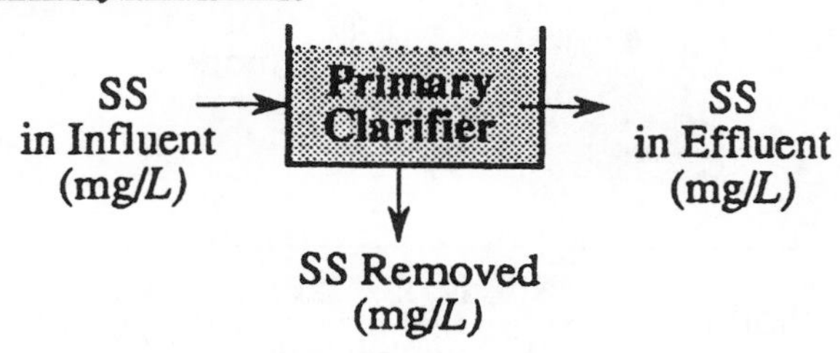

(mg/L) SS Removed	(MGD) flow	(8.34) lbs/gal	=	Solids lbs/day

For Secondary Clarifier:

A Y-value, bacteria growth rate, is used to determine SS produced in a secondary clarifier. The Y-value is the lbs/day of SS generated as a result of each lb of BOD consumed, or removed, during treatment. Typical Y-values are established for each plant based on the average ratio of lbs SS/lb BOD removed by the secondary system . The equation below assumes a Y-value, or growth rate, of 0.45 lbs SS produced/lb BOD removed. The Y-value is established for each plant.

First calculate lbs/day BOD removed:

BOD in Influent (mg/L) → Secondary Clarifier → BOD in Effluent (mg/L)

↓

BOD Removed (mg/L)

(mg/L) BOD Removed	(MGD) flow	(8.34) lbs/gal	=	BOD Removed lbs/day

Then use the *Y*-value as a ratio to determine lbs/day SS produced:

$$\frac{0.45 \text{ lbs SS}}{1 \text{ lb BOD Removed}} = \frac{x \text{ lbs SS}}{\text{lbs/day BOD Removed}}$$

SUMMARY—Cont'd

2. Percent Solids and Sludge Pumping

The same equation can be used for percent solids and lbs/day sludge calculations:

$$\text{\% Solids} = \frac{\text{Solids, lbs/day}}{\text{Sludge, lbs/day}} \times 100$$

This equation is sometimes rearranged for use in lbs/day sludge calculations:

$$\text{Sludge, lbs/day} = \frac{\text{Solids, lbs/day}}{\frac{\text{\% Solids}}{100}}$$

3. Thickening and Sludge Volume Changes

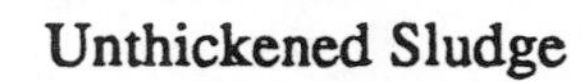

Sludge

Water

Solids

Simplified Equation:

$$\text{lbs Solids} = \text{lbs Solids}$$

Expanded Equations:

$$\left(\begin{array}{c}\text{Prim. or Sec.}\\ \text{Sludge,}\\ \text{lbs/day}\end{array}\right)\left(\frac{\text{\% Solids}}{100}\right) = \left(\begin{array}{c}\text{Thickened}\\ \text{Sludge,}\\ \text{lbs/day}\end{array}\right)\left(\frac{\text{\%Solids}}{100}\right)$$

Or

$$\left(\begin{array}{c}\text{Prim. or Sec}\\ \text{Sludge, gpd}\end{array}\right)\left(\begin{array}{c}8.34\\ \text{lbs/gal}\end{array}\right)\left(\frac{\text{\% Solids}}{100}\right) = \left(\begin{array}{c}\text{Thickened}\\ \text{Sludge, gpd}\end{array}\right)\left(\begin{array}{c}8.34\\ \text{lbs/gal}\end{array}\right)\left(\frac{\text{\%Solids}}{100}\right)$$

SUMMARY—Cont'd

4. Gravity Thickening Calculations

- **Hydraulic Loading Rate**

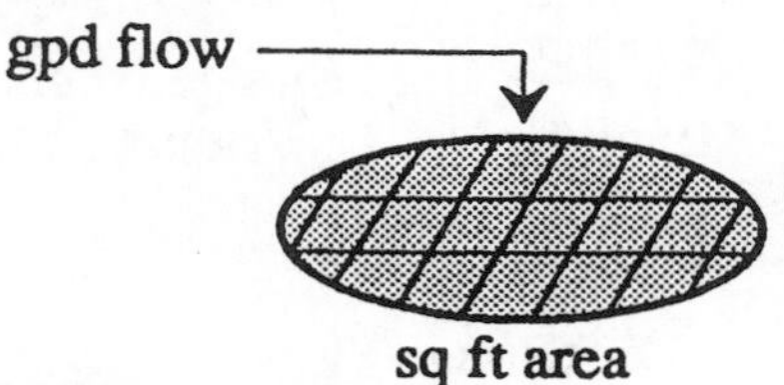

Simplified Equation:

$$\text{Hydraulic Loading Rate, gpd/sq ft} = \frac{\text{Total flow, gpd}}{\text{Area, sq ft}}$$

Expanded Equation:

$$\text{Hydraulic Loading Rate, gpd/sq ft} = \frac{\text{(gpm flow)(1440 min/day)}}{\text{Area, sq ft}}$$

- **Solids Loading Rate**

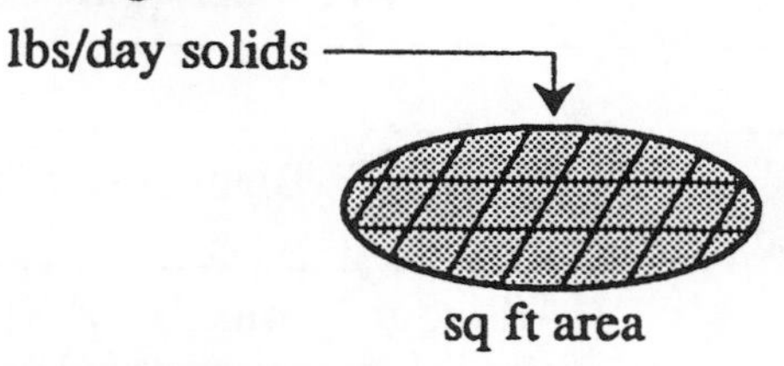

Simplified Equation:

$$\text{Solids Loading Rate, lbs/day/sq ft} = \frac{\text{Solids, lbs/day}}{\text{Area, sq ft}}$$

Expanded Equation:

$$\text{Solids Loading Rate, lbs/day/sq ft} = \frac{\underset{\text{gpd}}{\text{(Sludge)}}\ \underset{\text{lbs/gal}}{(8.34)}\ \left(\frac{\text{\% Solids}}{100}\right)}{\text{Area, sq ft}}$$

Or

$$\text{Solids Loading Rate, lbs/day/sq ft} = \frac{\underset{\text{gpm}}{\text{(Sludge)}}\ \underset{\text{min/day}}{(1440)}\ \underset{\text{lbs/gal}}{(8.34)}\ \left(\frac{\text{\% Solids}}{100}\right)}{\text{Area, sq ft}}$$

SUMMARY—Cont'd

- **Sludge Detention Time (Sludge-Volume Ratio)**

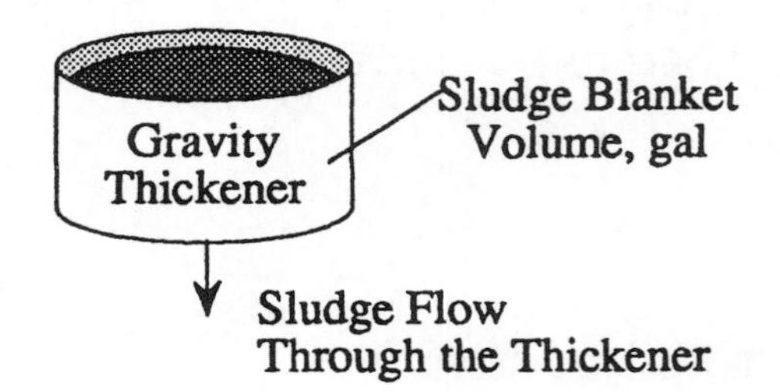

Simplified Equation:

$$\text{Sludge Detention Time, days} = \frac{\text{Sludge Blanket Vol., gal}}{\text{Sludge Pumped from Thickener, gpd}}$$

Expanded Equation:

$$\text{Sludge Detention Time, days} = \frac{(0.785)\,(D^2)\,(\text{Blanket Depth, ft})\,(7.48\ \text{gal/cu ft})}{\text{Sludge Pumped from Thickener, gpd}}$$

Or

$$\text{Sludge Detention Time, days} = \frac{(0.785)\,(D^2)\,(\text{Blanket Depth, ft})\,(7.48\ \text{gal/cu ft})}{(\text{Sludge Pumped from Thickener, gpm})\,(1440\ \frac{\text{min}}{\text{day}})}$$

- **Gravity Thickener Efficiency**

mg/*L* or % SS in Influent → Gravity Thickener → mg/*L* or % SS in Effluent

mg/*L* or % SS Removed

$$\text{Efficiency, \%} = \frac{\text{SS Removed, \%}}{\text{SS in Influent, \%}} \times 100$$

Or

$$\text{Efficiency, \%} = \frac{\text{SS Removed, mg/}L}{\text{SS in Influent, mg/}L} \times 100$$

SUMMARY—Cont'd

- **Concentration Factor**

$$\text{Concentration Factor} = \frac{\text{Thickened Sludge, \%}}{\text{Influent Sludge, \%}}$$

- **Solids Balance at the Thickener**

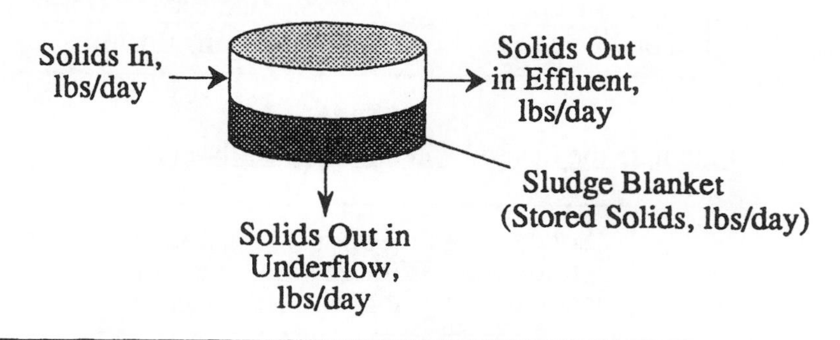

$$\text{Solids In, lbs/day} = \text{Solids Out in Underflow, lbs/day} + \text{Solids Out in Effluent, lbs/day} + \text{Solids Stored lbs/day}$$

- **Time Required for Sludge Blanket Rise or Fall**

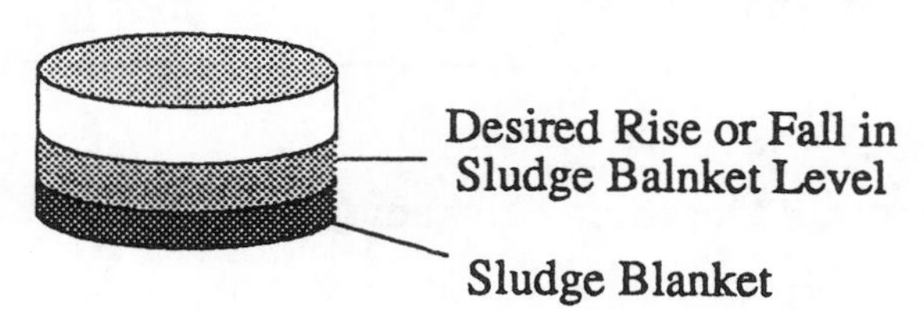

Simplified Equation*:

$$\text{Rise or Fall Time, hrs} = \frac{\text{Solids in Rise or Fall, lbs}}{\text{Solids Storage Rate, lbs/hr}}$$

Expanded Equation:

$$\text{Rise or Fall Time, hrs} = \frac{(0.785)(D^2)\ \binom{\text{Rise or Fall}}{\text{Depth, ft}}\ \binom{7.48}{\text{gal/cu ft}}\ \binom{8.34}{\text{lbs/gal}}\ \left(\frac{\%\ \text{Sol.}}{100}\right)}{\text{Solids Storage Rate, lbs/hr}}$$

* This equation uses solids rather than sludge simply because solids data is generally required for other thickener calculations. The rise or fall time could be calculated using lbs sludge in rise or fall divided by the sludge storage rate, lbs/hr.

SUMMARY—Cont'd

- **Adjusting Withdrawal Rates**

Three calculations are part of determining appropriate withdrawal rates during non-steady-state conditions:

Calculate the desired solids storage rate:

$$\frac{\text{Storage Depth, ft}}{\text{Total Depth, ft}} = \frac{\text{Solids Storage Rate, lbs/min}}{\text{Solids Entering, lbs/min}}$$

Calculate the desired solids withdrawal rate:

$$\begin{array}{c}\text{Solids}\\\text{Entering,}\\\text{lbs/min}\end{array} = \begin{array}{c}\text{Solids}\\\text{Withdrawal}\\\text{lbs/min}\end{array} + \begin{array}{c}\text{Solids}\\\text{Storage,}\\\text{lbs/min}\end{array}$$

Calculate the desired gpm sludge withdrawal rate:

$$\begin{array}{c}\text{(Sludge)}\\\text{Withd.}\\\text{Rate, gpm}\end{array} \begin{array}{c}(8.34)\\\text{lbs/gal}\end{array} \left(\frac{\text{\% Solids of Thick. Sludge}}{100}\right) = \begin{array}{c}\text{Solids}\\\text{Withdr.,}\\\text{lbs/min}\end{array}$$

5. Dissolved Air Flotation Thickening Calculations

- **Hydraulic Loading Rate**

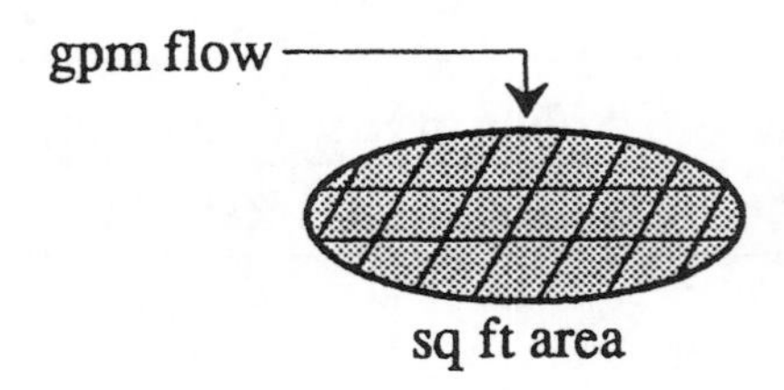

$$\begin{array}{c}\text{Hydraulic}\\\text{Loading Rate,}\\\text{gpm/sq ft}\end{array} = \frac{\text{Flow, gpm}}{\text{Surface Area, sq ft}}$$

SUMMARY—Cont'd

- **Solids Loading Rate**

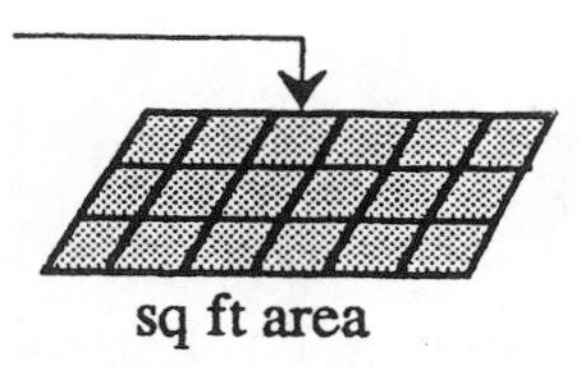

Simplified Equation:

$$\text{Solids Loading Rate, lbs/hr/sq ft} = \frac{\text{Solids, lbs/hr}}{\text{Area, sq ft}}$$

Expanded Equation:

$$\text{Solids Loading Rate, lbs/hr/sq ft} = \frac{(\text{gpm})(60\ \frac{\text{min}}{\text{hr}})(8.34\ \text{lbs/gal})(\frac{\%\text{SS}}{100})}{\text{Area, sq ft}}$$

(gpm = sludge flow)

- **Air Applied**

$$\text{Air, lbs/hr} = (\text{Air, cfm})(60\ \frac{\text{min}}{\text{hr}})(0.075\ \frac{\text{lbs}}{\text{cu ft}})$$

- **Air/Solids Ratio**

Simplified Equation:

$$\text{Air/Solids Ratio} = \frac{\text{Air, lbs/min}}{\text{Solids, lbs/min}}$$

Expanded Equation:

$$\text{Air/Solids Ratio} = \frac{(\text{cfm Air})(0.075\ \text{lbs/cu ft})}{(\text{gpm Sludge})(8.34\ \text{lbs/gal})(\frac{\%\ \text{SS}}{100})}$$

- **Percent Recycle Rate**

$$\text{Recycle, \%} = \frac{\text{Recycle Flow, gpm}}{\text{Sludge Flow to DAF Unit, gpm}} \times 100$$

SUMMARY—Cont'd

- **Solids Removal Efficiency**

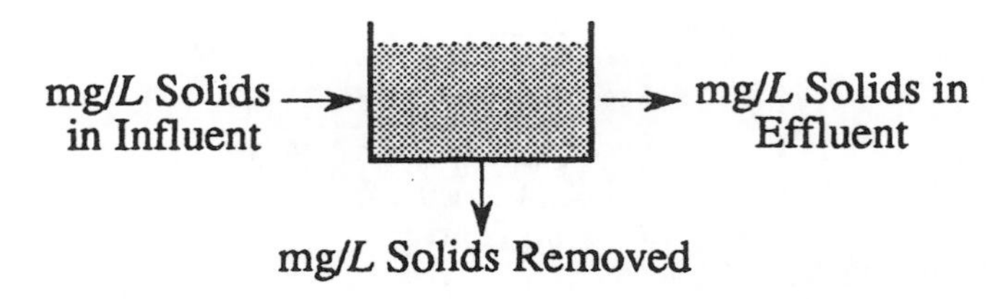

$$\text{Solids Removal Efficiency, \%} = \frac{\text{Solids Removed, mg/L}}{\text{Solids in Influent, mg/L}} \times 100$$

- **Concentration Factor**

$$\text{Concentration Factor} = \frac{\text{Thickened Sludge Conc., \%}}{\text{Influent Sludge Conc., \%}}$$

6. Centrifuge Thickening Calculations

- **Hydraulic Loading Rate—Scroll or Disc Centrifuges**

$$\text{Hydraulic Loading, gph} = \frac{\text{Flow, gpd}}{\text{24 hrs/day}}$$

Or

$$\text{Hydraulic Loading, gph} = (\text{Flow, gpm})\left(60\ \frac{\text{min}}{\text{hr}}\right)$$

- **Hydraulic Loading Rate—Basket Centrifuges**

Simplified Equation:

$$\text{Hydraulic Loading, gph} = \underset{\text{flow}}{(\text{gph})}\ \frac{(\text{Duration of Sludge Flow})}{\text{Time in Operation}}$$

Expanded Equation:

$$\text{Hydraulic Loading, gph} = \underset{\text{flow}}{(\text{gpm})}\ \left(60\ \frac{\text{min}}{\text{hr}}\right) \frac{(\text{Dur. of Sludge Flow})}{\text{Time in Operation}}$$

Or

$$\text{Hydraulic Loading, gph} = \frac{(\text{gpd flow})}{\text{24 hrs/day}} \cdot \frac{(\text{Dur. of Sludge Flow})}{\text{Time in Operation}}$$

SUMMARY—Cont'd

- **Solids Loading Rate—Scroll or Disc Centrifuges**

The equation used depends on how the flow rate is expressed:

$$\text{Solids Loading, lbs/hr} = (\text{gph sludge flow})(8.34 \text{ lbs/gal})\frac{(\%\ \text{Solids})}{100}$$

Or

$$\text{Solids Loading, lbs/hr} = \frac{(\text{gpd sludge flow})}{24 \text{ hr/day}}(8.34 \text{ lbs/gal})\frac{(\%\ \text{Solids})}{100}$$

Or

$$\text{Solids Loading, lbs/hr} = (\text{gpm sludge flow})\left(60\ \frac{\text{min}}{\text{hr}}\right)(8.34 \text{ lbs/gal})\frac{(\%\ \text{Solids})}{100}$$

- **Solids Loading Rate—Basket Centrifuges**

Solids loading rates for basket centrifuges are calculated the same as for scroll or disc centrifuges <u>except that the flow rate must be adjusted</u> for the actual duration of sludge flow to the unit.

- **Feed Time (Fill Time) for Basket Centrifuges**

<u>Simplified Equation:</u>

$$\text{Fill Time, min} = \frac{\text{Stored Solids Capacity, lbs}}{\text{Solids Entering, lbs/min}}$$

<u>Expanded Equation:</u>

$$\text{Fill Time, min} = \frac{(\text{Basket Capac., cu ft})(7.48 \text{ gal/cu ft})(8.34 \text{ lbs/gal})\frac{(\%\ \text{Basket Solids})}{100}}{(\text{gpm sludge flow})(8.34 \text{ lbs/gal})\frac{(\%\ \text{Influent Solids})}{100}}$$

SUMMARY—Cont'd

- **Removal Efficiency**

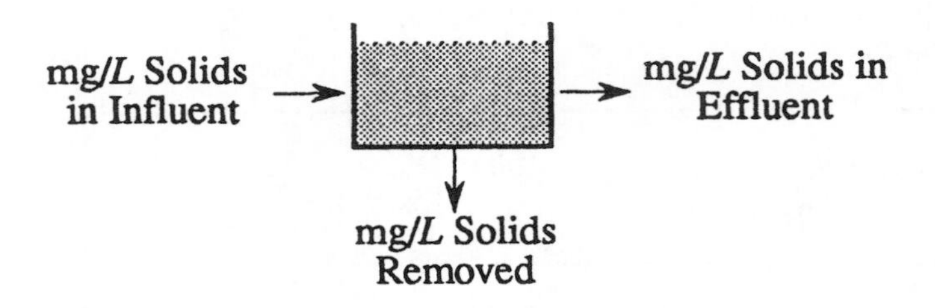

$$\text{Efficiency, \%} = \frac{\text{Solids Removed, \%}}{\text{Solids in Influent, \%}} \times 100$$

Or

$$\text{Efficiency, \%} = \frac{\text{Solids Removed, mg/}L}{\text{Solids in Influent, mg/}L} \times 100$$

- **Average Total Solids Concentration—Basket Centrifuge**

Simplified Equation:

$$\begin{array}{c}\text{\% Solids of}\\ \text{Sludge Mixture}\end{array} = \frac{\text{Solids in Mixture, lbs}}{\text{Sludge Mixture, lbs}} \times 100$$

Expanded Equations:

$$\begin{array}{c}\text{\% Solids of}\\ \text{Sludge Mixture}\end{array} = \frac{\begin{array}{c}\text{lbs Solids in}\\ \text{Skimmed Sludge}\end{array} + \begin{array}{c}\text{lbs Solids in}\\ \text{Knifed Sludge}\end{array}}{\begin{array}{c}\text{lbs Skimmed}\\ \text{Sludge}\end{array} + \begin{array}{c}\text{lbs Knifed}\\ \text{Sludge}\end{array}} \times 100$$

Or

$$\begin{array}{c}\text{\% Solids}\\ \text{of}\\ \text{Sludge}\\ \text{Mixture}\end{array} = \frac{\begin{pmatrix}\text{Skimmed}\\ \text{Sludge,}\\ \text{cu ft}\end{pmatrix} \begin{pmatrix}62.4\\ \text{lbs/cu ft}\end{pmatrix} \frac{(\text{\% Sol.})}{100} + \begin{pmatrix}\text{Knifed}\\ \text{Sludge,}\\ \text{cu ft}\end{pmatrix} \begin{pmatrix}62.4\\ \text{lbs/cu ft}\end{pmatrix} \frac{(\text{\% Sol.})}{100}}{\begin{pmatrix}\text{Skimmed}\\ \text{Sludge,}\\ \text{cu ft}\end{pmatrix} \begin{pmatrix}62.4\\ \text{lbs/cu ft}\end{pmatrix} + \begin{pmatrix}\text{Knifed}\\ \text{Sludge,}\\ \text{cu ft}\end{pmatrix} \begin{pmatrix}62.4\\ \text{lbs/cu ft}\end{pmatrix}} \times 100$$

NOTES:

15.1 PRIMARY & SECONDARY CLARIFIER SOLIDS PRODUCTION

CALCULATING SOLIDS PRODUCED—PRIMARY SYSTEM

The solids produced during primary treatment depends on the solids that settle in, or are removed by, the clarifier.

Use the mg/*L* to lbs/day equation* to determine the lbs/day solids removed by the clarifier:

(mg/*L*) SS Rem.	(MGD) flow	(8.34) lbs/gal	= lbs/day SS Rem.

When making calculations pertaining to solids and sludge, remember that the term "solids" refers to dry solids and the term "sludge" refers to the solids and water.

Example 1: (Solids Production)

❑ A primary clarifier receives a flow of 1.82 MGD with a suspended solids concentration of 345 mg/*L*. If the clarifier effluent has a suspended solids concentration of 190 mg/*L*, how many pounds of solids are generated daily?

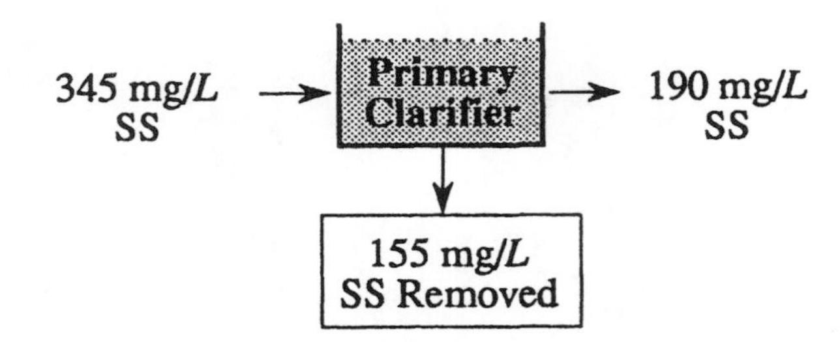

(mg/*L* SS Removed) (MGD flow) (8.34 lbs/gal) = lbs/day SS Removed

(155 mg/*L*) (1.82 MGD) (8.34 lbs/gal) = **2353 lbs/day Solids**

Example 2: (Solids Production)

❑ The suspended solids content of the primary influent is 360 mg/*L* and the primary effluent is 208 mg/*L*. How many pounds of solids are produced during a day that the flow is 4,220,000 gpd?

360 mg/*L* SS → Primary Clarifier → 208 mg/*L* SS

↓

152 mg/*L* SS Removed

Complete the lbs/day calculation:

(mg/*L* SS Removed) (MGD flow) (8.34 lbs/gal) = lbs/day SS Removed

(152 mg/*L*) (4.22 MGD) (8.34 lbs/gal) = **5350 lbs/day Solids Removed**

---

* Refer to Chapter 3 for a review of mg/*L* to lbs/day calculations.

Example 3: (Solids Production)

❑ The 1.6 MGD influent to the secondary system has a BOD concentration of 178 mg/*L*. The secondary effluent contains 24 mg/*L* BOD. If the bacteria growth rate, *Y*-value, for this plant is 0.45 lbs SS/lb BOD removed, how many pounds of dry sludge solids are produced each day by the secondary system?

First calculate the lbs/day BOD removed:

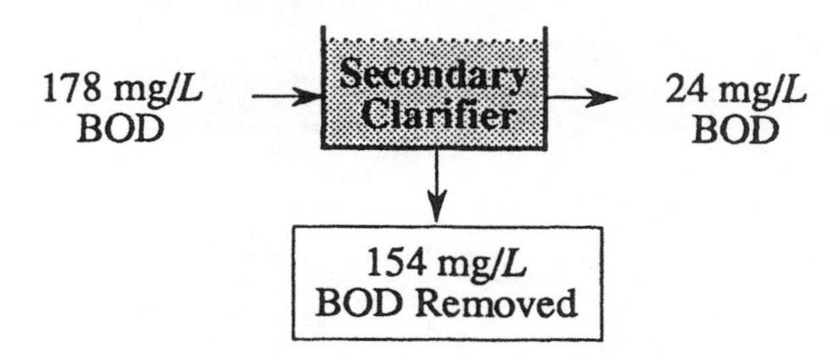

$$(\text{mg/}L\ \text{BOD Removed})\ (\text{MGD flow})\ (8.34\ \text{lbs/gal}) = \text{lbs/day BOD Removed}$$

$$(154\ \text{mg/}L)\ (1.6\ \text{MGD})\ (8.34\ \text{lbs/gal}) = \boxed{\begin{array}{l}2055\ \text{lbs/day}\\ \text{BOD Removed}\end{array}}$$

Then use the *Y*-value to determine lbs/day solids produced.

$$\frac{0.45\ \text{lbs SS Produced}}{1\ \text{lb BOD Removed}} = \frac{x\ \text{lbs SS Produced}}{2055\ \text{lbs/day BOD Rem.}}$$

$$\frac{(0.45)\ (2055)}{1} = x$$

$$\boxed{\begin{array}{c}925\ \text{lbs/day}\\ \text{Solids Produced}\end{array}} = x$$

CALCULATING SOLIDS PRODUCED—SECONDARY SYSTEM

The solids produced during secondary treatment depends on many factors, including the amount of organic matter removed by the system and the growth rate of the bacteria. Although precise calculations of sludge production is rather complex, a rough estimate of solids production can be obtained using an estimated growth rate (*Y*) value.

For most treatment plants, every pound of food consumed (BOD removed) by the bacteria, produces between 0.3 and 0.7 lbs of new bacteria cells; these are solids that have to be removed from the system.

There are, therefore, two steps in calculating lbs/day solids production for the secondary system:

1. Calculate the lbs/day BOD removed from the system.

2. Use the *Y*-value to determine lbs/day solids produced.

Example 3 illustrates this calculation.

15.2 PERCENT SOLIDS AND SLUDGE PUMPING

PERCENT SOLIDS

Sludge is comprised of water and solids. The vast majority of sludge is water—usually in the range of 93% to 97%.

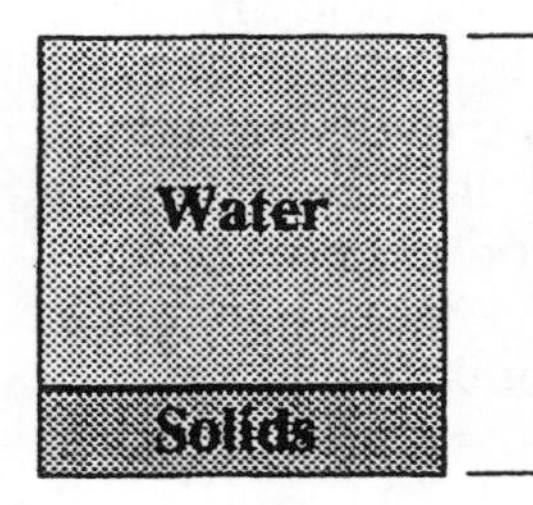

To determine the solids content of a sludge, a sample of sludge is dried overnight in an oven at 103°- 105° C. **The solids that remain after drying represent the total solids content of the sludge.** This solids content may be expressed as a percent or as a mg/*L* concentration.** Two equations are used to calculate percent solids, depending on whether lab data or plant data is used in the calculation. In both cases, the calculation is **on the basis of solids and sludge weight**:

$$\% \text{ Solids} = \frac{\text{Total Solids, g}}{\text{Sludge Sample, g}} \times 100$$

Or

$$\% \text{ Solids} = \frac{\text{Solids, lbs/day}}{\text{Sludge, lbs/day}} \times 100$$

Example 1: (Percent Solids and Sludge Pumping)

❑ The total weight of a sludge sample is 20 grams. (Sludge sample only, not the dish.) If the weight of the solids after drying is 0.58 grams, what is the percent total solids of the sludge?

$$\% \text{ Solids} = \frac{\text{Total Solids, grams}}{\text{Sludge Sample, grams}} \times 100$$

$$= \frac{0.58 \text{ grams}}{20 \text{ grams}} \times 100$$

$$= \boxed{2.9\%}$$

Example 2: (Percent Solids and Sludge Pumping)

❑ A total of 4100 gallons of sludge is pumped to a digester daily. If the sludge has a 5% solids content, how many lbs/day solids are pumped to the digester? (Assume the sludge weighs 8.34 lbs/gal.)

First, write the percent solids equation and fill in the given information:

$$\% \text{ Solids} = \frac{\text{Solids, lbs/day}}{\text{Sludge, lbs/day}} \times 100$$

$$5 = \frac{x \text{ lbs/day Solids}}{(4100 \text{ gal})(8.34 \text{ lbs/gal})} \times 100$$

$$\frac{(4100)(8.34)(5)}{100} = x$$

$$\boxed{\text{1710 lbs/day Solids}} = x$$

* This graphic is to illustrate the composition of sludge. Under normal circumstances, of course, the solids and water are mixed.

** 1% solids = 10,000 mg/*L*. For a review of % to mg/*L* conversions, refer to Chapter 8.

Example 3: (Percent Solids and Sludge Pumping)
❑ A total of 8,700 lbs/day SS are removed from a primary clarifier and pumped to a sludge thickener. If the sludge has a solids content of 2.5%, how many lbs/day sludge are pumped to the thickener?

$$\% \text{ Solids} = \frac{\text{Solids, lbs/day}}{\text{Sludge, lbs/day}} \times 100$$

$$2.5 = \frac{8{,}700 \text{ lbs/day Solids}}{x \text{ lbs/day Sludge}} \times 100$$

$$x = \frac{(8{,}700)(100)}{2.5}$$

$$= \boxed{348{,}000 \text{ lbs/day Sludge}}$$

CALCULATING OTHER UNKNOWN VARIABLES

The three variables in percent solids calculations are percent solids, lbs/day solids, and lbs/day sludge. In Example 1 percent solids was the unknown variable. Examples 2-4 illustrate calculations when other variables are unknown. To solve these problems, write the equation as usual, fill in the known information, then solve for the unknown value.*

Example 4: (Percent Solids and Sludge Pumping)
❑ It is anticipated that 320 lbs/day SS will be pumped from the primary clarifier of a new plant. If the primary clarifier sludge has a solids content of 4.5%, how many gpd sludge will be pumped from the clarifier? (Assume the sludge weighs 8.34 lbs/gal.)

First calculate lbs/day sludge to be pumped using the % solids equation, then convert lbs/day sludge to gpd sludge:

$$\text{lbs/day Sludge} = \frac{\text{Solids, lbs/day}}{\frac{\% \text{ Solids}}{100}}$$

$$x \text{ lbs/day Sludge} = \frac{320 \text{ lbs/day Solids}}{0.045}$$

$$x = 7111 \text{ lbs/day Sludge}$$

Converting lbs/day sludge to gpd sludge:

$$\frac{7111 \text{ lbs/day Sludge}}{8.34 \text{ lbs/gal}} = \boxed{853 \text{ gpd Sludge}}$$

SLUDGE TO BE PUMPED

As mentioned above, one of the variables in the percent solids calculation is lbs/day sludge. Example 3 illustrates a calculation of lbs/day sludge. Example 4 illustrates the calculation of gpd sludge to be pumped.

Because lbs/day sludge is calculated relatively frequently, the percent solids equation is often rearranged as follows:

$$\boxed{\text{Sludge, lbs/day} = \frac{\text{Solids, lbs/day}}{\frac{\% \text{ Solids}}{100}}}$$

* Refer to Chapter 2 in *Basic Math Concepts* for a review of solving for the unknown factor.

15.3 THICKENING AND SLUDGE VOLUME CHANGES

Sludge thickening calculations are based on the concept that **the solids in the primary or secondary sludge are equal to the solids in the thickened sludge**. The solids are the same.* It is primarily water that has been removed in order to thicken the sludge and result in a higher percent solids. In the unthickened sludge, the solids might represent 1 or 4% of the total pounds of sludge. But when some of the water is removed, that same amount of solids might represent 5-7% of the total pounds of sludge.

Because the pounds of solids remain unchanged, when making sludge thickening calculations, the lbs/day solids of the unthickened sludge may be set **equal to** the lbs/day solids of the thickened sludge, as shown to the right.

Normally the lbs/day solids are not known directly but must be calculated, based on lbs/day (or gpd) sludge and percent solids content of that sludge, as shown in the expanded equation.

Note that both the primary or secondary sludge and the thickened sludge are assumed to have a density of 8.34 lbs/gal). **The actual density of the sludge should be determined** and used if it is different than 8.34 lbs/gal. The density of the primary or secondary sludge may be different than the thickened sludge density. Refer to Chapter 7, Section 7.1, "Density and Specific Gravity".

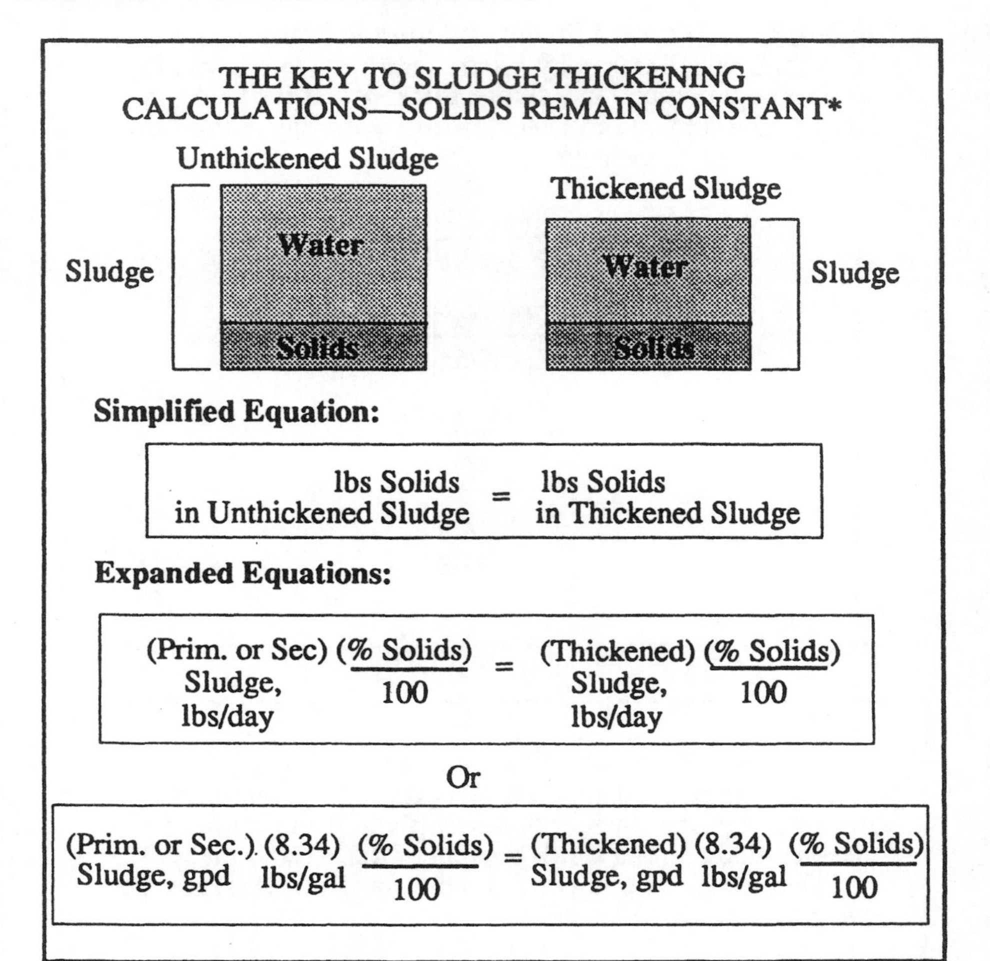

THE KEY TO SLUDGE THICKENING CALCULATIONS—SOLIDS REMAIN CONSTANT*

Simplified Equation:

$$\text{lbs Solids in Unthickened Sludge} = \text{lbs Solids in Thickened Sludge}$$

Expanded Equations:

$$(\text{Prim. or Sec) Sludge, lbs/day})\left(\frac{\%\ \text{Solids}}{100}\right) = (\text{Thickened) Sludge, lbs/day})\left(\frac{\%\ \text{Solids}}{100}\right)$$

Or

$$(\text{Prim. or Sec.) Sludge, gpd})(8.34\ \text{lbs/gal})\left(\frac{\%\ \text{Solids}}{100}\right) = (\text{Thickened) Sludge, gpd})(8.34\ \text{lbs/gal})\left(\frac{\%\ \text{Solids}}{100}\right)$$

Example 1: (Sludge Thickening)

❑ A primary clarifier sludge has a 3% solids content. A total of 25,400 lbs/day sludge is pumped to a thickener. If the sludge has been concentrated to 5% solids, what will be the expected lbs/day sludge flow from the thickener?

$$(\text{Primary) Sludge, lbs/day})\left(\frac{\%\ \text{Solids}}{100}\right) = (\text{Thickened) Sludge, lbs/day})\left(\frac{\%\ \text{Solids}}{100}\right)$$

$$(25{,}400\ \text{lbs/day})\left(\frac{3}{100}\right) = (x\ \text{lbs/day})\left(\frac{5}{100}\right)$$

$$\frac{(25{,}400)(0.03)}{0.05} = x$$

$$\boxed{\text{15,240 lbs/day Thickened Sludge}} = x$$

* This assumes a negligible amount of solids are lost in the thickener overflow.

Example 2: (Sludge Thickening)

❑ A primary clarifier sludge has a 3.2% solids content. If 3140 gpd primary sludge is pumped to a thickener and the thickened sludge has a solids content of 5.8%, what would be the expected gpd flow of thickened sludge? (Assume both sludges weigh 8.34 lbs/gal.)

Since sludge data is given as gpd, the second expanded equation is used:

$$(\text{Prim. or Sec.})\ \text{Sludge, gpd}\ (8.34\ \text{lbs/gal}) \frac{(\%\ \text{Solids})}{100} = (\text{Thickened})\ \text{Sludge, gpd}\ (8.34\ \text{lbs/gal}) \frac{(\%\ \text{Solids})}{100}$$

Fill in the given information and solve for the unknown value:

$$(3140\ \text{gpd})\ (8.34\ \text{lbs/gal})\ (0.032) = (x\ \text{gpd})\ (8.34\ \text{lbs/gal})\ (0.058)$$

$$\frac{(3140)\ (\cancel{8.34})\ (0.032)}{(\cancel{8.34})\ (0.058)} = x$$

1732 gpd Thickened Sludge	= x

Example 3: (Sludge Thickening)

❑ The sludge from a primary clarifier has a solids content of 2.8%. The primary sludge is pumped at a rate of 4510 gpd to a thickener. If the thickened sludge has a solids content of 5.2%, what is the anticipated gpd sludge flow from the thickener? (Assume both sludges weigh 8.34 lbs/gal.)

$$(\text{Prim. or Sec.})\ \text{Sludge, gpd}\ (8.34\ \text{lbs/gal}) \frac{(\%\ \text{Solids})}{100} = (\text{Thickened})\ \text{Sludge, gpd}\ (8.34\ \text{lbs/gal}) \frac{(\%\ \text{Solids})}{100}$$

$$(4510\ \text{gpd})\ (8.34\ \text{lbs/gal})\ (0.028) = (x\ \text{gpd})\ (8.34\ \text{lbs/gal})\ (0.052)$$

$$\frac{(4510\ \text{gpd})\ (\cancel{8.34})\ (0.028)}{(\cancel{8.34})\ (0.052)} = x$$

2428 gpd Thickened Sludge	= x

15.4 GRAVITY THICKENING CALCULATIONS

HYDRAULIC LOADING RATE

A gravity thickener is designed to thicken or further concentrate sludges before they are sent to additional sludge handling and treatment processes, such as digestion, conditioning, and dewatering. By thickening the sludge, there is a reduced load on these subsequent processes.

As with many other processes, the calculation of hydraulic loading is important in determining whether the process is underloaded or overloaded. The equation used is as follows:

$$\text{Hydraulic Loading Rate, gpd/sq ft} = \frac{\text{Flow, gpd}}{\text{Area, sq ft}}$$

The flow rate to the thickener is often given as gpm. The equation incorporating gpm flow rate is:

$$\text{Hydraulic Loading Rate, gpd/sq ft} = \frac{(\text{gpm flow})(1440 \text{ min/day})}{\text{Area, sq ft}}$$

Gravity thickeners, like clarifiers, are typically circular in shape. Therefore, the square feet area is a calculation of the area of a circle*: $(0.785)(D^2)$.

Example 1: (Gravity Thickening)

❑ A gravity thickener 20 ft in diameter receives a flow of 35 gpm primary sludge combined with a 75 gpm secondary effluent flow. What is the hydraulic loading on the thickener in gpd/sq ft?

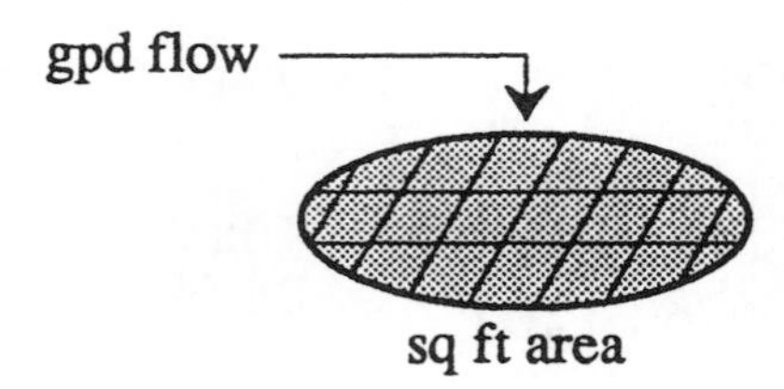

$$\text{Hydraulic Loading Rate, gpd/sq ft} = \frac{\text{Total Flow, gpd}}{\text{Area, sq ft}}$$

$$= \frac{(35 \text{ gpm} + 75 \text{ gpm})(1440 \text{ min/day})}{(0.785)(20 \text{ ft})(20 \text{ ft})}$$

$$= \frac{158{,}400 \text{ gpd}}{314 \text{ sq ft}}$$

$$= \boxed{504 \text{ gpd/sq ft}}$$

Example 2: (Gravity Thickening)

❑ The primary sludge flow to gravity thickener is 60 gpm. This is blended with a 80 gpm secondary effluent flow. If the thickener has a diameter of 25 ft, what is the hydraulic loading rate, in gpd/sq ft?

gpd flow

sq ft area

$$\text{Hydraulic Loading Rate, gpd/sq ft} = \frac{\text{Total Flow, gpd}}{\text{Area, sq ft}}$$

$$= \frac{(140 \text{ gpm})(1440 \text{ min/day})}{(0.785)(25 \text{ ft})(25 \text{ ft})}$$

$$= \boxed{411 \text{ gpd/sq ft}}$$

---

* For a review of area calculations, refer to Chapter 10 in *Basic Math Concepts*.

SOLIDS LOADING RATE

Solids loading on the gravity thickener is calculated as pounds of solids entering daily per square feet area:

$$\text{Solids Loading Rate, lbs/day/sq ft} = \frac{\text{Solids, lbs/day}}{\text{Area, sq ft}}$$

Many times the lbs/day solids is not given directly, but must be calculated using lbs/day sludge and percent solids data. The equation used in this case is:

$$\text{Solids Loading Rate, lbs/day/sq ft} = \frac{(\underset{\text{lbs/day}}{\text{Sludge,}})\left(\frac{\text{\%Solids}}{100}\right)}{\text{Area, sq ft}}$$

The sludge flow rate to the thickener may be expressed as gpd or gpm. The corresponding equations are given below. Remember, each equation is simply an expanded form of the basic equation lbs/day solids per square feet area.

$$\text{S.L.R.} = \frac{(\underset{\text{gpd}}{\text{Sludge,}})\,(\underset{\text{lbs/gal}}{8.34^*})\left(\frac{\text{\%Solids}}{100}\right)}{\text{Area, sq ft}}$$

$$\text{S.L.R.} = \frac{(\underset{\text{gpm}}{\text{Sludge,}})\,(\underset{\text{min/day}}{1440})\,(\underset{\text{lbs/gal}}{8.34^*})\left(\frac{\text{\%Sol.}}{100}\right)}{\text{Area, sq ft}}$$

Example 3: (Gravity Thickening)

❑ A primary sludge flow equivalent to 115,200 gpd is pumped to a 40-ft diameter gravity thickener. If the solids concentration of the sludge is 3.7%, what is the solids loading in lbs/day/sq ft?

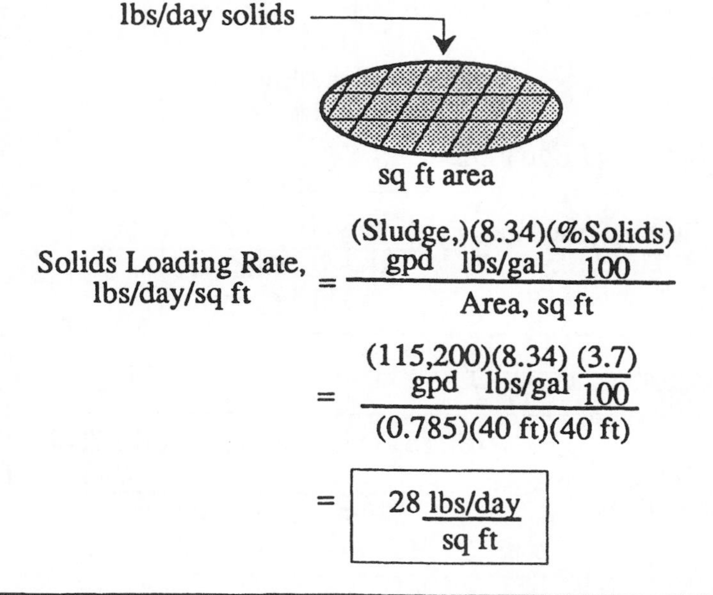

$$\text{Solids Loading Rate, lbs/day/sq ft} = \frac{(\underset{\text{gpd}}{\text{Sludge,}})(\underset{\text{lbs/gal}}{8.34})\left(\frac{\text{\%Solids}}{100}\right)}{\text{Area, sq ft}}$$

$$= \frac{(\underset{\text{gpd}}{115{,}200})(\underset{\text{lbs/gal}}{8.34})\left(\frac{3.7}{100}\right)}{(0.785)(40\text{ ft})(40\text{ ft})}$$

$$= \boxed{28\ \frac{\text{lbs/day}}{\text{sq ft}}}$$

Example 4: (Gravity Thickening)

❑ What is the solids loading on a gravity thickener (in lbs/day/sq ft), if the primary sludge flow to the 35-ft diameter gravity thickener is 55 gpm, with a solids concentration of 3.5%?

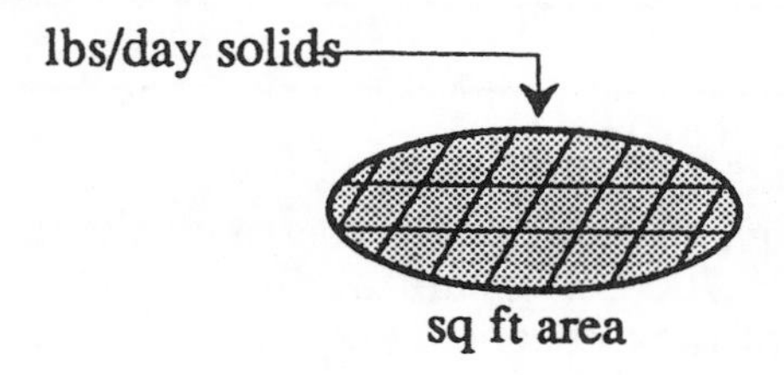

$$\text{Solids Loading Rate, lbs/day/sq ft} = \frac{(\underset{\text{gpd}}{\text{Sludge,}})(\underset{\text{lbs/gal}}{8.34})\left(\frac{\text{\%Solids}}{100}\right)}{\text{Area, sq ft}}$$

$$= \frac{(55\text{ gpm})(\underset{\text{min/day}}{1440})\,(\underset{\text{lbs/gal}}{8.34})\left(\frac{3.5}{100}\right)}{(0.785)(35\text{ ft})(35\text{ ft})}$$

$$= \boxed{24\ \frac{\text{lbs/day}}{\text{sq ft}}}$$

* If the sludge has a density greater than water, a larger number would be required here. Refer to Chapter 7, Section 7.1, Density and Specific Gravity.

SLUDGE DETENTION TIME (SLUDGE-VOLUME RATIO)

The **sludge detention time*** refers to the length of time the solids remain in the gravity thickener. This length of time depends on the solids added to the thickener, the depth of the sludge blanket, and the solids pumped from the thickener. The sludge detention time is sometimes referred to as the **sludge-volume ratio.**

When detention time is calculated for the wastewater flow, such as for a sedimentation tank or clarifier, or for an oxidation pond, the flow rate of interest is the **flow rate into the tank.**

In contrast, however, when detention time is calculated for the solids in the tank, such as for this calculation, the flow rate of interest is the **flow rate of solids pumped from the tank.****

These two types of detention time calculations are sometimes distinguished by the terms **hydraulic detention time** and **solids detention time,** respectively.

The equations and the associated graphic to be used in the calculation of sludge detention time are shown in the box to the right.

SLUDGE DETENTION TIME IS BASED ON SLUDGE PUMPED FROM THE GRAVITY THICKENER

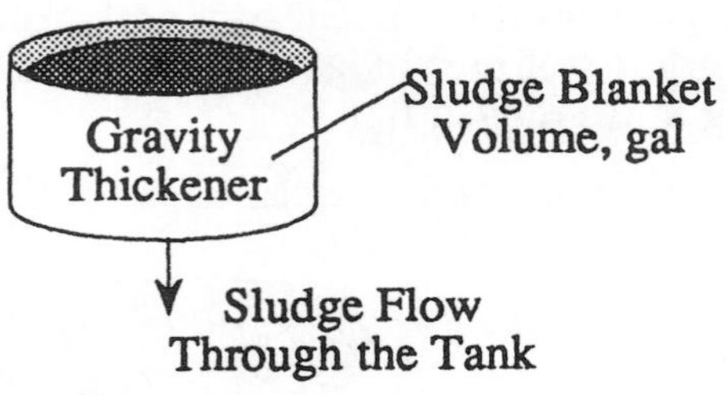

Simplified Equation:

$$\text{Sludge Detention Time, days} = \frac{\text{Sludge Blanket Vol., gal}}{\text{Sludge Pumped from Thickener, gpd}}$$

Expanded Equation:

$$\text{Sludge Detention Time, days} = \frac{(0.785)(D^2 \text{ ft})(\text{Blanket Depth})(7.48 \text{ gal/cu ft})}{\text{Sludge Pumped from Thickener, gpd}}$$

Or

$$\text{Sludge Detention Time, days} = \frac{(0.785)(D^2 \text{ ft})(\text{Blanket Depth})(7.48 \text{ gal/cu ft})}{(\text{Sludge Pumped from Thickener, gpm})(1440 \frac{\text{min}}{\text{day}})}$$

Example 5: (Gravity Thickening)

❑ The sludge blanket volume in a gravity thickener has been calculated to be 21,700 gallons. If the sludge pumping rate from the bottom of the thickener is equal to 36,000 gpd, what is the sludge detention time, in days?

$$\text{Sludge Detention Time, days} = \frac{\text{Sludge Blanket Vol., gal}}{\text{Sludge Pumped from Thickener, gpd}}$$

$$= \frac{21{,}700 \text{ gal}}{36{,}000 \text{ gpd}}$$

$$= \boxed{0.6 \text{ days}}$$

* This calculation is similar in form to many other detention time calculations. Refer to Chapter 5.

** The denominator of the sludge detention time calculation includes only the solids pumped from the thickener. Note that the solids retention time calculation (see Chapter 5) includes both the solids pumped from the tank as well as the solids in the tank effluent.

Example 6: (Gravity Thickening)

❑ A gravity thickener 40-ft in diameter has a sludge blanket depth of 3.2 ft. If sludge is pumped from the bottom of the thickener at the rate of 20 gpm, what is the sludge detention time (in days) in the thickener? (Assume the pumping rate is continuous.*)

$$\text{Sludge Detention Time, days} = \frac{\text{Sludge Blanket Vol., gal}}{\text{Sludge Pumped from Thickener, gpd}}$$

$$= \frac{(0.785)(40\text{ ft})(40\text{ ft})(3.2\text{ ft})(7.48\text{ gal/cu ft})}{(20\text{ gpm})(1440\text{ min/day})}$$

$$= \frac{30{,}064\text{ gal}}{28{,}800\text{ gpd}}$$

$$= \boxed{1.0\text{ days}}$$

Example 7: (Gravity Thickening)

❑ A gravity thickener 35 ft in diameter has a sludge blanket depth of 3.8 ft. If the sludge is pumped from the bottom of the thickener at a rate of 25 gpm, what is the sludge detention time, in hours, in the thickener? (Assume the pumping rate is continuous.)

$$\text{Sludge Detention Time, hrs} = \frac{\text{Sludge Blanket Vol., gal}}{\text{Sludge Pumped from Thickener, gph}}$$

$$= \frac{(0.785)(35\text{ ft})(35\text{ ft})(3.8\text{ ft})(7.48\text{ gal/cu ft})}{(25\text{ gpm})(60\text{ min/hr})}$$

$$= \frac{27{,}333\text{ gal}}{1500\text{ gph}}$$

$$= \boxed{18.2\text{ hrs}}$$

SLUDGE DETENTION TIME EXPRESSED IN HOURS

For detention times less than one day, it may be desirable to express the **detention time in hours**. If so, the pumping rate must also be expressed in terms of gallons **per hour**:

$$\text{Sludge Detention Time, hours} = \frac{\text{Sludge Blanket Vol., gal}}{\text{Pumping Rate, gph}}$$

Or

$$\text{Sludge Detention Time, hours} = \frac{\text{Sludge Blanket Vol., gal}}{(\text{Pumping Rate, gpm})(60\text{ min/hr})}$$

Example 7 illustrates this type of calculation.

* If the pumping rate were not continuous, you would use a different conversion factor than 1440 min/day to convert to gpd pumped. The factor to be used would be the total number of minutes pumped during the day.

EFFICIENCY

The efficiency of a gravity thickener is a measure of the effectiveness in removing suspended solids from the flow. As with other calculations of efficiency*, the calculation includes amount removed divided by the total amount:

$$\text{Efficiency, \%} = \frac{\text{Part Removed}}{\text{Total}} \times 100$$

In calculations of gravity thickener efficiency, the percent sludge solids (% SS) removed is determined using the equations given below.

$$\text{Effic., \%} = \frac{\dfrac{\text{\% SS Removed}}{100}}{\dfrac{\text{\% SS in Influent}}{100}} \times 100$$

This equation can be simplified by dividing out the hundreds in the numerator and denominator of the fraction:

$$\text{Effic., \%} = \frac{\text{SS Removed, \%}}{\text{SS in Influent, \%}} \times 100$$

The percent sludge solids may also be expressed as mg/*L*** to calculate efficiency:

$$\text{Effic., \%} = \frac{\text{SS Removed, mg/L}}{\text{SS in Influent, mg/L}} \times 100$$

Examples 8 and 9 illustrate efficiency calculations.

Example 8: (Gravity Thickening)

❑ The sludge flow entering a gravity thickener contains 3.1% solids. The effluent from the thickener contains 0.18% solids. What is the efficiency of the gravity thickener in removing solids?

3.1% Solids → [thickener] → 0.18 % Solids

2.92 % Solids Removed

$$\text{Efficiency, \%} = \frac{\text{\% SS Removed}}{\text{\% SS in Influent}} \times 100$$

$$= \frac{2.92}{3.1} \times 100$$

$$= \boxed{94\%}$$

Example 9: (Gravity Thickening)

❑ What is the efficiency of the gravity thickener if the influent flow to the thickener has a solids concentration of 2.8%, and the effluent flow has a solids concentration of 0.7%? Calculate the efficiency using mg/*L* solids concentration.

First convert the data from percent to mg/*L* (1% = 10,000 mg/*L*):

2.8% = 28,000 mg/*L*
0.7% = 7,000 mg/*L*

28,000 mg/*L* → [thickener] → 7,000 mg/*L*

21,000 mg/*L*

$$\text{Efficiency, \%} = \frac{21{,}000 \text{ mg/L}}{28{,}000 \text{ mg/L}} \times 100$$

$$= \boxed{75\%}$$

---

* For a review of other types of efficiency calculations, refer to Chapter 6.

** For a discussion of mg/*L* to percent conversions, refer to Chapter 8 in *Basic Math Concepts*.

CONCENTRATION FACTOR

The concentration factor is another means of determining the effectiveness of the gravity thickening process.

The concentration factor indicates how much the sludge has been concentrated as a result of the thickening process. For example, a concentration factor of 2 means that after thickening the sludge is twice as concentrated as when it entered the thickener; a concentration factor of 3 indicates that the thickened sludge is three times as concentrated as when it entered the thickener, etc.

To determine the concentration factor, simply compare the thickened sludge concentration with the influent sludge concentration:

$$\text{Concent. Factor} = \frac{\text{Thickened Sludge, \%}}{\text{Influent Sludge, \%}}$$

Example 10: (Gravity Thickening)

❑ The sludge solids concentration of the influent flow to a gravity thickener is 3.2%. If the sludge withdrawn from the bottom of the thickener has a sludge solids concentration of 8%, what is the concentration factor?

$$\text{Concentration Factor} = \frac{\text{Thickened Sludge, \%}}{\text{Influent Sludge, \%}}$$

$$= \frac{8}{3.2}$$

$$= \boxed{2.5}$$

Example 11: (Gravity Thickening)

❑ The influent flow to a gravity thickener has a sludge solids concentration of 3%. What is the concentration factor if the sludge solids concentration of the sludge withdrawn from the thickener is 8.7%?

$$\text{Concentration Factor} = \frac{\text{Thickened Sludge, \%}}{\text{Influent Sludge, \%}}$$

$$= \frac{8.7}{3}$$

$$= \boxed{2.9}$$

SOLIDS BALANCE AT THE THICKENER

Monitoring the solids entering and leaving the gravity thickener, as well as those retained in the thickener (sludge blanket) is essential in the effective operation of the process.

When determining the **solids balance** (mass balance) at the thickener, the solids entering the thickener are compared with those leaving the system (either through the underflow as thickened sludge, or in the effluent). The balance between incoming and outgoing solids determines whether the sludge blanket will increase, decrease, or stay the same.

- **If the solids going in equal the solids going out**, the sludge blanket level will stay the same. For example, if 40,000 lbs/day solids leave the thickener (in underflow and overflow), there are no additional solids being stored:

$$\text{Solids In} = \text{Solids Out} + \text{Solids Stored}$$

$$40{,}000 \text{ lbs/day} = 40{,}000 \text{ lbs/day} + 0 \text{ lbs/day}$$

- **If the solids going in are greater than the solids going out**, the sludge blanket level will begin to rise. For example, if 40,000 lbs/day solids enter the thickener and only 30,000 lbs/day leave the thickener, then 10,000 lbs/day will be stored in the sludge blanket:

$$\text{Solids In} = \text{Solids Out} + \text{Solids Stored}$$

$$40{,}000 \text{ lbs/day} = 30{,}000 \text{ lbs/day} + 10{,}000 \text{ lbs/day}$$

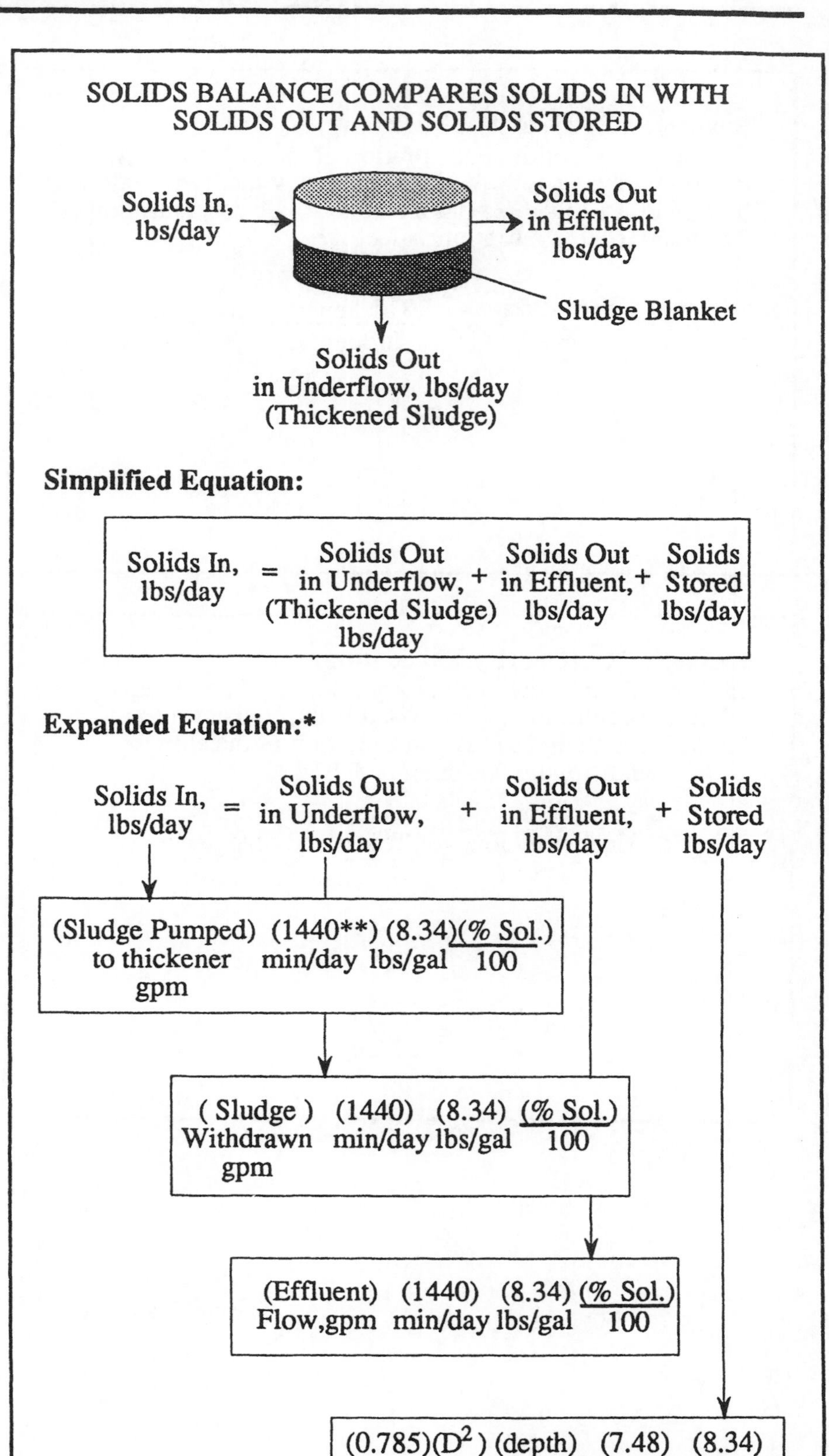

* Note that each of these lbs/day quantities may be calculated using mg/L concentration rather than percent. In that case, the calculation would be (mg/L solids)(MGD flow)(8.34 lbs/gal). The result will be the same. Refer to Chapter 8 in *Basic Math Concepts* for a review of mg/L and percent conversions.

** If the pumping is not continuous, use the total number of minutes pumped per day rather than the 1440 min/day factor.

Example 12: (Gravity Thickening)

❑ Given the data below, (a) determine whether the sludge blanket in the gravity thickener will increase, decrease, or remain the same; and (b) if there is an increase or decrease, how many lbs/day solids change will this be?

Sludge Pumped to Thickener = 110 gpm
Thickened Sludge Pumped from Thickener = 50 gpm
Primary Sludge Solids = 3%
Thickened Sludge Solids = 8%
Thickener Effluent Suspended Solids = 600 mg/*L*

a) First, determine **lbs/day sludge solids entering the thickener.**

$$(110 \text{ gpm})(1440 \text{ min/day})(8.34 \text{ lbs/gal})\left(\frac{3}{100}\right) = 39{,}632 \text{ lbs/day}$$

Next, calculate lbs/day sludge solids leaving the thickener via the underflow:

$$(50 \text{ gpm})(1440 \text{ min/day})(8.34 \text{ lbs/gal})\left(\frac{8}{100}\right) = 48{,}038 \text{ lbs/day}$$

And the lbs/day sludge solids leaving the thickener via the effluent flow:* (The effluent flow is 110 gpm – 50 gpm = 60 gpm. Also, 600 mg/*L* = 0.06%)

$$(60 \text{ gpm})(1440 \text{ min/day})(8.34 \text{ lbs/gal})\left(\frac{0.06}{100}\right) = 432 \text{ lbs/day}$$

To summarize the sludge solids entering and leaving:

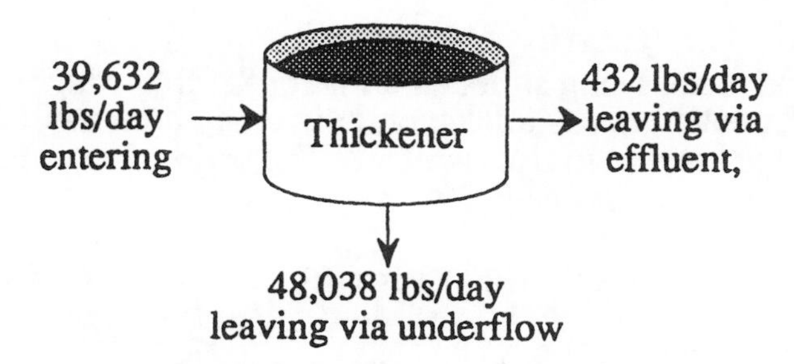

As indicated by the diagram, there are more solids leaving the thickener than entering. Therefore the sludge blanket level will drop.

b) The lbs/day drop in level depends on the difference between solids entering and leaving the thickener:

$$\text{48,470 Total lbs/day leaving} - \text{39,632 lbs/day entering} = \boxed{\text{8838 lbs/day total}}$$

• **If the solids going in are less than the solids going out,** the sludge blanket level will begin to fall. For example, if 35,000 lbs/day solids enter the thickener and 40,000 lbs/day leave the thickener, then the sludge blanket will fall an equivalent of 5,000 lbs/day:

Solids In = Solids Out + Solids Stored

$$\text{35,000 lbs/day} = \text{40,000 lbs/day} + \text{(–5,000) lbs/day}$$

In Example 12, there is a comparison of solids entering and leaving the thickener. The effect on the sludge blanket level is also calculated.

Although the equation shown in Example 12 includes solids in the thickener effluent, **these solids are sometimes considered negligible** (that is, they have little effect on the overall calculation) provided the effluent contains a solids concentration less than 500 mg/*L* and there is little solids carry over in the effluent.

* The lbs/day solids may be calculated using the flow and percent solids, as shown in the calculation of solids entering the digester; or, the lbs/day solids may be calculated using mg/*L* concentration—(mg/*L*)(Flow, MGD)(8.34 lbs/gal) = lbs/day.

TIME REQUIRED FOR SLUDGE BLANKET RISE OR FALL

Once you have determined the sludge storage rate, as illustrated in Example 12, you can determine the time required for the sludge blanket to reach a desired level.

This calculation is similar to a detention time or fill-time calculation,* as shown to the right. As with all detention time calculations, if the time is desired in hours, the flow rate (or storage rate in this calculation) must be expressed in corresponding time units—lbs/hr.

Flow conversions from one unit to another may be required in these calculations. For a review of flow conversions, refer to Chapter 8 in *Basic Math Concepts*.

RISE OR FALL TIME IS SIMILAR TO A DETENTION TIME CALCULATION

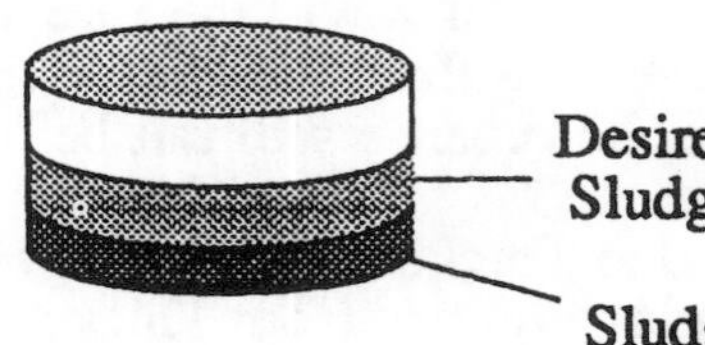

For detention time, the volume and flow rate are generally expressed as gallons and gal/hr (or gal/min, etc.). For rise or fall time, the solids and flow rate are expressed as lbs and lbs/hr, respectively:

Simplified Equation:**

$$\text{Rise or Fall Time, hrs} = \frac{\text{Solids in Rise or Fall, lbs}}{\text{Solids Storage Rate, lbs/hr}}$$

Expanded Equation:

$$\text{Rise or Fall Time, hrs} = \frac{(0.785)(D^2)(\text{Rise or Fall Depth, ft})\ (7.48\ \text{gal/cu ft})\ (8.34\ \text{lbs/gal})\ \left(\frac{\%\ \text{Sol.}}{100}\right)}{\text{Solids Storage Rate, lbs/hr}}$$

Example 13: (Gravity Thickening)

❑ If solids are being stored at a rate of 8800 lbs/day in a 20-ft diameter gravity thickener, how many hours will it take the sludge blanket to rise two feet? The solids concentration of the thickened sludge is 7%.

$$\text{Rise Time, hrs} = \frac{(0.785)(D^2)(\text{Rise, ft})\ (7.48\ \text{gal/cu ft})\ (8.34\ \text{lbs/gal})\ \left(\frac{\%\ \text{Sol.}}{100}\right)}{\text{Solids Storage Rate, lbs/hr}}$$

Before filling in the equation, the storage rate must be expressed as lbs/hr: 8800 lbs/day ÷ 24 hrs/day = 367 lbs/hr.

$$\text{Rise Time, hrs} = \frac{(0.785)\ (20\ \text{ft})\ (20\ \text{ft})\ (2\ \text{ft})\ (7.48)\ (8.34)\ \left(\frac{7}{100}\right)}{367\ \text{lbs/hr}}$$

$$= \boxed{7.5\ \text{hrs}}$$

* For a review of detention time calculations, refer to Chapter 5.

** Note that the sludge blanket volume, rise or fall, must be expressed as pounds of solids (rather than pounds of sludge) since the storage rate, lbs/min or lbs/hr is expressed in terms of solids.

Example 14: (Gravity Thickening)

☐ After several hours of startup of a gravity thickener, the sludge blanket level is measured at 2.5 ft. The desired sludge blanket level is 6 ft. If the sludge solids are entering the thickener at a rate of 40 lbs/min, what is the desired sludge withdrawal rate, in gpm? The thickened sludge solids concentration is 6%.

First calculate the desired solids storage rate, using the ratio of stored (or current) to total sludge blanket depth:

$$\frac{\text{Storage Depth, ft}}{\text{Total Depth, ft}} = \frac{\text{Solids Storage Rate, lbs/min}}{\text{Solids Entering, lbs/min}}$$

$$\frac{2.5 \text{ ft}}{6 \text{ ft}} = \frac{x \text{ lbs/min}}{40 \text{ lbs/min}}$$

$$x = \frac{(2.5)(40)}{6}$$

$$x = \boxed{\text{17 lbs/min Solids Storage Rate}}$$

Next, calculate the solids withdrawal rate necessary to achieve the 17 lbs/min solids storage rate:

$$\begin{matrix}\text{Solids} \\ \text{Entering} \\ \text{lbs/min}\end{matrix} = \begin{matrix}\text{Solids} \\ \text{Withdrawal,} \\ \text{lbs/min}\end{matrix} + \begin{matrix}\text{Solids} \\ \text{Storage,} \\ \text{lbs/min}\end{matrix}$$

$$40 \text{ lbs/min} = x \text{ lbs/min} + 17 \text{ lbs/min}$$

$$40 \text{ lbs/min} - 17 \text{ lbs/min} = x \text{ lbs/min}$$

$$\boxed{\text{23 lbs/min Solids Withdrawal Rate}} = x$$

Then calculate the gpm sludge withdrawal rate that will result in withdrawing 23 lbs/min solids withdrawal:

$$\begin{matrix}\text{(Sludge)} \\ \text{Withd.,} \\ \text{gpm}\end{matrix} \begin{matrix}\text{(8.34)} \\ \text{lbs/gal}\end{matrix} \frac{\text{(\% Sol. of) Thickened Sl.}}{100} = \begin{matrix}\text{Solids} \\ \text{Withd.,} \\ \text{lbs/min}\end{matrix}$$

$$(x \text{ gpm}) \underset{\text{lbs/gal}}{(8.34)} \left(\frac{6}{100}\right) = 23 \text{ lbs/min}$$

$$x = \frac{23}{(8.34)(0.06)}$$

$$x = \boxed{\text{46 gpm Sludge Withdrawal Rate}}$$

ADJUSTING WITHDRAWAL RATES

Under continuous operation, the sludge blanket should be maintained at a relatively constant level. The sludge withdrawal rates are generally not adjusted more than about 20% at a time.

During such times as startup, however, the withdrawal rate is sometimes adjusted significantly more than 20 percent to achieve an increase in sludge blanket depth. An appropriate sludge withdrawal rate in this situation may be determined using three steps:

1. Calculate the **desired solids storage rate**, lbs/min, using the ratio* of the storage depth (current sludge blanket depth) to the total desired depth:

$$\frac{\text{Storage Depth, ft}}{\text{Total Depth, ft}} = \frac{\text{Solids Storage Rate, lbs/min}}{\text{Solids Entering, lbs/min}}$$

2. Once the desired solids storage rate has been calculated, the **desired solids withdrawal rate**, lbs/min, may be determined using the equation, solids in equal solids out plus solids stored:

$$\begin{matrix}\text{Solids} \\ \text{Entering} \\ \text{lbs/min}\end{matrix} = \begin{matrix}\text{Solids} \\ \text{Withdrawal,} \\ \text{lbs/min}\end{matrix} + \begin{matrix}\text{Solids} \\ \text{Storage,} \\ \text{lbs/min}\end{matrix}$$

3. This withdrawal rate may then be expressed as **gpm sludge withdrawal rate**:

$$\begin{matrix}\text{(Sludge)} \\ \text{Withd.,} \\ \text{gpm}\end{matrix} \begin{matrix}\text{(8.34)} \\ \text{lbs/gal}\end{matrix} \frac{\text{(\% Sol. of) Thick. Sl.}}{100} = \begin{matrix}\text{Solids} \\ \text{Withd.,} \\ \text{lbs/min}\end{matrix}$$

Withdrawal rate calculations are often made every few hours to verify sludge storage and sludge blanket changes.

* For a review of ratios and proportions, refer to Chapter 7 in *Basic Math Concepts*.

15.5 DISSOLVED AIR FLOTATION THICKENING CALCULATIONS

HYDRAULIC LOADING RATE

The hydraulic loading calculation for dissolved air flotation (DAF) thickeners is similar to other calculations of hydraulic loading—flow rate per surface area. For DAF thickeners, hydraulic loading is expressed as gpm/sq ft:

$$\text{Hydraulic Loading Rate, gpm/sq ft} = \frac{\text{Flow, gpm}}{\text{Surface Area, sq ft}}$$

Example 1: (DAF Thickening)

❑ A dissolved air flotation unit receives a sludge flow of 800 gpm. If the DAF unit is 50 ft in diameter, what is the hydraulic loading rate, in gpm/sq ft?

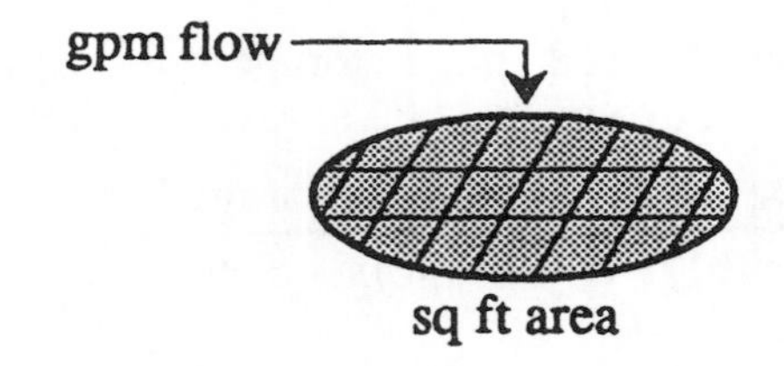

$$\text{Hydraulic Loading Rate, gpm/sq ft} = \frac{\text{Flow, gpm}}{\text{Area, sq ft}}$$

$$= \frac{800 \text{ gpm}}{(0.785)(50 \text{ ft})(50 \text{ ft})}$$

$$= \boxed{0.4 \text{ gpm/sq ft}}$$

Example 2: (DAF Thickening)

❑ A dissolved air flotation unit is 45 ft long and 15 ft wide. If the unit receives a sludge flow of 170,000 gpd, what is the hydraulic loading, in gpm/sq ft?

$$\frac{170{,}000 \text{ gpd}}{1440 \text{ min/day}}$$

$$= 118 \text{ gpm}$$

sq ft area

$$\text{Hydraulic Loading Rate, gpm/sq ft} = \frac{\text{Flow, gpm}}{\text{Area, sq ft}}$$

$$= \frac{118 \text{ gpm}}{(45 \text{ ft})(15 \text{ ft})}$$

$$= \boxed{0.2 \text{ gpm/sq ft}}$$

Example 3: (DAF Thickening)

❑ The sludge flow to a dissolved air flotation thickener is 120 gpm, with a suspended solids concentration of 0.75%. If the DAF unit is 50 ft long and 15 ft wide, what is the solids loading rate in lbs/hr/sq ft?

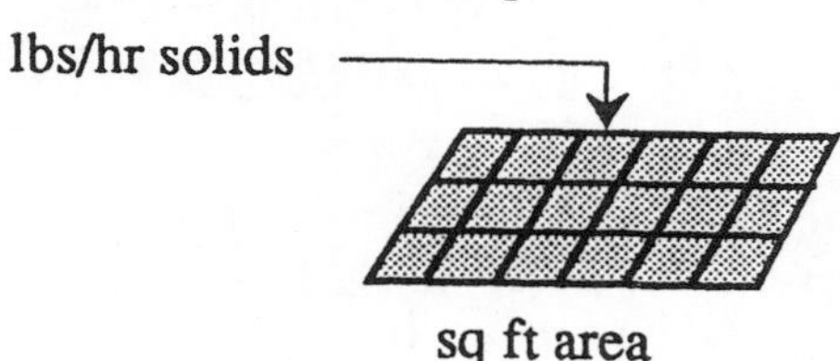

$$\text{Solids Loading Rate, lbs/hr/sq ft} = \frac{\text{Solids, lbs/hr}}{\text{Area, sq ft}}$$

$$= \frac{(120\text{ gpm})\left(60\frac{\text{min}}{\text{hr}}\right)\left(8.34\frac{}{\text{lbs/gal}}\right)\left(\frac{0.75}{100}\right)}{(50\text{ ft})(15\text{ ft})}$$

$$= \boxed{0.6\text{ lbs/hr/sq ft}}$$

Example 4: (DAF Thickening)

❑ The 0.13-MGD sludge flow to a 30-ft diameter dissolved air thickener has a suspended solids concentration of 8200 mg/*L*, what is the solids loading rate in lbs/hr/sq ft?

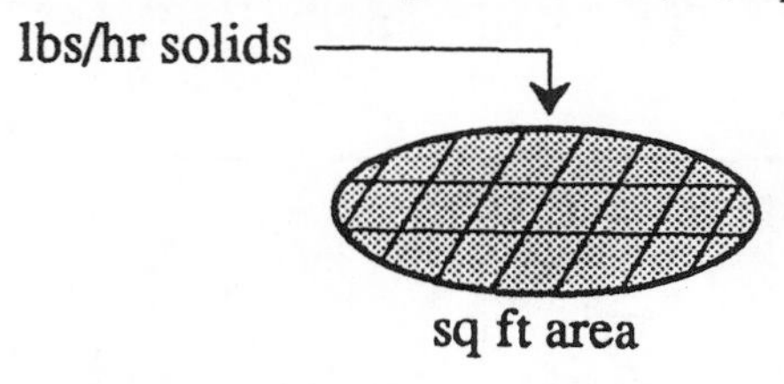

$$\text{Solids Loading Rate, lbs/hr/sq ft} = \frac{\text{Solids, lbs/hr}}{\text{Area, sq ft}}$$

First calculate lbs/day/sq ft solids loading:

$$\text{Solids Loading Rate, lbs/day/sq ft} = \frac{(8200\text{ mg/L})(0.13\text{ MGD})(8.34\text{ lbs/gal})}{(0.785)(30\text{ ft})(30\text{ ft})}$$

$$= 12.6\text{ lbs/day/sq ft}$$

Then convert to lbs/hr/sq ft solids loading:

$$\frac{12.6\text{ lbs/day/sq ft}}{24\text{ hrs/day}} = \boxed{0.5\text{ lbs/hr/sq ft}}$$

SOLIDS LOADING RATE

The solids loading rate for gravity thickeners is expressed in lbs/day/sq ft. For DAF thickeners, however, the solids loading rate is expressed as lbs/hr/sq ft.

Simplified Equation:

$$\text{Solids Loading Rate, lbs/hr/sq ft} = \frac{\text{Solids, lbs/hr}}{\text{Area, sq ft}}$$

Expanded Equation:

$$\text{SLR, lbs/hr/sq ft} = \frac{(\text{gpm sludge flow})\left(60\frac{\text{min}}{\text{hr}}\right)\left(8.34\frac{}{\text{lbs/gal}}\right)\left(\frac{\%SS}{100}\right)}{\text{Area, sq ft}}$$

Example 3 illustrates this calculation.

If the suspended solids concentration is expressed as mg/*L** rather than percent, the following equation may be used to calculate solids loading rate as lbs/day/sq ft:

$$\text{SLR, lbs/day/sq ft} = \frac{(\text{mg/L SS})(\text{MGD flow})(8.34\text{ lbs/gal})}{\text{Area, sq ft}}$$

The solids loading rate can then be converted to lbs/hr/sq ft as follows:

$$\frac{\text{lbs/day/sq ft}}{24\text{ hrs/day}} = \text{lbs/hr/sq ft}$$

Example 4 illustrates this calculation.

* For a review of mg/*L* to lbs/day calculations refer to Chapter 3. For a discussion of mg/*L* and percent conversions, refer to Chapter 8 in *Basic Math Concepts*.

AIR APPLIED

The pounds of air applied to the DAF unit is an essential part of its operation. It is the dissolved air that floats the solids to the surface for removal.

The air supplied to the system is measured by a rotameter in cubic feet per minute (cfm). The air-to-solids ratio requires that the applied air be expressed as lbs/hr.

When converting from cu ft water to lbs of water, you must multiply cu ft of water by the density* of water (62.4 lbs/cu ft):

$$(\text{cu ft water})\ (7.48\ \text{gal/cu ft})\ (8.34\ \text{lbs/gal}) = \text{lbs water}$$

or

$$(\text{cu ft water})\ (62.4\ \text{lbs/gal}) = \text{lbs water}$$

When converting cu ft of air to pounds of air, the density of air must be used:

1 cu ft air = 0.075 lbs air

This is the density of air at standard conditions (20°C, 14.7 psia).** To convert cfm air to lbs/hr, use the following equation:

$$\text{Air, lbs/hr} = (\text{Air, cfm})\left(\frac{60\ \text{min}}{\text{hr}}\right)\left(\frac{0.075\ \text{lbs}}{\text{cu ft}}\right)$$

Example 5: (DAF Thickening)

❑ The air rotameter indicates 7 cfm is supplied to the dissolved air flotation thickener. What is this air supply expressed as lbs/hr?

$$\text{Air Supplied, lbs/hr} = (\text{Air supplied, cfm})\left(\frac{60\ \text{min}}{\text{hr}}\right)\left(\frac{0.075\ \text{lbs}}{\text{cu ft}}\right)$$

$$= (7\ \text{cfm})\left(\frac{60\ \text{min}}{\text{hr}}\right)\left(\frac{0.075\ \text{lbs}}{\text{cu ft}}\right)$$

$$= \boxed{32\ \text{lbs/hr}}$$

Example 6: (DAF Thickening)

❑ The air supplied to a dissolved air flotation thickener is 9 cfm. What is this air supply expressed as lbs/hr?

$$\text{Air Supplied, lbs/hr} = (\text{Air supplied, cfm})\left(\frac{60\ \text{min}}{\text{hr}}\right)\left(\frac{0.075\ \text{lbs}}{\text{cu ft}}\right)$$

$$= (9\ \text{cfm})\left(\frac{60\ \text{min}}{\text{hr}}\right)\left(\frac{0.075\ \text{lbs}}{\text{cu ft}}\right)$$

$$= \boxed{41\ \text{lbs/hr}}$$

---

* Density and specific gravity are described in Chapter 7, Section 7.1.

** Air density may vary considerably, depending on the surrounding (ambient) temperature, pressure, and humidity.

Example 7: (DAF Thickening)

❑ A dissolved air flotation thickener receives a 95 gpm flow of waste activated sludge with a suspended solids concentration of 8600 mg/*L*. If air is supplied at a rate of 6 cfm, what is the air-to-solids ratio?

$$\text{Air/Solids Ratio} = \frac{(\text{cfm air})\ (0.075\ \text{lbs/cu ft})}{\underset{\text{Sludge}}{(\text{gpm})}\ \underset{\text{lbs/gal}}{(8.34)}\ \left(\frac{\%\ \text{SS}}{100}\right)}$$

The solids concentration of 8600 mg/*L* can be expressed as a percent (0.86%):*

$$= \frac{(6\ \text{cfm})\ (0.075\ \text{lbs/cu ft})}{(95\ \text{gpm})\ \underset{\text{lbs/gal}}{(8.34)}\ \left(\frac{0.86}{100}\right)}$$

$$= \boxed{0.07}$$

Example 8: (DAF Thickening)

❑ The sludge flow to a DAF thickener is 110 gpm. The solids concentration of the sludge is 0.8%. If the air supplied to the DAF unit is 4 cfm, what is the air-to-solids ratio?

$$\text{Air/Solids Ratio} = \frac{(\text{cfm air})\ (0.075\ \text{lbs/cu ft})}{\underset{\text{Sludge}}{(\text{gpm})}\ \underset{\text{lbs/gal}}{(8.34)}\ \left(\frac{\%\ \text{SS}}{100}\right)}$$

$$= \frac{(4\ \text{cfm})\ (0.075\ \text{lbs/cu ft})}{(110\ \text{gpm})\ \underset{\text{lbs/gal}}{(8.34)}\ \left(\frac{0.8}{100}\right)}$$

$$= \boxed{0.04}$$

AIR/SOLIDS RATIO

The air supplied to a dissolved air flotation thickener must be in balance with the solids in the system. Generally the air-to-solids ratio falls in the range of 0.01 to 0.10.

To calculate the air-to-solids ratio, use the equations shown below.

Simplified Equation:

$$\text{Air/Solids Ratio} = \frac{\text{Air, lbs/min}}{\text{Solids, lbs/min}}$$

Expanded Equation:

$$\text{A/S Ratio} = \frac{\underset{\text{Air}}{(\text{cfm})}\ \underset{\text{lbs/cu ft}}{(0.075)}}{\underset{\text{Sludge}}{(\text{gpm})}\ \underset{\text{lbs/gal}}{(8.34)}\ \left(\frac{\%\ \text{SS}}{100}\right)}$$

* For a review of mg/*L* and percent conversions, refer to Chapter 8 in *Basic Math Concepts*.

PERCENT RECYCLE RATE

The recycle rate is one of the operation control techniques for the dissolved air flotation thickener.

The percent recycle is the recycle flow rate compared to the sludge flow rate to the thickener:

$$\text{Recycle, \%} = \frac{\text{Recycle Flow, gpm}}{\text{Sludge Flow to DAF unit, gpm}} \times 100$$

Example 9: (DAF Thickening)

❑ A dissolved air flotation thickener receives a sludge flow of 75 gpm. If the recycle rate is 80 gpm, what is the percent recycle rate?

$$\text{\% Recycle} = \frac{\text{Recycle Flow, gpm}}{\text{Sludge Flow, gpm}} \times 100$$

$$= \frac{80 \text{ gpm}}{75 \text{ gpm}} \times 100$$

$$= \boxed{107\%}$$

Example 10: (DAF Thickening)

❑ The desired percent recycle rate for a DAF unit is 125%. If the sludge flow to the thickener is 55 gpm, what should the recycle flow be, in MGD?

$$\text{\% Recycle} = \frac{\text{Recycle Flow, gpm}}{\text{Sludge Flow, gpm}} \times 100$$

$$125 = \frac{x \text{ gpm}}{55 \text{ gpm}} \times 100$$

$$x = \boxed{69 \text{ gpm}}$$

Example 11: (DAF Thickening)

❑ An 80-ft diameter DAF thickener receives a sludge flow with a solids concentration of 7800 mg/L. If the effluent solids concentration is 70 mg/L, what is the solids removal efficiency?

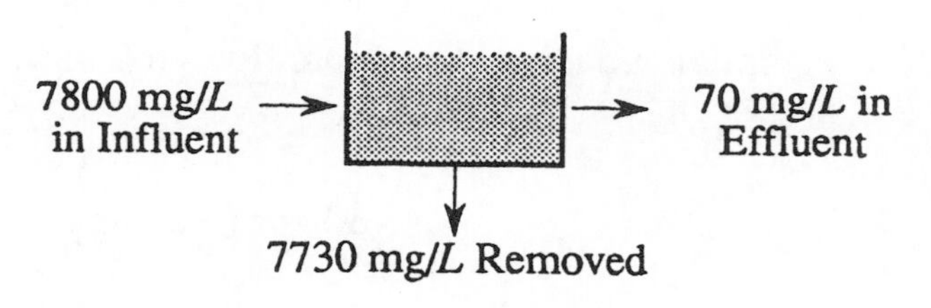

$$\text{Solids Removal Efficiency, \%} = \frac{\text{Solids Removed, mg/L}}{\text{Influent Solids, mg/L}} \times 100$$

$$= \frac{7730 \text{ mg/L}}{7800 \text{ mg/L}} \times 100$$

$$= \boxed{99.1\% \text{ Solids Removal}}$$

Example 12: (DAF Thickening)

❑ The solids concentration of the influent sludge to a dissolved air flotation unit is 8100 mg/L. If the thickened sludge solids concentration is 4%, what is the concentration factor?

$$\text{Concentration Factor} = \frac{4}{0.81}$$

$$= \boxed{5}$$

SOLIDS REMOVAL EFFICIENCY AND CONCENTRATION FACTOR

The solids removal efficiency and concentration factor for the dissolved air flotation thickener are calculated the same as described for the gravity thickener. The equations to be used in these calculations are:

$$\text{Solids Removal Effic., \%} = \frac{\text{Solids Removed, mg/L}}{\text{Influent Solids, mg/L}} \times 100$$

$$\text{Concentration Factor} = \frac{\text{Thickened Sludge Conc., \%}}{\text{Influent Sludge Conc., \%}}$$

15.6 CENTRIFUGE THICKENING CALCULATIONS

HYDRAULIC LOADING RATE—SCROLL OR DISC CENTRIFUGES

Hydraulic loading is normally measured as flow rate per unit of area. However, due to variety of sizes and designs, hydraulic loading to centrifuges does not include area considerations. It is expressed only as gallons per hour.

Since scroll and disc centrifuges are continuous feed, the hydraulic loading is simply the flow rate to the unit expressed in gallons per hour.* The equations to be used if the flow rate to the centrifuge is given as gallons per day or gallons per minute are:

$$\text{Hydraulic Loading, gph} = \frac{\text{Flow, gpd}}{\text{24 hrs/day}}$$

Or

$$\text{Hydraulic Loading, gph} = (\text{gpm flow})\left(60\ \frac{\text{min}}{\text{hr}}\right)$$

Example 1: (Centrifuge Thickening)

❑ A disc centrifuge receives a waste activated sludge flow of 30 gpm. What is the hydraulic loading on the unit, in gal/hr?

$$\text{Hydraulic Loading, gph} = (\text{gpm flow})\left(60\ \frac{\text{min}}{\text{hr}}\right)$$

$$= (30\ \text{gpm})\left(60\ \frac{\text{min}}{\text{hr}}\right)$$

$$= \boxed{1800\ \text{gph}}$$

Example 2: (Centrifuge Thickening)

❑ The waste activated sludge flow to a scroll centrifuge thickener is 90,000 gpd. What is the hydraulic loading on the thickener, in gph?

$$\text{Hydraulic Loading, gph} = \frac{\text{gpd flow}}{\text{24 hrs/day}}$$

$$= \frac{90{,}000\ \text{gpd}}{\text{24 hrs/day}}$$

$$= \boxed{3750\ \text{gph}}$$

* Flow conversions are discussed in Chapter 8 of *Basic Math Concepts*.

Example 3: (Centrifuge Thickening)

❑ The waste activated sludge flow to a basket centrifuge is 60 gpm. The basket run time is 25 minutes until the basket is full of solids. If it takes 1-1/2 minutes to skim the solids out of the unit, what is the hydraulic loading on the unit in gal/hr?

$$\text{Hydraulic Loading, gph} = (\text{gpm flow})\left(60\ \frac{\text{min}}{\text{hr}}\right)\frac{\text{Duration of Sludge Flow, min}}{\text{Time in Operation, min}}$$

$$= (60\ \text{gpm})\left(60\ \frac{\text{min}}{\text{hr}}\right)\frac{25\ \text{min}}{26.5\ \text{min}}$$

$$= \boxed{3396\ \text{gal/hr}}$$

Example 4: (Centrifuge Thickening)

❑ The sludge flow to a basket centrifuge is 79,000 gpd. The basket run time is 22 minutes until the flow to the unit must be stopped for the skimming operation. If skimming takes 1-1/2 minutes, what is the hydraulic loading on the unit in gal/hr?

$$\text{Hydraulic Loading, gph} = \frac{\text{gpd flow}}{24\ \text{hrs/day}}\cdot\frac{\text{Duration of Sludge Flow, min}}{\text{Time in Operation, min}}$$

$$= \frac{79{,}000\ \text{gpd}}{24\ \text{hrs/day}}\cdot\frac{22\ \text{min}}{23.5\ \text{min}}$$

$$= \boxed{3082\ \text{gal/hr}}$$

HYDRAULIC LOADING RATE—BASKET CENTRIFUGES

Hydraulic loading for basket centrifuges is calculated slightly differently than that for scroll or disc centrifuges.

Since scroll or disc centrifuges operate on a **continuous feed** basis, the hydraulic loading on the unit is simply the flow to the unit expressed as gallons per hour. Basket centrifuges, however, operate on a **batch feed** basis. Sludge is fed to the unit until the effluent (called centrate) begins to deteriorate. The flow is then stopped and the concentrated sludge is removed from the unit.

During this "down time," therefore, the basket centrifuge is in operation but there is no flow (no hydraulic load) to the unit. You can calculate the hydraulic loading by simply **adjusting the flow rate**: multiply the flow rate by the fraction of the time sludge was flowing to the unit (the time sludge was fed divided by the total time in operation), as shown below.

Simplified Equation:

$$\text{Hydr. Loading, gph} = (\text{gph flow})\frac{\text{Duration of Sludge Flow}}{\text{Time in Operation}}$$

Expanded Equation:

$$\text{Hyd. Load., gph} = (\text{gpm flow})\left(60\ \frac{\text{min}}{\text{hr}}\right)\frac{\text{Dur. of Sl. Flow}}{\text{Time in Op.}}$$

Or

$$\text{Hyd. Load., gph} = \frac{\text{gpd flow}}{24\ \text{hrs/day}}\cdot\frac{\text{Dur. of Sl. Flow}}{\text{Time in Op.}}$$

SOLIDS LOADING RATE—SCROLL OR DISC CENTRIFUGES

As with hydraulic loading, the solids loading rate for centrifuges does not include area considerations. It is simply pounds of solids per hour. The calculation of solids loading depends on the expression of sludge flow rate, as shown below.*

$$\text{Solids Load., lbs/hr} = (\text{gph sludge flow})(8.34\ \text{lbs/gal})\frac{(\%\ \text{Sol.})}{100}$$

$$\text{Solids Load., lbs/hr} = \frac{(\text{gpd sludge flow})}{24\ \text{hrs/day}}(8.34\ \text{lbs/gal})\frac{(\%\ \text{Sol.})}{100}$$

$$\text{Solids Load., lbs/hr} = (\text{gpm sludge flow})\left(60\ \frac{\text{min}}{\text{hr}}\right)(8.34\ \text{lbs/gal})\frac{(\%\ \text{Sol.})}{100}$$

SOLIDS LOADING RATE—BASKET CENTRIFUGES

Solids loading rates for basket centrifuges are calculated the same as that shown above for scroll or disc centrifuges <u>except</u> that the flow rate (and therefore loading rate) **must be adjusted for the actual duration of sludge flow to the unit**. This is the same adjustment as described for hydraulic loading to basket centrifuges.

The same three equations may be used as shown above, with the inclusion of the "adjustment factor." Example 6 illustrates this calculation.

Example 5: (Centrifuge Thickening)

❑ A scroll centrifuge receives a waste activated sludge flow of 110,000 gpd with a suspended solids concentration of 7900 mg/L. What is the solids loading to the centrifuge?

To complete the calculation, 7900 mg/L solids must be expressed as percent. (7900 mg/L = 0.79% solids)**

$$\text{Solids Loading, lbs/hr} = \frac{(\text{gpd flow})}{24\ \text{hrs/day}}(8.34\ \text{lbs/gal})\frac{(\%\ \text{Sl. Sol.})}{100}$$

$$= \frac{(110{,}000\ \text{gpd})}{24\ \text{hrs/day}}(8.34\ \text{lbs/gal})\frac{(0.79)}{100}$$

$$= \boxed{302\ \text{lbs/hr}}$$

Example 6: (Centrifuge Thickening)

❑ The sludge flow to a basket thickener is 75 gpm with a suspended solids concentration of 8300 mg/L. The basket operates 20 minutes before the flow must be stopped to the unit during the 1-1/2 minute skimming operation. What is the solids loading to the centrifuge?

$$\text{Solids Load., lbs/hr} = (\text{gpm flow})\left(60\ \frac{\text{min}}{\text{hr}}\right)(8.34\ \text{lbs/gal})\frac{(\%\ \text{Sol.})}{100}\underbrace{\frac{(\text{Dur. of Sl. Flow})}{\text{Time in Oper.}}}_{\text{Adjustment Factor}}$$

$$= (75\ \text{gpm})\left(60\ \frac{\text{min}}{\text{hr}}\right)(8.34\ \text{lbs/gal})\frac{(0.83)}{100}\frac{(20\ \text{min})}{21.5\ \text{min}}$$

$$= \boxed{290\ \text{lbs/hr}}$$

---

* The density of the sludge is assumed to be 8.34 lbs/gal. If it is different than this, the new density figure should be used in place of 8.34 lbs/gal.

** To review mg/L and percent conversions refer to Chapter 8 in *Basic Math Concepts*.

Example 7: (Centrifuge Thickening)

❑ A basket centrifuge with a 25 cu ft capacity receives a flow of 65 gpm with a suspended solids concentration of 7800 mg/*L*.. The average solids concentration within the basket is 7%. What is the feed time for the centrifuge?

$$\text{Fill Time, min} = \frac{(\text{Basket Capac., cu ft})(7.48\ \text{gal/cu ft})(8.34\ \text{lbs/gal})\left(\frac{\%\ \text{Bask. Sol.}}{100}\right)}{(\text{gpm flow})(8.34\ \text{lbs/gal})\left(\frac{\%\ \text{Inf. Sol.}}{100}\right)}$$

$$= \frac{(25\ \text{cu ft})(7.48\ \text{gal/cu ft})(8.34\ \text{lbs/gal})\left(\frac{7}{100}\right)}{(65\ \text{gpm flow})(8.34\ \text{lbs/gal})\left(\frac{0.78^{**}}{100}\right)}$$

$$= \boxed{26\ \text{min}}$$

Example 8: (Centrifuge Thickening)

❑ A basket centrifuge thickener has a capacity of 16 cu ft. The 55 gpm sludge flow to the thickener has a solids concentration of 8100 mg/*L*. The average solids concentration within the basket is 9%. What is the feed time for the centrifuge?

$$\text{Fill Time, min} = \frac{(\text{Basket Capac., cu ft})(7.48\ \text{gal/cu ft})(8.34\ \text{lbs/gal})\left(\frac{\%\ \text{Bask. Sol.}}{100}\right)}{(\text{gpm flow})(8.34\ \text{lbs/gal})\left(\frac{\%\ \text{Inf. Sol.}}{100}\right)}$$

$$= \frac{(16\ \text{cu ft})(7.48\ \text{gal/cu ft})(8.34\ \text{lbs/gal})\left(\frac{9}{100}\right)}{(55\ \text{gpm flow})(8.34\ \text{lbs/gal})\left(\frac{0.81}{100}\right)}$$

$$= \boxed{24\ \text{min}}$$

FEED TIME (FILL TIME) FOR BASKET CENTRIFUGES

The time required for a batch-fed basket centrifuge to fill with solids depends on several factors:

- The cubic feet volume of the basket,
- The gpm flow to the thickener, and
- The solids concentration of the flow to the thickener.

The feed time (or fill time) calculation is a detention time type of problem.*

Simplified Equation:

$$\text{Fill Time, min} = \frac{\text{Stored Sol. Capac., lbs}}{\text{Sol. Entering, lbs/min}}$$

Expanded Equation:

$$\text{Fill Time, min} = \frac{(\text{Bask. Cap., cu ft})(7.48\ \text{gal/cu ft})(8.34\ \text{lbs/gal})\left(\frac{\%\ \text{Bask. Solids}}{100}\right)}{(\text{gpm sludge flow})(8.34\ \text{lbs/gal})\left(\frac{\%\ \text{Inf. Sol.}}{100}\right)}$$

Examples 7 and 8 illustrate the calculation of feed time.

---

* Refer to Chapter 5.

** 7800 mg/*L* is equivalent to 0.78%. For a review of mg/*L* and percent conversions, refer to Chapter 8, *Basic Math Concepts*.

REMOVAL EFFICIENCY

The solids removal efficiency is calculated like other efficiency calculations.* The solids concentration may be expressed as a percent or as mg/L.

$$\text{Effic., \%} = \frac{\text{Sol. Rem., \%}}{\text{Sol. in Infl., \%}} \times 100$$

Or

$$\text{Effic., \%} = \frac{\text{Solids Rem., mg/L}}{\text{Sol. in Infl., mg/L}} \times 100$$

Example 9: (Centrifuge Thickening)

❑ The influent sludge solids concentration to a disc centrifuge is 7700 mg/L. If the suspended solids concentration of the centrifuge effluent (centrate) is 1500 mg/L, what is the suspended solids removal efficiency?

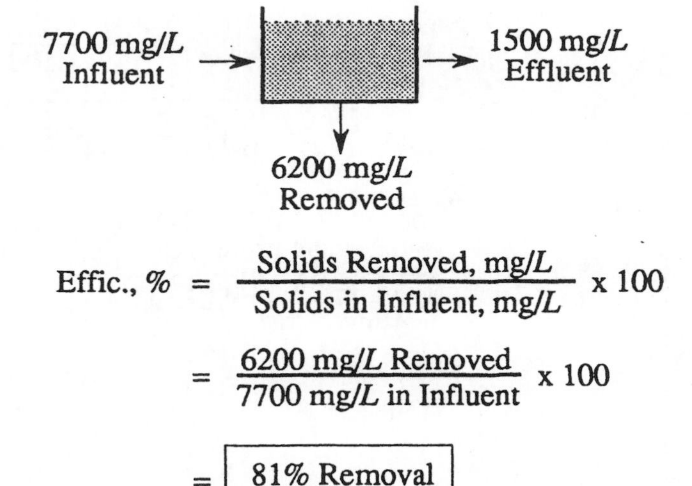

$$\text{Effic., \%} = \frac{\text{Solids Removed, mg/L}}{\text{Solids in Influent, mg/L}} \times 100$$

$$= \frac{\text{6200 mg/L Removed}}{\text{7700 mg/L in Influent}} \times 100$$

= **81% Removal Efficiency**

Example 10: (Centrifuge Thickening)

❑ The influent sludge to a scroll centrifuge has a suspended solids concentration of 8300 mg/L. If the centrifuge effluent has a suspended solids concentration of 0.2%, what is the solids removal efficiency?

8300 mg/L = 0.83% → [centrifuge] → 0.2%

0.63% Removed

$$\text{Effic., \%} = \frac{\text{Solids Removed, \%}}{\text{Solids in Influent, \%}} \times 100$$

$$= \frac{0.63}{0.83} \times 100$$

= **76% Removal Efficiency**

* For a review of other efficiency calculations, refer to Chapter 6.

MIXING SLUDGES WITH DIFFERENT PERCENT SOLIDS

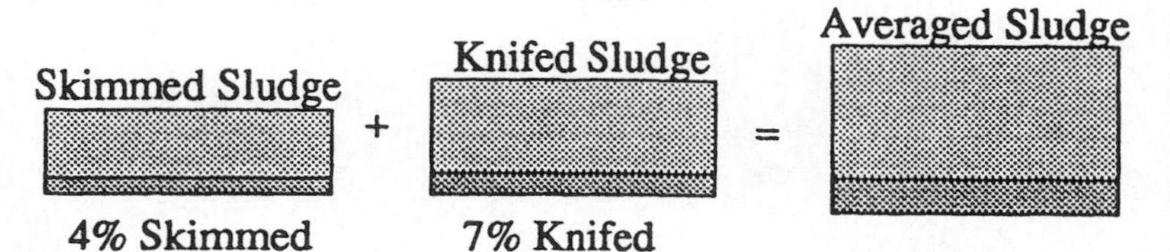

Blended Sludge (% Solids Somewhere Between 4% and 7%)

Simplified Equation:

$$\text{\% Solids of Sludge Mixture} = \frac{\text{Solids in Mixture, lbs}}{\text{Sludge Mixture, lbs}} \times 100$$

Expanded Equation:

$$\text{\% Solids of Sludge Mixture} = \frac{\text{lbs Solids in Skimmed Solids} + \text{lbs Solids in Knifed Solids}}{\text{lbs Skimmed Sludge} + \text{lbs Knifed Sludge}} \times 100$$

$$\text{\% Solids of Sludge Mixture} = \frac{(\text{Skim. Sludge, cu ft})(62.4\text{ lbs/cu ft})\left(\frac{\text{\% Sol.}}{100}\right) + (\text{Knifed Sludge, cu ft})(62.4\text{ lbs/cu ft})\left(\frac{\text{\% Sol.}}{100}\right)}{(\text{Skimmed Sludge, cu ft})(62.4\text{ lbs/cu ft}) + (\text{Knifed Sludge, cu ft})(62.4\text{ lbs/cu ft})} \times 100$$

Example 11: (Centrifuge Thickening)

❑ A total of 12 cu ft of skimmed sludge and 4 cu ft of knifed sludge is removed from a basket centrifuge. If the skimmed sludge has a solids concentration 3.8% and the knifed sludge has a solids concentration of 8%, what is the solids concentration of the sludge mixture?

$$\text{\% Solids of Sludge Mixture} = \frac{(\text{Skim. Sludge, cu ft})(62.4\text{ lbs/cu ft})\left(\frac{\text{\% Sol.}}{100}\right) + (\text{Knifed Sludge, cu ft})(62.4\text{ lbs/cu ft})\left(\frac{\text{\% Sol.}}{100}\right)}{(\text{Skimmed Sludge, cu ft})(62.4\text{ lbs/cu ft}) + (\text{Knifed Sludge, cu ft})(62.4\text{ lbs/cu ft})} \times 100$$

$$= \frac{(12\text{ cu ft})(62.4\text{ lbs/cu ft})\left(\frac{3.8}{100}\right) + (4\text{ cu ft})(62.4\text{ lbs/cu ft})\left(\frac{8}{100}\right)}{(12\text{ cu ft})(62.4\text{ lbs/cu ft}) + (4\text{ cu ft})(62.4\text{ lbs/cu ft})} \times 100$$

$$= \boxed{4.9\%}$$

AVERAGE TOTAL SOLIDS CONCENTRATION —BASKET CENTRIFUGE

The solids concentration of thickened sludge can be determined for scroll and disc centrifuges using the total solids test of thickened sludge samples.

The thickened sludge solids concentration for a basket centrifuge, however, is the **average of the skimmed and knifed solids removed from the basket.**

In effect, to determine the average solids concentration, you must determine the percent solids concentration resulting from the mixture of two different percent sludges. Similar calculations to this are presented in Chapter 14, Section 14.5 and Chapter 16, Section 16.1.

When making these calculations, solids and sludge should be expressed in pounds in the event there are differences in sludge densities*.

* Refer to Chapter 7, Section 7.1, for a discussion of densities.

** This number is a product of two numbers (7.48 gal/cu ft)(8.34 lbs/gal) = 62.4 lbs/cu ft.

NOTES:

16 Sludge Digestion

SUMMARY

1. Mixing Different Percent Solids Sludges

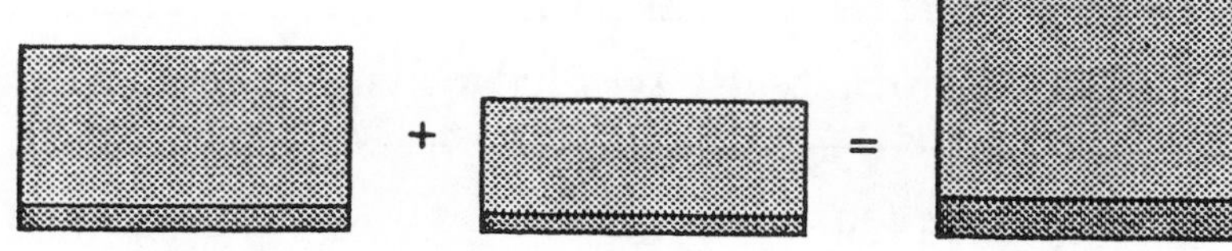

5% Primary Sludge — *3% Thickened Secondary Sludge* — *Blended Sludge (% Solids Somewhere Between 3% and 5%)*

Simplified Equation:

$$\text{\% Solids of Sludge Mixture} = \frac{\text{Solids in Mixture, lbs/day}}{\text{Sludge Mixture, lbs/day}} \times 100$$

Expanded Equation:

$$\text{\% Solids of Sludge Mixture} = \frac{\text{Prim. Sol., lbs/day} + \text{Sec. Sol., lbs/day}}{\text{Prim. Sludge, lbs/day} + \text{Sec. Sludge, lbs/day}} \times 100$$

2. Sludge Volume Pumped

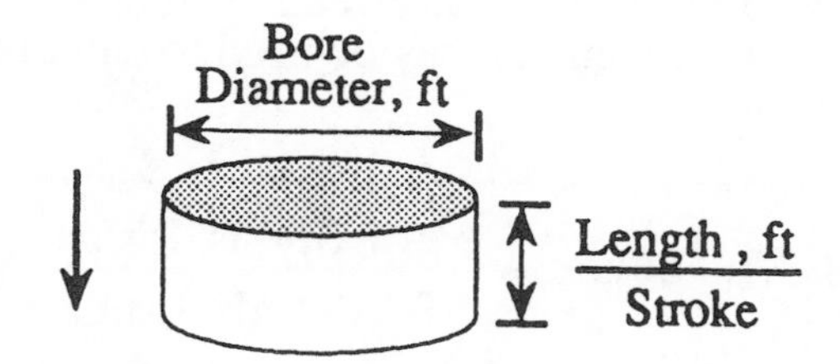

Simplified Equation:

$$\begin{matrix}\text{Volume of} \\ \text{Sludge Pumped} \\ \text{(gal/min)}\end{matrix} = \begin{matrix}\text{(Gallons pumped)} \\ \text{each Stroke}\end{matrix} \begin{matrix}\text{(No. of Strokes)} \\ \text{each Minute}\end{matrix}$$

Expanded Equation:

$$\begin{matrix}\text{Volume of} \\ \text{Sludge Pumped} \\ \text{(gal/min)}\end{matrix} = \left[(0.785)(D^2)\frac{\text{(Length)}}{\text{Stroke}}\ \begin{matrix}(7.48) \\ \text{gal/cu ft}\end{matrix}\right]\left[\begin{matrix}\text{No. of} \\ \text{Strokes/min}\end{matrix}\right]$$

SUMMARY—Cont'd

3. Sludge Pump Operating Time

Simplified Equation:

$$\text{Sludge Removed, lbs/day} = \text{Sludge Pumped, lbs/day}$$

Expanded Equation:

$$\frac{\underset{\text{SS Rem.}}{(\text{mg}/L)}\ \underset{\text{flow}}{(\text{MGD})}\ \underset{\text{lbs/gal}}{(8.34)}}{\frac{\%\text{ Solids}}{100}} = \underset{\text{Sludge Pumping Rate}}{(\text{gpm})}\ \left(\frac{x\text{ min}}{\text{day}}\right)\ \underset{\text{lbs/gal}}{(8.34)}$$

4. Volatile Solids to the Digester, lbs/day

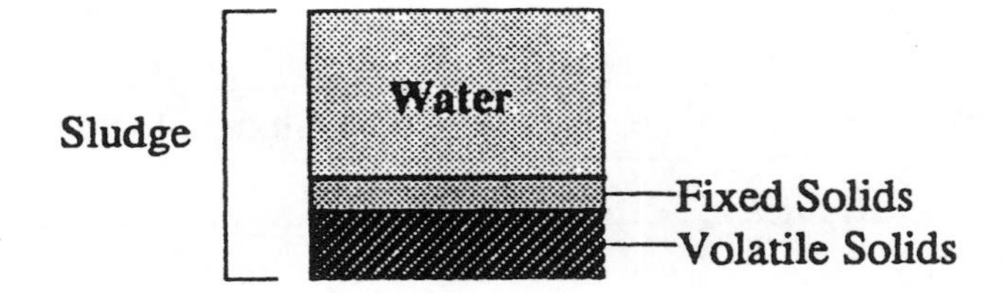

If lbs/day solids have already been calculated, either of the following two equations may be used to calculate lbs/day volatile solids:

$$\%\text{ Vol. Solids} = \frac{\text{Vol. Solids, lbs/day}}{\text{Tot. Solids, lbs/day}} \times 100$$

$$\underset{\text{lbs/day}}{(\text{Tot. Sol.})}\ \frac{(\%\text{ Vol.) Solids}}{100} = \underset{\text{lbs/day}}{\text{Vol. Sol.}}$$

If lbs/day solids have not been determined yet, the following equation can be used to calculate lbs/day volatile solids:

$$\underset{\text{lbs/day}}{(\text{Sludge})}\ \frac{(\%)\text{ Solids}}{100}\ \frac{(\%\text{ Vol.) Solids}}{100} = \underset{\text{lbs/day}}{\text{Vol. Sol.}}$$

SUMMARY—Cont'd

5. Seed Sludge Based on Digester Capacity

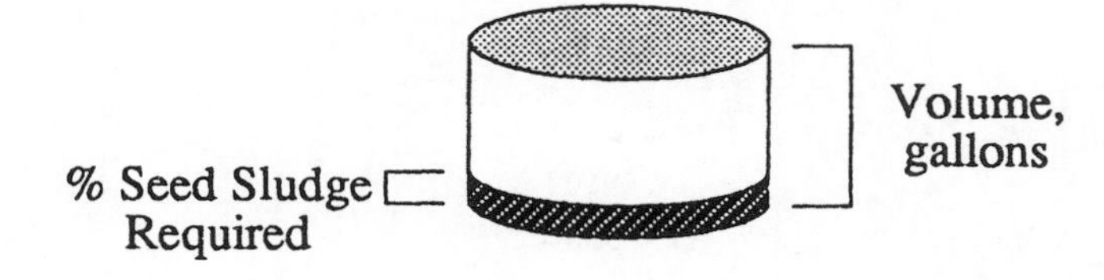

$$\%\ \text{Seed Sludge} = \frac{\text{Seed Sludge, gal}}{\text{Total Digester Capacity, gal}} \times 100$$

6. Seed Sludge Based on Volatile Solids Loading

VOLATILE SOLIDS LOADING RATIO COMPARES
VS ADDED WITH VS IN THE DIGESTER

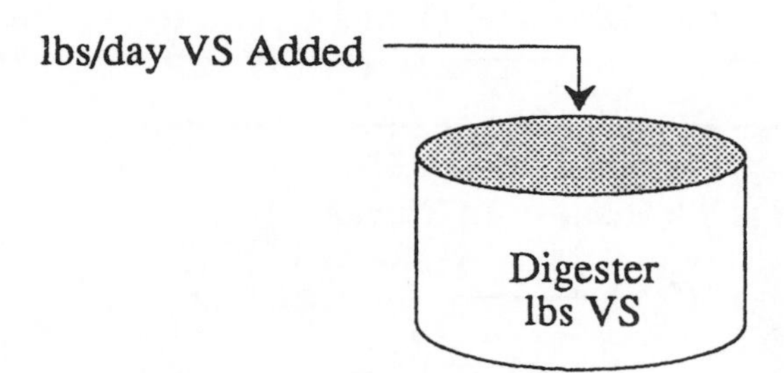

Simplified Equation:

$$\text{VS Loading Ratio} = \frac{\text{VS Added, lbs/day}}{\text{VS in Digester, lbs}}$$

Expanded Equation:

$$\text{VS Loading Ratio} = \frac{(\text{Sludge Added, lbs/day})\left(\frac{\%\ \text{Sol}}{100}\right)\left(\frac{\%\ \text{VS}}{100}\right)}{(\text{Sludge in Dig., lbs})\left(\frac{\%\ \text{Sol}}{100}\right)\left(\frac{\%\ \text{VS}}{100}\right)}$$

SUMMARY—Cont'd

7. Digester Loading Rate, lbs VS Added/day/cu ft

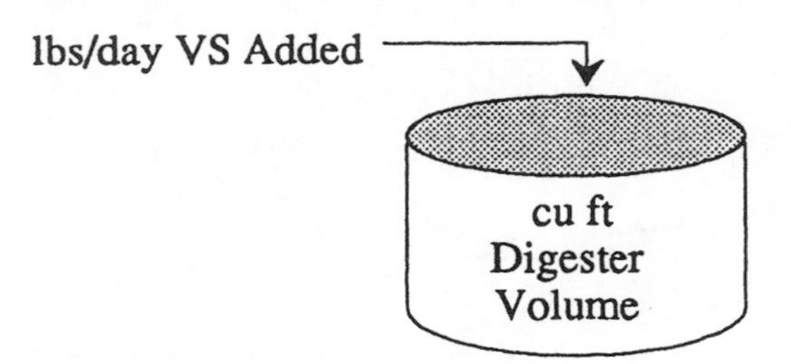

Simplified Equation:

$$\text{Digester Loading} = \frac{\text{VS Added, lbs/day}}{\text{Volume, cu ft}}$$

Expanded Equation:

$$\text{Digester Loading} = \frac{(\text{Sludge, lbs/day})\ \frac{(\%\ \text{Solids})}{100}\ \frac{(\%\ \text{VS})}{100}}{(0.785)\ (D^2)\ (\text{Water Depth, ft})}$$

8. Digester Sludge To Remain In Storage

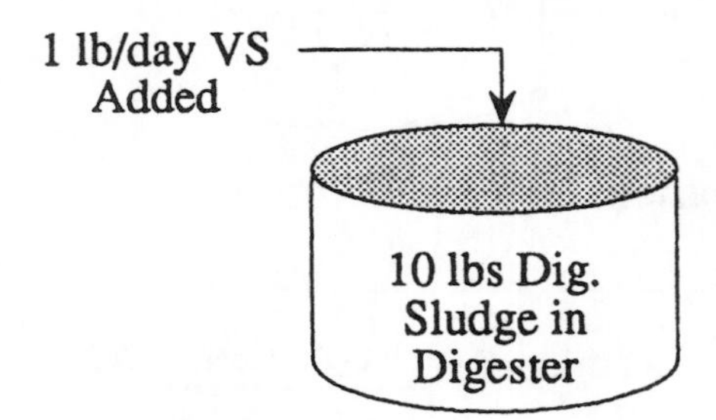

Simplified Equation:

$$\frac{\text{1 lb/day VS Added}}{\text{10 lbs Dig. Sludge in Storage}} = \frac{\text{VS Added, lbs/day}}{\text{Dig. Sludge in Storage, lbs}}$$

Expanded Equation:

$$\frac{\text{1 lb/day VS Added}}{\text{10 lbs Dig. Sludge in Storage}} = \frac{\underset{\text{Sludge}}{(\text{gpd})}\ \underset{\text{lbs/gal}}{(8.34)}\ \frac{(\%\ \text{Sol.})}{100}\ \frac{(\%\ \text{VS})}{100}}{\text{Dig. Sludge in Storage, lbs}}$$

SUMMARY—Cont'd

9. Volatile Acids/Alkalinity Ratio

$$\text{Volatile Acids/Alkalinity Ratio} = \frac{\text{Volatile Acids, mg/}L}{\text{Alkalinity, mg/}L}$$

10. Lime Required for Neutralization

$$\underset{\text{lbs}}{\text{Volatile Acids}} = \underset{\text{lbs}}{\text{Lime Required}}$$

$$\underset{\text{Volatile Acids}}{(\text{mg/}L)}\ \underset{\text{MG}}{(\text{Dig. Vol.})}\ \underset{\text{lbs/gal}}{(8.34)} = \text{lbs Volatile Acids}$$

$$= \text{lbs Lime Required}$$

11. Percent Volatile Solids Reduction

When Volatile Solids Are Expressed as Pounds:

$$\text{\% VS Reduction} = \frac{\text{VS Reduced, lbs}}{\text{Total VS Entering, lbs}} \times 100$$

When Volatile Solids Are Expressed As Percents:

There are two equations that may be used to calculate percent volatile solids reduction.

$$\text{\% VS Reduction} = \frac{(\text{\% VS In} - \text{\% VS Out})}{\text{\% VS}_{\text{In}} - \dfrac{(\text{\% VS})_{\text{In}}(\text{\% VS})_{\text{Out}}}{100}} \times 100$$

In the equation shown below, the "In" and "Out" data are written as **decimal fractions.**

$$\text{\% VS Reduction} = \frac{\text{In} - \text{Out}}{\text{In} - (\text{In} \times \text{Out})} \times 100$$

SUMMARY—Cont'd

12. Volatile Solids Destroyed, lbs VS/day/cu ft

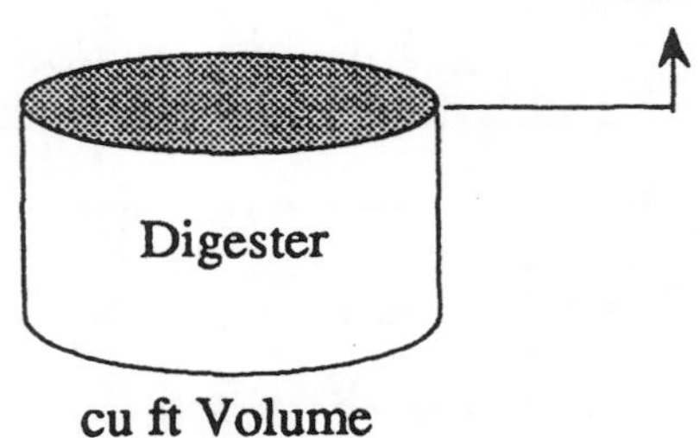

Simplified Equation:

$$\text{Volatile Solids Destroyed, lbs VS/day/cu ft} = \frac{\text{VS Destroyed, lbs/day}}{\text{Dig. Volume, cu ft}}$$

Expanded Equation:

$$\text{Volatile Solids Destroyed, lbs VS/day/cu ft} = \frac{\underset{\text{Sludge}}{(\text{gpd})}\ \underset{\text{lbs/gal}}{(8.34)}\ \left(\frac{\%\ \text{Sol.}}{100}\right)\left(\frac{\%\ \text{VS}}{100}\right)\left(\frac{\%\ \text{VS Red.}}{100}\right)}{(0.785)(D^2)(\text{depth, ft})}$$

13. Digester Gas Production

Simplified Equation:

$$\text{Gas Produced cu ft/lb VS Destroyed} = \frac{\text{Gas, cu ft}}{\text{VS Destroyed, lbs}}$$

Expanded Equation:

$$\text{Digester Gas Production} = \frac{\text{Gas Produced, cu ft/day}}{\underset{\text{lbs/day}}{(\text{Vol. Sol. to Dig.})}\left(\frac{\%\ \text{VS Reduction}}{100}\right)}$$

SUMMARY—Cont'd

14. Digester Solids Balance

SOLIDS BALANCE FOR DIGESTER
"WHAT GOES IN, MUST COME OUT"

SLUDGE IN = SLUDGE OUT + GAS + H_2O

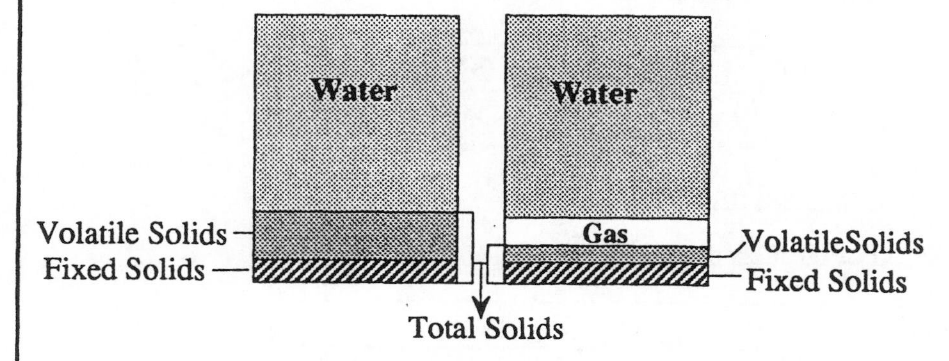

Sludge In, lbs = Sludge Out, lbs + Gas, lbs

Total Solids + Water = Total Solids + Water + Gas

Vol. Solids Fixed Solids Vol. Solids Fixed Solids

Calculations 1-14 are applicable to anaerobic and aerobic digestion. (Calculation 13 is a calculation used for anaerobic digestion.) Calculations 15-17 apply particularly to aerobic digestion.

15. Digestion Time

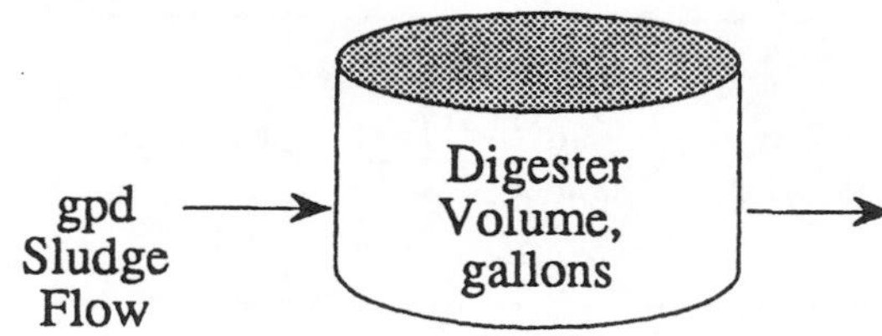

Simplified Equation:

$$\text{Digestion Time, days} = \frac{\text{Digester Volume, gal}}{\text{Sludge Flow Rate, gpd}}$$

Expanded Equation:

$$\text{Digestion Time, days} = \frac{(0.785)(D^2)(\text{depth, ft})(7.48 \text{ gal/cu ft})}{\text{Sludge Flow Rate, gpd}}$$

SUMMARY—Cont'd

16. Air Requirements and Oxygen Uptake

Air Requirements:

Simplified Equation:

$$\frac{0.02 \text{ cfm Air}}{1 \text{ cu ft Dig. Vol}} = \frac{\text{Air Req., cfm}}{\text{Dig. Vol., cu ft}}$$

Expanded Equation:

For circular digesters

$$\frac{0.02 \text{ cfm Air}}{1 \text{ cu ft}} = \frac{\text{Air Req., cfm}}{(0.785)(D^2)(\text{depth, ft})}$$

For rectangular digesters

$$\frac{0.02 \text{ cfm Air}}{1 \text{ cu ft}} = \frac{\text{Air Req., cfm}}{(\text{length, ft})(\text{width, ft})(\text{depth, ft})}$$

Oxygen Uptake:

Simplified Equation:

$$\text{O}_2 \text{ Uptake, mg/}L\text{/hr} = \frac{\text{mg/}L \text{ DO used during Test}}{\text{Min. during Measurement}} \times \frac{60 \text{ min}}{\text{hr}}$$

Expanded Equation:

$$\text{O}_2 \text{ Up., mg/}L\text{/hr} = \frac{\text{mg/}L \text{ DO at Time}_1 - \text{mg/}L \text{ DO at Time}_2}{\text{Time}_2\text{, min} - \text{Time}_1\text{, min}} \times \frac{60 \text{ min}}{\text{hr}}$$

SUMMARY—Cont'd

17. pH Adjustment Using Jar Tests

$$\underset{\text{Lime or Caustic}}{(\text{mg}/L)}\ \underset{\text{Dig. Vol.}}{(\text{MG})}\ \underset{\text{lbs/gal}}{(8.34)} = \text{lbs Lime or Caustic}$$

NOTES:

16.1 MIXING DIFFERENT PERCENT SOLIDS SLUDGES

When sludges with different percent solids content are mixed, the resulting sludge has a percent solids content somewhere **between** the solids contents of the original sludges. For example, if a primary sludge with a 4% solids content were mixed with a secondary sludge with a 1% solids content, the resulting sludge might have a solids content of about 2 or 3%. The actual percent solids content will depend on how much (lbs) of each sludge is mixed together. If, in the example, most of the sludge was from the secondary sludge (1% solids) and very little from the primary sludge (4% solids), then the resulting sludge would be closer to a 1% sludge— perhaps a 1.5% sludge. If, on the other hand, most of the sludge was primary sludge and very little was secondary sludge, then the resulting sludge mixture might have a solids content closer to 4%—such as 3 or 3.5%.

The actual solids content of a mixture of two or more sludges depends on:

- the pounds of sludge contributed from each source, and
- the percent solids of each sludge.

As with the sludge thickening equation, remember that if the thickened sludge has a density different than 8.34 lbs/gal, it should be used in the equation instead of 8.34 lbs/gal.*

WHEN SLUDGES ARE MIXED, THE MIXTURE HAS A % SOLIDS CONTENT BETWEEN THE TWO ORIGINAL % SOLIDS VALUES

+

=

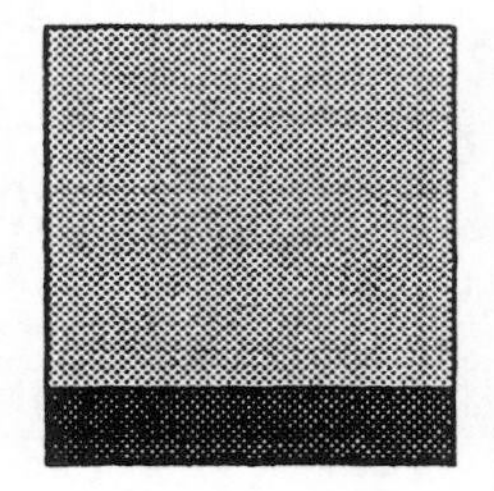

Primary Sludge with 5% Solids	Thickened Secondary Sludge with 3% Solids	Blended Sludge (% Solids Somewhere Between 3% and 5%)

Simplified Equation:

$$\text{\% Solids of Sludge Mixture} = \frac{\text{Solids in Mixture, lbs/day}}{\text{Sludge Mixture, lbs/day}} \times 100$$

Expanded Equation:

$$\text{\% Solids of Sludge Mixture} = \frac{\text{Prim. Sol., lbs/day} + \text{Sec. Sol., lbs/day}}{\text{Prim. Sludge lbs/day} + \text{Sec. Sludge lbs/day}} \times 100$$

Example 1: (Mixing Sludges)

❑ A primary sludge (5% solids) flow of 7540 gpd is mixed with a thickened secondary sludge (3% solids) flow of 3220 gpd. What is the percent solids content of the mixed sludge flow?

$$\text{\% Solids of Sludge Mixture} = \frac{\text{Prim. Sl. Sol., lbs/day} + \text{Sec. Sl. Sol., lbs/day}}{\text{Prim. Sludge, lbs/day} + \text{Sec. Sludge, lbs/day}} \times 100$$

$$= \frac{\underset{\text{Prim. Sl.}}{(7540\text{ gpd})}\ \underset{\text{lbs/gal}}{(8.34)}\ \left(\frac{5}{100}\right) + \underset{\text{Sec. Sl.}}{(3220\text{ gpd})}\ \underset{\text{lbs/gal}}{(8.34)}\ \left(\frac{3}{100}\right)}{\underset{\text{Prim. Sludge}}{(7540\text{ gpd})}\ \underset{\text{lbs/gal}}{(8.34)} + \underset{\text{Sec. Sludge}}{(3220\text{ gpd})}\ \underset{\text{lbs/gal}}{(8.34)}} \times 100$$

$$= \frac{3144\text{ lbs/day Prim Sol} + 806\text{ lbs/day Sec. Sol.}}{62{,}884\text{ lbs/day} + 26{,}855\text{ lbs/day}} \times 100$$

$$= \frac{3950\text{ lbs/day Solids}}{89{,}739\text{ lbs/day Sludge}} \times 100$$

$$= \boxed{4.4\%\text{ Solids}}$$

* Refer to Chapter 7, Section 7.1, for a review of density and specific gravity.

Example 2: (Mixing Sludges)

❑ Primary and thickened secondary sludges are to be mixed and sent to the digester. The 5930 gpd primary sludge has a solids content of 4.8%; the 2660 gpd thickened secondary sludge has a solids content of 3.8%. What would be the percent solids content of the mixed sludge?

$$\text{\% Solids of Sludge Mixture} = \frac{\text{Prim. Sol., lbs/day} + \text{Sec. Sol., lbs/day}}{\text{Prim. Sludge, lbs/day} + \text{Sec. Sludge, lbs/day}} \times 100$$

$$= \frac{\underset{\text{Prim. Sl.}}{(5930\text{ gpd})}\ \underset{\text{lbs/gal}}{(8.34)}\left(\frac{4.8}{100}\right) + \underset{\text{Sec. Sl.}}{(2660\text{ gpd})}\ \underset{\text{lbs/gal}}{(8.34)}\left(\frac{3.8}{100}\right)}{\underset{\text{Prim. Sludge}}{(5930\text{ gpd})}\ \underset{\text{lbs/gal}}{(8.34)} + \underset{\text{Sec. Sludge}}{(2660\text{ gpd})}\ \underset{\text{lbs/gal}}{(8.34)}} \times 100$$

$$= \frac{2374\text{ lbs/day} + 843\text{ lbs}}{49{,}456\text{ lbs/day} + 22{,}184} \times 100$$

$$= \frac{3217\text{ lbs/day Solids}}{71{,}640\text{ lbs/day}} \times 100$$

$$= \boxed{4.5\%\text{ Solids}}$$

Example 3: (Mixing Sludges)

❑ A primary sludge flow (3.8% solids) of 6720 gpd is mixed with a thickened secondary sludge flow (5% solids) of 3670 gpd. What is the percent solids of the combined sludge flow?

$$\text{\% Solids of Sludge Mixture} = \frac{\text{lbs/day Prim. Sol.} + \text{lbs/day Sec. Sol.}}{\text{lbs/day Prim. Sludge} + \text{lbs/day Sec. Sludge}} \times 100$$

$$= \frac{\underset{\text{Prim. Sl.}}{(6720\text{ gpd})}\ \underset{\text{lbs/gal}}{(8.34)}\left(\frac{3.8}{100}\right) + \underset{\text{Sec. Sl.}}{(3670)}\ \underset{\text{lbs/gal}}{(8.34)}\left(\frac{5}{100}\right)}{(6720\text{ gpd})(8.34\text{ lbs/gal}) + (3670)(8.34\text{ lbs/gal})} \times 100$$

$$= \frac{2130\text{ lbs/day} + 1530\text{ lbs/day}}{56{,}045\text{ lbs/day} + 30{,}608\text{ lbs/day}} \times 100$$

$$= \frac{3660\text{ lbs/day Solids}}{86{,}653\text{ lbs/day Sludge}} \times 100$$

$$= \boxed{4.2\%\text{ Solids}}$$

16.2 SLUDGE VOLUME PUMPED

CAPACITY FOR POSITIVE DISPLACEMENT PUMPS

One of the most common types of sludge pumps is the piston pump.* This type of pump operates on the principle of positive displacement. This means that it displaces, or pushes out, a volume of sludge equal to the volume of the piston. The length of the piston, called the stroke, can be lengthened or shortened to increase or decrease the gpm sludge delivered by the pump.

EACH STROKE OF A PISTON PUMP "DISPLACES" OR PUSHES OUT SLUDGE

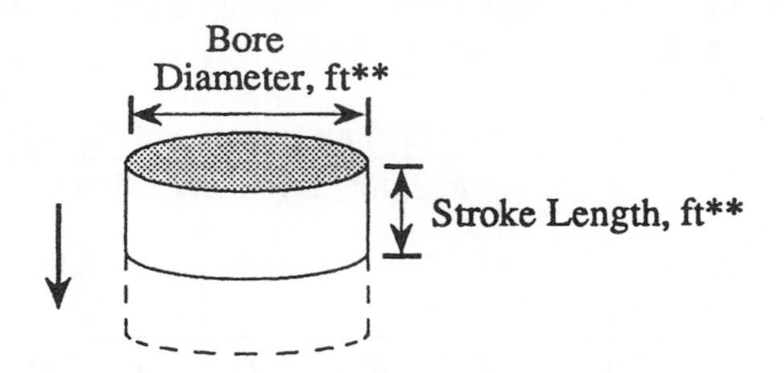

Simplified Equation:

$$\text{Volume of Sludge Pumped (gal/min)} = \frac{\text{(Gallons Pumped)}}{\text{Stroke}} \frac{\text{(No. of Strokes)}}{\text{Minute}}$$

Expanded Equation:

$$\text{Volume of Sludge Pumped (gal/min)} = \left[(0.785)(D^2)\frac{\text{(Length)}}{\text{Stroke}}\frac{(7.48)}{\text{gal/cu ft}}\right]\left[\frac{\text{Strokes}}{\text{min}}\right]$$

Example 1: (Pump Capacities)

❑ A piston pump discharges a total of 0.75 gallons per stroke (or revolution). If the pump operates at 25 revolutions per minute, what is the gpm pumping rate? (Assume the piston is 100% efficient and displaces 100% of its volume each stroke)

0.75 gal

$$\text{Vol. of Sludge Pumped} = \frac{\text{(Gallons Pumped)}}{\text{Stroke}} \frac{\text{(No. of Strokes)}}{\text{Minute}}$$

$$= \frac{(0.75\text{ gal})}{\text{stroke}} \frac{(25\text{ strokes})}{\text{min}}$$

$$= \boxed{19\text{ gpm}}$$

* This type pump is also known as a plunger type pump or positive displacement pump.

** Since the cu ft or gallon volume of sludge pumped is to be calculated, the bore diameter and length/stroke should be expressed in terms of feet.

Example 2: (Pump Capacities)

❑ A sludge pump has a bore of 10 inches and a stroke of 4 inches. If the pump operates at 35 strokes (or revolutions) per minute, how many gpm are pumped? (Assume the piston is 100% efficient and displaces 100% of its volume each stroke.)

$$\frac{10 \text{ in}}{12 \text{ in/ft}} = 0.83 \text{ ft}$$

← 0.83 ft →

$$\frac{4 \text{ in.}}{12 \text{ in./ft}} = 0.33 \text{ ft}$$

$$\text{Vol. of Sludge Pumped} = \frac{\text{(Gallons Pumped)}}{\text{Stroke}} \frac{\text{(No. of Strokes)}}{\text{Minute}}$$

$$= \left[(0.785)(D^2)\frac{\text{(Length)}}{\text{Stroke}}\frac{(7.48)}{\text{gal/cu ft}}\right]\left[\text{Strokes/min}\right]$$

$$= \left[(0.785)(0.83 \text{ ft})(0.83 \text{ ft})\frac{(0.33 \text{ ft})}{\text{Stroke}}\frac{(7.48 \text{ gal})}{\text{cu ft}}\right]\left[\frac{35 \text{ Strokes}}{\text{min}}\right]$$

$$= \frac{(1.33 \text{ gal})}{\text{Stroke}}\frac{(35 \text{ Strokes})}{\text{min}}$$

$$= \boxed{47 \text{ gpm}}$$

Example 3: (Pump Capacities)

❑ A sludge pump has a bore of 8 inches and a stroke setting of 3 inches. The pump operates at 50 revolutions per minute. If the pump operates a total of 90 minutes during a 24-hour period, what is the gpd pumping rate? (Assume the piston is 100% efficient.)

$$\frac{8 \text{ in}}{12 \text{ in/ft}} = 0.67 \text{ ft}$$

← 0.67 ft →

$$\frac{3 \text{ in.}}{12 \text{ in./ft}} = 0.25 \text{ ft}$$

First calculate the gpm pumping rate:

$$\text{Vol. Pumped gpm} = \frac{\text{(Gallons Pumped)}}{\text{Stroke}} \frac{\text{(No. of Strokes)}}{\text{Minute}}$$

$$= \left[(0.785)(0.67 \text{ ft})(0.67 \text{ ft})\frac{(0.25 \text{ ft})}{\text{Stroke}}\frac{(7.48 \text{ gal})}{\text{cu ft}}\right]\left[\frac{50 \text{ Strokes}}{\text{min}}\right]$$

$$= \frac{(0.66 \text{ gal})}{\text{stroke}}\frac{(50 \text{ strokes})}{\text{min}}$$

$$= 33 \text{ gpm}$$

Then convert gpm to gpd pumping rate, based on total minutes pumped during 24-hours:

$$(33 \text{ gpm})\frac{(90 \text{ min})}{\text{day}} = \boxed{2970 \text{ gpd}}$$

CALCULATING GPD PUMPED

There are two methods to determine gpd pumping rate:

- Calculate the gpm pumping rate, then multiply by the total minutes operation during the 24-hour period:

$$\text{Pumping Rate, gpd} = \text{(Pumping) Rate, gpm} \times \text{(Total min) pumping in 24 hrs}$$

- Calculate the gallons pumped each revolution, then multiply by the total revolutions during the 24-hour period:

$$\text{Pumping Rate, gpd} = \frac{\text{(Gallons)}}{\text{Revolution}}\frac{\text{(Total Revol.)}}{\text{day}}$$

16.3 SLUDGE PUMP OPERATING TIME

The sludge pump operating time depends on the amount of sludge to be removed. As shown to the right, the basis of this calculation is simply—the lbs/day sludge to be removed is set equal to the lbs/day sludge to be pumped. Examples 1-3 illustrate this calculation.

REQUIRED SLUDGE PUMPING RATE DEPENDS ON SLUDGE REMOVED PER DAY

Simplified Equation:

Sludge Removed, lbs/day = Sludge Pumped, lbs/day

Expanded Equation:

$$\frac{(\text{mg}/L)_{\text{SS Rem.}}\ (\text{MGD})_{\text{flow}}\ (8.34)_{\text{lbs/gal}}}{\frac{\%\ \text{Solids}}{100}} = (\text{gpm})_{\text{Sludge Pumping Rate}}\ \left(\frac{x\ \text{min}}{\text{day}}\right)\ (8.34)_{\text{lbs/gal}}$$

Example 1: (Sludge Pump Operating Time)

❑ The flow to a primary clarifier is 2 MGD. The influent suspended solids concentration is 210 mg/*L* and the effluent suspended solids concentration is 108 mg/*L*. If the sludge to be removed from the clarifier has a solids content of 3.5% and the sludge pumping rate is 30 gpm, how many minutes per hour should the pump operate?

First calculate min/day pumping rate required:

Sludge Removed, lbs/day = Sludge Pumped, lbs/day

$$\frac{(\text{mg}/L)_{\text{SS Rem.}}\ (\text{MGD})_{\text{flow}}\ (8.34)_{\text{lbs/gal}}}{\frac{\%\ \text{Solids}}{100}} = (\text{gpm})_{\text{Sludge Pumping Rate}}\ \left(\frac{x\ \text{min}}{\text{day}}\right)\ (8.34)_{\text{lbs/gal}}$$

$$\frac{(102\ \text{mg}/L)\ (2\ \text{MGD})\ (8.34\ \text{lbs/gal})}{0.035} = (30\ \text{gpm})\ \left(\frac{x\ \text{min}}{\text{day}}\right)\ (8.34\ \text{lbs/gal})$$

$$\frac{(102)(2)\cancel{(8.34)}}{(0.035)(30)\cancel{(8.34)}} = x$$

$$194\ \text{min/day} = x$$

Then convert min/day to min/hr:

$$\frac{194\ \text{min/day}}{24\ \text{hrs/day}} = \boxed{8.1\ \text{min/hr}}$$

* For a review of calculating lbs/day sludge to be removed (the left side of the equation), refer to Chapter 15, Section 15.2.

Example 2: (Sludge Pump Operating Time)

❑ A primary clarifier receives a flow of 3,600,000 gpd with a suspended solids concentration of 225 mg/L. The clarifier effluent has a suspended solids concentration of 95 mg/L. If the sludge to be removed from the clarifier has a solids content of 4% and the sludge pumping rate is 45 gpm, how many minutes per hour should the pump operate?

First calculate min/day pumping rate required:

$$\text{Sludge Removed, lbs/day} = \text{Sludge Pumped, lbs/day}$$

$$\frac{(130 \text{ mg/}L)(3.6 \text{ MGD})(8.34 \text{ lbs/gal})}{0.04} = (45 \text{ gpm})\left(\frac{x \text{ min}}{\text{day}}\right)(8.34 \text{ lbs/gal})$$

$$\frac{(130)(3.6)(\cancel{8.34})}{(0.04)(45)(\cancel{8.34})} = x$$

$$260 \text{ min/day} = x$$

Then convert min/day to min/hr:

$$\frac{260 \text{ min/day}}{24 \text{ hrs/day}} = \boxed{10.8 \text{ min/hr}}$$

Example 3: (Sludge Pump Operating Time)

❑ The flow to a primary clarifier is 1.7 MGD, with a suspended solids concentration of 205 mg/L. The clarifier effluent suspended solids concentration is 90 mg/L. The sludge to be removed from the clarifier has a solids content of 3%. If the sludge pumping rate is 30 gpm, how many minutes per hour should the pump operate?

First determine the min/day pumping rate required:

$$\text{Sludge Removed, lbs/day} = \text{Sludge Pumped, lbs/day}$$

$$\frac{(115 \text{ mg/}L)(1.7 \text{ MGD})(8.34 \text{ lbs/gal})}{0.03} = (30 \text{ gpm})\left(\frac{x \text{ min}}{\text{day}}\right)(8.34 \text{ lbs/gal})$$

$$\frac{(115)(1.7)(\cancel{8.34})}{(0.03)(30)(\cancel{8.34})} = x$$

$$217 \text{ min/day} = x$$

Then convert min/day to min/hr:

$$\frac{217 \text{ min/day}}{24 \text{ hrs/day}} = \boxed{9 \text{ min/hr}}$$

16.4 VOLATILE SOLIDS TO THE DIGESTER

Sludge solids are comprised of organic matter (from plant or animal sources) and inorganic matter (material from mineral sources, such as sand and grit). The organic matter is called **volatile solids**, and the inorganic matter is called **fixed solids**. Together, the volatile solids and fixed solids make up the **total solids**.

When calculating percent volatile solids, it is essential to remember the general concept of percent:

$$\text{Percent} = \frac{\text{Part}}{\text{Whole}} \times 100$$

When calculating percent volatile solids, the "part" of interest is the weight of the volatile solids; the "whole" is the weight of total solids:

$$\%\ \text{Vol. Solids} = \frac{\text{Vol. Sol., lbs/day}}{\text{Tot. Sol., lbs/day}} \times 100$$

To calculate lbs/day volatile solids to the digester, this equation is often rearranged as:

$$\frac{(\text{Tot. Sol. lbs/day})(\%\ \text{Vol. Sol.})}{100} = \text{Vol. Sol. lbs/day}$$

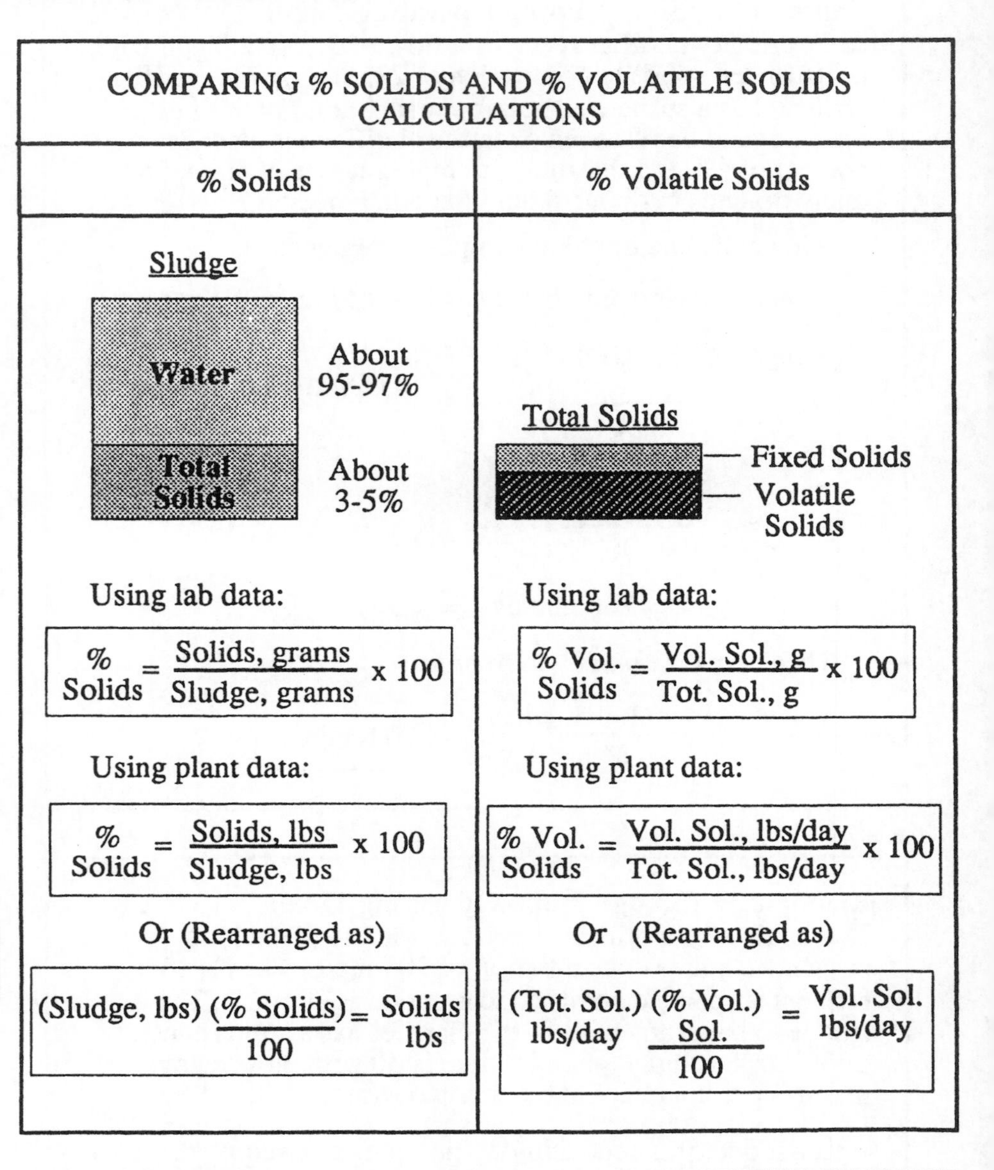

Example 1: (% Volatile Solids)

❑ If 1480 lbs/day solids are sent to the digester, with a volatile solids content of 70%, how many lbs/day volatile solids are sent to the digester?

Write the equation, then fill in the known information:

$$\frac{(\text{Total Solids lbs/day})(\%\ \text{Vol. Solids})}{100} = \text{Vol. Solids lbs/day}$$

$$(1480\ \text{lbs/day})\left(\frac{70}{100}\right) = \boxed{1036\ \text{lbs/day Vol. Solids}}$$

* The calculation of volatile solids using laboratory data described in greater detail in Chapter 18.

Example 2: (% Volatile Solids)

❑ A total of 3240 gpd sludge is to be pumped to the digester. If the sludge has a 5% solids content with 68% volatile solids, how many lbs/day volatile solids are pumped to the digester?

$$(\text{Sludge, lbs/day}) \frac{(\%\ \text{Solids})}{100} \frac{(\%\ \text{Vol. Sol.})}{100} = \text{Vol. Sol., lbs/day}$$

Since sludge is given in gpd, it must be multiplied by 8.34 lbs/gal to convert gpd sludge to lbs/day sludge:*

$$\underbrace{(\text{Sludge, gpd})(8.34\ \text{lbs/gal})}_{\text{lbs/day Sludge}} \frac{(\%\ \text{Solids})}{100} \frac{(\%\ \text{Vol. Sol.})}{100} = \text{Vol. Sol., lbs/day}$$

$$(3240\ \text{gpd})(8.34\ \text{lbs/gal})(0.05)(0.68) = \boxed{919\ \text{lbs/day Vol. Sol.}}$$

Example 3: (% Volatile Solids)

❑ A total of 5480 gpd sludge is to be pumped to the digester. If the sludge has a total solids content of 4.5% and a volatile solids content of 72%, how many lbs/day volatile solids are pumped to the digester? (Assume the sludge weighs 8.34 lbs/gal.)**

$$\underbrace{(\text{Sludge, gpd})(8.34\ \text{lbs/gal})}_{\text{lbs/day Sludge}} \frac{(\%\ \text{Solids})}{100} \frac{(\%\ \text{Vol. Sol.})}{100} = \text{Vol. Sol., lbs/day}$$

$$(5480\ \text{gpd})(8.34\ \text{lbs/gal})(0.045)(0.72) = \boxed{1481\ \text{lbs/day Vol. Sol.}}$$

CALCULATING VOLATILE SOLIDS GIVEN SLUDGE DATA

Sometimes you will have lbs/day sludge information and will want to calculate lbs/day volatile solids. **When this is the case, you must include the percent solids factor in the equation as well,** shown in the equation below. In effect, you are calculating lbs/day solids first, (using the percent solids factor), then the lbs/day volatile solids (using the % volatile solids factor):

$$(\text{Sludge, lbs/day}) \frac{(\%\ \text{Solids})}{100} \frac{(\%\ \text{Vol. Solids})}{100} = \text{Vol. Solids, lbs/day}$$

---

* For a review of flow conversions, refer to Chapter 8 in *Basic Math Concepts*.

** Remember, if the sludge has a density greater than than of water, a factor greater than 8.34 lbs/gal must be used. Refer to Chapter 7 for a discussion of density and specific gravity.

16.5 SEED SLUDGE BASED ON DIGESTER CAPACITY

There are many methods to determine seed sludge required to start a new digester. One method is to calculate seed sludge required based on the volume of the digester. Examples 1-4 illustrate this calculation.

Most digesters have cone-shaped bottoms. For simplicity, however, the side water depth (SWD) is commonly used to represent the average digester depth.

Although determining seed sludge requirements based on digester volume can give you a quick estimate of seed sludge required, it is not sensitive to the balance between the volatile solids in the seed sludge and the volatile solids in the incoming sludge. A calculation of seed sludge requirements based on volatile solids loading is given in the next section.

Example 1: (% Seed Sludge)

❑ A digester has a capacity of 250,000 gallons. If the digester seed sludge is to be 25% of the digester capacity, how many gallons of seed sludge will be required?*

25%

250,000 gallons

$$\text{\% Seed Sludge} = \frac{\text{Seed Sludge, gal}}{\text{Total Digester Capacity, gal}} \times 100$$

$$25 = \frac{x \text{ gal Seed Sludge}}{250{,}000 \text{ gal Capacity}} \times 100$$

$$\frac{(250{,}000 \text{ gal})(25)}{100} = x$$

$$\boxed{\begin{array}{c}62{,}500 \text{ gal} \\ \text{Seed Sludge}\end{array}} = x$$

Example 2: (% Seed Sludge)

❑ A 50-ft diameter digester has a typical water depth of 20 ft. If the seed sludge to be used is 20% of the tank capacity, how many gallons of seed sludge will be required?*

50 ft

20 ft

20%

$$\text{\% Seed Sludge} = \frac{\text{Seed Sludge, gal}}{\text{Total Digester Capacity, gal}} \times 100$$

$$20 = \frac{x \text{ gal Seed Sludge}}{(0.785)(50 \text{ ft})(50 \text{ ft})(20 \text{ ft})(7.48 \text{ gal/cu ft})} \times 100$$

$$\frac{(0.785)(50 \text{ ft})(50 \text{ ft})(20 \text{ ft})(7.48)(20)}{100} = x$$

$$\boxed{\begin{array}{c}58{,}718 \text{ gal} \\ \text{Seed Sludge}\end{array}} = x$$

* For a review of volume calculations, refer to Chapter 11 in *Basic Math Concepts*.

Example 3: (% Seed Sludge)

❑ A digester 40 ft in diameter has a side water depth of 20 ft. If the digester seed sludge is to be 22% of the digester capacity, how many gallons of seed sludge will be required?

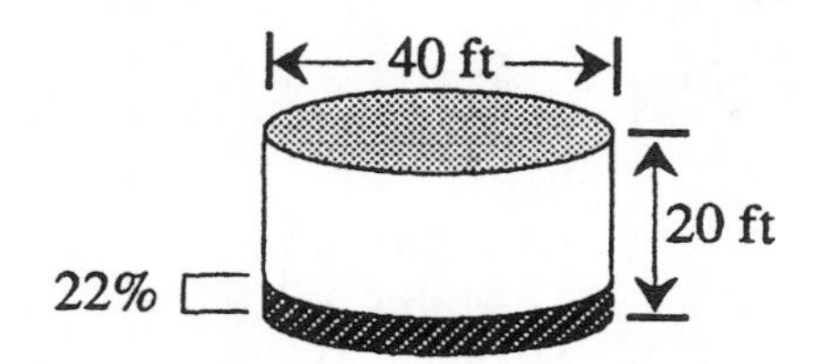

$$\text{\% Seed Sludge} = \frac{\text{Seed Sludge, gal}}{\text{Total Digester Capacity, gal}} \times 100$$

$$22 = \frac{x \text{ gal Seed Sludge}}{(0.785)\ (40 \text{ ft})\ (40 \text{ ft})\ (20 \text{ ft})\ (7.48 \text{ gal/cu ft})} \times 100$$

$$\frac{(0.785)\ (40 \text{ ft})\ (40 \text{ ft})\ (20 \text{ ft})\ (7.48)\ (22)}{100} = x$$

$$\boxed{\text{41,337 gal Seed Sludge}} = x$$

Example 4: (% Seed Sludge)

❑ A 40-ft diameter digester has a typical side water depth of 18 ft. If 52,100 gallons seed sludge are to be used in starting up the digester what percent of the digester volume will be seed sludge?

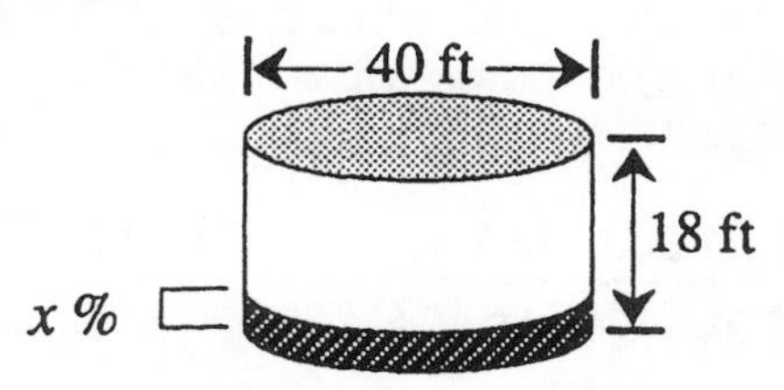

$$\text{\% Seed Sludge} = \frac{\text{Seed Sludge, gal}}{\text{Total Digester Capacity, gal}} \times 100$$

$$x = \frac{\text{52,100 gal Seed Sludge}}{(0.785)\ (40 \text{ ft})\ (40 \text{ ft})\ (18 \text{ ft})\ (7.48 \text{ gal/cu ft})} \times 100$$

$$x = \boxed{31\%}$$

CALCULATING OTHER UNKNOWN FACTORS

There are three variables in percent seed sludge calculations: percent seed sludge, gallons seed sludge, and total gallons digester capacity.

In Examples 1-3, the unknown factor was seed sludge gallons. However, the same equation can be used to calculate either one of the other two variables. In Example 4 the variable is percent seed sludge.

16.6 SEED SLUDGE BASED ON VOLATILE SOLIDS LOADING

One way of calculating seed sludge requirements was described in the previous section—seed sludge based on a percent of the digester volume. Another way to express digester loading is based on lbs/day volatile solids* added daily per each lb of volatile solids under digestion (in the digesters).

VOLATILE SOLIDS LOADING RATIO COMPARES VS ADDED DAILY WITH VS IN THE DIGESTER

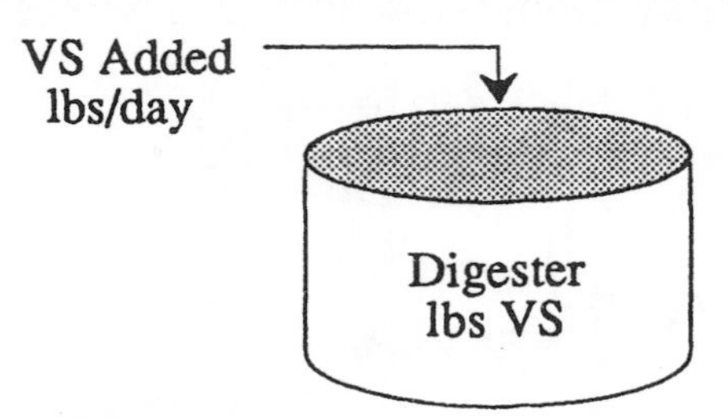

Simplified Equation:

$$\text{VS Loading Ratio} = \frac{\text{VS Added, lbs/day}}{\text{VS in Digester, lbs}}$$

Expanded Equation:

$$\text{VS Loading Ratio} = \frac{(\text{Sludge Added, lbs/day})\left(\frac{\%\ \text{Sol}}{100}\right)\left(\frac{\%\ \text{VS}}{100}\right)}{(\text{Sludge in Dig., lbs})\left(\frac{\%\ \text{Sol}}{100}\right)\left(\frac{\%\ \text{VS}}{100}\right)}$$

Example 1: (Volatile Solids Loading Ratio)

❑ A total of 64,900 lbs/day sludge is pumped to a 100,000-gallon digester. The sludge being pumped to the digester has total solids content of 5.5% and volatile solids content of 67%. The sludge in the digester has a solids content of 6% with a 56% volatile solids content. What is the volatile solids loading on the digester in lbs VS added/day/lb VS in digester?

$$\text{VS Loading Ratio} = \frac{\text{VS Added, lbs/day}}{\text{VS in Digester, lbs}}$$

$$= \frac{(64{,}900\ \text{lbs/day})\left(\frac{5.5}{100}\right)\left(\frac{67}{100}\right)}{\underbrace{(100{,}000\ \text{gal})(8.34\ \text{lbs/gal})}_{\text{This is lbs digester sludge}}\left(\frac{6}{100}\right)\left(\frac{56}{100}\right)}$$

$$= \boxed{\frac{0.085\ \text{lbs VS Added/day}}{\text{lbs VS in Digester}}}$$

* For a review of calculating percent solids and percent volatile solids, refer to Chapters 6 (Section 6.4) and Chapter 18 (Section 18.6).

Example 2: (Volatile Solids Loading Ratio)

❑ A total of 20,700 gal of digested sludge is in a digester. The digested sludge contains 6% total solids and 58% volatile solids. If the desired VS loading ratio is 0.05 lbs VS added/day/lb VS under digestion, what is the desired lbs VS/day to enter the digester?

$$\text{VS Loading Ratio} = \frac{\text{VS Added, lbs/day}}{\text{VS in Digester, lbs}}$$

$$0.05 = \frac{x \text{ lbs VS Added/day}}{\underbrace{(20{,}700 \text{ gal})(8.34 \text{ lbs/gal})}_{\text{This is lbs digester sludge}}\left(\frac{6}{100}\right)\left(\frac{58}{100}\right)}$$

Now solve for the unknown value:*

$$(20{,}700)(8.34)\left(\frac{6}{100}\right)\left(\frac{58}{100}\right)(0.05) = x$$

$$\boxed{\text{300 lbs/day VS Added}} = x$$

Example 3: (Volatile Solids Loading Ratio)

❑ A new digester 50 ft in diameter is operating at an average depth of 25 ft. The raw sludge flow to the digester is expected to be 820 gpd. The raw sludge contains 6% solids and 72% volatile solids. The desired VS loading ratio is 0.07 lbs VS/day/lb VS in digester. How many gallons of seed sludge will be required if the seed sludge contains 9% solids with a 52% volatile solids content? (Assume the seed sludge weighs 9 lbs/gal.)

$$\text{VS Loading Ratio} = \frac{(\text{Sludge Added, lbs/day})\left(\frac{\%\text{ Sol}}{100}\right)\left(\frac{\%\text{ VS}}{100}\right)}{(\text{Seed Sludge in Dig., lbs S})\left(\frac{\%\text{ Sol}}{100}\right)\left(\frac{\%\text{ VS}}{100}\right)}$$

$$0.07 = \frac{\overbrace{(820 \text{ gpd})(8.34 \text{ lbs/gal})}^{\text{This is lbs/day sludge added}}(0.06)(0.72)}{\underbrace{(x \text{ gal seed})(9 \text{ lbs/gal})}_{\text{This is lbs seed sludge}}(0.09)(0.52)}$$

Now solve for x*:

$$x = \frac{(820)(8.34)(0.06)(0.72)}{(0.07)(9)(0.09)(0.52)}$$

$$x = \boxed{\text{10,020 gal Seed Sludge}}$$

CALCULATING OTHER UNKNOWN VALUES

Volatile solids loading ratio calculations have three variables: VS loading ratio, lbs VS added daily, and lbs VS in the digester.

Given a **desired VS loading ratio**, you can calculate either of the other two variables, as shown in Examples 2 and 3.

* Refer to Chapter 2 in *Basic Math Concepts* for a review of solving for the unknown value.

16.7 DIGESTER LOADING RATE, lbs VS Added/day/cu ft

Sludge is sent to a digester in order to break down or stabilize the organic portion of the sludge. Therefore, it is the organic part of the sludge (the volatile solids) that are of interest when calculating solids loading on a digester.

Digester loading rate is a measure of the lbs volatile solids/day* entering each cubic foot of digester volume, as illustrated in the diagram to the right.

DIGESTER LOADING RATE

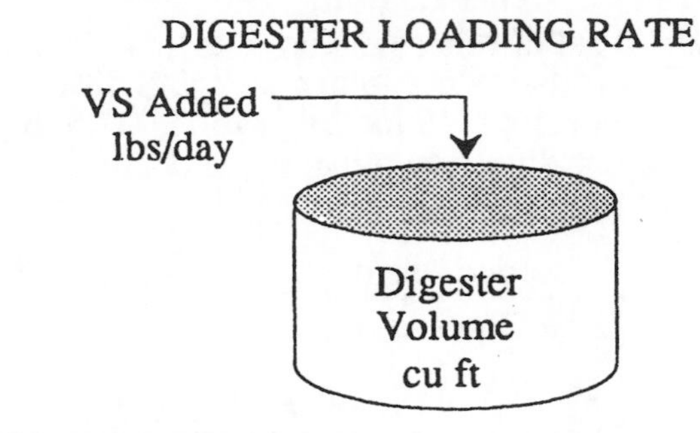

Simplified Equation:

$$\text{Digester Loading} = \frac{\text{VS Added, lbs/day}}{\text{Volume, cu ft}}$$

Expanded Equation:

$$\text{Digester Loading} = \frac{(\text{Sludge, lbs/day})\left(\frac{\%\text{ Solids}}{100}\right)\left(\frac{\%\text{ VS}}{100}\right)}{(0.785)(D^2)(\text{Water Depth, ft})}$$

Example 1: (Digester Loading Rate)

❑ A digester 40 ft in diameter with a water depth of 20 ft receives 84,000 lbs/day raw sludge. If the sludge contains 6.5% solids with 70% volatile matter, what is the digester loading in lbs VS added/day /cu ft volume?

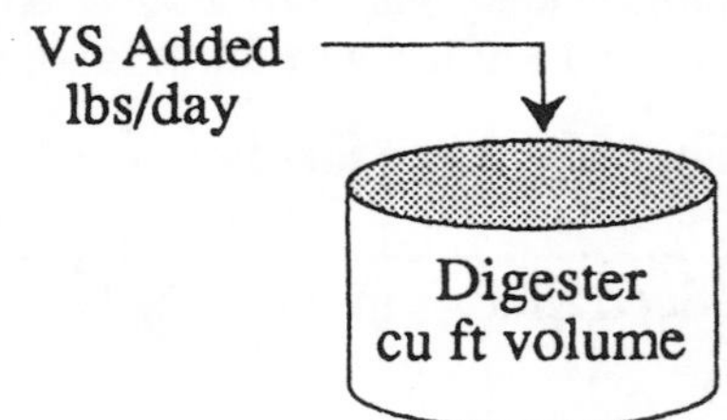

$$\text{Digester Loading} = \frac{(\text{Sludge, lbs/day})\left(\frac{\%\text{ Solids}}{100}\right)\left(\frac{\%\text{ VS}}{100}\right)}{(0.785)(D^2)(\text{Water Depth, ft})}$$

$$= \frac{(84{,}000\text{ lbs/day})(0.065)(0.70)}{(0.785)(40\text{ ft})(40\text{ ft})(20\text{ ft})}$$

$$= 0.15\ \frac{\text{lbs VS/day}}{\text{cu ft}}$$

* For a review of calculating percent solids and percent volatile solids, refer to Chapters 6 (Section 6.4) and Chapter 18, Section 18.6.

Example 2: (Digester Loading Rate)

❑ A digester 50 ft in diameter with a liquid level of 20 ft receives 36,900 gpd sludge with 5.5% solids and 70% volatile solids. What is the digester loading in lbs VS added/day /cu ft volume?

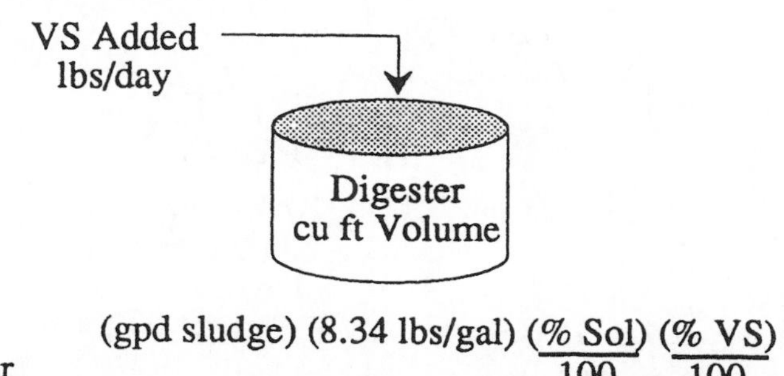

$$\text{Digester Loading} = \frac{(\text{gpd sludge})(8.34 \text{ lbs/gal})\left(\frac{\% \text{ Sol}}{100}\right)\left(\frac{\% \text{ VS}}{100}\right)}{(0.785)(D^2)(\text{Water Depth, ft})}$$

$$= \frac{(36{,}900 \text{ gpd})(8.34 \text{ lbs/gal})(0.055)(0.70)}{(0.785)(50 \text{ ft})(50 \text{ ft})(20 \text{ ft})}$$

$$= \boxed{0.30 \ \frac{\text{lbs VS/day}}{\text{cu ft}}}$$

GIVEN GPD OR GPM SLUDGE PUMPED TO DIGESTER

In Example 1, sludge pumped to the digester was expressed as lbs/day. Most times, however, sludge pumped to the digester will be expressed as gpd or gpm. When this is the case, convert the gpd or gpm pumping rate to lbs/day* and continue as in Example 1. You can make the gpd to lbs/day conversion a separate calculation, or you can incorporate it into the numerator of the equation as shown in Example 2.

Example 3: (Digester Loading Rate)

❑ A digester 40 ft in diameter with a liquid level of 18 ft receives 175,000 lbs/day sludge with 5% total solids and 72% volatile solids. What is the digester loading in lbs VS added/day per 1000 cu ft?

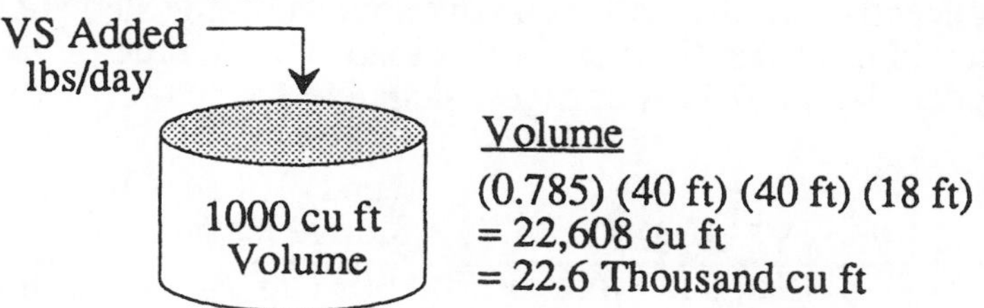

Volume

(0.785) (40 ft) (40 ft) (18 ft)
= 22,608 cu ft
= 22.6 Thousand cu ft

$$\text{Digester Loading} = \frac{(\text{lbs/day Sludge})\left(\frac{\% \text{ Solids}}{100}\right)\left(\frac{\% \text{ VS}}{100}\right)}{(0.785)(D^2)(\text{Water Depth, ft})}$$

$$= \frac{(175{,}000 \text{ lbs/day})(0.05)(0.72)}{22.6 \quad \text{1000-cu ft}}$$

$$= \boxed{279 \ \frac{\text{lbs VS/day}}{\text{1000-cu ft}}}$$

DIGESTER LOADING RATE AS LB/DAY VS/1000 CU FT

Digester loading is sometimes expressed as lbs volatile solids added/day per 1000 cu ft digester volume:

$$\boxed{\text{Digester Loading} = \frac{\text{VS Added, lbs/day}}{\text{Volume, 1000-cu ft}}}$$

When this is the case, express the cu ft volume as 1000 cu ft **before using it in the denominator**. For example, if the digester volume were 35,000 cu ft, then you would use 35 in the denominator of the equation. Remember that the "1000", as part of the "1000-cu ft" in the denominator, is a unit of measure and is **not part of the calculation**. Example 3 illustrates this calculation.

* To review flow conversions, refer to Chapter 8 in *Basic Math Concepts*.

16.8 DIGESTER SLUDGE TO REMAIN IN STORAGE

The ratio of the lbs volatile solids entering the digester daily to the lbs digested sludge in storage is an important consideration in maintaining a proper volatile acids/alkalinity balance.

Using a ratio of 1 lb volatile solids added/day to 10 lbs of digested sludge in storage, you can calculate the pounds of digested sludge that should remain in storage.*

FOR EACH POUND OF VOLATILE SOLIDS ADDED DAILY, AT LEAST 10 POUNDS OF DIGESTED SLUDGE SHOULD BE IN STORAGE

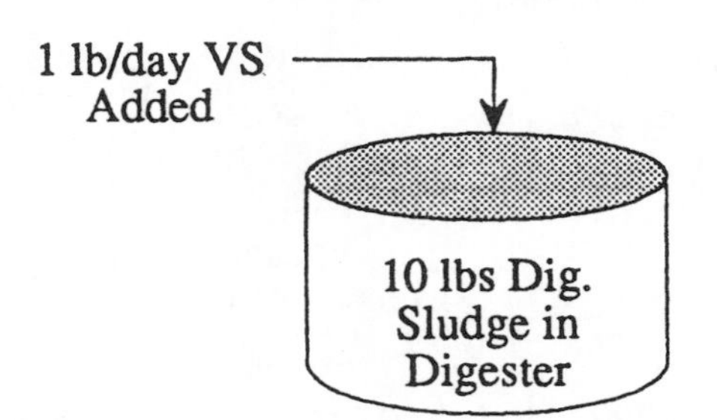

Simplified Equation:

$$\frac{\text{1 lb/day VS Added}}{\text{10 lbs Dig. Sludge in Storage}} = \frac{\text{lbs/day VS Added}}{\text{lbs Dig. Sludge in Storage}}$$

Expanded Equation:

$$\frac{\text{1 lb/day VS Added}}{\text{10 lbs Dig. Sludge in Storage}} = \frac{(\text{gpd})\ \underset{\text{Sludge lbs/gal}}{(8.34)}\ \frac{(\%\ \text{Sol.})}{100}\ \frac{(\%\ \text{VS})}{100}}{\text{lbs Dig. Sludge in Storage}}$$

Example 1: (Digester Sludge to Remain in Storage)

❑ A total of 2500 gpd sludge is pumped to a digester. If the sludge has a total solids content of 6% and a volatile solids concentration of 70% how many pounds of digested sludge should be in the digester for this load? (Use a ratio of 1 lb VS added/day per 10 lbs of digested sludge.)

$$\frac{\text{1 lb/day VS Added}}{\text{10 lbs Dig. Sludge in Storage}} = \frac{(\text{gpd})\ \underset{\text{Sludge lbs/gal}}{(8.34)}\ \frac{(\%\ \text{Sol.})}{100}\ \frac{(\%\ \text{VS})}{100}}{\text{lbs Dig. Sludge in Storage}}$$

$$\frac{\text{1 lb/day VS}}{\text{10 lbs Dig. Sludge}} = \frac{(2500\ \text{gpd})\ \underset{\text{lbs/gal}}{(8.34)}\ \frac{(6)}{100}\ \frac{(70)}{100}}{x\ \text{lbs Dig. Sludge in Storage}}$$

$$x^{*} = (2500)(8.34)(0.06)(0.70)(10)$$

$$x = \boxed{\text{8757 lbs Dig. Sludge in Storage}}$$

---

* For a review of of ratio and proportion problems, refer to Chapter 7 in *Basic Math Concepts.*

** For a review of solving for the unknown value, refer to Chapter 2 in *Basic Math Concepts.*

Example 2: (Digester Sludge to Remain in Storage)

❑ A total of 5000 gpd sludge is pumped to a digester. The sludge has a solids concentration of 6.5% and a volatile solids content of 67%. How many pounds of digested sludge should be in the digester for this load? (Use the ratio of 1 lb VS added/day per 10 lbs digested sludge.)

$$\frac{\text{1 lb/day VS Added}}{\text{10 lbs Dig. Sludge in Storage}} = \frac{\underset{\text{Sludge}}{\text{(gpd)}}\ \underset{\text{lbs/gal}}{(8.34)}\ \frac{(\text{\% Sol.})}{100}\ \frac{(\text{\% VS})}{100}}{\text{lbs Dig. Sludge in Storage}}$$

$$\frac{\text{1 lb/day VS}}{\text{10 lbs Dig. Sludge}} = \frac{(\text{5000 gpd})\underset{\text{lbs/gal}}{(8.34)}\ \frac{(6.5)}{100}\ \frac{(67)}{100}}{x\ \text{lbs Dig. Sludge in Storage}}$$

$$x = (5000)(8.34)(0.065)(0.67)(10)$$

$$x = \boxed{\text{18,160 lbs Digested Sludge}}$$

Example 3: (Digester Sludge to Remain in Storage)

❑ A digester receives a flow of 3000 gallons of sludge during a 24-hour period. If the sludge has a solids content of 7% solids and a volatile solids concentration of 72%, how many pounds of digested sludge should be in the digester for this load? (Use the ratio of 1 lb VS added/day per 10 lbs digested sludge.)

$$\frac{\text{1 lb/day VS Added}}{\text{10 lbs Dig. Sludge in Storage}} = \frac{\underset{\text{Sludge}}{\text{(gpd)}}\ \underset{\text{lbs/gal}}{(8.34)}\ \frac{(\text{\% Sol.})}{100}\ \frac{(\text{\% VS})}{100}}{\text{lbs Dig. Sludge in Storage}}$$

$$\frac{\text{1 lb/day VS}}{\text{10 lbs Dig. Sludge}} = \frac{(\text{3000 gpd})\underset{\text{lbs/gal}}{(8.34)}\ \frac{(7)}{100}\ \frac{(72)}{100}}{x\ \text{lbs Dig. Sludge in Storage}}$$

$$x = (3000)(8.34)(0.07)(0.72)(10)$$

$$x = \boxed{\text{12,610 lbs Digested Sludge}}$$

16.9 VOLATILE ACIDS/ALKALINITY RATIO

The process of anaerobic digestion occurs in two basic stages, both of which are in intricate balance. The first phase of digestion is that of acid fermentation and is related to the first stage in the digestion of new volatile solids entering the digester. The second stage of digestion, methane fermentation, occurs in a more alkaline environment and is thus indicative of advanced stages of digestion.

The volatile acid/alkalinity ratio is therefore an indicator of the progress of digestion and the balance between the two stages. **Though the ratio varies among different treatment plants, it is normally below 0.1.** If the ratio begins to increase, due to an overabundance of acid fermenters, this is the first indication of trouble in the digestion process. Because acid fermenters are associated with the digestion of new volatile solids entering the digester, an increase in the VA/Alkalinity ratio indicates a possible excessive feeding of raw sludge to the digester. It may also indicate a removal of too much digested sludge (the alkaline portion) thus leaving the digester with an overbalance of volatile acids.

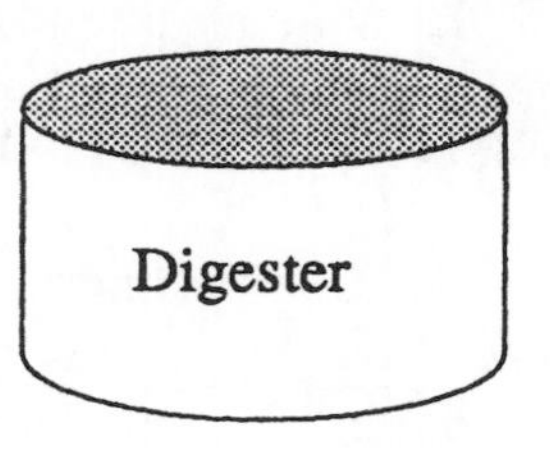

$$\text{VA/Alkalinity Ratio} = \frac{\text{Volatile Acids, mg}/L}{\text{Alkalinity, mg}/L}$$

Example 1: (VA/Alkalinity Ratio)

❑ The volatile acids concentration of the sludge in the anaerobic digester is 160 mg/*L*. If the alkalinity is measured as 2280 mg/*L*, what is the VA/Alkalinity ratio?

$$\text{VA/Alkalinity Ratio} = \frac{\text{Volatile Acids, mg}/L}{\text{Alkalinity, mg}/L}$$

$$= \frac{160\text{ mg}/L\text{ Volatile Acids}}{2280\text{ mg}/L\text{ Alkalinity}}$$

$$= \boxed{0.07\text{ VA/Alkalinity Ratio}}$$

Example 2: (VA/Alkalinity Ratio)

❑ The volatile acid concentration of the sludge in an anaerobic digester is 155 mg/*L*. If the alkalinity is measured as 2460 mg/*L*, what is the VA/Alkalinity ratio?

$$\text{VA/Alkalinity Ratio} = \frac{\text{Volatile Acids, mg}/L}{\text{Alkalinity, mg}/L}$$

$$= \frac{155 \text{ mg}/L \text{ Volatile Acids}}{2460 \text{ mg}/L \text{ Alkalinity}}$$

$$= \boxed{0.06}$$

Example 3: (VA/Alkalinity Ratio)

❑ The desired VA/Alkalinity ratio for the anaerobic digester at a particular plant is 0.05. If the alkalinity is found to be 2800 mg/*L*, what is the desired volatile acids concentration of the digester sludge?

$$\text{VA/Alkalinity Ratio} = \frac{\text{Volatile Acids, mg}/L}{\text{Alkalinity, mg}/L}$$

$$0.05 = \frac{x \text{ Volatile Acids}}{2800 \text{ mg/L Alkalinity}}$$

$$(2800 \text{ mg}/L)(0.05) = x \text{ Volatile Acids}$$

$$\boxed{\begin{array}{c}140 \text{ mg}/L \\ \text{Volatile Acids}\end{array}} = x$$

CALCULATING OTHER UNKNOWN VALUES

There are three variables in the volatile acid/alkalinity ratio: the VA/Alkalinity ratio, the mg/*L* volatile acids, and the mg/*L* Alkalinity. In Examples 1 and 2, the unknown variable was the ratio. In Example 3, a different variable is unknown.

Set up the equation as usual, filling in the known data, then solve for the unknown value.

16.10 LIME REQUIRED FOR NEUTRALIZATION

When the volatile acid/alkalinity ratio of an anaerobic digester increases above 0.8, the pH of the digester begins to drop, resulting in a sour digester. Although it is preferable to take corrective action and allow the digester to recover naturally, this is not always possible due to limited digester capacity and/or recovery time. Under these conditions, lime neutralization of the sour digester may be necessary.

Should lime neutralization be required, the dosage of lime is based on the volatile acids content of the digester sludge. Each mg/*L* volatile acids requires a lime dosage of 1 mg/*L*.

LIME DOSAGE REQUIRED DEPENDS ON VOLATILE ACIDS CONTENT OF DIGESTER SLUDGE

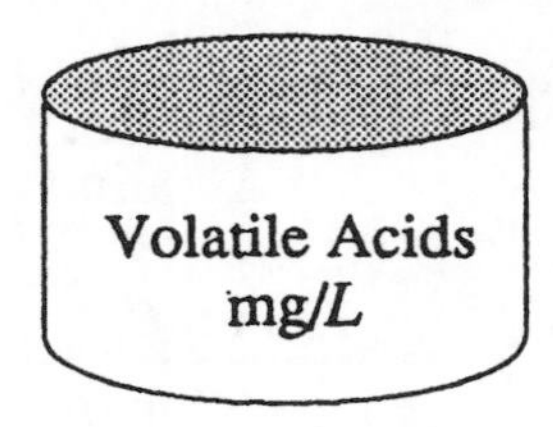

$$\underset{\text{Volatile Acids}}{1\ \text{mg}/L} = \underset{\text{Lime Dosage}}{1\ \text{mg}/L}$$

$$\underset{\text{Lime}}{(\text{mg}/L)}\ \underset{\text{Dig. Vol.}}{(\text{MG})}\ \underset{\text{lbs/gal}}{(8.34)} = \text{lbs Lime Required}$$

Example 1: (Lime for Neutralization)

❑ The digester sludge is found to have a volatile acids content of 2160 mg/*L*. If the digester volume is 150,000 gallons, how many pounds of lime will be required for neutralization?

$$\underset{\text{Volatile Acids}}{2160\ \text{mg}/L} = \underset{\text{Lime Dosage}}{2160\ \text{mg}/L}$$

Now calculate lbs/day lime required:

$$\underset{\text{Lime}}{(\text{mg}/L)}\ \underset{\text{Dig. Vol.}}{(\text{MG})}\ \underset{\text{lbs/gal}}{(8.34)} = \text{lbs Lime Required}$$

$$(2160\ \text{mg}/L)\ (0.15\ \text{MG})\ (8.34\ \text{lbs/gal}) = \boxed{2702\ \text{lbs Lime Required}}$$

Example 2: (Lime for Neutralization)

❑ To neutralize a sour digester, one mg/*L* of lime is to be added for every mg/*L* of volatile acids in the digester sludge. If the digester contains 250,000 gal of sludge with a volatile acid (VA) level of 2300 mg/*L*, how many pounds of lime should be added?

Since the VA concentration is 2300 mg/*L*, the lime concentration should also be 2300 mg/*L*:

$$\underset{\text{Lime}}{(\text{mg/}L)}\ \underset{\text{Dig. Vol..}}{(\text{MG})}\ \underset{\text{lbs/gal}}{(8.34)} = \begin{array}{c}\text{lbs Lime}\\ \text{Required}\end{array}$$

$$(2300\ \text{mg/}L)\ (0.25\ \text{MG})\ (8.34\ \text{lbs/gal}) = \boxed{\begin{array}{c}\text{4796 lbs Lime}\\ \text{Required}\end{array}}$$

Example 3: (Lime for Neutralization)

❑ To neutralize a sour digester, one mg/*L* of lime is to be added for every mg/*L* of volatile acids in the digester sludge. If the digester contains 180,000 gal of sludge with a volatile acid (VA) level of 1820 mg/*L*, how many pounds of lime should be added?

Since the VA concentration is 1820 mg/*L*, the lime concentration should also be 1820 mg/*L*:

$$\underset{\text{Lime}}{(\text{mg/}L)}\ \underset{\text{Dig. Vol..}}{(\text{MG})}\ \underset{\text{lbs/gal}}{(8.34)} = \begin{array}{c}\text{lbs Lime}\\ \text{Required}\end{array}$$

$$(1820\ \text{mg/}L)\ (0.18\ \text{MG})\ (8.34\ \text{lbs/gal}) = \boxed{\begin{array}{c}\text{2732 lbs Lime}\\ \text{Required}\end{array}}$$

16.11 PERCENT VOLATILE SOLIDS REDUCTION

One of the best indicators of the effectiveness of the digestion process is the volatile solids content of the digested sludge. This volatile content may be compared with the original volatile content of the influent sludge, and from the two values the **percent volatile solids reduction** due to digestion may be calculated. This reduction may be as high as 70%. When volatile solids data is given in lbs or lbs/day, the percent volatile solids reduction calculation is similar to any other percent removal calculation:

$$\text{\% VS Rem.} = \frac{\text{VS Removed, lbs}}{\text{VS Tot. Entering, lbs}} \times 100$$

In most volatile solids reduction calculations, however, the volatile solids data is given as percents. **This creates a problem since the percents are based on different wholes** (during digestion some of the volatile solids are converted to gases and water).

To remedy this problem, the equations shown to the right are used to calculate percent volatile solids reduction.

WHEN VOLATILE SOLIDS ARE EXPRESSED AS POUNDS, PERCENT VS REDUCTION IS A USUAL PERCENT CALC.

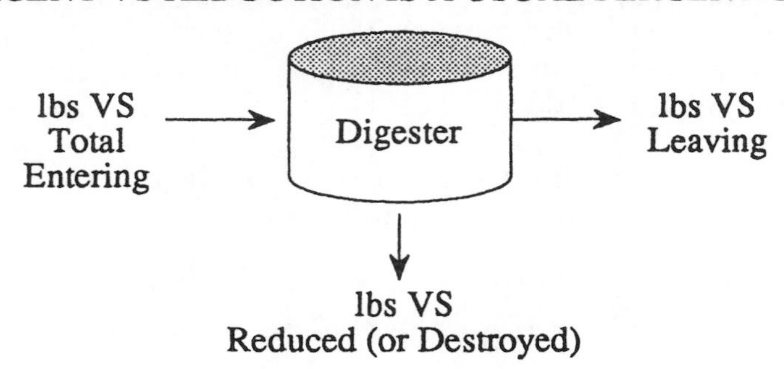

$$\text{\% VS Reduction} = \frac{\text{VS Reduced, lbs}}{\text{Total VS Entering, lbs}} \times 100$$

WHEN VOLATILE SOLIDS ARE EXPRESSED AS PERCENTS A <u>DIFFERENT EQUATION</u> MUST BE USED

SOLIDS ENTERING DIGESTION

If the VS entering the digester are 70%, the remaining solids (fixed solids) must be 30%

SOLIDS LEAVING DIGESTION

If the VS leaving the digester are 40%, the remaining solids (fixed) must be 60%

Fixed solids normally remain relatively unchanged by digestion. Yet fixed solids are 30% in one case above and 60% in the other. This is because **the percents are based on different wholes.**

There are two equations that may be used to calculate percent volatile solids reduction when volatile solids are expressed as percents:.

$$\text{\% VS Reduction} = \frac{(\text{\% VS In} - \text{\% VS Out})}{\text{\% VS}_{\text{In}} - \dfrac{(\text{\% VS})_{\text{In}}(\text{\% VS})_{\text{Out}}}{100}} \times 100$$

In the equation shown below, the "In" and "Out" data are written as **decimal fractions.***

$$\text{\% VS Reduction} = \frac{\text{In} - \text{Out}}{\text{In} - (\text{In} \times \text{Out})} \times 100$$

* For example, 70% VS entering the digester would be written as 0.70 and 56% VS leaving the digester would be written as 0.56.

Example 1: (% VS Reduction)

❑ The sludge entering a digester has a volatile solids content of 70%. The sludge leaving the digester has a volatile solids content of 52%. What is the percent volatile solids reduction?

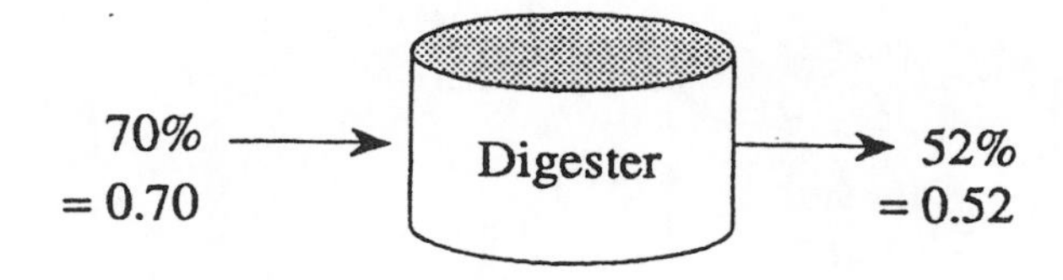

$$\text{\% VS Reduction} = \frac{\text{In} - \text{Out}}{\text{In} - (\text{In} \times \text{Out})} \times 100$$

To use this equation, express percents as decimal fractions:

$$\text{\% VS Reduction} = \frac{0.70 - 0.52}{0.70 - (0.70 \times 0.52)} \times 100$$

$$= \boxed{\text{54\% VS Reduction}}$$

Example 2: (% VS Reduction)

❑ The raw sludge to a digester has a volatile solids content of 72%. The digested sludge volatile solids content is 46%. What is the percent volatile solids reduction?

72% → Digester → 46%

= 0.72 = 0.46

$$\text{\% VS Reduction} = \frac{\text{In} - \text{Out}}{\text{In} - (\text{In} \times \text{Out})} \times 100$$

To use this equation, express percents as decimal fractions:

$$\text{\% VS Reduction} = \frac{0.72 - 0.46}{0.72 - (0.72 \times 0.46)} \times 100$$

$$= \boxed{\text{67\% VS Reduction}}$$

16.12 VOLATILE SOLIDS DESTROYED, lbs VS/day/cu ft

One measure of digester effectiveness is pounds of volatile solids reduced or destroyed per cubic feet of digester volume. The equations to be used in these calculations are shown to the right.

LBS/DAY VOLATILE SOLIDS DESTROYED PER CU FT DIGESTER VOLUME

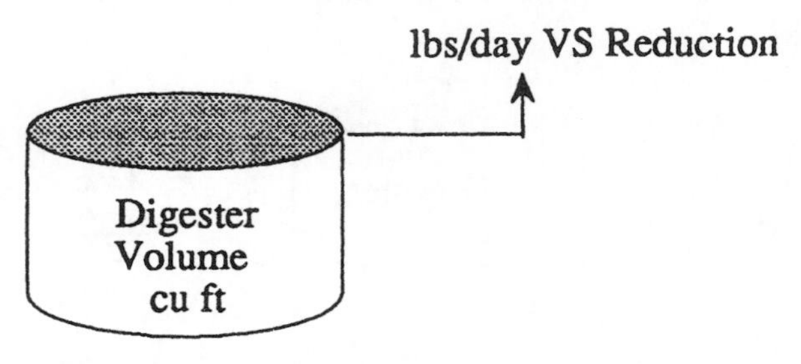

Simplified Equation:

$$\text{Volatile Solids Destroyed, lbs VS/day /cu ft} = \frac{\text{VS Destroyed, lbs/day}}{\text{Dig. Volume, cu ft}}$$

Expanded Equation:

$$\text{Volatile Solids Destroyed, lbs VS/day /cu ft} = \frac{(\text{gpd Sludge})\,(8.34\ \text{lbs/gal})\left(\frac{\%\ \text{Sol.}}{100}\right)\left(\frac{\%\ \text{VS}}{100}\right)\left(\frac{\%\ \text{VS Red.}}{100}\right)}{(0.785)(D^2)(\text{depth, ft})}$$

Example 1: (Volatile Solids Destroyed)

❑ A flow of 3300 gpd sludge is pumped to a 32,000 cu ft digester. The solids concentration of sludge is 6.3%, with a volatile solids content of 72%. If the volatile solids reduction during digestion is 54%, how many lbs/day volatile solids are destroyed per cu ft of digester capacity?

$$\text{Volatile Solids Destroyed, lbs/day VS/cu ft} = \frac{(\text{gpd Sludge})\,(8.34\ \text{lbs/gal})\left(\frac{\%\ \text{Sol.}}{100}\right)\left(\frac{\%\ \text{VS}}{100}\right)\left(\frac{\%\ \text{VS Red.}}{100}\right)}{\text{Digester Volume, cu ft}}$$

$$= \frac{(3300\ \text{gpd})\,(8.34\ \text{lbs/gal})\left(\frac{6.3}{100}\right)\left(\frac{72}{100}\right)\left(\frac{54}{100}\right)}{32{,}000\ \text{cu ft}}$$

$$= \boxed{0.021\ \frac{\text{lbs VS/day}}{\text{cu ft}}}$$

Example 2: (Volatile Solids Destroyed)

❑ A 50-ft diameter digester receives a sludge flow of 2600 gpd, with a solids content of 6% and a volatile solids concentration of 71%. The volatile solids reduction during digestion is 53%. The digester operates at a level of 21 ft. What is the lbs/day volatile solids reduction per cu ft of digester capacity?

$$\text{Volatile Solids Destroyed, lbs VS/day /cu ft} = \frac{(\text{gpd Sludge})\ (8.34\ \text{lbs/gal})\ \frac{(\%\ \text{Sol.})}{100}\ \frac{(\%\ \text{VS})}{100}\ \frac{(\%\ \text{VS Red.})}{100}}{\text{cu ft Digester Volume}}$$

$$= \frac{(2600\ \text{gpd})\ (8.34\ \text{lbs/gal})\ \frac{(6)}{100}\ \frac{(71)}{100}\ \frac{(53)}{100}}{(0.785)(50\ \text{ft})(50\ \text{ft})(21\ \text{ft})}$$

$$= \boxed{\frac{0.01\ \text{lbs/day VS Destroyed}}{\text{cu ft Volume}}}$$

Example 3: (Volatile Solids Destroyed)

❑ The sludge flow to a 40-ft diameter digester is 2800 gpd, with a solids concentration of 6.5% and a volatile solids concentration of 67%. The digester is operated at a depth of 20 ft. If the volatile solids reduction during digestion is 55%, what is the lbs/day volatile solids reduction per 1000 cu ft of digester capacity?

$$\text{Volatile Solids Destroyed, lbs VS/day /cu ft} = \frac{(\text{gpd Sludge})\ (8.34\ \text{lbs/gal})\ \frac{(\%\ \text{Sol.})}{100}\ \frac{(\%\ \text{VS})}{100}\ \frac{(\%\ \text{VS Red.})}{100}}{\text{Digester Volume, 1000-cu ft}}$$

$$= \frac{(2800\ \text{gpd})\ (8.34\ \text{lbs/gal})\ \frac{(6.5)}{100}\ \frac{(67)}{100}\ \frac{(55)}{100}}{\frac{(0.785)(40\ \text{ft})(40\ \text{ft})(20\ \text{ft})}{1000}}$$

$$= \frac{559\ \text{lbs/day VS}}{25.1\ \ 1000\text{-cu ft}}$$

$$= \boxed{\frac{22\ \text{lbs/day VS}}{1000\text{-cu ft}}}$$

VOLATILE SOLIDS DESTROYED PER 1000 CU FT VOLUME

The volatile solids destroyed calculation is sometimes expressed as lbs volatile solids destroyed/day per 1000 cu ft digester volume:

$$\boxed{\text{VS Destroyed} = \frac{\text{VS Destr., lbs/day}}{\text{Volume, 1000-cu ft}}}$$

For such a calculation, express the cu ft digester volume as 1000 cu ft **before using it in the denominator**. Remember that the "1000" in the denominator is a unit of measure and is **not part of the calculation**. Example 3 illustrates this calculation.

16.13 DIGESTER GAS PRODUCTION

The volume and composition of gas produced during anaerobic digestion is important not only as an indicator of the progress of digestion, but also in its utilization as a fuel for heating digesters and buildings, for driving gas engines, etc.

A decrease in the rate of gas production usually indicates that the digestion process is slowing down, and thus perhaps the removal of sludge is overdue. If a sharp increase in gas production occurs, this may indicate the presence of a high organic content of the sludge under digestion. Normally, the gas production is approximately 12-18 cu ft of gas per lb of volatile solids destroyed, though the industrial wastes in the sludge can affect this range, depending upon the composition.

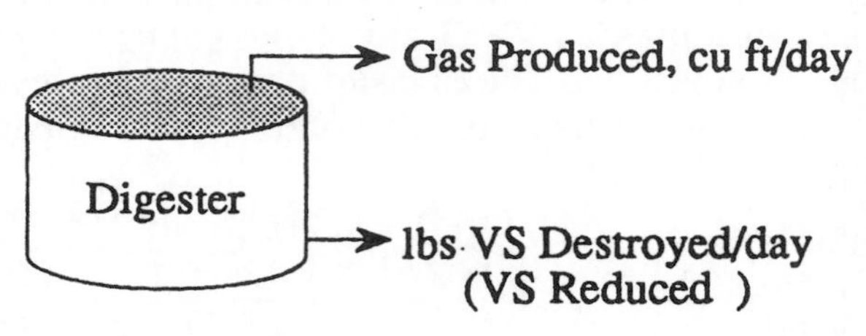

Simplified Equation:

$$\text{Digester Gas Production} = \frac{\text{Gas Produced, cu ft/day}}{\text{VS Destroyed, lbs/day}}$$

Expanded Equations:

$$\text{Digester Gas Production} = \frac{\text{Gas Produced, cu ft/day}}{(\text{Vol. Sol. to Dig.})\ (\frac{\text{\% VS Reduction}}{100})}$$

(Vol. Sol. to Dig. in lbs/day)

Or

$$\text{Digester Gas Production} = \frac{\text{Gas Produced, cu ft/day}}{(\text{gpd raw sludge})\ (8.34\ \text{lbs/day})\ (\frac{\text{\% Solids}}{100})\ (\frac{\text{\% VS}}{100})\ (\frac{\text{\% VS Red.}}{100})}$$

This is the lbs/day VS entering the digester

Example 1: (Digester Gas Production)

❑ A digester gas meter reading indicates an average of 6340 cu ft of gas is produced per day. If a total of 490 lbs/day volatile solids are destroyed, what is the digester gas production in cu ft gas/lb VS destroyed?

$$\text{Digester Gas Production} = \frac{\text{Gas Produced, cu ft/day}}{\text{VS Destroyed, lbs/day}}$$

$$= \frac{6340\ \text{cu ft/day}}{490\ \text{lbs VS Destroyed/day}}$$

$$= \boxed{12.9\ \frac{\text{cu ft Gas Produced}}{\text{lb VS Destroyed}}}$$

Example 2: (Digester Gas Production)

❑ A total of 1900 lbs of volatile solids are pumped to the digester daily. If the percent reduction of volatile solids due to digestion is 60% and the average gas production for the day is 18,240 cu ft, what is the daily gas production in cu ft/lb VS destroyed daily?

$$\text{Digester Gas Production} = \frac{\text{Gas Produced, cu ft/day}}{(\text{Vol. Sol. to Dig.})\ \text{lbs/day}\ \left(\frac{\%\ \text{VS Reduction}}{100}\right)}$$

$$= \frac{\text{18,240 cu ft Gas Produced/day}}{(\text{1900 lbs/day})\left(\frac{60}{100}\right)}$$

$$= \boxed{16\ \frac{\text{cu ft Gas Produced}}{\text{lb VS Destroyed}}}$$

Example 3: (Digester Gas Production)

❑ The anaerobic digester at a plant receives a total of 11,400 gpd of raw sludge. This sludge has a solids content of 5.4%, of which 62% is volatile. If the digester yields a volatile solids reduction of 58%, and the average digester gas production is 25,850 cu ft, what is the daily gas production in cu ft/lb VS destroyed?

$$\text{Digester Gas Production} = \frac{\text{Gas Produced, cu ft/day}}{(\text{gpd raw sludge})(8.34\ \text{lbs/day})\left(\frac{\%\ \text{Solids}}{100}\right)\left(\frac{\%\ \text{VS}}{100}\right)\left(\frac{\%\ \text{VS Red.}}{100}\right)}$$

This is the lbs/day VS entering the digester

$$= \frac{\text{25,850 cu ft Gas Produced/day}}{(\text{11,400 gpd raw sludge})(8.34^{*}\ \text{lbs/gal})\left(\frac{5.4}{100}\right)\left(\frac{62}{100}\right)\left(\frac{58}{100}\right)}$$

$$= \boxed{14\ \frac{\text{cu ft Gas Produced}}{\text{lb VS Destroyed}}}$$

* If the sludge has a higher density than water, a number greater than 8.34 lbs/gal should be used here. Refer to Chapter 7 for a discussion of density and specific gravity.

16.14 DIGESTER SOLIDS BALANCE

A solids balance (sometimes called "mass balance") can be calculated for a single process, such as digestion, or it can be calculated for the plant as a whole. These calculations help verify the many calculations to control and optimize the wastewater treatment system.

When calculating a mass balance for a treatment system, you should be able to account for about 90% of the material entering and leaving the system (solids, water and gases).

The sludge entering the digester is comprised of solids and water. After digestion, the products are solids and water plus gas (such as methane and carbon dioxide), as shown in the diagram to the right.

When calculating a solids balance, **first determine the total solids, volatile solids and fixed solids entering the digester**:

- Total Solids Entering, lbs
- Volatile Solids Entering, lbs
- Fixed Solids Entering, lbs

Then determine the lbs water entering.

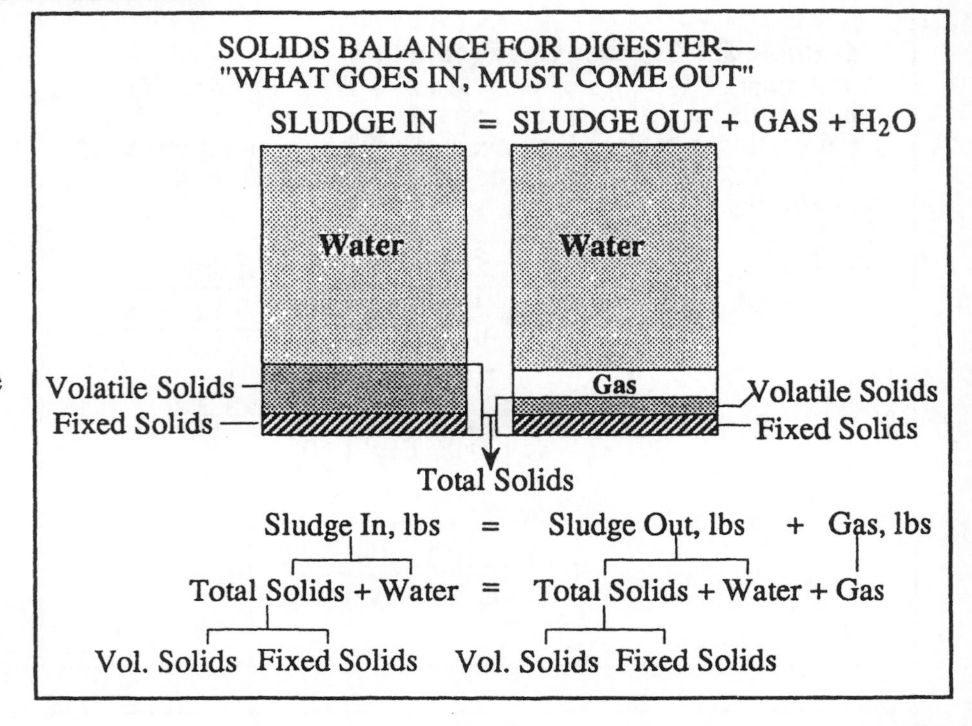

Example 1: (Solids Balance)

❑ Given the following data, calculate the solids balance for the digester.

Sludge to Digester	Sludge After Digestion
Raw Sludge—26,700 lbs/day	Digested Sludge
% Solids—6.5%	% Solids—4.5%
% Volatile Solids—70%	% Volatile Solids—55%

First calculate solids and water entering the digester:

- ***Total Solids Entering, lbs/day***

$$(\text{Sludge, lbs/day})\left(\frac{\%\ \text{Solids}}{100}\right) = \text{Total Solids, lbs/day}$$

$$(26{,}700\ \text{lbs/day})\ (0.065) = \boxed{1736\ \text{Total Solids, lbs/day}}$$

- ***Volatile Solids Entering, lbs/day***

$$(\text{Tot. Sol., lbs/day})\left(\frac{\%\ \text{VS}}{100}\right) = \text{Vol. Solids, lbs/day}$$

$$(1736\ \text{lbs/day})\ (0.70) = \boxed{1215\ \text{Vol. Sol., lbs/day}}$$

- ***Fixed Solids Entering, lbs/day***

$$\text{Total Solids, lbs/day} - \text{Vol. Sol., lbs/day} = \text{Fixed Solids, lbs/day}$$

$$1736\ \text{lbs/day} - 1215\ \text{lbs/day} = \boxed{521\ \text{Fixed Solids, lbs/day}}$$

- ***Water Entering, lbs/day***

$$\text{Sludge, lbs/day} - \text{Total Solids, lbs/day} = \text{Water, lbs/day}$$

$$26{,}700\ \text{lbs/day} - 1736\ \text{lbs/day} = \boxed{24{,}964\ \text{Water, lbs/day}}$$

Then calculate solids, water and gas leaving the digester:
(To begin these calculations, you must first determine the percent volatile solids reduction during digestion.

- ***% Volatile Solids Reduction****
  (Write percents as decimal fractions)

$$\%\ \text{VS Reduction} = \frac{\text{In} - \text{Out}}{\text{In} - (\text{In} \times \text{Out})} \times 100$$

$$= \frac{0.70 - 0.55}{0.70 - (0.70 \times 0.55)} \times 100$$

$$= \boxed{47.6\%\ \text{VS Reduction}}$$

- ***Gas Produced, lbs/day***
  (VS destroyed or reduced = lbs gas produced)

$$\frac{(\text{VS Entering Digester, lbs/day})(\%\ \text{VS Reduction})}{100} = \text{Gas Produced, lbs/day}$$

$$(1215\ \text{lbs/day Vol. Sol.})(0.476) = \boxed{578\ \text{lbs/day Gas Produced}}$$

- ***Volatile Solids in Digested Sludge, lbs/day***

VS Entering Digester, lbs/day – VS Destroyed, lbs/day = VS Leaving Digester, lbs/day

$$1215\ \text{lbs/day} - 578\ \text{lbs/day} = \boxed{637\ \text{lbs/day VS in Dig. Slud.}}$$

- ***Total Solids in Digested Sludge, lbs/day***

$$\frac{\text{Vol. Sol., lbs/day}}{\frac{\%\ \text{VS}}{100}} = \text{Total Solids, lbs/day}$$

$$\frac{637\ \text{lbs/day VS}}{0.55} = \boxed{1158\ \text{lbs/day Total Solids}}$$

- ***Fixed Solids in Digested Sludge, lbs/day***

Tot. Sol., lbs/day – VS, lbs/day = Fixed Solids, lbs/day

$$1158\ \text{lbs/day} - 637\ \text{lbs/day} = \boxed{521\ \text{lbs/day}}$$

- ***Digested Sludge, lbs/day***

$$\frac{\text{Total Solids, lbs/day}}{\frac{\%\ \text{Solids}}{100}} = \text{Digested Sludge, lbs/day}$$

$$\frac{1158\ \text{lbs/day Tot. Sol.}}{0.045} = \boxed{25{,}733\ \text{lbs/day Digested Sludge}}$$

- ***Water in Digested Sludge, lbs/day***

Sludge, lbs/day – Total Solids, lbs/day = Water, lbs/day

$$25{,}733\ \text{lbs/day} - 1158\ \text{lbs/day} = \boxed{24{,}575\ \text{lbs/day}}$$

To account for solids leaving the digester, you will make the same basic calculations but in a slightly different order. You must begin by determining the % volatile solids reduction, then calculate the solids, water, and gas, as shown in the example calculation.

Summary Comparison

Sludge Entering	Sludge Leaving
Tot. Sol.—1736 lbs	Tot. Sol.—1158 lbs
(VS—1215 lbs)	(VS—637 lbs)
(FS— 521 lbs)	(FS— 521 lbs)
Water—24,964 lbs	Water—24,575 lbs
	Gas— 578 lbs
26,700 lbs	26,311 lbs

* For a review of the percent volatile solids reduction calculation, refer to Section 16.11.

16.15 DIGESTION TIME

The digestion time calculation is simply a detention time calculation.* The most common calculation of digestion time is a determination of the flow-through time in the digester. This is sometimes referred to as the **"hydraulic digestion time."** The equations for this calculation are shown to the right.

Another calculation of digestion time is the **"solids digestion time"**. This calculation is based on the amount of time solids remain in the digester and is calculated using the equation shown below.

$$\text{Digestion Time, days} = \frac{\text{Digester Solids, lbs}}{\text{Sol. Wasted, lbs/day}}$$

DIGESTION TIME (HYDRAULIC) IS FLOW-THROUGH TIME

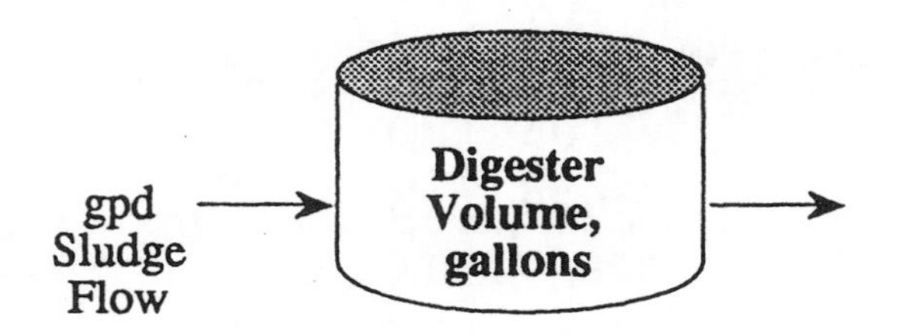

Simplified Equation:

$$\text{Digestion Time, days} = \frac{\text{Digester Volume, gal}}{\text{Sludge Flow Rate, gpd}}$$

Expanded Equation:

$$\text{Digestion Time, days} = \frac{(0.785)\ (D^2)\ (\text{Depth, ft})\ (7.48\ \text{gal/cu ft})}{\text{Sludge Flow Rate, gpd}}$$

Example 1: (Digestion Time)

❑ A 50-ft diameter aerobic digester has a side water depth (SWD) of 12 ft. The sludge flow to the digester is 9000 gpd. Calculate the hydraulic digestion time, in days.

$$\text{Digestion Time, days} = \frac{(0.785)\ (D^2)\ (\text{Depth, ft})\ (7.48\ \text{gal/cu ft})}{\text{Sludge Flow Rate, gpd}}$$

$$= \frac{(0.785)(50\ \text{ft})\ (50\ \text{ft})\ (12\ \text{ft})\ (7.48\ \text{gal/cu ft})}{9000\ \text{gpd flow}}$$

$$= \boxed{19.6\ \text{days}}$$

* Detention time calculations are discussed in Chapter 5.

Example 2: (Digestion Time)

❑ A sludge flow of 9500 gpd has a solids concentration of 2.7%. If the solids concentration is increased to 3.8% as a result of thickening, what is the anticipated flow rate of the thickened sludge to the digester? (Assume both sludges have a density of 8.34 lbs/gal.)

lbs/day Solids = lbs/day Solids

$$(\text{gpd})_{\text{Unthick. Sludge}}\ (8.34\ \text{lbs/gal})\ \frac{(\%\ \text{Sol.})}{100} = (\text{gpd})_{\text{Thickened Sludge}}\ (8.34\ \text{lbs/gal})\ \frac{(\%\ \text{Sol.})}{100}$$

$$(9500\ \text{gpd})\ (8.34\ \text{lbs/gal})\ \frac{(2.7)}{100} = (x\ \text{gpd})_{\text{Thickened Sludge}}\ (8.34\ \text{lbs/gal})\ \frac{(3.8)}{100}$$

$$\frac{(9500)\cancel{(8.34)}(0.027)}{\cancel{(8.34)}(0.038)} = x$$

$$\boxed{6750\ \text{gpd}} = x$$

Example 3: (Digestion Time)

❑ For a digester 35-ft in diameter with a side water depth (SWD) of 10 ft, what is the difference in digestion time for each of the two sludge flow rates in Example 2 (9500 gpd and 6750 gpd)?

For the unthickened sludge flow:

$$\text{Digestion Time, days} = \frac{(0.785)\ (35\ \text{ft})\ (35\ \text{ft})\ (10\ \text{ft})\ (7.48\ \text{gal/cu ft})}{9500\ \text{gpd}}$$

$$= \boxed{7.6\ \text{days}}$$

For the thickened sludge flow:

$$\text{Digestion Time, days} = \frac{(0.785)\ (35\ \text{ft})\ (35\ \text{ft})\ (10\ \text{ft})\ (7.48\ \text{gal/cu ft})}{6750\ \text{gpd}}$$

$$= \boxed{10.7\ \text{days}}$$

THE EFFECT OF THICKENING ON DIGESTION TIME

The digestion time depends on sludge flow, gpd. When a sludge flow is thickened, a smaller volume of sludge is pumped to the digester, resulting in a longer detention time:

Increase in % Solids (thicker sludge) —results in→ Increase in Detent. Time (longer d.t.)

And conversely,

Decrease in % solids (less thick) sludge —results in→ Decrease in Detent. Time (shorter d.t.)

To calculate the effect of thickening on digestion time, first **determine the change in sludge flow** resulting from the percent solids sludge change*. The equation used for this calculation is based on the concept that the pounds of solids in the unthickened sludge is equal to the pounds of solids in the thickened sludge:**

Simplified Equation:

$$\boxed{\text{Solids, lbs/day} = \text{Solids, lbs/day}}$$

Expanded Equation:

$$(\text{lbs/day})_{\text{Unth. Sl.}}\ \frac{(\%\ \text{Sol.})}{100} = (\text{lbs/day})_{\text{Thick. Sludge}}\ \frac{(\%\ \text{Sol.})}{100}$$

Or

$$(\text{gpd})_{\text{Unth. Sl.}}\ (8.34\ \text{lbs/gal})\ \frac{(\%\ \text{Sol.})}{100} = (\text{gpd})_{\text{Thick. Sludge}}\ (8.34\ \text{lbs/gal})\ \frac{(\%\ \text{Sol.})}{100}$$

Example 2 illustrates this calculation. The digestion times for both the unthickened and thickened sludge flows can then be calculated and compared, as shown in Example 3.

* This calculation is also discussed in Chapter 15, Section 15.3.

** There is an assumption here that a negligible amount of solids are lost in the thickener effluent.

16.16 AIR REQUIREMENTS AND OXYGEN UPTAKE

AIR REQUIREMENTS

The specific air supply required for a digester depends on several factors, such as sludge volatile solids content, temperature, and biomass activity, and must be determined experimentally at the plant. To determine the total cfm air required, use the air supply rate, cfm/cu ft, and set up a proportion.* For example, suppose the desired air supply rate for a digester has been determined as 0.02 cfm/cu ft digester volume, the proportion would be set up as:

Simplified Equation:

$$\frac{0.02 \text{ cfm Air}}{1 \text{ cu ft Dig. Vol}} = \frac{\text{Air Req., cfm}}{\text{Dig. Vol., cu ft}}$$

Expanded Equation:
For circular digesters

$$\frac{0.02 \text{ cfm Air}}{1 \text{ cu ft}} = \frac{\text{cfm Air Req.}}{(0.785)(D^2)(\text{depth, ft})}$$

For rectangular digesters

$$\frac{0.02 \text{ cfm Air}}{1 \text{ cu ft}} = \frac{\text{Air Req., cfm}}{(\text{length})(\text{width})(\text{depth})}$$
ft ft ft

Sometimes the air supply rate is expressed as cfm per 1000 cu ft of digester volume. This calculation is demonstrated in Example 2. The denominator of the equation (cu ft volume) must a be divided by 1000, since cfm per 1000-cu ft volume is required. Remember that the 1000-cu ft shown in the denominator of the left side of the equation is a unit of measure only and is **not to be used in the calculation of the answer.**

Example 1: (Air Requirements and Oxygen Uptake)
❑ The desired air supply rate for an aerobic digester was determined to be 0.03 cfm/cu ft digester capacity. What is the total cfm air required if the digester is 80 ft long, 20 ft wide with a side water depth of 10 ft?

$$\frac{0.03 \text{ cfm}}{1 \text{ cu ft Dig. Vol}} = \frac{x \text{ cfm Air Required}}{(\text{length, ft})(\text{width, ft})(\text{depth, ft})}$$

$$\frac{0.03 \text{ cfm}}{1 \text{ cu ft}} = \frac{x \text{ cfm}}{(80 \text{ ft})(20 \text{ ft})(10 \text{ ft})}$$

$$(0.03)(80)(20)(10) = x$$

$$\boxed{480 \text{ cfm Air}} = x$$

Example 2: (Air Requirements and Oxygen Uptake)
❑ An aerobic digester is 50 ft in diameter, with a side water depth of 10 ft. If the desired air supply for this digester was determined to be 40 cfm/1000-cu ft digester capacity, what is the total cfm air required for this digester?

This problem can be calculated as usual using the new air supply rate:

$$\frac{40 \text{ cfm}}{\text{Dig. Vol, 1000-cu ft}} = \frac{x \text{ cfm Air Required}}{\frac{(0.785)(D^2)(\text{depth, ft})}{1000}}$$

$$\frac{40 \text{ cfm}}{1000\text{-cu ft}} = \frac{x \text{ cfm}}{\frac{(0.785)(50 \text{ ft})(50 \text{ ft})(10 \text{ ft})}{1000}}$$

$$\frac{40 \text{ cfm}}{1000\text{-cu ft}} = \frac{x \text{ cfm}}{19.6 \text{ 1000-cu ft}}$$

$$(40)(19.6) = x$$

$$\boxed{784 \text{ cfm Air}} = x$$

* Refer to Chapter 7 in *Basic Math Concepts* for a review of ratios and proportions. Note in this calculation, since the ratio is expressed as cfm air/cu ft vol., the proportion is set up as cfm/cu ft rather than grouping like terms (cfm/cfm = cu ft/cu ft).

Example 3: (Air Requirements and Oxygen Uptake)

❑ The dissolved air concentrations recorded during a 5-minute test of an air-saturated sample of aerobic digester sludge are given below. Calculate the oxygen uptake, in mg/*L*/hr.

Elapsed Time, min	D.O., mg/*L*	Elapsed Time, min	D.O., mg/*L*
At Start	6.9	3 min	4.3
1 min	5.8	4 min	3.7
2 min	5.0	5 min	2.9

$$\text{O}_2\text{ Uptake, mg/}L\text{/hr} = \frac{\text{mg/}L\text{ DO at 2 min.} - \text{mg/}L\text{ DO at 5 min.}}{5\text{ min} - 2\text{ min}} \times 60\,\frac{\text{min}}{\text{hr}}$$

$$= \frac{5.0\text{ mg/}L - 2.9\text{ mg/}L}{3\text{ min}} \times 60\,\frac{\text{min}}{\text{hr}}$$

$$= \boxed{42\text{ mg/}L\text{/hr}}$$

Example 4: (Air Requirements and Oxygen Uptake)

❑ Dissolved air concentrations are taken at one-minute intervals on an air-saturated sample of digester sludge. Given the results below, calculate the oxygen uptake, in mg/*L*/hr.

Elapsed Time, min	D.O., mg/*L*	Elapsed Time, min	D.O., mg/*L*
At Start	7.3	3 min	4.9
1 min	6.4	4 min	4.1
2 min	5.6	5 min	3.2

$$\text{O}_2\text{ Uptake, mg/}L\text{/hr} = \frac{\text{mg/}L\text{ DO at 2 min.} - \text{mg/}L\text{ DO at 5 min.}}{5\text{ min} - 2\text{ min}} \times 60\,\frac{\text{min}}{\text{hr}}$$

$$= \frac{5.6\text{ mg/}L - 3.2\text{ mg/}L}{3\text{ min}} \times 60\,\frac{\text{min}}{\text{hr}}$$

$$= \boxed{48\text{ mg/}L\text{/hr}}$$

OXYGEN UPTAKE

Another measurement of the aerobic digestion system is oxygen air uptake. The oxygen uptake is an indication of microbiological (biomass) activity. There is an increase in oxygen uptake with increased microorganism activity; and there is a decrease in oxygen uptake when the biomass activity slows (such as during upset conditions).

To determine oxygen uptake, a one-liter sample of digested sludge is tested for dissolved oxygen (DO) levels. The DO measurement is recorded at the start of the test and at one-minute intervals for the duration of the five-minute test.

The DO measurements at 2 minutes and 5 minutes are generally used to calculate the oxygen uptake.

Simplified Equation:

$$\text{O}_2\text{ Uptake, mg/}L\text{/hr} = \frac{\text{mg/}L\text{ DO used during Test}}{\text{Min. during Measurement}} \times 60\,\frac{\text{min}}{\text{hr}}$$

Expanded Equation:

$$\text{O}_2\text{ Up., mg/}L\text{/hr} = \frac{\text{mg/}L\text{ DO at 2 min.} - \text{mg/}L\text{ DO at 5 min.}}{5\text{ min} - 2\text{ min}} \times 60\,\frac{\text{min}}{\text{hr}}$$

For measurements taken at times other than 2 and 5 minutes, use the following general equation:

Expanded Equation:

$$\text{O}_2\text{ Up., mg/}L\text{/hr} = \frac{\text{mg/}L\text{ DO at Time}_1 - \text{mg/}L\text{ DO at Time}_2}{\text{Time}_2\text{, min} - \text{Time}_1\text{, min}} \times 60\,\frac{\text{min}}{\text{hr}}$$

16.17 pH ADJUSTMENT USING JAR TESTS

The pH of aerobic digesters should not be allowed to drop below 6.0. As the pH approaches 6.0, the digester should be neutralized to adjust the pH upward. The quantity of lime ($Ca(OH)_2$), caustic (NaOH), or bicarbonate (HCO_3^-) required for neutralization may be determined using jar tests.

A one-liter sample of digested sludge is tested for the quantity of lime or caustic required to raise the pH to the desired level. Based on this quantity then, the pounds of chemical required for the entire digester is calculated using the mg/L to lbs equation:*

(mg/L) Lime or Caustic	(MG) Dig. Vol.	(8.34) lbs/gal	= lbs Lime or Caustic

Example 1: (pH Adjustment Using Jar Tests)

❑ A jar test indicates that 22 mg of caustic are required to raise the pH of the one-liter sludge sample to 7.0. If the digester volume is 100,000 gallons, how many pounds of caustic will be required for pH adjustment?

(mg/L) Caustic Req'd	(MG) Dig. Vol.	(8.34) lbs/gal	= lbs Caustic Required

(22 mg/L) (0.1 MG) (8.34 lbs/gal) = **18.3 lbs Caustic**

Example 2: (pH Adjustment Using Jar Tests)

❑ Jar testing indicates that 18 mg of caustic are required to raise the pH of the one-liter sludge sample to 7.0. If the digester volume is 90,000 gallons, how many pounds of caustic will be required for pH adjustment?

(mg/L) Caustic Req'd	(MG) Dig. Vol.	(8.34) lbs/gal	= lbs Caustic Required

(18 mg/L) (0.09 MG) (8.34 lbs/gal) = **13.5 lbs Caustic**

* For a review of mg/L calculations, refer to Chapter 3.

Example 3: (pH Adjustment Using Jar Tests)
❑ A two-liter sample of digested sludge is used to determine the required caustic dosage for pH adjustment. If 56 mg of caustic are required for pH adjustment in the jar test, and the digester volume is 94,000 gallons, how many pounds of caustic will be required for pH adjustment?

First determine the mg/*L* caustic dosage required:

$$\frac{56 \text{ mg}}{2\,L} = \boxed{28 \text{ mg}/L}$$

The pounds caustic required can now be calculated:

$$(28 \text{ mg}/L)\,(0.094 \text{ MG})\,(8.34 \text{ lbs/gal}) = \boxed{22.0 \text{ lbs Caustic}}$$

Example 4: (pH Adjustment Using Jar Tests)
❑ A 2-liter sample of digested sludge is used to determine the required caustic dosage for pH adjustment. A total of 62 mg caustic were used in the jar test. The aerobic digester is 45 feet in diameter with a side water depth of 10 ft. How many pounds of caustic are required for pH adjustment of the digester?

To complete this calculation, the required dosage and the gallon volume of the digester must be determined:

$$\frac{62 \text{ mg}}{2 \text{ liters}} = 31 \text{ mg}/L$$

$$(0.785)\,(45 \text{ ft})\,(45 \text{ ft})\,(10 \text{ ft})\,(7.48) = 118{,}904 \text{ gal}$$

The pounds caustic may now be calculated:

$$(31 \text{ mg}/L)\,(0.119 \text{ MG})\,(8.34 \text{ lbs/gal}) = \boxed{30.8 \text{ lbs Caustic}}$$

USING DIFFERENT SAMPLE VOLUMES

In Examples 1 and 2, the sample volume of digested sludge was one-liter. When a different sample volume is used, such as a half-liter (500 m*L*) or two liters (2000 m*L*), the quantity of chemical required for one liter is first determined, then the calculation is completed as usual.

For example, if 11 mg chemical were required for desired pH adjustment of a 500 ml sample, what is the chemical required per liter?

One way to determine the answer is to use a proportion:

$$\frac{11 \text{ mg Chemical}}{0.5\,L \text{ Sample}} = \frac{x \text{ mg}}{1 \text{ liter}}$$

$$x = 22 \text{ mg}$$

Note that the answer may be obtained by simply dividing the number of grams by the number of liters. Using another example, if 55 mg chemical are required for pH adjustment of a 2-liter sample, how many milligrams per liter is this?

$$\frac{55 \text{ mg}}{2\,L} = 27.5 \text{ mg}/L$$

Example 3 and 4 illustrate calculations where the sample volume was different than one liter.

NOTES:

17 Sludge Dewatering and Disposal

SUMMARY

1. Filter Press Dewatering

• Solids Loading Rate—Plate and Frame Filter Press

Simplified Equation:

$$\text{Solids Loading Rate, lbs/hr/sq ft} = \frac{\text{Solids, lbs/hr}}{\text{Plate Area, sq ft}}$$

Expanded Equation:

$$\text{Solids Loading Rate, lbs/hr/sq ft} = \frac{(\text{gph Sludge})(8.34 \text{ lbs/gal})\left(\frac{\% \text{ Sol.}}{100}\right)}{\text{Plate Area, sq ft}}$$

• Net Filter Press—Plate and Frame Filter Press

Simplified Equation:

$$\text{Net Filter Yield, lbs/hr/sq ft} = \frac{\text{(lbs/hr Sol.)}}{\text{sq ft}} \cdot \frac{\text{(Filtration Run Time, hrs)}}{\text{Total Cycle Time, hrs}}$$

Expanded Equation:

$$\text{Net Filter Yield, lbs/hr/sq ft} = \frac{(\text{gph Sludge})(8.34 \text{ lbs/gal})\left(\frac{\% \text{ Sol.}}{100}\right)}{\text{Plate Area, sq ft}} \cdot \frac{\text{(Filt. Run Time, hrs)}}{\text{Total Cycle Time, hrs}}$$

2. Belt Filter Press Dewatering

• Hydraulic Loading Rate

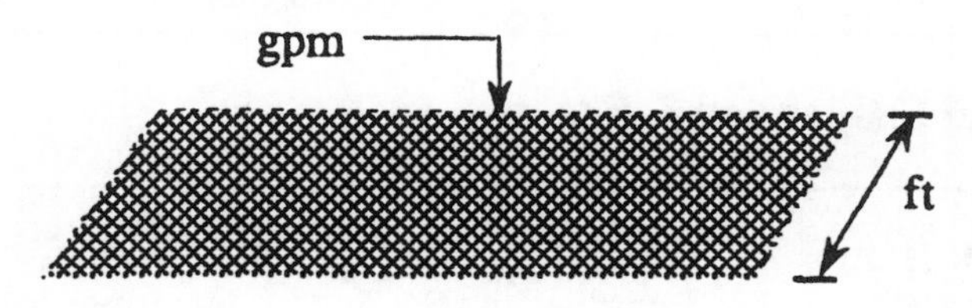

$$\text{Hydraulic Loading Rate, gpm/ft} = \frac{\text{Flow, gpm}}{\text{Belt Width, ft}}$$

SUMMARY—Cont'd

- **Sludge Feed Rate**

$$\text{Sludge Feed Rate, lbs/hr} = \frac{\text{Sludge to be Dewatered, lbs/day}}{\text{Operating Time, hrs/day}}$$

- **Solids Loading Rate**

If the TSS data is given as percent:

$$\text{Solids Loading Rate, lbs/hr} = (\text{Sludge Feed, gpm})\left(\frac{60\ \text{min}}{\text{hr}}\right)\left(\frac{8.34\ \text{lbs}}{\text{gal}}\right)\left(\frac{\%\ \text{TSS}}{100}\right)$$

If the TSS data is given as mg/*L*:

$$\text{Solids Loading Rate, lbs/hr} = \frac{(\text{mg/}L\ \text{TSS})(\text{MGD Sludge Feed})\left(\frac{8.34\ \text{lbs}}{\text{gal}}\right)}{24\ \text{hrs/day}}$$

Solids loading rate is sometimes expressed as tons/hr:

$$\text{Solids Loading Rate, tons/hr} = \frac{(\text{Sludge Feed gpm})\left(\frac{60\ \text{min}}{\text{hr}}\right)\left(\frac{8.34\ \text{lbs}}{\text{gal}}\right)\left(\frac{\%\ \text{TSS}}{100}\right)}{2000\ \text{lbs/ton}}$$

- **Flocculant Feed Rate**

$$\text{Flocculant Feed, lbs/hr} = \frac{(\text{mg/}L\ \text{Flocc.})(\text{MGD Feed Rate})\left(\frac{8.34\ \text{lbs}}{\text{gal}}\right)}{24\ \text{hrs/day}}$$

- **Total Suspended Solids**

$$\text{Total Residue, mg/}L - \text{Total Filterable Residue, mg/}L = \text{Total Non-Filterable Residue, mg/}L$$

SUMMARY—Cont'd

3. Vacuum Filter Dewatering

- **Filter Loading**

Simplified Equation:

$$\text{Filter Loading, lbs/hr/sq ft} = \frac{\text{Solids to Filter, lbs/hr}}{\text{Surface Area, sq ft}}$$

Expanded Equation:

$$\text{Filter Loading, lbs/hr/sq ft} = \frac{\dfrac{\text{Sol. to Filter, lbs/day}}{\text{Filter Oper., hrs/day}}}{(\pi D)(\text{width, ft})}$$

- **Filter Yield**

If lbs/hr wet cake flow, percent cake solids, and filter area are used to calculate filter yield:

Simplified Equation:

$$\text{Filter Yield, lbs/hr/sq ft} = \frac{\text{Dry Solids in Cake, lbs/hr}}{\text{Filter Area, sq ft}}$$

Expanded Equation:

$$\text{Filter Yield, lbs/hr/sq ft} = \frac{(\text{Wet Cake Flow, lbs/hr})\left(\dfrac{\text{\% Solids in Cake}}{100}\right)}{\text{Filter Area, sq ft}}$$

If filter loading (lbs/hr/sq ft) and percent solids recovery are used to calculate filter yield:

Simplified Equation:

$$\text{Filter Yield, lbs/hr/sq ft} = (\text{Filter Loading, lbs/hr/sq ft})\left(\frac{\text{\% Recovery}}{100}\right)$$

Expanded Equation:

$$\text{Filter Yield, lbs/hr/sq ft} = \frac{\left(\dfrac{\text{Sol. to Filter, lbs/day}}{\text{Fil. Oper., hrs/day}}\right)}{\text{Filter Area, sq ft}} \cdot \frac{(\text{\% Recov.})}{100}$$

SUMMARY—Cont'd

- **Percent Solids Recovery***

Simplified Equation:

$$\text{\% Solids Recovery} = \frac{\text{Solids in Cake, lbs/hr}}{\text{Solids in Feed, lbs/hr}} \times 100$$

Expanded Equation:

$$\text{\% Solids Recovery} = \frac{(\text{Wet Cake Flow, lbs/hr})\left(\frac{\text{\% Sol. in Cake}}{100}\right)}{(\text{Sludge Feed, lbs/hr})\left(\frac{\text{\% Sol. in Feed}}{100}\right)} \times 100$$

4. Sand Drying Beds

- **Total Sludge Applied**

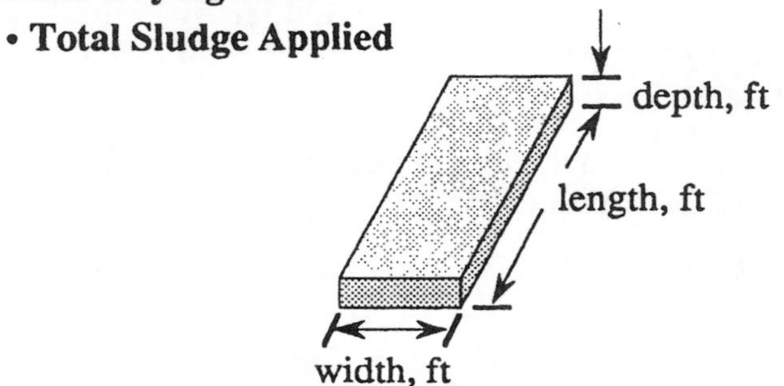

$$\text{Volume, gal} = (\text{length, ft})(\text{width, ft})(\text{depth, ft})(7.48\ \text{gal/cu ft})$$

- **Solids Loading Rate**

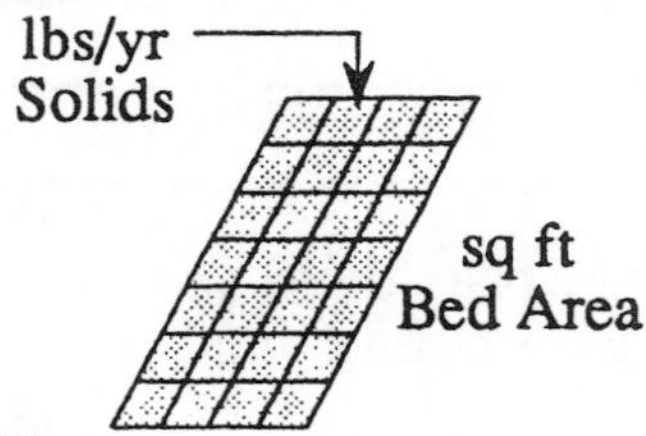

Simplified Equation:

$$\text{Solids Loading Rate, lbs/yr/sq ft} = \frac{(\text{Sludge Applied, lbs/yr})}{\text{Bed Area, sq ft}} \frac{(\text{\% Sol.})}{100}$$

Expanded Equation:

$$\text{Solids Loading Rate, lbs/yr/sq ft} = \frac{\frac{(\text{Sludge) Applied, lbs}}{\text{days of Applic.}} \frac{(365\ \text{days})}{\text{yr}} \frac{(\text{\% Sol.})}{100}}{(\text{length, ft})(\text{width, ft})}$$

* Centrifugation is another method of dewatering. The calculations associated with this process are given in Chapter 15 Thickening. In addition to those calculations, percent solids recovery is determined for centrifuge dewatering. This is calculated as shown for the vacuum filtration process.

SUMMARY—Cont'd

- **Sludge Withdrawal to Drying Beds**

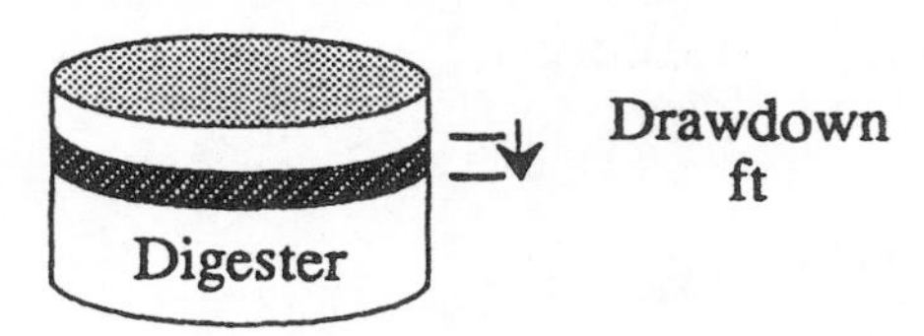

$$\text{Sludge Withdrawn, cu ft} = (0.785)(D^2)(\text{Drawdown, ft})$$

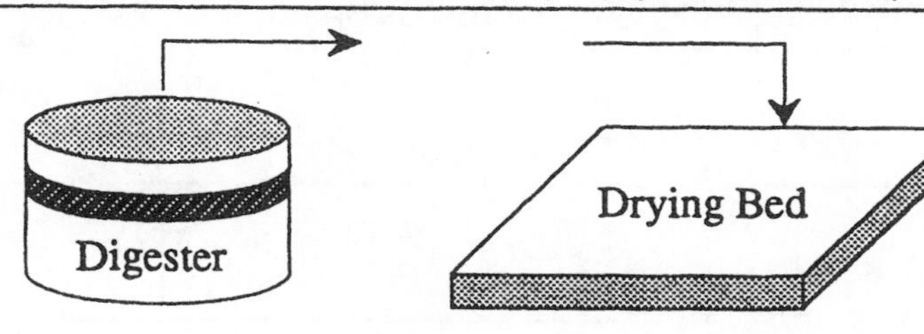

Sludge Withdrawn, cu ft = Sludge to Drying Beds, cu ft

$$(0.785)(D^2)(\text{Drawdown, ft}) = (\text{length, ft})(\text{width, ft})(\text{depth, ft})$$

5. Composting

- **Compost Blending—Dewatered Sludge with Composted Sludge**

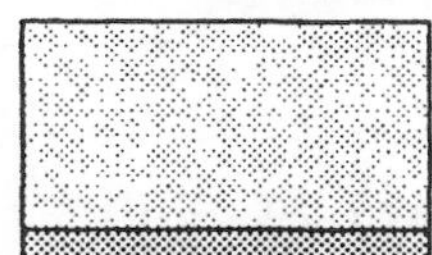 + 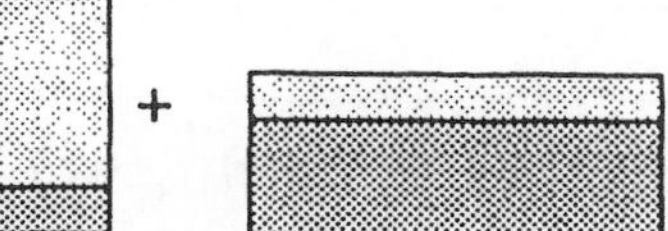 = 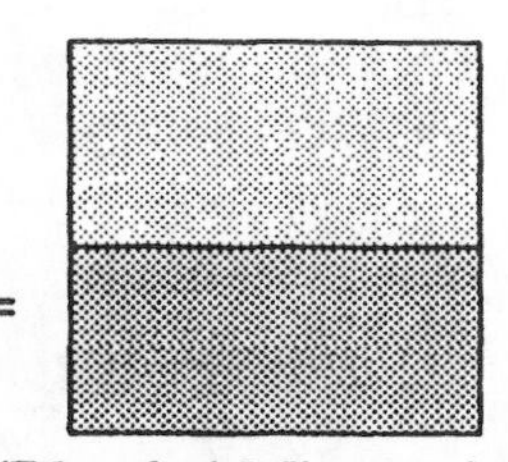

70% Moisture 30% Moisture (Blended Mixture in this example will have a moisture content between 70% and 30%)

Simplified Equation:

$$\text{\% Moisture of Compost Mixture} = \frac{\text{Moisture in Mixture, lbs/day}}{\text{Compost Mixture, lbs/day}} \times 100$$

Expanded Equation:

$$\text{\% Moisture of Compost Mixture} = \frac{\text{Moisture in Sludge, lbs/day} + \text{Moisture in Compost, lbs/day}}{\text{Sludge, lbs/day} + \text{Compost, lbs/day}} \times 100$$

$$\text{\% Moist. of Mixture} = \frac{(\text{Sludge lbs/day})\left(\frac{\text{\% Moist.}}{100}\right) + (\text{Compost lbs/day})\left(\frac{\text{\% Moist.}}{100}\right)}{\text{Sludge, lbs/day} + \text{Compost, lbs/day}} \times 100$$

SUMMARY—Cont'd

5. Composting—Cont'd

- **Compost Blending—Dewatered Sludge with Wood Chips**

Simplified Equation:

$$\text{\% Solids of Compost Blend} = \frac{\text{Solids in Compost Blend, lbs}}{\text{Compost Blend, lbs}} \times 100$$

Expanded Equations:

$$\text{\% Solids of Compost Blend} = \frac{\text{Solids in Sludge, lbs} + \text{Solids in Wood Chips, lbs}}{\text{Sludge, lbs} + \text{Wood Chips, lbs}} \times 100$$

$$\text{\% Solids of Compost Blend} = \frac{(\text{Solids in Sludge, cu yds})\left(\frac{\text{lbs}}{\text{cu yd}}\right) + (\text{Solids in Wood Chips, cu yds})\left(\frac{\text{lbs}}{\text{cu yd}}\right)}{(\text{Sludge, cu yds})\left(\frac{\text{lbs}}{\text{cu yd}}\right) + (\text{Wood Chips, cu yds})\left(\frac{\text{lbs}}{\text{cu yd}}\right)} \times 100$$

$$\text{\% Solids of Compost Blend} = \frac{(\text{Sludge, cu yds})\left(\frac{\text{lbs}}{\text{cu yd}}\right)\left(\frac{\text{\% Sol.}}{100}\right) + (\text{Wood Chips, cu yds})\left(\frac{\text{lbs}}{\text{cu yd}}\right)\left(\frac{\text{\% Sol.}}{100}\right)}{(\text{Sludge, cu yds})\left(\frac{\text{lbs}}{\text{cu yd}}\right) + (\text{Wood Chips, cu yds})\left(\frac{\text{lbs}}{\text{cu yd}}\right)} \times 100$$

Cu yds wood chips depends on the mix ratio of wood chips to sludge*

$$\text{\% Solids of Compost Blend} = \frac{(\text{Sludge, cu yds})\left(\frac{\text{lbs}}{\text{cu yd}}\right)\left(\frac{\text{\% Sol. of Sludge}}{100}\right) + (\text{Sludge, cu yds})(\text{Mix Ratio})\left(\frac{\text{lbs}}{\text{cu yd}}\right)\left(\frac{\text{\% Sol. of W.Chips}}{100}\right)}{(\text{Sludge cu yds})\left(\frac{\text{lbs}}{\text{cu yd}}\right) + (\text{Sludge cu yds})(\text{Mix Ratio})\left(\frac{\text{lbs}}{\text{cu yd}}\right)} \times 100$$

SUMMARY—Cont'd

- **Compost Site Capacity**

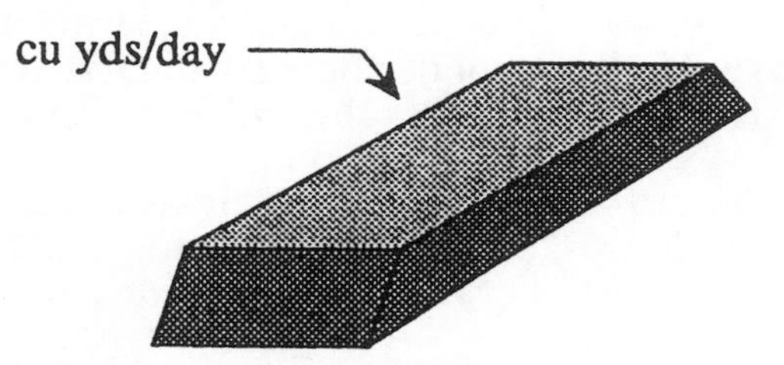

Simplified Equation:

$$\text{Fill Time, days} = \frac{\text{Total Available Capacity, cu yds}}{\text{Wet Compost, cu yds/day}}$$

Expanded Equations:

$$\text{Fill Time, days} = \frac{\text{Total Available Capacity, cu yds}}{\dfrac{\text{Wet Compost, lbs/day}}{\text{Compost Bulk Density, lbs/cu yd}}}$$

$$\text{Fill Time, days} = \frac{\text{Total Available Capacity, cu yds}}{\dfrac{\text{Wet Sludge lbs/day} + \text{Wet Wood Chips lbs/day}}{\text{Compost Bulk Density, lbs/cu yd}}}$$

$$\text{Fill Time, days} = \frac{(\text{Avail. Capac., cu yds})(\text{Comp. Bulk Dens., lbs/cu yd})}{\dfrac{\text{Dry Solids lbs/day}}{\frac{\%\ \text{Sol}}{100}} + \dfrac{(\text{Dry Solids lbs/day})}{\frac{\%\ \text{Sol}}{100}}(\text{Mix Ratio})\dfrac{(\text{Bulk Density of W.C. lbs/cu yd})}{\text{Slud. Bulk Dens., lb/cu yd}}}$$

Compost site capacity can also be calculated by the use of a nomograph, as described in the section.

17.1 FILTER PRESS DEWATERING CALCULATIONS (PLATE AND FRAME FILTER PRESS)

SOLIDS LOADING RATE

The solids loading rate is an important calculation in plate and frame filter press operation. It is a measure of the lbs/hr solids applied per square foot of plate area, as shown in the box to the right.

Simplified Equation:

$$\text{Solids Loading Rate, lbs/hr/sq ft} = \frac{\text{Solids, lbs/hr}}{\text{Plate Area, sq ft}}$$

Expanded Equation:

$$\text{Sol. Load. Rate, lbs/hr/sq ft} = \frac{(\text{gph Sludge})\,(8.34\text{ lbs/gal})\,\left(\frac{\%\text{ Sol.}}{100}\right)}{\text{Plate Area, sq ft}}$$

Examples 1 and 2 illustrate this calculation.

Example 1: (Filter Press Dewatering)

❑ A filter press used to dewater digested primary sludge receives a flow of 720 gallons during a 2-hr period. The sludge has a solids content of 3.5%. If the plate surface area is 125 sq ft, what is the solids loading rate in lbs/hr/sq ft?

The flow rate is given as gallons per 2 hours. First express this flow rate as gallons per hour: 720 gal/2 hrs = 360 gal/hr

$$\text{Solids Loading Rate, lbs/hr/sq ft} = \frac{(\text{Sludge, gph})\,(8.34\text{ lbs/gal})\,\left(\frac{\%\text{ Sol.}}{100}\right)}{\text{Plate Area, sq ft}}$$

$$= \frac{(360\text{ gph})\,(8.34\text{ lbs/gal})\,\left(\frac{3.5}{100}\right)}{125\text{ sq ft}}$$

$$= \boxed{0.84\text{ lbs/hr/sq ft}}$$

Example 2: (Filter Press Dewatering)

❑ A filter press used to dewater digested primary sludge receives a flow of 800 gallons of sludge during a 2-hour period. The solids content of the sludge is 3.2%. If the plate surface area is 110 sq ft, what is the solids loading rate in lbs/hr/sq ft?

The flow rate is given as gallons per 2 hours. First express this flow rate as gallons per hour: 800 gal/2 hrs = 400 gal/hr

$$\text{Solids Loading Rate, lbs/hr/sq ft} = \frac{(\text{Sludge, gph})\,(8.34\text{ lbs/gal})\,\left(\frac{\%\text{ Sol.}}{100}\right)}{\text{Plate Area, sq ft}}$$

$$= \frac{(400\text{ gph})\,(8.34\text{ lbs/gal})\,\left(\frac{3.2}{100}\right)}{110\text{ sq ft}}$$

$$= \boxed{0.97\text{ lbs/hr/sq ft}}$$

Example 3: (Filter Press Dewatering)

❑ A plate and frame filter press receives a solids loading of 0.8 lbs/hr/sq ft. If the filtration time is 2 hours and the time required to remove the sludge cake and begin sludge feed to the press is 25 minutes, what is the net filter yield in lbs/hr/sq ft? (25 min ÷ 60 min/hr = 0.42 hrs)

$$\text{Net Filter Yield, lbs/hr/sq ft} = \frac{\text{(lbs/hr)}}{\text{sq ft}} \frac{\text{(Filtration Run Time)}}{\text{Total Cycle Time}}$$

$$= \frac{\text{(0.8 lbs/hr)}}{\text{sq ft}} \frac{\text{(2 hrs)}}{\text{2.42 hrs}}$$

$$= \boxed{\text{0.66 lbs/hr/sq ft Net Filter Yield}}$$

Example 4: (Filter Press Dewatering)

❑ A plate and frame filter press receives a flow of 680 gallons of sludge during a 2-hour period. The solids concentration of the sludge is 3.4%. The surface area of the plate is 100 sq ft. If the down time for sludge cake discharge is 20 minutes, what is the net filter yield?

A simple way to calculate net filter yield is to calculate solids loading rate then multiply that number by the corrected time factor:

$$\text{Solids Loading Rate, lbs/hr/sq ft} = \frac{\text{(Sludge, gph) (8.34 lbs/gal)} \frac{\text{(\% Sol.)}}{100}}{\text{Plate Area, sq ft}}$$

$$= \frac{\text{(340 gph) (8.34 lbs/gal)} \frac{(3.4)}{100}}{\text{100 sq ft}}$$

$$= \boxed{\text{0.96 lbs/hr/sq ft}}$$

Now calculate net filter yield, using the corrected time factor:

$$\text{Net Filter Yield, lbs/hr/sq ft} = \text{(0.96 lbs/hr/sq ft)} \frac{\text{(2 hrs)}}{\text{2.33 hrs}}$$

$$= \boxed{\text{0.82 lbs/hr/sq ft}}$$

NET FILTER YIELD

The plate and frame filter press operates in a batch mode. Sludge is fed to the press until the space between the plates is completely filled with solids. The sludge flow to the press is then stopped and the plates are separated, allowing the sludge cake to fall into a hopper or conveyor below.

The **solids loading rate** measures the lbs/hr of solids applied to each sq ft of plate surface area. However, this does not reflect the "down time," the time when sludge feed to the press is stopped.

The **net filter yield**, measured in lbs/hr/sq ft, reflects the run time as well as the down time of the plate and frame filter press.* To calculate the net filter yield, simply multiply the solids loading rate (in lbs/hr/sq ft) by the ratio of filter run time to total cycle time as follows:

Simplified Equation:

$$\text{Net Filter Yield, lbs/hr/sq ft} = \frac{\text{(lbs/hr)}}{\text{sq ft}} \frac{\text{(Filt. Run Time)}}{\text{Tot. Cycle Time}}$$

Expanded Equation:

$$\text{N.F.Y.} = \frac{\text{(gph) Sludge (8.34) lbs/gal} \frac{\text{(\% Sol.)}}{100}}{\text{Plate Area, sq ft}} \frac{\text{(Filt.) Run Time}}{\text{Tot. Cycle Time}}$$

* This same type of calculation, one that reflects run time as well as down time, is described for hydraulic and solids loading of a basket centrifuge, Chapter 15, Section 15.6.

17.2 BELT FILTER PRESS DEWATERING

HYDRAULIC LOADING RATE

Hydraulic loading for belt filters is a measure of gpm flow per foot of belt width. The diagram and associated equation are shown to the right.

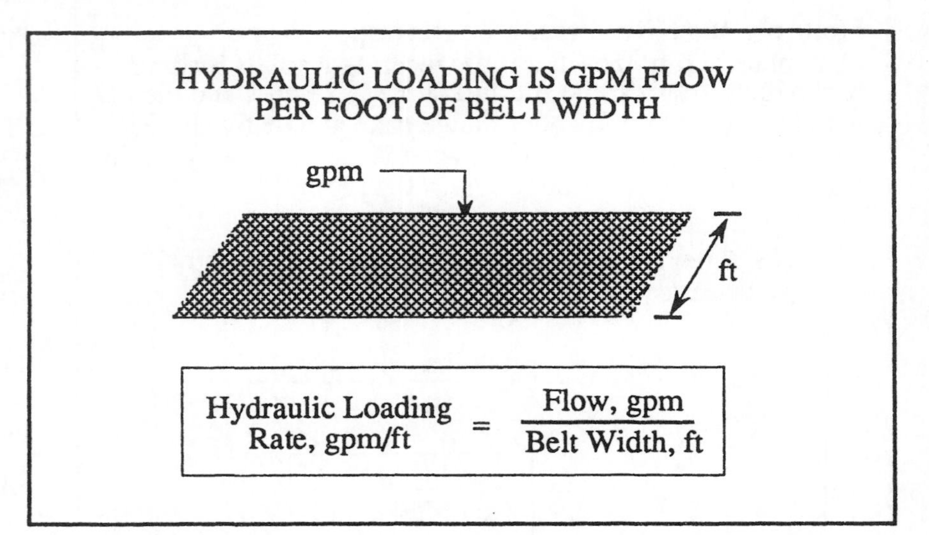

$$\text{Hydraulic Loading Rate, gpm/ft} = \frac{\text{Flow, gpm}}{\text{Belt Width, ft}}$$

Example 1: (Belt Filter Press Dewatering)

❑ A 5-feet wide belt press receives a flow of 120 gpm of primary sludge. What is the hydraulic loading rate in gpm/ft?

$$\text{Hydraulic Loading Rate, gpm/ft} = \frac{\text{Flow, gpm}}{\text{Belt Width, ft}}$$

$$= \frac{120 \text{ gpm}}{5 \text{ ft}}$$

$$= \boxed{24 \text{ gpm/ft}}$$

Example 2: (Belt Filter Press Dewatering)

❑ A belt filter press 6 ft wide receives a primary sludge flow of 155 gpm. What is the hydraulic loading rate in gpm/sq ft?

$$\text{Hydraulic Loading Rate, gpm/ft} = \frac{\text{Flow, gpm}}{\text{Belt Width, ft}}$$

$$= \frac{155 \text{ gpm}}{6 \text{ ft}}$$

$$= \boxed{26 \text{ gpm/ft}}$$

Example 3: (Belt Filter Press Dewatering)

❑ The amount of sludge to be dewatered by the belt filter press is 20,400 lbs/day. If the belt filter press is to be operated 12 hours each day, what should the sludge feed rate in lbs/hr be to the press?

$$\text{Sludge Feed Rate, lbs/hr} = \frac{\text{Sludge to be Dewatered, lbs/day}}{\text{Operating Time, hrs/day}}$$

$$= \frac{20{,}400 \text{ lbs/day}}{12 \text{ hrs/day}}$$

$$= \boxed{\begin{array}{l}\text{1700 lbs/hr}\\ \text{Feed Rate}\end{array}}$$

Example 4: (Belt Filter Press Dewatering)

❑ The amount of sludge to be dewatered by a belt filter press is 24,200 lbs/day. If the maximum feed rate that still provides an acceptable cake is 1800 lbs/hr, how many hours per day should the belt remain in operation?

Use the sludge feed rate calculation, filling in the known information. Then solve for the unknown value.

$$\text{Sludge Feed Rate, lbs/hr} = \frac{\text{Sludge to be Dewatered, lbs/day}}{\text{Operating Time, hrs/day}}$$

$$1800 \text{ lbs/hr} = \frac{24{,}200 \text{ lbs/day Sludge to be Dewatered}}{x \text{ hrs/day Operating Time}}$$

$$= \frac{24{,}200 \text{ lbs/day}}{1800 \text{ lbs/hr}}$$

$$= \boxed{\begin{array}{l}\text{13.4 hrs/day}\\ \text{Operating Time}\end{array}}$$

SLUDGE FEED RATE

The sludge feed rate to the belt filter press depends on several factors, including:

- The sludge, lbs/day, that must be dewatered,
- The maximum solids feed rate, lbs/hr, that will produce an acceptable cake dryness, and
- The number of hours per day the belt press is in operation.

The equation used in calculating sludge feed rate is:

$$\text{Sludge Feed Rate, lbs/hr} = \frac{\text{Sludge to be Dewatered, lbs/day}}{\text{Operating Time, hrs/day}}$$

In effect, in this calculation you are converting from lbs/day sludge to be processed to lbs/hr sludge to be processed. However, because the belt press may not operate 24 hour per day, instead of using 24 hrs/day to make this conversion, the operating time (perhaps 8 or 10 hrs/day) is used to make the conversion.

Example 3 illustrates the calculation of sludge feed rate. In Example 4, the unknown factor* is the desired operating time.

* For a review of solving for the unknown value, refer to Chapter 2 in *Basic Math Concepts*.

SOLIDS LOADING RATE

The solids loading rate may be expressed as lbs/hr or as tons/hr, as desired. In either case, the calculation is based on sludge flow (or feed) to the belt press and percent or mg/*L* concentration of total suspended solids (TSS) in the sludge.

$$\text{Sol. Load. Rate, lbs/hr} = (\text{Slud.) Feed, gpm}) \left(60\ \frac{\text{min}}{\text{hr}}\right) (8.34\ \text{lbs/gal}) \left(\frac{\%\ \text{TSS}}{100}\right)$$

If the TSS information is given as mg/*L*, the lbs/day solids loading rate can be calculated as a mg/*L* to lbs/day problem, as shown below. By dividing by 24 hours/day, the lbs/day solids loading is converted from lbs/day to lbs/hr solids loading:

$$\text{Solids Load. Rate, lbs/hr} = \frac{(\text{mg/}L\ \text{TSS})(\text{MGD Sludge Feed})\left(8.34\ \frac{\text{lbs}}{\text{gal}}\right)}{24\ \text{hrs/day}}$$

To express the solids loading rate as tons/hr, simply convert lbs/hr (as calculated using the equation above) to tons/hr solids loading rate:

$$\text{Sol. Load. Rate, tons/hr} = \frac{(\text{Slud. Feed, gpm})\left(60\ \frac{\text{min}}{\text{hr}}\right)(8.34\ \text{lbs/gal})\left(\frac{\%\ \text{TSS}}{100}\right)}{2000\ \text{lbs/ton}}$$

Example 5: (Belt Filter Press Dewatering)

❑ The sludge feed to a belt filter press is 130 gpm. If the total suspended solids concentration of the feed is 4%, what is the solids loading rate, in lbs/hr?

Since TSS concentration is expressed as a percent, the equation using percents will be used:

$$\text{Solids Loading Rate, lbs/hr} = (\text{Sludge Feed gpm})\left(60\ \frac{\text{min}}{\text{hr}}\right)\left(8.34\ \frac{\text{lbs}}{\text{gal}}\right)\left(\frac{\%\ \text{TSS}}{100}\right)$$

$$= (130\ \text{gpm})\left(60\ \frac{\text{min}}{\text{hr}}\right)\left(8.34\ \frac{\text{lbs}}{\text{gal}}\right)\left(\frac{4}{100}\right)$$

= **2602 lbs/hr Solids Loading Rate**

Example 6: (Belt Filter Press Dewatering)

❑ The sludge feed to a belt filter press is 150 gpm, If the total suspended solids concentration of the feed is 45,000 mg/*L*, what is the lbs/hr solids loading rate on the belt filter press?

Since TSS concentration is expressed as mg/*L* in this problem, the equation using mg/*L* will be used:

$$\text{Solids Loading Rate, lbs/hr} = \frac{(\text{mg/}L\ \text{TSS})(\text{MGD Sludge Feed})\left(8.34\ \frac{\text{lbs}}{\text{gal}}\right)}{24\ \text{hrs/day}}$$

Note that the feed rate, 150 gpm, must be expressed in terms of MGD*: $\frac{(150\ \text{gpm})(1440\ \text{min/day})}{1{,}000{,}000} = 0.216\ \text{MGD}$

$$\text{Solids Loading Rate, lbs/hr} = \frac{(45{,}000\ \text{mg/}L)(0.216\ \text{MGD})(8.34\ \text{lbs/gal})}{24\ \text{hrs/day}}$$

= **3378 lbs/hr Solids Loading Rate**

* Refer to Chapter 8 in *Basic Math Concepts* for a review of flow conversions.

Example 7: (Belt Filter Press Dewatering)

❑ The flocculant concentration for a belt filter press is 1% (10,000 mg/*L*). If the flocculant feed rate is 2 gpm, what is the flocculant feed rate in lbs/hr?

First calculate lbs/day flocculant using the mg/*L* to lbs/day calculation. Note that the gpm feed flow must be expressed as MGD feed flow:

$$\frac{(2 \text{ gpm})(1440 \text{ min/day})}{1{,}000{,}000} = 0.00288 \text{ MGD}$$

$$\text{Flocculant Feed, lbs/day} = (\text{mg/}L \text{ Flocc.}) (\text{MGD Feed Rate}) (8.34 \text{ lbs/gal})$$

$$= (10{,}000 \text{ mg/}L) (0.00288 \text{ MGD}) (8.34 \text{ lbs/gal})$$

$$= 240 \text{ lbs/day}$$

Then convert lbs/day flocculant to lbs/hr:

$$\frac{240 \text{ lbs/day}}{24 \text{ hrs/day}} = \boxed{10 \text{ lbs/hr}}$$

Example 8: (Belt Filter Press Dewatering)

❑ Using sludge and flocculant data from Examples 6 and 7, respectively, calculate the flocculant dose in lbs per ton of solids treated. (3378 lbs/hr solids treated; 10 lbs/hr flocculant used.)

First convert lbs/hr solids loading to tons/hr solids loading:

$$\frac{3378 \text{ lbs/hr}}{2000 \text{ lbs/ton}} = 1.69 \text{ tons/hr}$$

Now calculate lbs flocculant per ton of solids treated:

$$\text{Flocculant Dosage, lbs/ton} = \frac{\text{Flocculant, lbs/hr}}{\text{Solids Treated, tons/hr}}$$

$$= \frac{10 \text{ lbs/hr}}{1.69 \text{ tons/hr}}$$

$$= \boxed{5.9 \text{ lbs/ton}}$$

FLOCCULANT FEED RATE

The flocculant feed rate may be calculated like all other mg/*L* to lbs/day calculations,* and then converted to lbs/hr feed rate, as follows:

$$(\text{mg/}L \text{ Flocc.}) (\text{MGD Feed Rate}) (8.34 \text{ lbs/gal}) = \text{Floccul. Feed, lbs/day}$$

Or

$$\frac{\text{Flocc. Feed, lbs/day}}{24 \text{ hrs/day}} = \text{Flocc. Feed, lbs/hr}$$

These two equations can be combined into one equation, as follows:

$$\frac{(\text{mg/}L \text{ Flocc.}) (\text{MGD Feed Rate}) (8.34 \text{ lbs/gal})}{24 \text{ hrs/day}} = \text{Floccul. Feed, lbs/day}$$

FLOCCULANT DOSAGE, lbs/ton

The flocculant dosage is sometimes expressed as lbs/ton (pounds of flocculant per ton of solids treated) for use in cost and filter performance considerations. Once the solids loading rate (tons/hr) and flocculant feed rate (lbs/hr) have been calculated, the flocculant dose in lbs/ton can be determined. Simply place lbs/hr flocculant in the numerator, and tons/hr solids in the denominator. The units of the answer, lbs/ton, can be verified by dimensional analysis.**

$$\text{Flocc. Dosage, lbs/ton} = \frac{\text{Floccul., lbs/hr}}{\text{Sol. Treated, ton/hr}}$$

---

* For a review of mg/*L* to lbs/day calculations, refer to Chapter 3.

** Dimensional analysis is described in Chapter 15 of *Basic Math Concepts*.

TOTAL SUSPENDED SOLIDS

The feed sludge solids are comprised of two types of solids:

- Suspended solids, and
- Dissolved solids.

Suspended solids either float on the surface or are suspended in the wastewater. Some of these particles would settle given quiet conditions such as a clarifier. Others are too tiny or light to settle. Most suspended solids can be removed by laboratory filtering. The solids removed by filtering are called **non-filterable residue** because they do not pass through the filter. Suspended solids can be flocculated and removed by belt filter press as cake.

Dissolved solids are the other type of solids in the sludge.* These solids are dissolved in the water and therefore do not settle out of the wastewater. Because they are dissolved, they pass right through a laboratory filter, and are termed **filterable residue**. Dissolved solids cannot be removed by a belt filter press.

Two simple laboratory tests can be used to estimate the total suspended solids concentration of the feed sludge to the belt filter press:**

1. **Total residue test**—This measures both suspended and dissolved solids concentrations.
2. **Total filterable residue test**—This measures only the dissolved solids concentration.

By subtracting the total filterable residue from the total residue, the result is the total non-filterable residue (total suspended solids), as shown in the box at the top of this page.

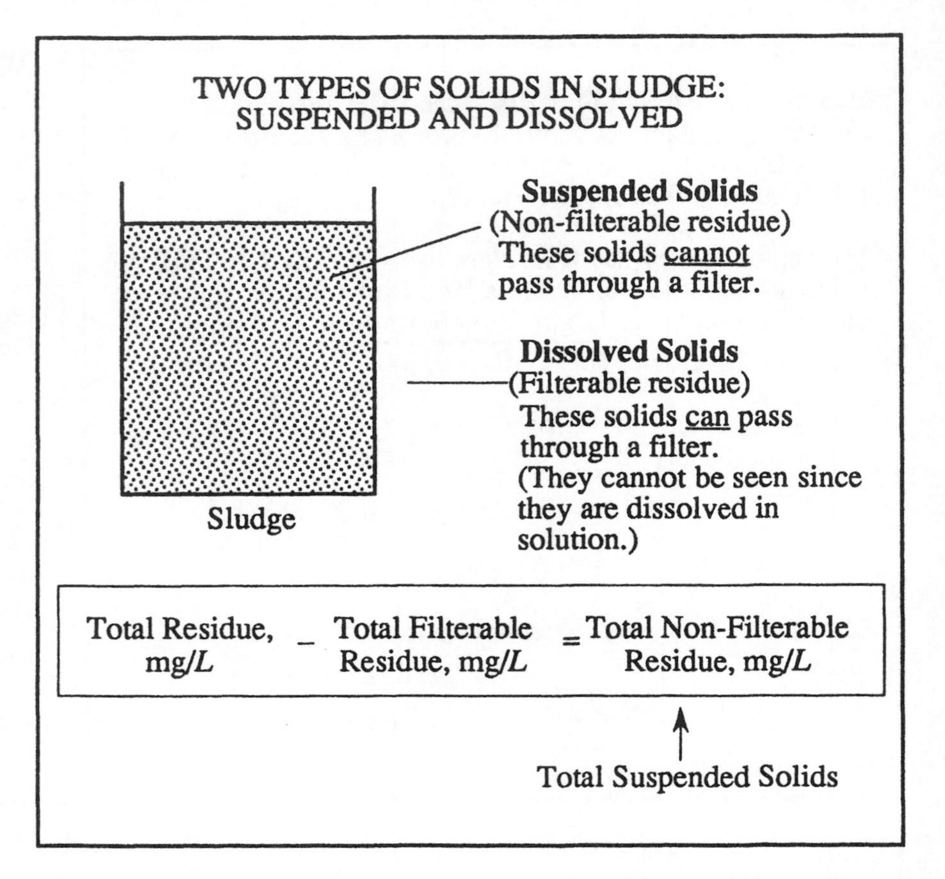

Example 9: (Belt Filter Press Dewatering)

❑ Laboratory tests indicate that the total residue portion of a feed sludge sample is 24,000 mg/L. The total filterable residue is 750 mg/L. On this basis what is the estimated total suspended solids concentration of the sludge sample?

Total Residue, mg/L − Total Filterable Residue, mg/L = Total Non-Filterable Residue, mg/L

24,000 mg/L − 750 mg/L = **23,250 mg/L Total SS**

* Dissolved solids are considered negligible in a normal dewatered sludge cake.

** Total non-filterable residue can be determined directly but the test takes longer than using the total residue and total filterable residue method.

Example 10: (Belt Filter Press Dewatering)

❑ Given the solids concentrations below, calculate the percent solids recovery of the belt filter press.

Feed Sludge TSS, %—3% : X_S
Return Flow TSS, %—0.04% : X_R
Cake TS, %—15% : X_C

$$\% \text{ Recovery} = \frac{(X_C)(X_S - X_R)}{(X_S)(X_C - X_R)} \times 100$$

$$= \frac{(15)(3 - 0.04)}{(3)(15 - 0.04)} \times 100$$

$$= \frac{44.4}{44.88} \times 100$$

$$= \boxed{99\%}$$

Example 11: (Belt Filter Press Dewatering)

❑ Given the solids concentrations below, calculate the percent solids recovery of the belt filter press.

Feed Sludge TSS, %—2.5% : X_S
Return Flow TSS, %—0.09% : X_R
Cake TS, %—14% : X_C

$$\% \text{ Recovery} = \frac{(X_C)(X_S - X_R)}{(X_S)(X_C - X_R)} \times 100$$

$$= \frac{(14)(2.5 - 0.09)}{(2.5)(14 - 0.09)} \times 100$$

$$= \frac{33.74}{34.78} \times 100$$

$$= \boxed{97\%}$$

PERCENT RECOVERY

The percent solids recovery is one measure of belt filter press efficiency. It is a measure of the percent of solids removed from the sludge feed and "captured" in the cake. This calculation is sometimes referred to as percent solids capture.

If the return flow rate can be measured, the percent recovery can be calculated as follows:

Simplified Equation:

$$\% \text{ Recov.} = \frac{\text{Cake Solids, lbs/hr}}{\text{Feed Solids, lbs/hr}} \times 100$$

Expanded Equation:

$$\% \text{ Recov.} = \frac{\text{Feed Sol., lbs/hr} - \text{Return Sol., lbs/hr}}{\text{Feed Solids, lbs/hr}} \times 100$$

When return flow rate information is not available, the percent recovery can still be calculated using percent solids information, as using the following equation:

$$\% \text{ Recovery} = \frac{\frac{(X_C)}{100}\frac{(X_S - X_R)}{100}}{\frac{(X_S)}{100}\frac{(X_C - X_R)}{100}} \times 100$$

The hundreds in the numerator and denominator divide out, leaving:

$$\% \text{ Recovery} = \frac{(X_C)(X_S - X_R)}{(X_S)(X_C - X_R)} \times 100$$

Where:
X_S = Sludge TSS, %
X_R = Return Flow TSS, % and
X_C = Cake TS, %

* Flow rates can all be expressed as lbs/min, if desired. It is necessary, however, that flow rates are expressed in the same terms.

17.3 VACUUM FILTER DEWATERING

FILTER LOADING

The filter loading for vacuum filters is a measure of lbs/hr of solids applied per square foot of drum surface area*. The equation to be used in this calculation is shown to the right.

The vacuum filter is generally operated on an intermittent basis. Therefore, when converting from lbs/day solids to lbs/hr solids, as shown in the expanded equation, a different factor than 24 hrs/day is used. In fact, to convert from lbs/day to lbs/hr, **use the actual operating hours (hrs/day) instead of 24 hrs/day.**

The dry weight of the solids used in the filter loading calculation must include the weight of chemicals that are added to the sludge.

TO FIND THE DRUM SURFACE AREA THINK OF "UNROLLING" THE DRUM

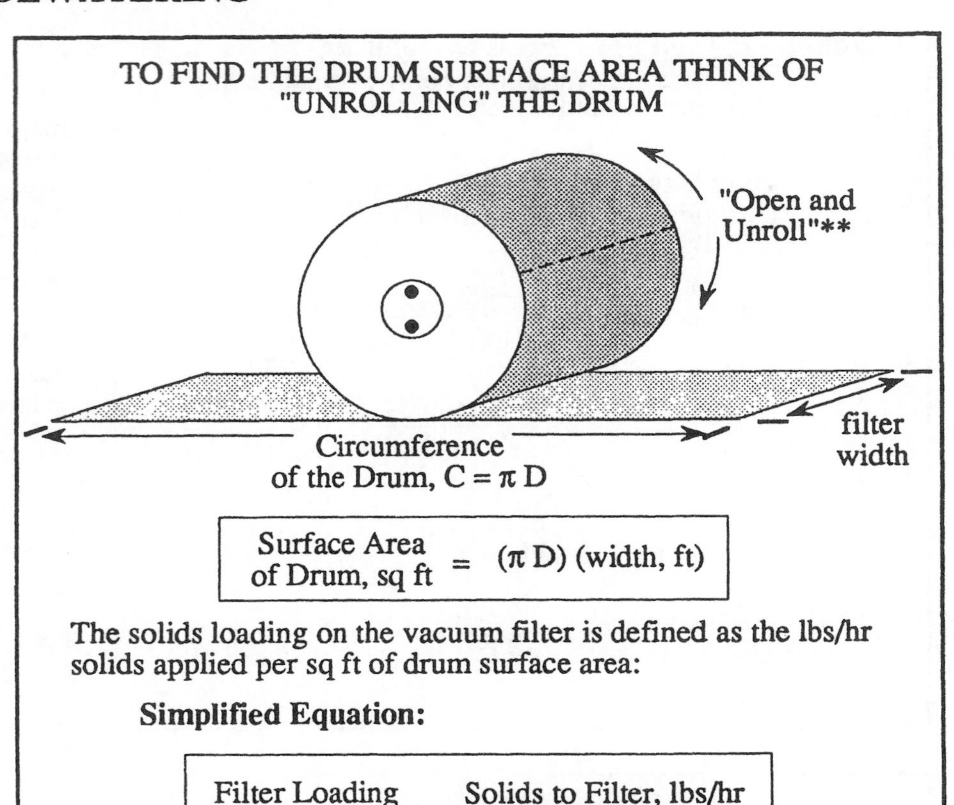

$$\text{Surface Area of Drum, sq ft} = (\pi D)(\text{width, ft})$$

The solids loading on the vacuum filter is defined as the lbs/hr solids applied per sq ft of drum surface area:

Simplified Equation:

$$\text{Filter Loading lbs/hr/sq ft} = \frac{\text{Solids to Filter, lbs/hr}}{\text{Surface Area, sq ft}}$$

Expanded Equation:

$$\text{Filter Loading lbs/hr/sq ft} = \frac{\dfrac{\text{Sol. to Filter, lbs/day}}{\text{Filter Oper., hrs/day}}}{(\pi D)(\text{width, ft})}$$

Example 1: (Vacuum Filter Dewatering)

❑ Digested sludge is applied to a vacuum filter at a rate of 80 gpm, with a solids concentration of 5%. If the vacuum filter has a surface area of 320 sq ft, what is the filter loading in lbs/hr/sq ft?

$$\text{Filter Loading lbs/hr/sq ft} = \frac{(\text{gpm sludge})\left(\dfrac{60\ \text{min}}{\text{hr}}\right)\left(\dfrac{8.34\ \text{lbs}}{\text{gal}}\right)\left(\dfrac{\%\ \text{Sol.}}{100}\right)}{\text{Surface Area, sq ft}}$$

$$= \frac{(80\ \text{gpm})\left(\dfrac{60\ \text{min}}{\text{hr}}\right)\left(\dfrac{8.34\ \text{lbs}}{\text{gal}}\right)\left(\dfrac{5}{100}\right)}{320\ \text{sq ft}}$$

$$= \boxed{6.3\ \frac{\text{lbs/hr}}{\text{sq ft}}}$$

---

* For a review of area calculations, refer to Chapter 10 in *Basic Math Concepts*.

** The concept of "opening and unrolling" the drum is for illustration purposes only, to demonstrate the basis of the equation.

FILTER YIELD

Filter yield is one of the most common measures of vacuum filter performance. It is the lbs/hr of dry solids in the dewatered sludge (cake) discharged per sq ft of filter area. It can be calculated directly, using lbs/hr wet cake flow, percent cake solids, and filter area data:

Simplified Equation:

$$\text{Filter Yield, lbs/hr/sq ft} = \frac{\text{Dry Sol. in Cake, lbs/hr}}{\text{Filter Area, sq ft}}$$

Expanded Equation:

$$\text{Filter Yield, lbs/hr/sq ft} = \frac{(\text{Wet Cake Flow, lbs/hr})\left(\frac{\text{\% Sol. in Cake}}{100}\right)}{\text{Filter Area, sq ft}}$$

Example 2 illustrates this type of calculation.

The filter yield may also be calculated using filter loading (lbs/hr/sq ft), and percent recovery* data. This is because filter loading indicates the lbs/hr/sq ft solids applied to the filter and the percent recovery indicates what percent of those solids are actually recovered in the sludge cake.

Simplified Equation:

$$\text{Fil. Yield, lbs/hr/sq ft} = (\text{Filter Loading, lbs/hr/sq ft})\left(\frac{\text{\% Recov.}}{100}\right)$$

Expanded Equation:

$$\text{Fil. Yield, lbs/hr/sq ft} = \frac{\frac{\text{Sol. to Fil., lbs/day}}{\text{Fil. Oper., hrs/day}}}{\text{Filter Area, sq ft}}\ \frac{(\text{\% Rec.})}{100}$$

Example 3 illustrates a calculation of this type.

Example 2: (Vacuum Filter Dewatering)

❑ The wet cake flow from a vacuum filter is 8000 lbs/hr. If the filter area is 310 sq ft and the percent solids in the cake is 30%, what is the filter yield in lbs/hr/sq ft?

$$\text{Filter Yield lbs/hr/sq ft} = \frac{(\text{Wet Cake Flow, lbs/hr})\left(\frac{\text{\% Solids in Cake}}{100}\right)}{\text{Filter Area, sq ft}}$$

$$= \frac{(8000\text{ lbs/hr})\left(\frac{30}{100}\right)}{310\text{ sq ft}}$$

$$= \boxed{7.7\ \frac{\text{lbs/hr}}{\text{sq ft}}}$$

Example 3: (Vacuum Filter Dewatering)

❑ The total pounds of dry solids pumped to a vacuum filter during a 24-hour period is 16,800 lbs/day. The vacuum filter is operated 8 hrs/day. If the percent solids recovery is 96% and the filter area is 300 sq ft, what is the filter yield in lbs/hr/sq ft?

$$\text{Filter Yield lbs/hr/sq ft} = \frac{\frac{\text{Sol. to Filter, lbs/day}}{\text{Fil. Oper., hrs/day}}}{\text{Filter Area, sq ft}}\ \frac{(\text{\% Recovery})}{100}$$

$$= \frac{\frac{16{,}800\text{ lbs/day}}{8\text{ hrs/day}}}{300\text{ sq ft}}\ \frac{(96)}{100}$$

$$= \frac{2100\text{ lbs/hr}}{300\text{ sq ft}}\ \frac{(96)}{100}$$

$$= \boxed{6.7\text{ lbs/hr/sq ft}}$$

* The percent recovery calculations are described on the following two-pages.

FILTER OPERATING TIME

The filter operating time required to process a given lbs/day solids can be calculated using the filter yield equation described on the previous page. Simply fill out the equation with the given data, leaving filter operation, hrs/day, as the unknown factor.* Examples 4 and 5 illustrate this calculation.

Example 4: (Vacuum Filter Dewatering)

❑ A total of 5000 lbs/day primary sludge solids are to be processed by a vacuum filter. The vacuum filter yield is 2.1 lbs/hr/sq ft. The solids recovery is 96%. If the area of the filter is 200 sq ft, how many hours per day must the vacuum filter remain in operation to process these solids?

$$\text{Filter Yield, lbs/hr/sq ft} = \frac{\dfrac{\text{Sol. to Filter, lbs/day}}{\text{Fil. Oper., hrs/day}}}{\text{Filter Area, sq ft}} \cdot \frac{(\%\ \text{Recovery})}{100}$$

$$2.1\ \text{lbs/hr/sq ft} = \frac{\dfrac{5000\ \text{lbs/day}}{x\ \text{hrs/day Oper.}}}{200\ \text{sq ft}} \cdot \frac{(96)}{100}$$

$$2.1\ \text{lbs/hr/sq ft} = \frac{(5000\ \text{lbs/day})}{x\ \text{hrs/day}} \cdot \frac{(1)}{200\ \text{sq ft}} \cdot \frac{(96)}{100}$$

$$x = \frac{(5000)(1)(96)}{(2.1)(200)(100)}$$

$$x = \boxed{11.4\ \text{hrs/day Operation}}$$

Example 5: (Vacuum Filter Dewatering)

❑ The total primary sludge solids to be processed by a vacuum filter is 4200 lbs/day. The vacuum filter yield is 1.8 lbs/hr/sq ft. The solids recovery is 92%. If the area of the filter is 250 sq ft, how many hours per day must the vacuum filter remain in operation to process the primary sludge?

$$\text{Filter Yield, lbs/hr/sq ft} = \frac{\dfrac{\text{Sol. to Filter, lbs/day}}{\text{Fil. Oper., hrs/day}}}{\text{Filter Area, sq ft}} \cdot \frac{(\%\ \text{Recovery})}{100}$$

$$1.8\ \text{lbs/hr/sq ft} = \frac{\dfrac{4200\ \text{lbs/day}}{x\ \text{hrs/day Oper.}}}{250\ \text{sq ft}} \cdot \frac{(92)}{100}$$

$$1.8\ \text{lbs/hr/sq ft} = \frac{(4200\ \text{lbs/day})}{x\ \text{hrs/day}} \cdot \frac{(1)}{250\ \text{sq ft}} \cdot \frac{(92)}{100}$$

$$x = \frac{(4200)(1)(92)}{(1.8)(250)(100)}$$

$$x = \boxed{8.6\ \text{hrs/day Operation}}$$

* For a review of solving for the unknown value, refer to Chapter 2 in *Basic Math Concepts*.

Example 6: (Vacuum Filter Dewatering)

❑ The sludge feed to a vacuum filter is 3540 lbs/hr, with a solids content of 5.5%. If the wet cake flow is 625 lbs/hr with a 30% solids content, what is the percent solids recovery?

$$\text{\% Solids Recovery} = \frac{(\text{Wet Cake Flow, lbs/hr})\left(\frac{\text{\% Solids in Cake}}{100}\right)}{(\text{Sludge Feed, lbs/day})\left(\frac{\text{\% Sol. in Feed}}{100}\right)} \times 100$$

$$= \frac{(625 \text{ lbs/hr})\left(\frac{30}{100}\right)}{(3540 \text{ lbs/hr})\left(\frac{5.5}{100}\right)} \times 100$$

$$= \frac{188 \text{ lbs/hr}}{195 \text{ lbs/hr}} \times 100$$

$$= \boxed{96\% \text{ Solids Recovery}}$$

Example 7: (Vacuum Filter Dewatering)

❑ The sludge feed to a vacuum filter is 98,000 lbs/day, with a solids content of 6%. If the wet cake flow is 19,500 lbs/day, with a 28% solids content, what is the percent solids recovery?

$$\text{\% Solids Recovery} = \frac{(\text{Wet Cake Flow, lbs/day})\left(\frac{\text{\% Solids in Cake}}{100}\right)}{(\text{Sludge Feed, lbs/day})\left(\frac{\text{\% Sol. in Feed}}{100}\right)} \times 100$$

$$= \frac{(19{,}500 \text{ lbs/day})\left(\frac{28}{100}\right)}{(98{,}000 \text{ lbs/day})\left(\frac{6}{100}\right)} \times 100$$

$$= \frac{5460 \text{ lbs/day}}{5880 \text{ lbs/day}} \times 100$$

$$= \boxed{93\% \text{ Solids Recovery}}$$

* Note that the 100 in the numerator and denominator of the fraction can be divided out. However the 100 shown to the right must not be cancelled.

** *Operations Manual—Sludge Handling and Conditioning*, EPA, 1978.

PERCENT SOLIDS RECOVERY

The function of the vacuum filtration process is to separate the solids from the liquids in the sludge being processed. Therefore, the efficiency of the process is the percent of feed solids "recovered" in the filter cake. This is sometimes referred to as the percent solids capture.

Simplified Equation:

$$\text{\% Solids Recovery} = \frac{\text{Sol. in Cake, lbs/hr}}{\text{Sol. in Feed, lbs/hr}} \times 100$$

Expanded Equation:*

$$\text{\% Sol. Rec.} = \frac{(\text{Wet Cake Flow, lbs/hr})\left(\frac{\text{\% Sol. in Cake}}{100}\right)}{(\text{Sl. Feed, lbs/hr})\left(\frac{\text{\% Sol. in Feed}}{100}\right)} \times 100$$

When the wet cake flow rate is not known, another equation may be used. This equation uses only suspended solids data to determine solids recovery:**

$$\text{\% Recov} = \frac{(\text{Cake Sol.,\%})(\text{Feed Sol.,\%} - \text{Filtrate Sol.,\%})}{(\text{Feed Sol.,\%})(\text{Cake Sol.,\%} - \text{Fil. Sol.,\%})} \times 100$$

If chemicals are added to the vacuum filter, filtrate solids must be corrected as shown below. This is because the filtrate has been diluted by additional water from the chemical and dilution water feeds. The measured filtrate solids, are actually less than they would have been had the additional water not been added.

First calculate the correction factor:

$$\frac{(\text{Feed rate, lbs/hr}) + (\text{Chem. Flow, lbs/hr}) + (\text{Dilution Wat., lbs/hr})}{\text{Feed Rate, lbs/hr}}$$

Then multiply the measured filtrate solids by the correction factor:

$$\text{Corrected Filtrate Sol.} = (\text{Meas. Filtrate Solids})(\text{Corr. Factor})$$

17.4 SAND DRYING BEDS CALCULATIONS

TOTAL SLUDGE APPLIED

The total gallons of sludge applied to sand drying beds may be calculated using the dimensions of the bed and depth of sludge applied, as shown in the diagram to the right.*

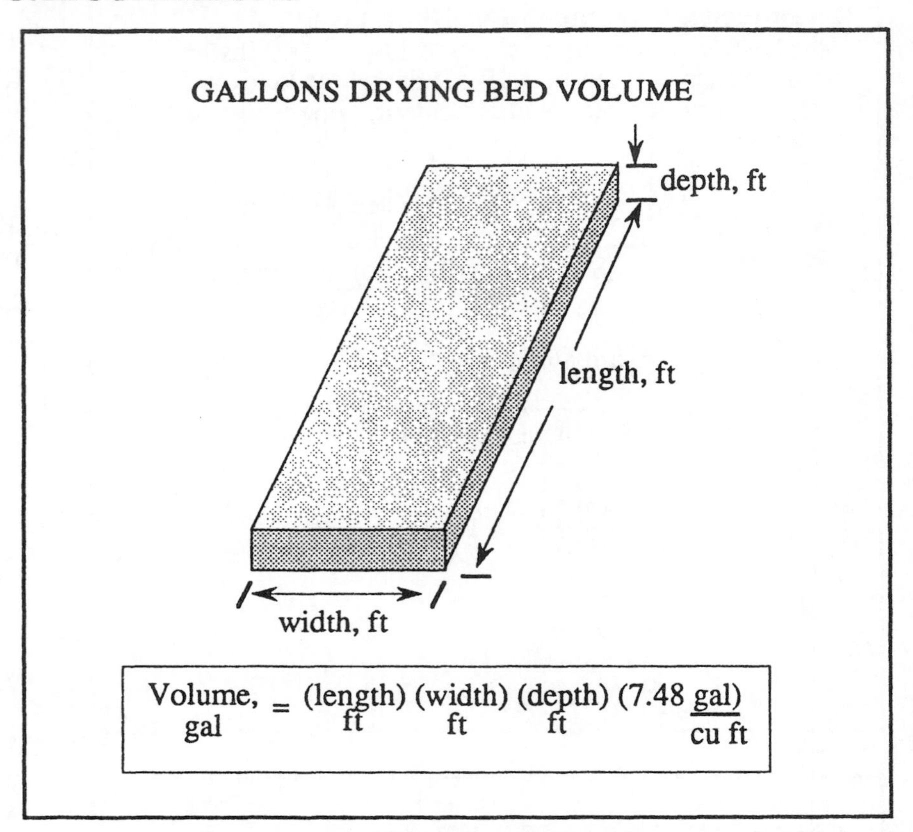

Example 1: (Sand Drying Beds)

❑ A drying bed is 225 ft long and 25 ft wide. If sludge is applied to a depth of 4 inches, how many gallons of sludge are applied to the drying bed?

$$\frac{4\text{ in}}{12\text{ in/ft}} = 0.33\text{ ft}$$

225 ft

25 ft

$$\text{Volume, gal} = (l)\,(w)\,(d)\,\left(7.48\,\frac{\text{gal}}{\text{cu ft}}\right)$$

$$= (225\text{ ft})\,(25\text{ ft})\,(0.33\text{ ft})\,\left(7.48\,\frac{\text{gal}}{\text{cu ft}}\right)$$

$$= \boxed{13{,}885\text{ gal}}$$

* For a review of volume calculations, refer to Chapter 11 in *Basic Math Concepts*.

Example 2: (Sand Drying Beds)

❑ A sludge bed is 200 ft long and 20 ft wide. A total of 171,600 lbs of sludge are applied each application of the sand drying bed. The sludge has a solids content of 5%. If the drying and removal cycle requires 21 days, what is the solids loading rate in lbs/yr/sq ft?

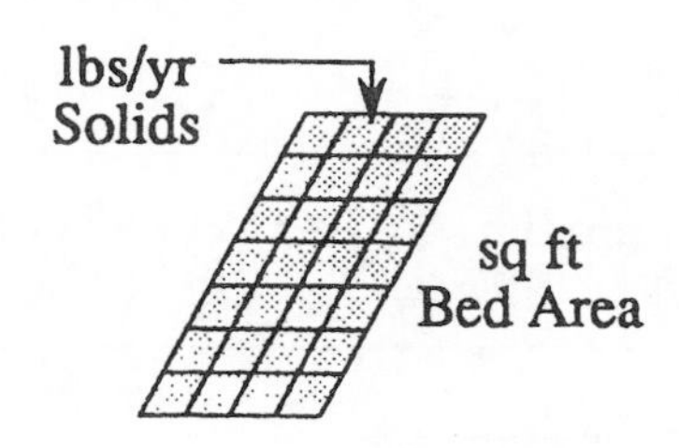

$$\text{Solids Loading Rate, lbs/yr/sq ft} = \frac{\frac{\text{(Sludge Applied, lbs)}}{\text{Days of Applic.}}\ \frac{(365\ \text{days})}{\text{yr}}\ \frac{(\%\ \text{Sol.})}{100}}{\text{Bed Area, sq ft}}$$

$$= \frac{\frac{(171{,}600\ \text{lbs})}{21\ \text{days}}\ \frac{(365\ \text{days})}{\text{yr}}\ \frac{(5)}{100}}{(200\ \text{ft})(20\ \text{ft})}$$

$$= \boxed{37.3\ \text{lbs/yr/sq ft}}$$

Example 3: (Sand Drying Beds)

❑ A sludge drying bed is 180 ft long and 25 ft wide. The sludge is applied to a depth of 8 inches. The solids concentration of the sludge is 4.5%. If the drying and removal cycle requires 18 days, what is the solids loading rate to the beds in lbs/yr/sq ft?

First calculate the lbs of sludge applied: (8 in. = 0.67 ft)

$$(180\ \text{ft})\ (25\ \text{ft})\ (0.67\ \text{ft})\ (7.48\ \frac{\text{gal}}{\text{cu ft}})\ (8.34\ \frac{\text{lbs}}{\text{gal}}) = \boxed{188{,}085\ \text{lbs}}$$

Then determine the solids loading rate:

$$\text{Solids Loading Rate, lbs/yr/sq ft} = \frac{\frac{(188{,}085\ \text{lbs})}{18\ \text{days}}\ \frac{(365\ \text{days})}{\text{yr}}\ \frac{(4.5)}{100}}{(180\ \text{ft})(25\ \text{ft})}$$

$$= \frac{171{,}628\ \text{lbs/yr}}{4500\ \text{sq ft}}$$

$$= \boxed{38.1\ \text{lbs/yr/sq ft}}$$

SOLIDS LOADING RATE

The sludge loading rate may be expressed as lbs/yr/sq ft. This loading rate is dependent on:

- Sludge applied per application, lbs,
- Percent solids concentration,
- Cycle length, and
- Square feet of sand bed area.

The equation for sludge loading rate is given below.

Simplified Equation:

$$\text{Solids Loading Rate, lbs/yr/sq ft} = \frac{\frac{\text{(Sludge Applied,)}}{\text{lbs/yr}}\ \frac{(\%\ \text{Sol.})}{100}}{\text{Bed Area, sq ft}}$$

Expanded Equation:

$$\text{Solids Loading Rate, lbs/yr/sq ft} = \frac{\frac{\text{lbs Sludge Applied}}{\text{Days of Applic.}}\ \frac{(365)}{\text{days/yr}}\ \frac{(\%\ \text{Sol.})}{100}}{(\text{length, ft})\ (\text{width, ft})}$$

Note that the first term in the numerator of the expanded equation is simply lbs/day sludge:

$$\frac{\text{lbs Sludge}}{\text{Days of Applic}} = \frac{\text{lbs Sludge}}{\text{day}}$$

The lbs/day sludge is then multiplied by 365 days per year and percent solids.

SLUDGE WITHDRAWAL TO DRYING BEDS

Pumping digested sludge to drying beds is one method among many for dewatering sludge, thus making the dried sludge useful as a soil conditioner or for other such commercial uses. Depending upon the climate of a region (and the related evaporation rate), the drying bed depth may range from 8 to 18 inches. Therefore, the area covered by these drying beds may be substantial. For this reason, the use of drying beds is more common for medium or small plants than for large municipal plants.

When calculating sludge withdrawal to drying beds, remember that the volume of sludge withdrawn from the digester will be pumped to the drying beds. Therefore, in the equation, set the volume of one equal to the volume of the other, as shown in the diagram to the right. **Be sure that the volume terms match on both sides of the equation (cu ft or gallons).**

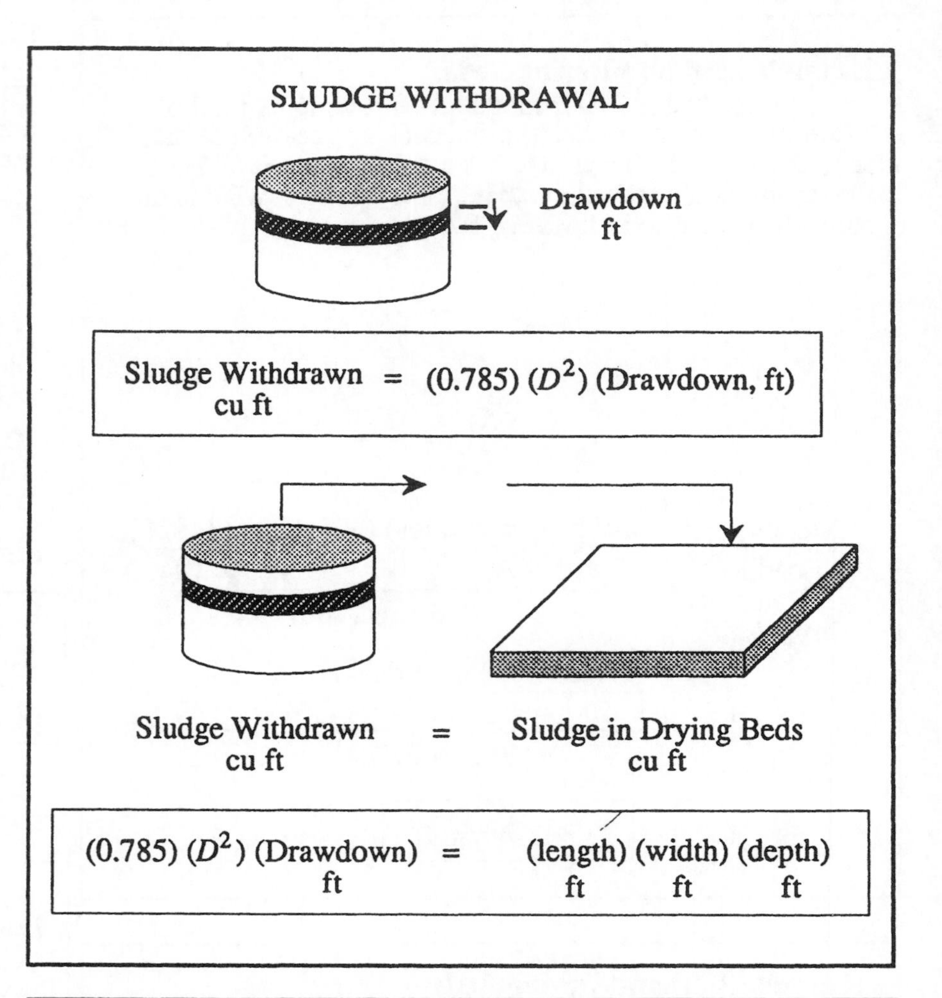

Example 4: (Sand Drying Beds)

❑ Sludge is withdrawn from a digester that has a diameter of 50 feet. If the sludge is drawn down 2 feet, how many cu ft will be sent to the drying beds?

50 ft

2 ft drop

Sludge Withdrawal, cu ft = (0.785) (D^2) (ft drop)

= (0.785) (50 ft) (50 ft) (2 ft)

= **3925 cu ft withdrawn**

---

* For a review of volume calculations, refer to Chapter 11 in *Basic Math Concepts*.

Example 5: (Sand Drying Beds)

❑ A 45-feet diameter digester has a drawdown of 1.5 feet. If the drying bed is 110 ft long and 25 feet wide, (a) how many feet deep will the drying bed be as a result of the drawdown?

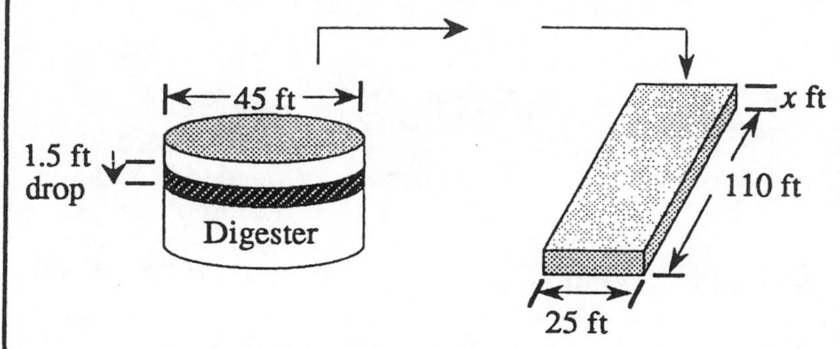

Sludge Withdrawn, cu ft = Sludge in Drying Beds, cu ft

$$(0.785)\,(45\text{ ft})\,(45\text{ ft})\,(1.5\text{ ft}) = (110\text{ ft})\,(25\text{ ft})\,(x\text{ ft})$$

$$\frac{(0.785)\,(45)\,(45)\,(1.5)}{(110)\,(25)} = x$$

$$\boxed{0.87\text{ ft}} = x$$

Example 6: (Sand Drying Beds)

❑ A sand drying bed is 150 ft long and 20 ft wide. If a 50-ft diameter digester has a drawdown of 1 ft (a) how many feet deep will the drying bed be as a result of the drawdown, and (b) how many inches is this?

(a)

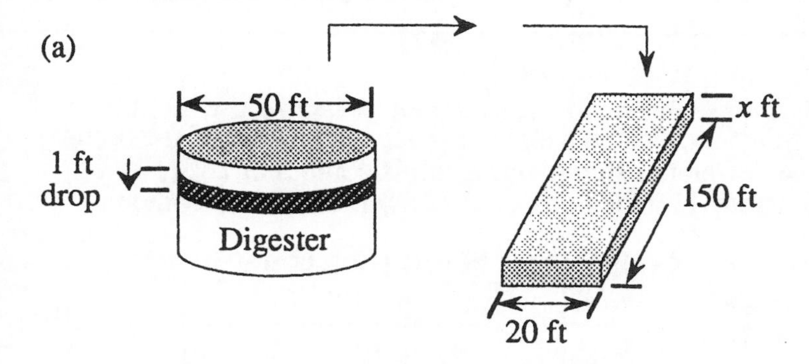

$$(0.785)(50\text{ ft})(50\text{ ft})(1\text{ ft}) = (150\text{ ft})(20\text{ ft})(x\text{ ft})$$

$$\frac{(0.785)(50)(50)(1)}{(150)(20)} = x$$

$$0.65\text{ ft} = x$$

(b) Convert 0.65 ft to inches:

$$(0.65\text{ ft})\left(\frac{12\text{ in.}}{\text{ft}}\right) = \boxed{8\text{ inches}}$$

17.5 COMPOSTING CALCULATIONS

BLENDING DEWATERED SLUDGE WITH COMPOSTED SLUDGE

Dewatered sludge is blended with previously composted sludge or a bulking agent such as straw, sawdust, or wood shavings to improve aeration during the composting process.

When blending composted material with dewatered sludge, it is similar to blending two different percent solids sludges. **The percent solids (or percent moisture) content of the mixture will always fall somewhere between the percent solids (or percent moisture) concentrations of the two materials being mixed.**

The equation used in calculating the percent solids of the mixture is shown to the right. This same basic equation has been described in Chapter 14 (mixing different percent solutions) and Chapter 15 (mixing different percent sludges). This calculation is the same with **two exceptions**:

- Rather than using percent solids in the equation, **percent moisture is used**. Given percent solids, the percent moisture can be determined by subtracting the solids content from 100%:

$$\text{Moisture Content, \%} = 100\% - \text{Solids Content, \%}$$

- In these calculations the desired moisture content of the mixture is known. The **unknown value is lbs/day compost**. It is recommended that the mixture equation be used as is, filling in given data, then solving for the unknown value. Example 1 illustrates the use of the equation when percent moisture of the mixture is unknown. Example 2 illustrates use of the equation when lbs/day compost is unknown.

MIXING DIFFERENT PERCENT MOISTURE MATERIALS

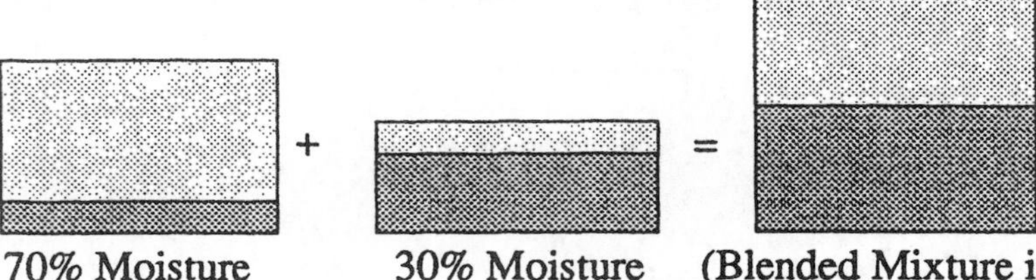

70% Moisture   30% Moisture   (Blended Mixture in this example will have a moisture content between 70% and 30%)

Simplified Equation:

$$\text{\% Moisture of Compost Mixture} = \frac{\text{Moisture in Mixture, lbs/day}}{\text{Compost Mixture, lbs/day}} \times 100$$

Expanded Equations:

$$\text{\% Moisture of Compost Mixture} = \frac{\text{Moisture in Sludge, lbs/day} + \text{Moisture in Compost, lbs/day}}{\text{Sludge, lbs/day} + \text{Compost, lbs/day}} \times 100$$

Or

$$\text{\% Moist. of Mixture} = \frac{(\text{Sludge) lbs/day}\ \frac{(\text{\% Moist})}{100} + (\text{Compost) lbs/day}\ \frac{(\text{\% Moist.})}{100}}{\text{Sludge, lbs/day} + \text{Compost, lbs/day}} \times 100$$

Example 1: (Composting)

❑ If 4000 lbs/day dewatered sludge is mixed with 3000 lbs/day compost, what is the percent moisture of the blend? The dewatered sludge has a solids content of 25% (75% moisture) and the compost has a 30% moisture content.

$$\text{\% Moist. of Mixture} = \frac{(\text{Sludge) lbs/day}\ \frac{(\text{\% Moist})}{100} + (\text{Compost) lbs/day}\ \frac{(\text{\% Moist.})}{100}}{\text{Sludge, lbs/day} + \text{Compost, lbs/day}} \times 100$$

$$= \frac{(4000\text{ lbs/day})\ \frac{(75)}{100} + (3000\text{ lbs/day})\ \frac{(30)}{100}}{4000\text{ lbs/day} + 3000\text{ lbs/day}} \times 100$$

$$= \frac{3000\text{ lbs/day} + 900\text{ lbs/day}}{7000\text{ lbs/day}}$$

$$= \boxed{56\%}$$

Example 2: (Composting)

☐ A treatment plant produces a total of 5500 lbs/day of dewatered digested primary sludge. The dewatered sludge has a solids concentration of 28%. Final compost to be used in blending has a moisture content of 32%. How much compost (lbs/day) must be blended with the dewatered sludge to produce a mixture with a moisture content of 50%?

The dewatered sludge has a solids content of 28%; therefore the moisture content is 72%.

$$\text{\% Moist. of Mixture} = \frac{(\text{Sludge, lbs/day})\left(\frac{\text{\% Moist}}{100}\right) + (\text{Compost, lbs/day})\left(\frac{\text{\% Moist.}}{100}\right)}{\text{Sludge, lbs/day} + \text{Compost, lbs/day}} \times 100$$

$$50 = \frac{(5500 \text{ lbs/day})\left(\frac{72}{100}\right) + (x \text{ lbs/day})\left(\frac{32}{100}\right)}{5500 \text{ lbs/day} + x \text{ lbs/day}} \times 100$$

First, move the 100 to the left side of the equation; then simplify terms before rearranging to solve for x*:

$$\frac{50}{100} = \frac{3960 \text{ lbs/day} + (x \text{ lbs/day})(0.32)}{5500 \text{ lbs/day} + x \text{ lbs/day}}$$

$$0.5\,(5500 + x) = (3960 + 0.32x)$$

$$2750 + 0.5x = 3960 + 0.32x$$

Next, group x terms on the left side of the equation, and numbers on the right side:

$$0.5x - 0.32x = 3960 - 2750$$

$$0.18x = 1210$$

$$x = \frac{1210}{0.18}$$

$$x = \boxed{\begin{array}{c}\text{6722 lbs/day} \\ \text{Compost Req'd}\end{array}}$$

---

* For a review of solving for the unknown value, refer to Chapter 2 in *Basic Math Concepts.*

BLENDING DEWATERED SLUDGE WITH WOOD CHIPS

When blending dewatered sludge and wood chips, the same basic "mixing equation" may be used. The basis of the percent solids calculation is simply the **pounds of solids** in the blended compost **divided by the total pounds of blended compost,** as shown by the simplified equation. The expanded equations are merely refinements of this basic concept.

The **first expanded equation** identifies the two sources of materials to be blended. Solids from the sludge and solids from the wood chips (numerator) constitute the solids in the blended product. The total pounds of compost blend is comprised of the pounds of sludge and the pounds of wood chips (denominator).

Since the sludge and compost is often mixed on the basis of volumes (cubic yards), the **second expanded equation** gives each component in terms of cubic yards. Then, because percent solids calculations should be made on the basis of weight, a lbs/cu yd factor must accompany each quantity of cubic yards. In this way, each component of the blend is still represented as pounds.

The lbs/cu ft density of the sludge and wood chips (**bulk density**) is generally given for wet sludge and wet wood chips, and not for dry solids, as shown in the second expanded equation. Therefore, the **third expanded equation** includes the sludge and wood chip data with the corresponding bulk density and percent solids factors.

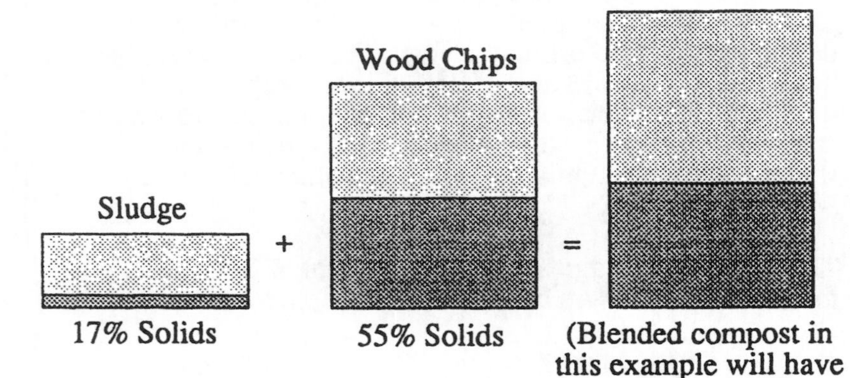

Simplified Equation:

$$\text{\% Solids of Compost Blend} = \frac{\text{Solids in Compost Blend, lbs}}{\text{Compost Blend, lbs}} \times 100$$

Expanded Equations:

$$\text{\% Solids of Compost Blend} = \frac{\text{Solids in Sludge, lbs} + \text{Solids in Wood Chips, lbs}}{\text{Sludge, lbs} + \text{Wood Chips, lbs}} \times 100$$

Or

$$\text{\% Solids of Compost Blend} = \frac{(\text{cu yds})\,\text{Solids in Sludge}\left(\frac{\text{lbs}}{\text{cu yd}}\right) + (\text{cu yds})\,\text{Solids in Wood Chips,}\left(\frac{\text{lbs}}{\text{cu yd}}\right)}{(\text{cu yds})\,\text{Sludge}\left(\frac{\text{lbs}}{\text{cu yd}}\right) + (\text{cu yds})\,\text{W. Chips}\left(\frac{\text{lbs}}{\text{cu yd}}\right)} \times 100$$

Or

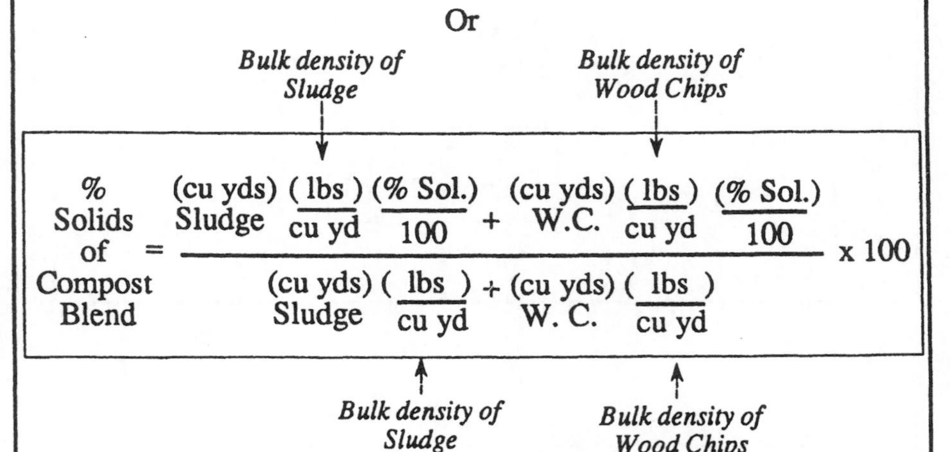

$$\text{\% Solids of Compost Blend} = \frac{(\text{cu yds})\,\text{Sludge}\left(\frac{\text{lbs}}{\text{cu yd}}\right)\left(\frac{\text{\% Sol.}}{100}\right) + (\text{cu yds})\,\text{W.C.}\left(\frac{\text{lbs}}{\text{cu yd}}\right)\left(\frac{\text{\% Sol.}}{100}\right)}{(\text{cu yds})\,\text{Sludge}\left(\frac{\text{lbs}}{\text{cu yd}}\right) + (\text{cu yds})\,\text{W. C.}\left(\frac{\text{lbs}}{\text{cu yd}}\right)} \times 100$$

Expanded Equations—Cont'd:

Cu yds wood chips depends on the mix ratio of wood chips to sludge*

$$\frac{\%\ \text{Solids of Compost Blend}}{} = \frac{(\text{cu yds Sludge})\left(\frac{\text{lbs}}{\text{cu yd}}\right)\left(\frac{\%\ \text{Sol.}}{100}\right) + (\text{cu yds Sludge})(\text{Mix Ratio})\left(\frac{\text{lbs}}{\text{cu yd}}\right)\left(\frac{\%\ \text{Sol.}}{100}\right)}{(\text{cu yds Sludge})\left(\frac{\text{lbs}}{\text{cu yd}}\right) + (\text{cu yds Sludge})(\text{Mix Ratio})\left(\frac{\text{lbs}}{\text{cu yd}}\right)} \times 100$$

Example 3: (Composting)

❑ Compost is to be blended from wood chips and dewatered sludge. The wood chips are to be mixed with 6.98 cu yds of dewatered sludge at a ratio of 3:1*. The solids content of the sludge is 17% and the solids content of the wood chips is 55%. If the bulk density of the sludge is 1685 lbs/cu yd and the bulk density of the wood chips is 750 lbs/cu yd, what is the percent solids of the compost blend?

Use the equation shown above, filling in given data:

$$\%\ \text{Sol. of Comp. Blend} = \frac{(6.98\ \text{cu yds Sludge})\left(1685\frac{\text{lbs}}{\text{cu yd}}\right)\left(\frac{17}{100}\right) + (6.98\ \text{cu yds Sludge})(3)\left(750\frac{\text{lbs}}{\text{cu yd}}\right)\left(\frac{55}{100}\right)}{(6.98\ \text{cu yds Sludge})\left(1685\frac{\text{lbs}}{\text{cu yd}}\right) + (6.98\ \text{cu yds Sludge})(3)\left(750\frac{\text{lbs}}{\text{cu yd}}\right)} \times 100$$

$$= \frac{\text{1999 lbs Solids from the Sludge} + \text{8638 lbs Solids from the Wood Chips}}{\text{11,761 lbs Sludge} + \text{15,705 lbs Wood Chips}} \times 100$$

$$= \frac{\text{10,637 lbs Solids}}{\text{27,466 lbs Compost Blend}} \times 100$$

= **38.7% Solids in Compost Blend**

* This is a volumetric mix ratio of wood chips to sludge.

The bulk density of each material varies, depending on the makeup of the material, including the amount of water in it. The bulk density commonly used for dewatered sludge is 1685 lbs/cu yd. (This is the same density as for water, 62.4 lbs/cu ft or 1685 lbs/cu yd.) The bulk density of wood chips varies from about 550 lbs/cu yd for new wood chips to 750 lbs/cu yd for recycled wood chips.

The percent solids for the dewatered sludge usually ranges from 15 to 25% and the percent solids of the wood chips ranges from 50 to 60%.

The **fourth expanded equation** incorporates the concept of a mix ratio—the ratio of wood chips to sludge. The mix ratio is by volume. A normal mix ratio of wood chips to sludge ranges from 2.6 to 3.6. A mix ratio of 3, for example, would be a mix of 3 cu yds of wood chips to 1 cu yd of sludge. To determine the cubic yards of wood chips, therefore, simply multiply the cubic yards of sludge by the mix ratio:

$$(\text{Sludge cu yds})(\text{Mix Ratio}) = \text{Wood Chips cu yds}$$

Note that from the third expanded equation to the fourth equation (shown above), the cubic yards wood chips has been replaced by cubic yards sludge times the mix ratio. All other factors remain unchanged.

COMPOST SITE CAPACITY CALCULATION

An important consideration in compost operation is the solids processing capability, lbs/day or lbs/wk.

This type of calculation is essentially a detention time or fill time problem*, as illustrated by the simplified equation shown to the right. Consistent with other fill time calculations, the volume is in the numerator of the equation and the fill rate is in the denominator.

The expanded equations replace wet compost, cu yds/day with equivalent expressions. For example, in the first expanded equation, note that the denominator of the equation (wet compost, lbs/day ÷ compost bulk density, lbs/cu yd) is equivalent to wet compost, cu yds/day.

There is substitution of terms in each succeeding expanded equation until **lbs/day dry solids** is included in the equation. This term is desirable since it is frequently the unknown variable in compost site capacity calculations.

THE SITE CAPACITY CALCULATION IS A DETENTION TIME OR "FILL TIME" CALCULATION*

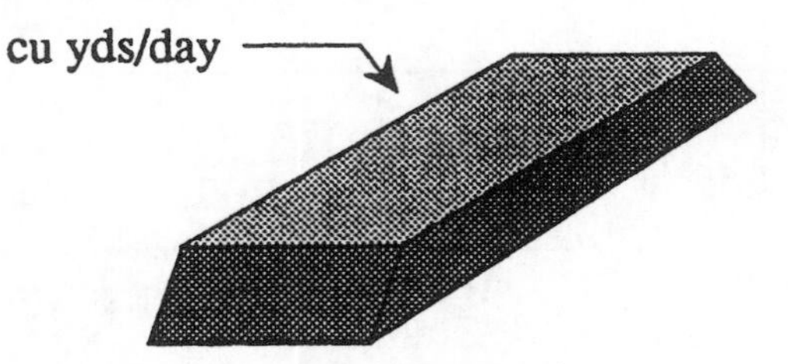

Simplified Equation:

$$\text{Fill Time, days} = \frac{\text{Total Available Capacity, cu yds}}{\text{Wet Compost, cu yds/day}}$$

Expanded Equation:

$$\text{Fill Time, days} = \frac{\text{Total Available Capacity, cu yds}}{\dfrac{\text{Wet Compost, lbs/day}}{\text{Compost Bulk Density, lbs/cu yd}}}$$

$$\text{Fill Time, days} = \frac{\text{Total Available Capacity, cu yds}}{\dfrac{\text{Wet Sludge lbs/day} + \text{Wet Wood Chips lbs/day}}{\text{Compost Bulk Density, lbs/cu yd}}}$$

$$\text{Fill Time, days} = \frac{\text{Total Available Capacity, cu yds}}{\dfrac{\dfrac{\text{Dry Solids lbs/day}}{\frac{\%\text{ Sol}}{100}} + \dfrac{\left(\dfrac{\text{Dry Solids lbs/day}}{\frac{\%\text{ Sol}}{100}}\right)}{\text{Slud. Bulk Dens., lbs/cu yd}}(\text{Mix Ratio})(\text{Bulk Dens. of Wood Chips, lbs/cu yd})}{\text{Compost Bulk Density, lbs/cu yd}}}$$

The last equation is then rearranged as:

$$\text{Fill Time, days} = \frac{(\text{Avail. Capac., cu yds})(\text{Comp. Bulk Dens., lbs/cu yd})}{\dfrac{\text{Dry Solids lbs/day}}{\frac{\%\text{ Sol}}{100}} + \left(\dfrac{\text{Dry Solids lbs/day}}{\frac{\%\text{ Sol}}{100}}\right)(\text{Mix Ratio})\left(\dfrac{\text{Bulk Density of W.C. lbs/cu yd}}{\text{Slud. Bulk Dens., lb/cu yd}}\right)}$$

* Refer to Chapter 5 for a review of detention and retention time calculations.

Example 4 : (Composting)

❑ A composting facility has an available capacity of 7800 cu yds. If the composting cycle is 21 days, how many lbs/day wet compost can be processed by this facility? Assume a compost bulk density of 1000 lbs/cu yd.*

$$\text{Fill Time, days} = \frac{\text{Total Available Capacity, cu yds}}{\dfrac{\text{Wet Compost, lbs/day}}{\text{Compost Bulk Density, lbs/cu yd}}}$$

$$21 \text{ days} = \frac{7800 \text{ cu yds}}{\dfrac{x \text{ lbs/day}}{1000 \text{ lbs/cu yd}}}$$

$$21 \text{ days} = \frac{(7800 \text{ cu yds})\,(1000 \text{ lbs/cu yd})}{x \text{ lbs/day}}$$

$$x \text{ lbs/day} = \frac{(7800 \text{ cu yds})\,(1000 \text{ lbs/cu yd})}{21 \text{ days}}$$

$$x = \boxed{371{,}429 \text{ lbs/day}}$$

Example 5 : (Composting)

❑ A composting facility has an available capacity of 5500 cu yds. If the composting cycle is 21 days, how many lbs/day wet compost can be processed by this facility? How many tons/day is this? Assume a compost bulk density of 950 lbs/cu yd.*

$$21 \text{ days} = \frac{5500 \text{ cu yds}}{\dfrac{x \text{ lbs/day}}{950 \text{ lbs/cu yd}}}$$

$$21 \text{ days} = \frac{(5500 \text{ cu yds})\,(950 \text{ lbs/cu yd})}{x \text{ lbs/day}}$$

$$x \text{ lbs/day} = \frac{(5500 \text{ cu yds})\,(950 \text{ lbs/cu yd})}{21 \text{ days}}$$

$$x = \boxed{248{,}810 \text{ lbs/day}}$$

The lbs/day compost can now be converted to tons/day:

$$\frac{248{,}810 \text{ lb/day}}{2{,}000 \text{ lbs/ton}} = \boxed{124 \text{ tons/day}}$$

* For a review of solving for the unknown value, refer to Chapter 2 in *Basic Math Concepts*.

In Examples 4 and 5, the unknown variable was wet compost, lbs/day. In Examples 6 and 7, the unknown variable is dry sludge, lbs/day. With this variable as the unknown, the most complex expanded equation given on the previous page must be used.

Example 6 : (Composting)

❑ Given the data listed below, calculate the dry sludge processing capability, lbs/day, of the compost operation.*

Cycle time—28 days
Total available capacity—7850 cu yds
% Solids of wet sludge—17%
Mix ratio (by volume) of wood chips to sludge—3
Wet Compost Bulk Density—1000 lbs/cu yd
Wet Sludge Bulk Density—1685 lbs/cu yd
Wet Wood Chips Bulk Density—750 lbs/cu yd

$$\text{Fill Time, days} = \frac{(\text{Avail. Capac., cu yds})(\text{Comp. Bulk Dens., lbs/cu yd})}{\dfrac{\text{Dry Solids lbs/day}}{\frac{\%\ \text{Sol}}{100}} + \dfrac{(\text{Dry Solids}) \text{ lbs/day}}{\frac{\%\ \text{Sol}}{100}}\ (\text{Mix Ratio})\ \dfrac{(\text{Bulk Density}) \text{ of W.C. lbs/cu yd}}{\text{Slud. Bulk Dens., lbs/cu yd}}}$$

$$28 \text{ days} = \frac{(7850 \text{ cu yds})(1000 \text{ lbs/cu yd})}{\dfrac{x \text{ lbs/day Dry Solids}}{0.17} + \dfrac{(x \text{ lbs/day}) \text{ Dry Solids}}{0.17}\ (3) \text{ Mix Ratio}\ \dfrac{(750 \text{ lbs/cu yd})}{1685 \text{ lbs/cu yd}}}$$

First simplify terms, as possible:

$$28 = \frac{7{,}850{,}000}{\dfrac{x}{0.17} + 7.85\,x}$$

The x term in the denominator can be further simplified:

$$28 = \frac{7{,}850{,}000}{\dfrac{1}{0.17}\,x + 7.85\,x}$$

$$28 = \frac{7{,}850{,}000}{5.88\,x + 7.85\,x}$$

$$28 = \frac{7{,}850{,}000}{13.73\,x}$$

$$13.73\,x = \frac{7{,}850{,}000}{28}$$

$$x = \frac{7{,}850{,}000}{(28)(13.73)}$$

$$x = \boxed{20{,}419 \text{ lbs/day Dry Sludge}}$$

* Refer to Chapter 2 in *Basic Math Concepts* for a review of solving for the unknown value.

Example 7 : (Composting)

❑ Given the data listed below, calculate the dry sludge, lbs/day, that can be processed at the compost facility.

Cycle time—28 days
Total available capacity—7850 cu yds
% Solids of wet sludge—17%
Mix ratio (by volume) of wood chips to sludge—3.45
Wet Compost Bulk Density—1000 lbs/cu yd
Wet Sludge Bulk Density—1685 lbs/cu yd
Wet Wood Chips Bulk Density—750 lbs/cu yd

$$\text{Fill Time, days} = \frac{(\text{Avail. Capac., cu yds})(\text{Comp. Bulk Dens., lbs/cu yd})}{\dfrac{\text{Dry Solids lbs/day}}{\dfrac{\%\text{ Sol}}{100}} + \dfrac{(\text{Dry Solids})\text{ lbs/day}}{\dfrac{\%\text{ Sol}}{100}}\,(\text{Mix Ratio})\,\dfrac{(\text{Bulk Density}) \text{ of W.C. lbs/cu yd}}{\text{Slud. Bulk Dens., lbs/cu yd}}}$$

$$28 \text{ days} = \frac{(7850 \text{ cu yds})(1000 \text{ lbs/cu yd})}{\dfrac{x \text{ lbs/day Dry Solids}}{0.17} + \dfrac{(x \text{ lbs/day})\text{ Dry Solids}}{0.17}\,(3.45)\text{ Mix Ratio}\,\dfrac{(750 \text{ lbs/cu yd})}{1685 \text{ lbs/cu yd}}}$$

First simplify terms, as possible:

$$28 = \frac{7{,}850{,}000}{\dfrac{x}{0.17} + 9.03\,x}$$

The x term in the denominator can be further simplified:

$$28 = \frac{7{,}850{,}000}{\dfrac{1}{0.17}\,x + 9.03\,x}$$

$$28 = \frac{7{,}850{,}000}{5.88\,x + 9.03\,x}$$

$$28 = \frac{7{,}850{,}000}{14.91\,x}$$

$$14.91\,x = \frac{7{,}850{,}000}{28}$$

$$x = \frac{7{,}850{,}000}{(28)(14.91)}$$

$$x = \boxed{18{,}803 \text{ lbs/day Dry Sludge}}$$

* Expressed in terms of tons/day, this is 18,803 lbs/day ÷ 2000 lbs/ton = 9.4 tons/day.

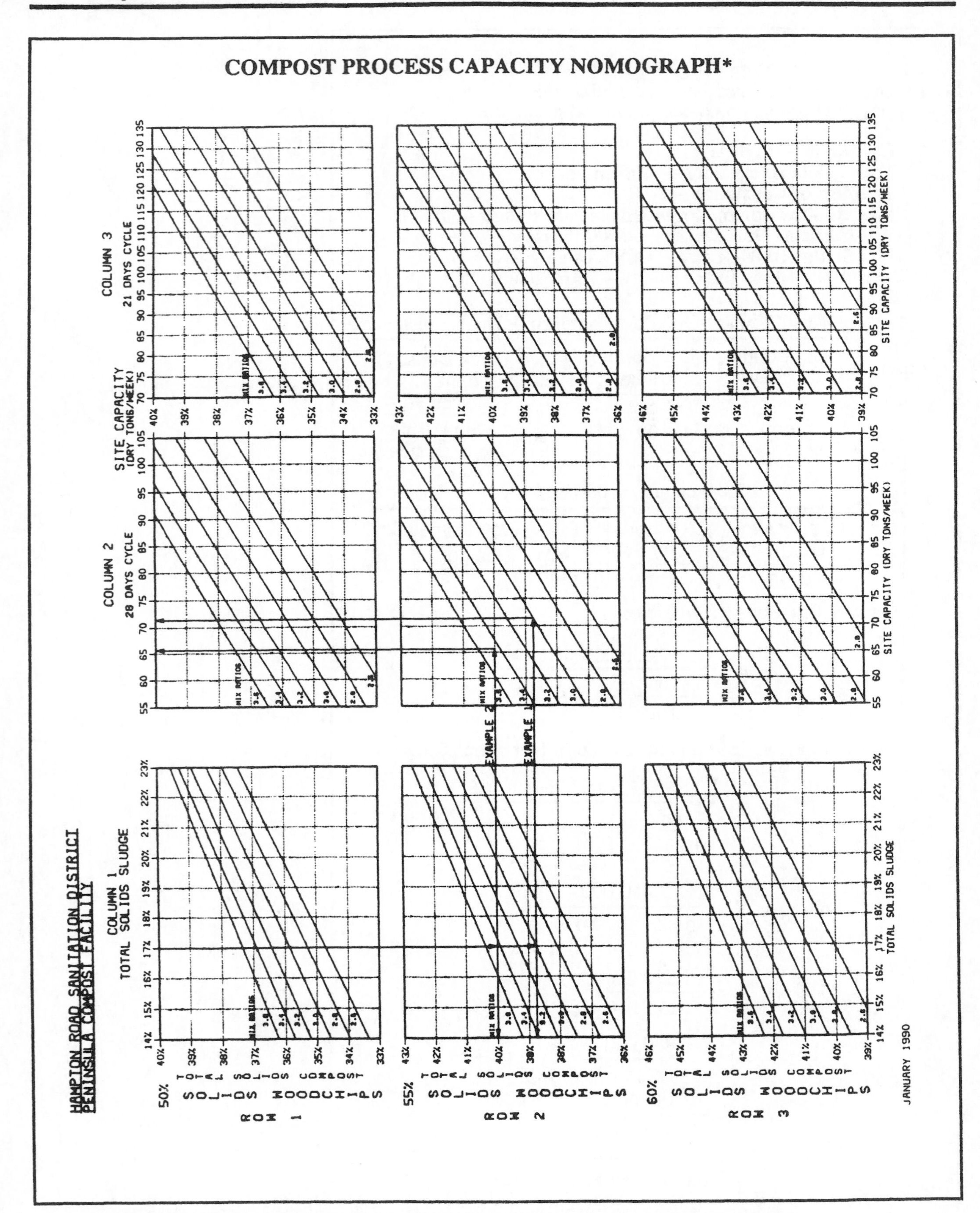

* Source for nomograph and Examples 8 and 9: "*Graphical Techniques for Quick and Comprehensive Evaluation of the Compost Process*," presented at the annual WPCF conference, October, 1990, by Alan B. Cooper, Regional Office Manager and Senior Project Manager for Black & Veatch, Gaithersburg, Maryland.

Example 8 : (Composting)

❑ Determine the resulting total solids compost and site capacity (dry solids, tons/wk) for the following conditions:

Total solid content of sludge—17%
Solids content of wood chips—55%
Mix ratio of wood chips to sludge—3
Cycle time—28 days

First, locate 17% solids at the top of column 1. Follow the line corresponding to 17% down to the row 2 nomograph (which represents 55% solids wood chips). At the 3.0 mix ratio line, read the scale directly left to determine the percent total solids of the compost:

38.7% solids compost

Move along the same horizontal line directly right to the column 2 nomograph (which represents a 28 day cycle). Then at the 3.0 mix ratio line, move directly up to the top scale and read the dry tons/week site capacity:

72 dry tons/wk

Example 9 : (Composting)

❑ Determine the required mix ratio to achieve the desired percent solids compost shown below. Then determine the resulting site capacity (dry solids/wk) for that mix ratio and a 28 day cycle time.

Desired total solids of compost—40%
Total solid content of sludge—17%
Solids content of wood chips—55%

First, locate 17% solids at the top of column 1. Follow the line corresponding to 17% down to the row 2 nomograph (which represents 55% solids wood chips). Find the point of intersection between the 17% solids vertical line and the horizontal line from the left scale at 40%. The mix ratio indicated falls between the 3.4 and 3.6 mix ratio lines. It can be estimated at about a 3.45 desired mix ratio:

3.45 mix ratio

Move along the same horizontal line directly right to the column 2 nomograph and at the estimated 3.45 mix ratio line, move directly up to the top scale and read the dry tons/week site capacity:

67 dry tons/wk

The calculation of percent solids of the compost and compost site capacity given in Examples 3-7 can be determined by the use of the nomograph given on the facing page.

Using the nomograph, you can quickly determine the effect on site capacity (solids processing rate) that results from changes in the percent solids content of either material being blended, changes in the the mix ratio of these materials, or changes in the cycle length (fill time, in the equation). Examples 8 and 9 illustrate the use the nomograph. Note that the problems used for Examples 8 and 9 are the same problems worked out "long hand" for Examples 6 and 7.

To use the nomograph:

1. Locate percent total solids sludge in Column 1.
2. Select % solids wood chips.
   - For: 50% solids wood chips, use nomographs from row 1,
   - For 55% solids wood chips, use row 2 nomographs,
   - For 60% solids wood chips, use row 3 nomographs.
3. Select cycle time.
   - For 28 day cycle, use column 2,
   - For 21 day cycle, use column 3.

NOTES:

18 Laboratory Calculations

SUMMARY

1. Biochemical Oxygen Demand (BOD)

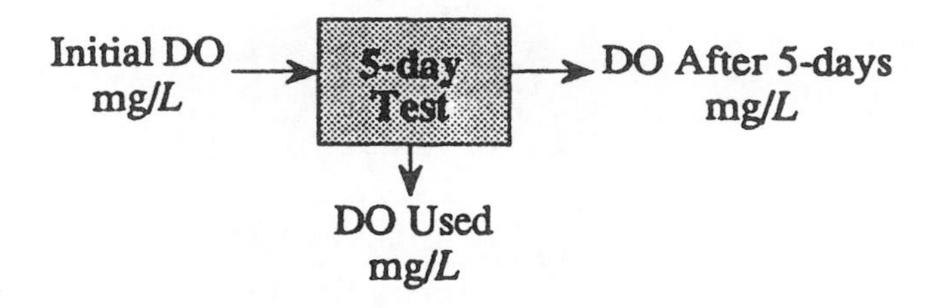

Simplified Equation:

$$\text{BOD, mg/}L = \frac{\text{DO Used During 5-day Test, mg/}L}{\text{\% Dilution of Sample}}$$

Expanded Equation:

$$\text{BOD, mg/}L = \frac{\text{Initial DO, mg/}L \quad \text{DO After 5-days, mg/}L}{\dfrac{\text{Sample Volume, ml}}{\text{BOD Bottle Volume, ml}}}$$

7-Day Average BOD:

$$\text{7-day Aver. BOD} = \frac{\text{BOD Day 1} + \text{BOD Day 2} + \text{BOD Day 3} + \text{BOD Day 4} + \text{BOD Day 5} + \text{BOD Day 6} + \text{BOD Day 7}}{7}$$

SUMMARY—Cont'd

2. Molarity and Moles

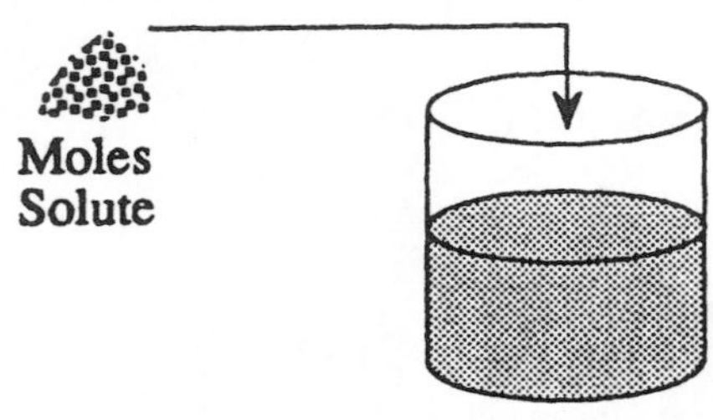

$$\text{Molarity} = \frac{\text{Moles Solute}}{\text{Liters Solution}}$$

$$\text{Molarity} = \frac{\text{Grams of Chemical}}{\text{Formula Wt. of the Chemical}}$$

3. Normality and Equivalents

$$\text{Normality} = \frac{\text{No. of Equivalents of Solute}}{\text{Liters of Solution}}$$

$$\text{Equivalent Weight} = \frac{\text{Formula Wt}}{\text{Net Valence}}$$

4. Settleability*

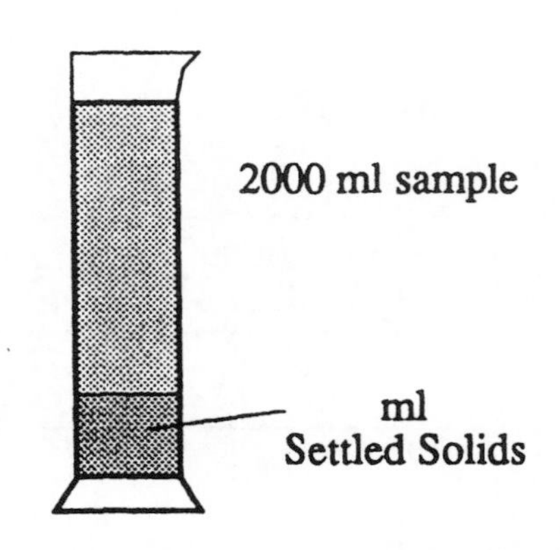

$$\%\ \text{Settleable Solids} = \frac{\text{ml Settled Solids}}{\text{2000 ml Sample}} \times 100$$

* Using a Mallory Direct Reading Settleometer.

SUMMARY—Cont'd

5 Settleable Solids

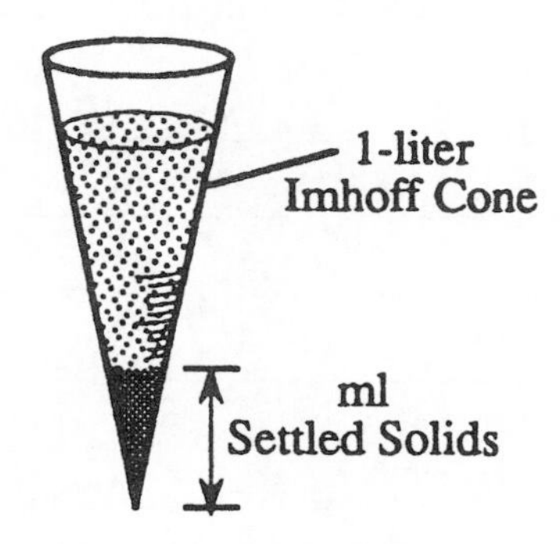

The most common calculation using settleable solids is percent removal of settleable solids:

$$\text{\% Removal of Set. Sol.} = \frac{\text{Set. Sol Removed, ml}/L}{\text{Set. Sol. in Influent, ml}/L} \times 100$$

Percent setteable solids may also be calculated:

$$\text{\% Settleable Solids} = \frac{\text{Settled Solids, ml}}{\text{1000 ml Sample}} \times 100$$

6. Sludge Total and Volatile Solids

$$\text{\% Total Solids} = \frac{\text{Wt. of Total Solids}}{\text{Wt. of Sludge Sample}} \times 100$$

$$\text{\% Volatile Solids} = \frac{\text{Wt. of Vol. Solids}}{\text{Wt. of Total Solids}} \times 100$$

To calculate mg/*L* Total or Volatile Solids:

$$\frac{\text{Solids, g}}{\text{100-ml Sample}} \times \frac{\text{1000 mg}}{\text{1 g}} \times \frac{10}{10} = \text{mg}/L \text{ Solids}$$

SUMMARY—Cont'd

7. Suspended Solids and Volatile Suspended Solids

To calculate mg/*L* SS, if a 25-m*L* sample is used:

$$\frac{\text{grams SS}}{\text{25-m}L\text{ Sample}} \times \frac{\text{1000 mg}}{\text{1 g}} \times \frac{40}{40} = \frac{\text{mg SS}}{\text{liter}} = \text{mg/}L\text{ SS}$$

To calculate mg/*L* SS, if a 50-m*L* sample is used:

$$\frac{\text{grams SS}}{\text{50-m}L\text{ Sample}} \times \frac{\text{1000 mg}}{\text{1 g}} \times \frac{20}{20} = \frac{\text{mg SS}}{\text{liter}} = \text{mg/}L\text{ SS}$$

To calculate percent volatile suspended solids:

$$\%\text{ VSS} = \frac{\text{Wt. of Volatile Solids}}{\text{Wt. of Suspended Solids}} \times 100$$

8. Sludge Volume Index and Sludge Density Index

$$\text{SVI} = \frac{\text{Volume, m}L}{\text{Density, g}}$$

$$\text{SDI} = \frac{\text{Density, g}}{\text{Volume, m}L} \times 100$$

SUMMARY—Cont'd

9. Temperature

Fahrenheit to Celsius : (Conventional Equation)

$$°C = \frac{5}{9}(°F - 32°)$$

Celsius to Fahrenheit : (Conventional Equation)

$$°F = \frac{9}{5}(°C) + 32°$$

3-Step Method: (Converting either direction)

1. Add 40°.
2. Multiply by 5/9 or 9/5.
   (Depends on the direction of conversion.)
3. Subtract 40°.

NOTES:

18.1 BIOCHEMICAL OXYGEN DEMAND (BOD) CALCULATIONS

The Biochemical Oxygen Demand (BOD) content of a wastewater is used as an indicator of the available food in the wastewater, and is therefore included in such calculations as organic loading and F/M ratio.

The BOD test measures the amount of oxygen used by the microorganisms as they breakdown food (complex organic compounds) in the wastewater.

The dissolved oxygen (DO) content of the sample is tested just prior to beginning the test (initial DO) and at the end of the test. Then by subtracting the second DO reading from the initial DO, **the amount of DO used during the test can be determined**:

$$\text{Initial DO, mg/}L - \text{DO After 5-day Test, mg/}L = \text{DO Used During, mg/}L$$

If the BOD test were conducted using a full-strength sample, the BOD content (mg/*L*) would be equal to the dissolved oxygen (DO) used or depleted during the 5-day test. For example, if the DO used during the 5-day BOD test was 75 mg/*L*, then the BOD would be the same—75 mg/*L*.

However, **the BOD test is conducted on a diluted sample**. Therefore, the percent dilution of the sample must be included in the calculation, as shown in the equations to the right.*

Depending on the percent dilution, the DO used in the diluted sample might represent only 1% to 10% of the DO used in the full-strength sample.

THE BOD TEST IS CONDUCTED ON A DILUTED SAMPLE

Diluted Sample

Simplified Equation:

$$\text{BOD mg/}L = \frac{\text{DO Used During 5-day Test, mg/}L}{\text{Dilution Fraction of Sample}}$$

Or

Expanded Equation:

$$\text{BOD mg/}L = \frac{\text{Initial DO, mg/}L - \text{DO After 5 days, mg/}L}{\dfrac{\text{Sample Volume, m}L}{\text{BOD Bottle Volume, m}L}}$$

Example 1: (BOD)

❑ Given the following information, determine the BOD of the wastewater:

Sample Volume—4 m*L*
BOD Bottle Volume—300 m*L*
Initial DO of Diluted Sample—7 mg/*L*
DO of Diluted Sample—4 mg/*L* (After 5 days)

7 mg/*L* DO (Initial) → **5-day Test** → 4 mg/*L* DO (After 5-days)
↓
3 mg/*L* DO Used in 5 days

$$\text{BOD mg/}L = \frac{\text{Initial DO, mg/}L - \text{DO After 5-days, mg/}L}{\dfrac{\text{Sample Volume, m}L}{\text{BOD Bottle Volume, m}L}}$$

$$= \frac{7\text{ mg/}L - 4\text{ mg/}L}{\dfrac{4\text{ m}L}{300\text{ m}L}}$$

$$= \frac{3\text{ mg/}L}{0.013}$$

$$= \boxed{231\text{ mg/}L\text{ BOD}}$$

* Note that this calculation is essentially a percent strength calculation—similar to a hypochlorite problem. Refer to Chapter 14, Section 14.3 to review similarities.

Example 2: (BOD)

❑ Results from a BOD test are given below. Calculate the BOD of the sample.

Sample Volume—30 m*L*
BOD Bottle Volume—300 m*L*
Initial DO of Diluted Sample—8 mg/*L*
DO of Diluted Sample—3.7 mg/*L* (After 5 days)

8 mg/*L* DO (Initial) → **5-day Test** → 3.7 mg/*L* DO (After 5-days)
↓
4.3 mg/*L* DO Used in 5 days

$$\text{BOD mg/}L = \frac{\text{Initial DO, mg/}L - \text{DO After 5 days, mg/}L}{\text{Dilution Fraction of Sample}}$$

$$= \frac{8 \text{ mg/}L - 3.7 \text{ mg/}L}{0.1}$$

$$= \boxed{43 \text{ mg/}L \text{ BOD}}$$

Example 3: (BOD)

❑ Given the information listed below, determine the BOD of the wastewater.

Sample Volume—7 m*L*
BOD Bottle Volume—300 m*L*
Initial DO of Diluted Sample—9 mg/*L*
DO of Diluted Sample—3.2 mg/*L* (After 5 days)

9 mg/*L* DO (Initial) → **5-day Test** → 3.2 mg/*L* DO (After 5-days)
↓
5.8 mg/*L* DO Used in 5 days

$$\text{BOD mg/}L = \frac{\text{Initial DO, mg/}L - \text{DO After 5 days, mg/}L}{\text{Dilution Fraction of Sample}}$$

$$= \frac{9 \text{ mg/}L - 3.2 \text{ mg/}L}{0.023}$$

$$= \boxed{252 \text{ mg/}L \text{ BOD}}$$

BOD 7-DAY MOVING AVERAGE

The BOD characteristic of wastewater varies from day to day, even hour-to-hour. However, operational control of the wastewater treatment system is most often accomplished based on trends in data rather than individual data points. The BOD 7-day moving average is a calculation of the BOD trend.

This calculation is called a moving average, since a new 7-day average is calculated each day, adding the new days value and the six previous days values.

BOD 7-DAY MOVING AVERAGE

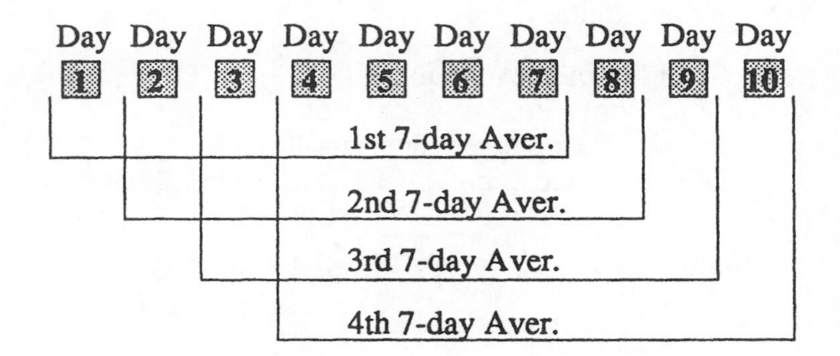

$$\text{7-day Aver. BOD} = \frac{\text{BOD Day 1} + \text{BOD Day 2} + \text{BOD Day 3} + \text{BOD Day 4} + \text{BOD Day 5} + \text{BOD Day 6} + \text{BOD Day 7}}{7}$$

Example 4: (7-day Moving Average)

❑ Given the following primary effluent BOD test results, calculate the 7-day average.

May 1—210 mg/*L*
May 2—218 mg/*L*
May 3—202 mg/*L*
May 4—207 mg/*L*
May 5—224 mg/*L*
May 6—216 mg/*L*
May 7—220 mg/*L*

$$\text{7-day Aver. BOD} = \frac{\text{BOD Day 1} + \text{BOD Day 2} + \text{BOD Day 3} + \text{BOD Day 4} + \text{BOD Day 5} + \text{BOD Day 6} + \text{BOD Day 7}}{7}$$

$$= \frac{210 + 218 + 202 + 207 + 224 + 216 + 220}{7}$$

$$= \boxed{214 \text{ mg/}L \text{ BOD}}$$

Example 5: (7-day Moving Average)

❑ Given the BOD test results shown below, calculate the 7-day average for influent BOD.

Sunday—280 mg/*L*
Monday—310 mg/*L*
Tuesday—314 mg/*L*
Wednesday—307 mg/*L*
Thursday—320 mg/*L*
Friday—303 mg/*L*
Saturday—292 mg/*L*

$$\text{7-day Aver. BOD} = \frac{\text{BOD Day 1} + \text{BOD Day 2} + \text{BOD Day 3} + \text{BOD Day 4} + \text{BOD Day 5} + \text{BOD Day 6} + \text{BOD Day 7}}{7}$$

$$= \frac{280 + 310 + 314 + 307 + 320 + 303 + 292}{7}$$

$$= \boxed{304 \text{ mg/}L \text{ BOD}}$$

Example 6: (7-day Moving Average)

❑ Calculate the 7-day average primary effluent BOD for March 12th, 13th, and 14th, given the BOD test results shown below.

Mar 1—180
Mar 2—174
Mar 3—179
Mar 4—185
Mar 5—189
Mar 6—172
Mar 7—178
Mar 8—184
Mar 9—195
Mar 10—181
Mar 11—169
Mar 12—175
Mar 13—187
Mar 14—174
Mar 15—184
Mar 16—191
Mar 17—187
Mar 18—172

To calculate the 7-day average BOD **for March 12th,** use the test result for March 12th and the 6 days previous (Mar 6-Mar 12):

$$\text{7-day Aver. BOD} = \frac{172 + 178 + 184 + 195 + 181 + 169 + 175}{7}$$

$$= \boxed{179 \text{ mg/}L \text{ BOD}}$$

For the next day's average, **keep the same string of numbers except** drop the "oldest" value (Mar 6) and add the new value (Mar 13):

$$\text{7-day Aver. BOD} = \frac{178 + 184 + 195 + 181 + 169 + 175 + 187}{7}$$

$$= \boxed{181 \text{ mg/}L \text{ BOD}}$$

For the March 14th average, repeat the same process—drop the oldest value and add the new value:

$$\text{7-day Aver. BOD} = \frac{184 + 195 + 181 + 169 + 175 + 187 + 174}{7}$$

$$= \boxed{181 \text{ mg/}L \text{ BOD}}$$

SHORTCUT METHOD FOR DETERMINING NEW TOTALS

If you keep track of th total of the 7 days BOD results (the numerator of the average calculation), you may use a shortcut in calculating a new total:

$$\text{Previous Total} - \text{Oldest Value} + \text{New Value} = \text{New Total}$$

In Example 6, the total in the numerator of the first 7-day average is 1254. To calculate the new total to be used for the next 7-day average:

$$\underset{\text{Prev. Total}}{1254} - \underset{\text{Oldest Value}}{172} + \underset{\text{New Value}}{187} = \underset{\text{New Total}}{1269}$$

18.2 MOLARITY AND MOLES

A **solution** is comprised of two parts:

- A **solvent** which is the dissolving medium (such as water), and

- A **solute** which is the substance dissolved (such as dry hypochlorite).

A concentrated solution is one that contains a relatively small amount of solute (e.g., chemical) per unit volume of solution. A dilute solution is one that contains a relatively small amount of solute per unit volume of solution.

The principal methods of expressing the concentrations of solutions are percent strength (see Chapter 14, Section 14.4), molarity (described in this section) and normality (described in the next section). Of these three methods, molarity is perhaps used least often in the field of water and wastewater. However, it has been included here since it may be required occasionally.

Molarity is one method devised to compare solution concentrations. For example, a solution of one molarity would be one mole of the solute dissolved in one liter. (The concept of "moles" will be discussed later in this section.) A one-molar solution (1M) might also be 2 moles of solute dissolved in 2 liters of solution; or 3 moles solute dissolved in 3 liters solution. It may also be a half mole solute dissolved in a half-liter solution (500 m*L*). Normally, however, we do not find such optimum conditions of 1:1, 2:2, etc. Examples 1-3 illustrate how to calculate the molarity of a solution, given the moles solute and solution volume.

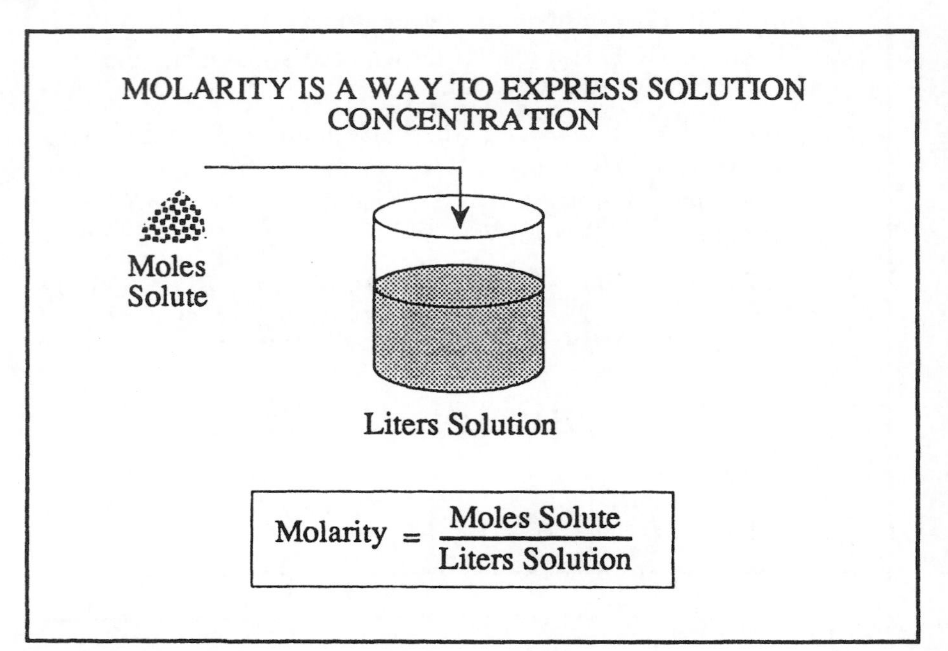

Example 1: (Molarity & Moles)

❑ If 2.5 moles of solute are dissolved in 0.5 liters solution, what is the molarity of the solution?

$$\text{Molarity} = \frac{\text{Moles Solute}}{\text{Liters Solution}}$$

$$= \frac{2.5 \text{ moles}}{0.5 \text{ liters}}$$

$$= \boxed{\text{5 Molarity or 5-Molar Solution}}$$

Example 2: (Molarity & Moles)

❑ What is the molarity of a solution that has 0.4 moles solute dissolved in 1250 ml solution?

$$\text{Molarity} = \frac{\text{Moles Solute}}{\text{Liters Solution}}$$

$$= \frac{0.4 \text{ moles}}{1.25 \text{ liters}}$$

$$= \boxed{\text{0.32 Molarity or 0.32-Molar Solution}}$$

Example 3: (Molarity & Moles)

❑ A 0.6-molar solution is to be prepared. If a total of 500 ml solution is to be prepared, how many moles solute will be required?

Use the same equation and fill in the known information:

$$\text{Molarity} = \frac{\text{Moles Solute}}{\text{Liters Solution}}$$

$$0.6 = \frac{x \text{ moles}}{0.5 \text{ liters}}$$

$$(0.5)\,(0.6) = x$$

$$\boxed{0.3 \text{ moles}} = x$$

WHEN SOLUTION VOLUME IS EXPRESSED AS MILLILITERS

Occasionally the solution volume will be expressed as milliliters rather than liters. Before calculating solution molarity, you will need to convert milliliters solution to liters solution (divide milliliters by 1000):

$$\boxed{\frac{mL}{1000\ mL/L}}$$

This can be done without any division—simply find the thousand's comma and place a decimal in that position:

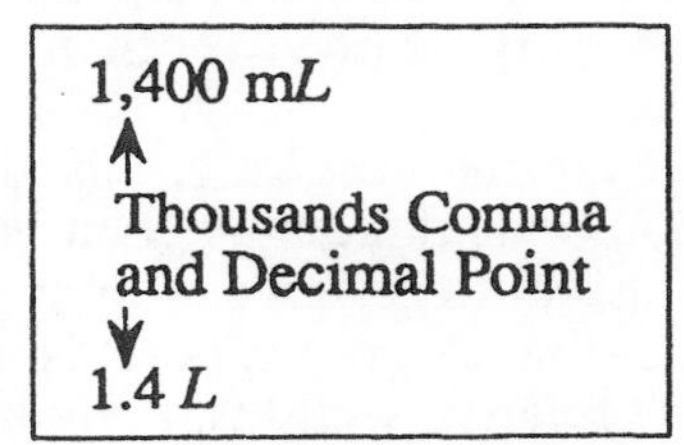

CALCULATING OTHER UNKNOWN VARIABLES

In molarity calculations there are three variables: molarity, moles solute, and liters solution. In Examples 1 and 2, molarity was the unknown variable. In Example 3 a different variable is unknown.

MOLES

Every chemical listed on the Periodic Table of Elements has a corresponding atomic weight listed. The number shown as an atomic weight is an arbitrary number designed to compare the densities of different elements.

The **relative densities** of the atoms of each elements are more convenient quantities to use than their absolute densities in grams. These relative densities or relative weights are based on an assigned value of exactly 12.0000 for the atomic weight of carbon. From this basis, the atomic weight of each element was assigned by comparing its weight to that of carbon and thus assigning a number higher or lower than 12.

Since atomic weights are used for various calculations in chemistry, it is useful to assign a weight value (that is, grams, pounds, etc.) to the relative weights. A mole is a quantity of a compound equal in weight to its formula weight.

For example, 12 grams of carbon would be one gram-mole of carbon. And 12 pounds of carbon, is a pound-mole of carbon. In water and wastewater calculations, we are primarily concerned with gram-moles. The term "mole" will be understood to mean "gram-mole".

To calculate the number of moles used in preparing a solution, use the following equation:

$$\text{Molarity} = \frac{\text{Grams of Chemical}}{\text{Formula Wt. of Chemical}}$$

A MOLE IS A QUANTITY OF A COMPOUND EQUAL IN WEIGHT TO ITS FORMULA WEIGHT

For example, the formula weight for water (H_2O) can be determined using the Periodic Table of Elements:

H_2O:

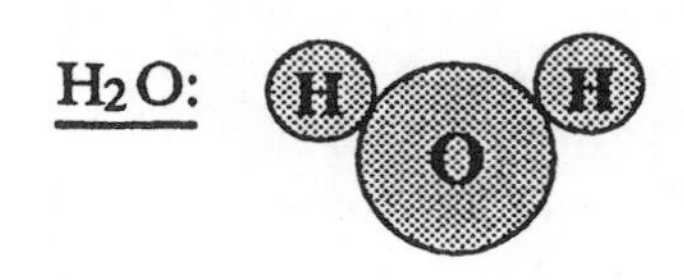

Hydrogen (1.008) x 2 = 2.016
Oxygen = 16.000
Formula weight of H_2O → 18.016

Since the formula weight of water is 18.016, a mole is 18.016 units of weight. A **gram-mole** is 18.016 grams of water. A **pound-mole** is 18.016 pounds of water.

Example 4: (Molarity & Moles)

❑ The atomic weight of calcium is 40. If 65 grams of calcium are used in making up a one-liter solution, how many moles are used?

$$\text{Moles} = \frac{\text{Grams of Chemical}}{\text{Formula wt.}}$$

$$= \frac{65 \text{ grams}}{40 \text{ grams/mole}}$$

$$= \boxed{1.6 \text{ moles}}$$

Example 5: (Molarity & Moles)

❑ If magnesium has a listed atomic weight of 24, how many moles is represented by 62 grams of magnesium?

$$\text{Moles} = \frac{\text{Grams of Chemical}}{\text{Formula wt.}}$$

$$= \frac{62 \text{ grams}}{24 \text{ grams/mole}}$$

$$= \boxed{2.6 \text{ moles}}$$

Example 6: (Molarity & Moles)

❑ The atomic weights listed for each element of sulfuric acid (H_2SO_4) are given below. How many grams make up a mole of sulfuric acid?

H = 1.008

S = 32.06

O = 16.000

To determine the formula weight of sulfuric acid, the total weight of each element must be calculated:

Element	Atomic Weight		Number of Atoms		Total Weight Represented
H:	1.008	x	2	=	2.016
S:	32.06	x	1	=	32.06
O:	16.000	x	4	=	64.000
					98.076

$\boxed{98.076 \text{ grams}}$

DETERMINING FORMULA WEIGHT OF A COMPOUND

To calculate the formula weight of a compound, first determine the total atomic weight represented by each element. Then total all these weights to obtain the formula weight. Example 6 illustrates this type of calculation.

18.3 NORMALITY AND EQUIVALENTS

The **molarity** of a solution refers to its **concentration** (the solute dissolved in the solution). The **normality** of the solution refers more specifically to the reacting power of the solution. One of the first concepts to be learned in a basic chemistry course is that the sharing or transfer of electrons is responsible for chemical activity or the reacting characteristics of an element or compound.

The concept of **equivalents** parallels this concept by relating the number of electrons available to be transferred or shared (valence) and the atomic weight associated with each of these valence electrons.*

Since the concept of equivalents is based upon the "reacting power" of an element or compound, it follows that **a specific number of equivalents of one substance will react with the same number of equivalents of another substance**. For example, two equivalents of a substance will react with two equivalents of another substance. If, however, one equivalent of Substance A is mixed with two equivalents of Substance B, only one equivalent of each substance will react, leaving an excess of one equivalent of Substance B.

Practically speaking, if the concept of equivalents is ignored when making up solutions, most likely chemicals will be wasted as excess amounts.

NORMALITY IS A MEASURE OF THE "REACTING POWER" OF A SOLUTION

1 Equivalent of a Substance ⟷ Reacts With **1 Equivalent of another Substance**

$$\text{Normality} = \frac{\text{No. of Equivalents of Solute}}{\text{Liters of Solution}}$$

This equation may be rearranged as:

$$(\text{Normality of Sol'n})(\text{Liters Sol'n}) = \text{Equivalents in Solution}$$

Example 1: (Normality and Equivalents)

❑ If 2.5 equivalents of a chemical are dissolved in 1.5 liters solution, what is the normality of the solution?

$$\text{Normality} = \frac{\text{No. of Equivalents of Solute}}{\text{Liters of Solution}}$$

$$= \frac{2.5 \text{ Equivalents}}{1.5 \text{ liters}}$$

$$= \boxed{1.67 \text{ N}}$$

* The valence number may be either positive or negative, depending upon whether electrons were added to or taken from the particular element or compound.

Example 2: (Normality and Equivalents)
❑ A 600-ml solution contains 1.8 equivalents of a chemical. What is the normality of the solution?

First convert 600 m*L* to liters:

$$\frac{600 \text{ m}L}{1000 \text{ m}L/L} = 0.6\ L$$

Then calculate the normality of the solution:

$$\text{Normality} = \frac{\text{No. of Equivalents of Solute}}{\text{Liters of Solution}}$$

$$= \frac{1.8 \text{ Equivalents}}{0.6 \text{ Liters}}$$

$$= \boxed{3 \text{ N}}$$

Example 3: (Normality and Equivalents)
❑ How many milliliters of 0.5 N NaOH will react with 500 ml of 0.01 N HCl?

Set the normality and volume of the first solution equal to the normality and volume of the second solution:

$$N_A V_A = N_B V_B$$

$$(0.5)\ (x \text{ m}L) = (0.01)\ (500 \text{ m}L)$$

$$x = \frac{(0.01)\ (500)}{0.5}$$

$$= \boxed{10 \text{ m}L \text{ NaOH}}$$

WHEN MILLILITERS VOLUME IS GIVEN

Many times the volume of solution is given as milliliters. To calculate normality, the volume must be expressed in liters. Therefore, convert the milliliters volume to liters:

$$\boxed{(\text{m}L)\ \frac{(1 \text{ liters})}{1000 \text{ m}L} = \text{liters}}$$

NORMALITY AND TITRATIONS

The second equation, given on the previous page, indicates that the normality of a solution times the volume of the solution is equal to the number of equivalents in the solution:

$$\boxed{\underset{\text{of Sol'n}}{(\text{Normality})}\ \underset{L}{(\text{Volume})} = \underset{\text{Equiv.}}{\text{No. of}}}$$

Because chemicals react on the basis of equivalents, this relationship is of great importance in understanding titrations (such as used in the COD test) or acid/base neutralizations. In general, where N = normality of the solution and V = volume of the solution:

$$\boxed{N_A V_A = N_B V_B}$$

When using this equation, the solution volume may be expressed as liters or milliliters. However, whichever term (ml or *L*) is used on one side of the equation, must also be used on the other side of the equation.

EQUIVALENT WEIGHT

Equivalents relate the number of valence electrons (electrons that may be shared or transferred) with a corresponding atomic weight. Although knowing how to determine the number of valence electrons is not required in water and wastewater calculations, a general understanding of how equivalents are calculated will help you better understand the concept of normality. Examples 4-7 illustrate the calculation of equivalents.

$$\text{Equivalent Weight} = \frac{\text{Atomic Weight}}{\text{Net Valence}}$$

This equation may be stated in more general terms to include compounds:

$$\text{Equivalent Weight} = \frac{\text{Formula Weight}}{\text{Net Valence}}$$

Example 4: (Normality and Equivalents)

❑ If oxygen has an atomic weight of 16 and has 2 valence electrons, what is the equivalent weight of oxygen?

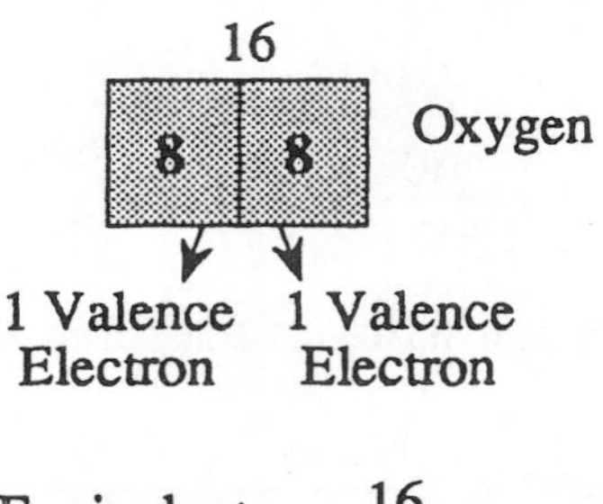

$$\text{Equivalent Weight} = \frac{16}{2}$$

$$= \boxed{8}$$

Example 5: (Normality and Equivalents)

❑ If aluminum has an atomic weight of 27 and has a valence of 3, what is the equivalent weight of aluminum?

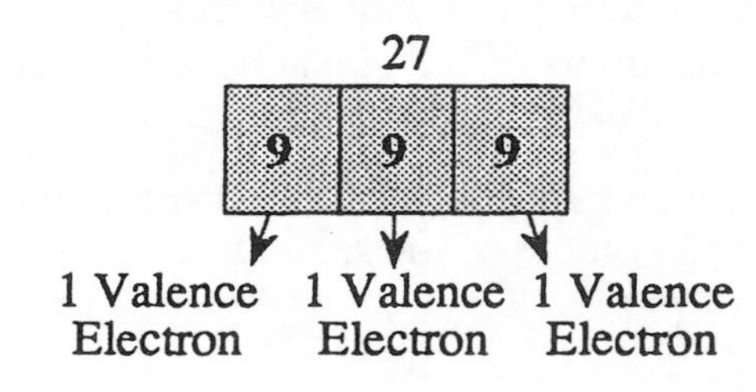

$$\text{Equivalent Weight} = \frac{27}{3}$$

$$= \boxed{9}$$

Example 6: (Normality and Equivalents)

❑ The molecular weight of Na_2CO_3 is 106. The net valence is 2. If 90 grams of Na_2CO_3 are dissolved in a solution, how many equivalents are dissolved in the solution?

The equivalent weight of Na_2CO_3 is $106 \div 2 = 53$. Calculate the number of equivalents as follows:

$$\text{Number of Equivalents} = \frac{\text{grams of NaOH}}{\text{Equivalent wt.}}$$

$$= \frac{90 \text{ grams}}{53 \text{ grams/equivalent}}$$

$$= \boxed{1.7 \text{ Equivalents}}$$

Example 7: (Normality and Equivalents)

❑ Given the atomic weights and net valence of NaOH, what is the normality of an 800-m*L* solution in which 45 grams of NaOH are dissolved?

Sodium (Na)..............23
Oxygen (O)................16
Hydrogen (H)..............1
Net Valence = 1

The equivalent weight of NaOH is $40 \div 1 = 40$. First calculate the number of equivalents dissolved in the solution:

$$\text{Number of Equivalents} = \frac{\text{grams NaOH}}{\text{Equivalent wt.}}$$

$$= \frac{45 \text{ grams}}{40 \text{ grams/equivalent}}$$

$$= \boxed{1.125}$$

The normality of the solution can now be determined:

$$\text{Normality} = \frac{\text{No. of Equivalents of NaOH}}{\text{Liters of Solution}}$$

$$= \frac{1.125 \text{ Equivalents}}{0.8 \text{ Liters}}$$

$$= \boxed{1.4 \text{ N}}$$

CALCULATING NUMBER OF EQUIVALENTS

To determine the number of equivalents used in a solution, you will need to know the grams of chemical used and the equivalent weight:

$$\text{Number of Equivalents} = \frac{\text{Chemical, grams}}{\text{Equivalent wt.}}$$

18.4 SETTLEABILITY

The settleability test is a test of the quality of the activated sludge solids (Mixed Liquor Suspended Solids). A suspended solids test should be run on the same sample so that the Sludge Volume Index (SVI) or Sludge Density Index (SDI) can also be calculated. (SVI and SDI calculations are described in Section 18.8.)

A sample of activated sludge is taken from the aeration tank, poured into a 2000-m*L* graduate, and allowed to settle for 60 minutes. The settling characteristics of the sludge in the graduate gives a general indication of the settling characteristics of the MLSS in the final clarifier.

From the settleability test the percent settleable solids can be calculated, as illustrated in Examples 1-3.

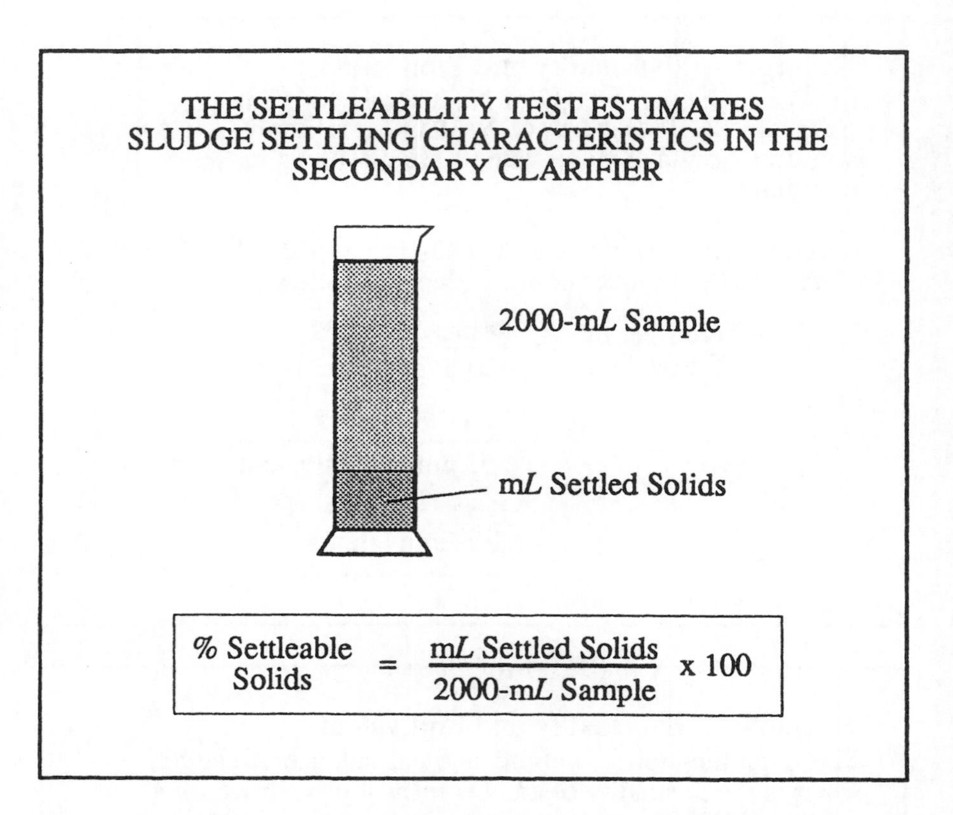

Example 1: (Settleability)

❑ The settleability test is conducted on a sample of MLSS. What is the percent settleable solids if 380 milliliters settle in the 2000-m*L* graduate?

$$\text{\% Settleable Solids} = \frac{\text{m}L\text{ Settled Solids}}{2000\text{ m}L\text{ Sample}} \times 100$$

$$= \frac{380\text{ m}L}{2000\text{ m}L} \times 100$$

$$= \boxed{19\%\text{ Settleable Solids}}$$

Example 2: (Settleability)

❑ A 2000-m*L* sample of activated sludge is tested for settleability. If the settled solids is measured as 350 milliliters, what is the percent settled solids?

$$\text{\% Settleable Solids} = \frac{\text{m}L\text{ Settled Solids}}{2000\text{-m}L\text{ Sample}} \times 100$$

$$= \frac{350\text{ m}L}{2000\text{ m}L} \times 100$$

$$= \boxed{17.5\%\text{ Settleable Solids}}$$

Example 3: (Settleability)

❑ The settleability test is conducted on a sample of MLSS. What is the percent settleable solids if 330 milliliters settle in the 2000-m*L* graduate?

$$\text{\% Settleable Solids} = \frac{\text{m}L\text{ Settled Solids}}{2000\text{-m}L\text{ Sample}} \times 100$$

$$= \frac{330\text{ m}L}{2000\text{ m}L} \times 100$$

$$= \boxed{16.5\%\text{ Settleable Solids}}$$

18.5 SETTLEABLE SOLIDS (IMHOFF CONE)

The settleable solids test, like the settleability test, measures the volume of solids that settle out from a wastewater sample during the 60-minute test. **This test differs from the settleability test in at least two respects**:

- The settleable solids test is conducted in a one-liter Imhoff Cone. (The settleability test or settled sludge volume test, is conducted in a one-liter or two-liter graduated cylinder.)
- The settleable solids test is conducted on samples from sedimentation tank or clarifier influent and effluent. (The settleability test or settled sludge volume test is conducted on samples from the activated sludge aeration basin.)

This test indicates the volume of solids removed by sedimentation in sedimentation tanks, clarifiers, or ponds.

By running a settleable solids test on the wastewater influent and effluent, the settleable solids test can be used to estimate the percent removal of settleable solids.

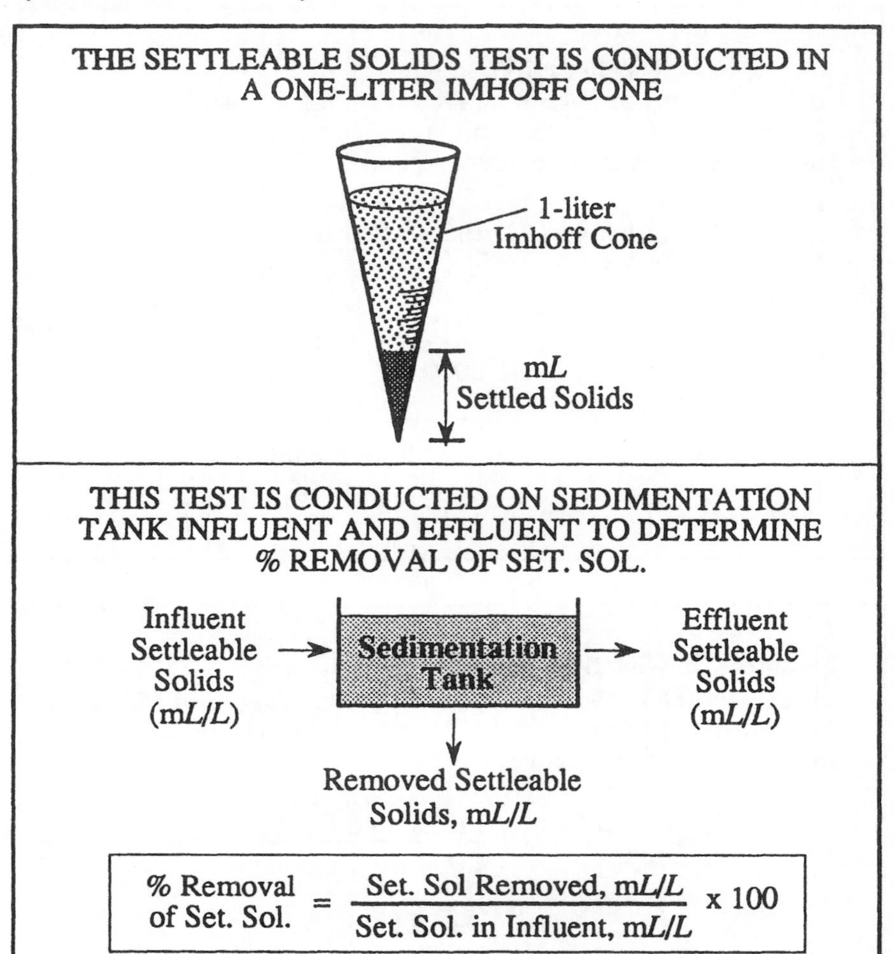

Example 1: (Settleable Solids)

❑ Calculate the percent removal of settleable solids if the settleable solids of the sedimentation tank influent is 17 mL/L and the settleable solids of the effluent is 0.3 mL/L.

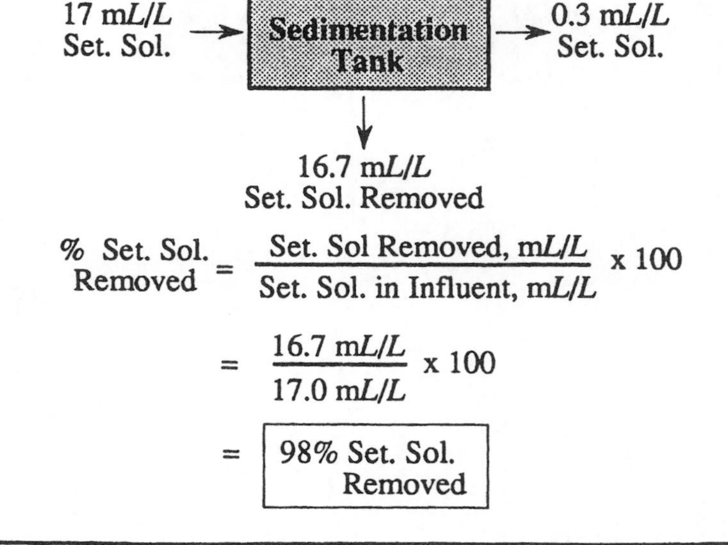

$$\text{\% Set. Sol. Removed} = \frac{\text{Set. Sol Removed, mL/L}}{\text{Set. Sol. in Influent, mL/L}} \times 100$$

$$= \frac{16.7 \text{ mL/L}}{17.0 \text{ mL/L}} \times 100$$

$$= \boxed{\text{98\% Set. Sol. Removed}}$$

Example 2: (Settleable Solids)

❑ The settleable solids of the raw wastewater is 14 mL/L. If the settleable solids of the clarifier effluent is 0.5 mL/L, what is the settleable solids removal efficiency of the clarifier?

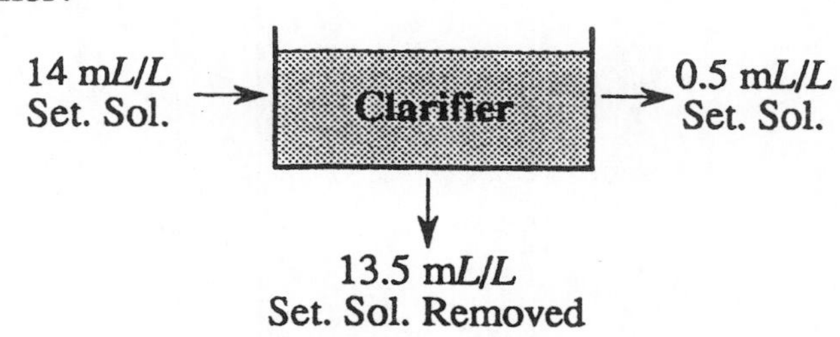

$$\frac{\%\text{ Set. Sol.}}{\text{Removed}} = \frac{\text{Set. Sol Removed, m}L/L}{\text{Set. Sol. in Influent, m}L/L} \times 100$$

$$= \frac{13.5 \text{ m}L}{14 \text{ m}L} \times 100$$

$$= \boxed{96\% \text{ Set. Sol. Removed}}$$

Example 3: (Settleable Solids)

❑ The settleable solids of the raw wastewater is 15 mL/L. If the settleable solids of the clarifier effluent is 0.3 mL/L, what is the settleable solids removal efficiency of the clarifier?

15 mL/L Set. Sol. → **Clarifier** → 0.3 mL/L Set. Sol.

↓

14.7 mL/L Set. Sol. Removed

$$\frac{\%\text{ Set. Sol.}}{\text{Removed}} = \frac{\text{Set. Sol Removed, m}L/L}{\text{Set. Sol. in Influent, m}L/L} \times 100$$

$$= \frac{14.7 \text{ m}L}{15 \text{ m}L} \times 100$$

$$= \boxed{98\% \text{ Set. Sol. Removed}}$$

18.6 SLUDGE TOTAL SOLIDS AND VOLATILE SOLIDS

Wastewater is comprised of both water and solids. The **total solids** may be further classified as either **volatile solids**, representing the organics, or **fixed solids**, representing the inorganics in the wastewater. This relationship must be clearly understood prior to any mathematical calculations regarding total solids content, fixed solids content, volatile solids content, or moisture content of any particular wastewater. These calculations may be expressed in terms of percent solids(by weight) or mg/*L* concentrations. Normally, total solids and volatile solids are expressed as percents; whereas suspended solids are generally expressed as mg/*L*.*

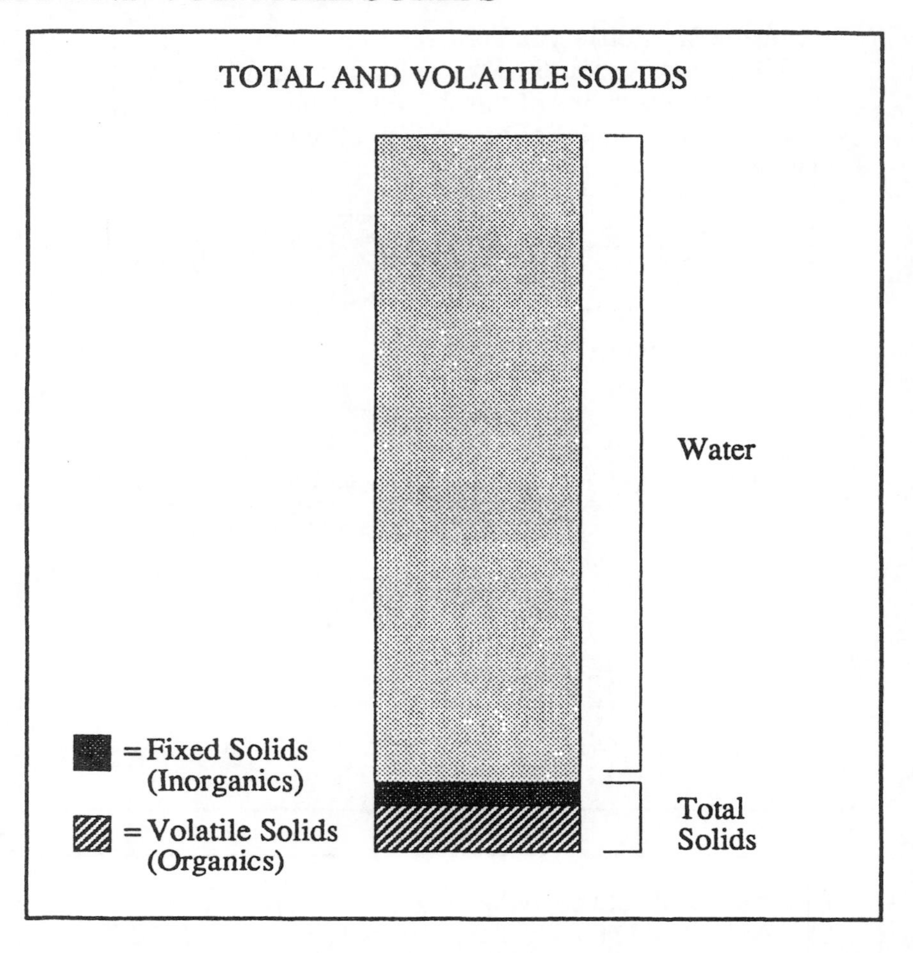

In calculating either percents or mg/*L* concentrations, certain concepts must be understood:

Total Solids
The **solids remaining after drying** wet sludge overnight at 103° - 105° C.

Fixed Solids
The **solids remaining after burning** for 1 hour at 600° C.

Volatile Solids
The **solids** which are destroyed or **lost through the 1-hour burning** period.

UNDERSTANDING THE TERMS IS ESSENTIAL

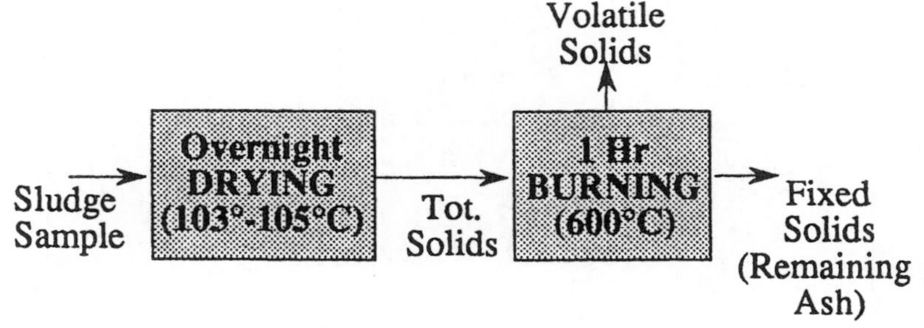

When the word "sludge" is used, it may be understood to mean a semi-liquid mass composed of **solids and water**. The term "solids" is used to mean dry solids after the evaporation of water.

This relationship may be expressed graphically as shown to the right. Percent total solids and volatile solids are calculated as follows:

$$\text{\% Total Solids} = \frac{\text{Total Solids Wt.}}{\text{Sludge Sample Wt.}} \times 100$$

$$\text{\% Volatile Solids} = \frac{\text{Vol. Solids Wt.}}{\text{Tot. Solids Wt.}} \times 100$$

* For a review of mg/*L* to % conversions, refer to Chapter 8.

Example 1: (Sludge Total and Volatile Solids)

❑ Given the information below, (a) determine the percent total solids in the sample, and (b) the percent of volatile solids in the sludge sample:

	Sludge (Total Sample)	After Drying	After Burning (Ash)
Wt. of Sample & Dish	71.82 g	24.57	22.95
Wt. of Dish (tare wt.)	22.08 g	22.08	22.08

a) To calculate % total solids, the grams total solids (solids after drying) and grams sludge sample must be determined:

Total Solids

24.57 g Total Solids & Dish
– 22.08 g Wt of Dish
2.49 g Total Solids

Sludge Sample

71.82 g Sludge & Dish
– 22.08 g Dish
49.74 g Sludge

$$\text{\% Total Solids} = \frac{\text{Wt. of Total Solids}}{\text{Wt. of Sludge Sample}} \times 100$$

$$= \frac{2.49 \text{ grams}}{49.74 \text{ grams}} \times 100$$

$$= \boxed{\text{5\% Total Solids}}$$

b) To calculate the % volatile solids, the grams total solids and grams volatile solids must be determined. Since total solids has already been calculated in Part a, only volatile solids must be calculated:

Volatile Solids

24.57 g Sample and Dish Before Burning
22.95 g Sample and Dish After Burning
1.62 g Solids Lost in Burning

$$\text{\% Vol. Solids} = \frac{\text{Wt. of Volatile Solids}}{\text{Wt. of Total Sample}} \times 100$$

$$= \frac{1.62 \text{ g}}{2.49 \text{ g}} \times 100$$

$$= \boxed{\text{65\% Volatile Solids}}$$

Example 2: (Sludge Total and Volatile Solids)

❑ Given the information below, calculate (a) the percent total solids and (b) percent volatile solids of the sludge sample.

	Sludge (Total Sample)	After Drying	After Burning (Ash)
Wt. of Sample & Dish	69.82 g	22.04	20.46
Wt. of Dish (tare wt.)	19.79 g	19.79	19.79

a) To calculate % total solids, the grams total solids (solids after drying) and grams sludge sample must be determined:

Total Solids

22.04 g Total Solids & Dish
− 19.79 g Wt of Dish
2.25 g Total Solids

Sludge Sample

69.82 g Sludge & Dish
− 19.79 g Dish
50.03 g Sludge

$$\text{\% Total Solids} = \frac{\text{Wt. of Total Solids}}{\text{Wt. of Sludge Sample}} \times 100$$

$$= \frac{2.25 \text{ grams}}{50.03 \text{ grams}} \times 100$$

$$= \boxed{4.5\% \text{ Total Solids}}$$

b) To calculate the % volatile solids, the grams total solids and grams volatile solids must be determined. Since total solids has already been calculated in Part a, only volatile solids must be calculated:

Volatile Solids

22.04 g Sample and Dish Before Burning
− 20.46 g Sample and Dish After Burning
1.58 g Solids Lost in Burning

$$\text{\% Total Solids} = \frac{\text{Wt. of Volatile Solids}}{\text{Wt. of Total Sample}} \times 100$$

$$= \frac{1.58 \text{ g}}{2.25 \text{ g}} \times 100$$

$$= \boxed{70\% \text{ Volatile Solids}}$$

Example 3: (Sludge Total and Volatile Solids)

❑ In Example 2, the 100-ml sludge sample was found to contain 1.58 grams volatile solids. What is this expressed as mg/*L*?

$$\frac{1.58\text{ g VS}}{100\text{ m}L\text{ Sample}} \times \frac{1000\text{ mg}}{1\text{ g}} = \frac{1580\text{ mg VS}}{100\text{ m}L} \frac{\times 10}{\times 10} = \frac{15800\text{ mg VS}}{1000\text{ m}L}$$

or = **15,800 mg/*L* VS**

Example 4: (Sludge Total and Volatile Solids)

❑ A 100 ml sludge sample has been dried and burned. Given the information below, a) determine the percent volatile solids content of the sample, and b) determine the mg/*L* concentration of the volatile solids.

	After Drying	After Burning (Ash)
Wt. of Sample & Crucible	22.0153 g	22.0067 g
Wt. of Crucible (tare wt.)	22.0021 g	22.0021 g

a) To calculate % volatile solids, you will first need to know grams volatile solids and grams total solids:

Total Solids

22.0153 g Sample & Crucible After Drying
– 22.0021 g Crucible Wt
0.0132 g Total Solids

Volatile Solids

22.0153 g Sample & Crucible Before Burning
– 22.0067 g Sample & Crucible After Burning
0.0086 g Volatile Solids

$$\%\text{ Volatile Solids} = \frac{\text{Wt. of Volatile Solids}}{\text{Wt. of Total Solids}} \times 100$$

$$= \frac{0.0086\text{ g}}{0.0132\text{ g}} \times 100$$

= **65% Volatile Solids**

b) Express the volatile solids content as mg/*L*:

$$\frac{0.0086\text{ g VS}}{100\text{ m}L\text{ Sample}} \times \frac{1000\text{ mg}}{1\text{ g}} = \frac{8.6\text{ mg VS}}{100\text{ m}L} \frac{\times 10}{\times 10} = \frac{86\text{ mg VS}}{1000\text{ m}L}$$

or = **86 mg/*L* VS**

CALCULATING mg/*L* TOTAL AND VOLATILE SOLIDS

The results of the total and volatile solids may be expressed as mg/*L*. To do this, the grams total or volatile solids must be known as well as the volume of sludge sample (milliliters). Since sludge samples are generally 100 milliliters, mg/*L* is calculated as shown below.

$$\frac{1\text{ g}}{100\text{ m}L} \times \underbrace{\frac{1000\text{ mg}}{1\text{ g}}}_{\text{Step 1}} \underbrace{\frac{\times 10}{\times 10}}_{\text{Step 2}}$$

$$= \frac{10{,}000\text{ mg}}{1000\text{ m}L} = 10{,}000\text{ mg/}L$$

In Step 1, grams solids (total or volatile) are converted to milligrams solids. Note that only the numerator is multiplied by 1000 since grams are simply being re-expressed as an equivalent number of milligrams. The numerator is now in the desired terms.

The denominator is desired in liters. To obtain liters, the denominator (100 ml) must be multiplied by 10. Since this is not a conversion of terms, if the denominator is multiplied by 10, the numerator must also be multiplied by 10. (Multiplying by 10/10 does not change the overall value of the fraction because 10/10 = 1.)

Examples 3 and 4 illustrate the calculation of mg/*L* total or volatile solids using sludge sample data.

18.7 SUSPENDED SOLIDS AND VOLATILE SUSPENDED SOLIDS (OF WASTEWATER)

The total and volatile solids of sludge are generally expressed as percents, by weight. The sludge samples are 100 m*L* and are unfiltered.

The suspended solids test is designed to measure the solids load or the strength of the wastewater by measuring the suspended particles in the water. Because the solids are filtered from the sample, often the sample is limited to 50 m*L* to prevent clogging of the filter.

Except for the required drying time, the suspended solids and volatile suspended solids tests of wastewater are similar to those of the total and volatile solids performed for sludges (described in the previous section).

Suspended solids are normally reported as milligrams per liter (mg/*L*) whereas volatile suspended solids are normally given as a percent (%). The equations for these calculations are described on the facing page. Note the similarity between these equations and those used in the previous section for sludge total solids and volatile solids.

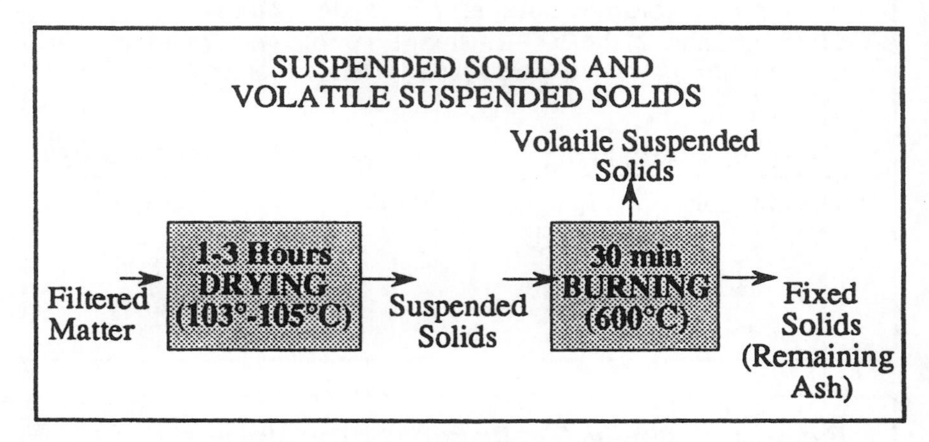

Example 1: (SS & VSS)

❑ Given the following information regarding a primary effluent sample, (a) calculate the mg/*L* suspended solids, and (b) the percent volatile suspended solids of the sample.

	After Drying (Before Burning)	After Burning (Ash)
Wt. of Sample & Dish	24.6862 g	24.6830 g
Wt. of Dish (tare wt.)	24.6820 g	24.6820 g

Sample Volume = 50 m*L*

a) To calculate the milligrams suspended solids per liter of sample (mg/*L*), you must first determine grams suspended solids:

24.6862 g Dish and Suspended Solids
− 24.6820 g Dish
0.0042 g Suspended Solids

Now mg/*L* suspended solids can be calculated:

$$\frac{0.0042 \text{ g SS}}{50 \text{ m}L} \times \frac{1000 \text{ mg}}{1 \text{ g}} \times \frac{20^*}{20} = \frac{84 \text{ mg}}{1000 \text{ m}L} = \boxed{84 \text{ mg}/L \text{ SS}}$$

b) To calculate percent volatile suspended solids, you must know the weight of both total suspended solids (calculated in part "a" above) and volatile suspended solids.

24.6862 g Dish & SS Before Burning
− 24.6830 g Dish & SS After Burning
0.0032 g Solids Lost in Burning

$$\% \text{ VSS} = \frac{\text{Wt. of Vol. Solids}}{\text{Wt. of Susp. Solids}} \times 100$$

$$= \frac{0.0032 \text{ g VSS}}{0.0042 \text{ g SS}} \times 100$$

$$= \boxed{76\% \text{ VSS}}$$

* Multiplication factor used to make the denominator equal to 1 liter (1000 m*L*). This number will vary with sample volume.

Example 2: (SS & VSS)

❑ Given the following information regarding a treatment plant influent sample, a) calculate the mg/*L* suspended solids, and b) the percent volatile suspended solids of the sample.

	After Drying (Before Burning)	After Burning (Ash)
Wt. of Sample & Dish	24.2048 g	24.2002 g
Wt. of Dish (tare wt.)	24.1987 g	24.1987 g

Sample Volume = 25 m*L*

a) To determine mg/*L* suspended solids, first calculate the grams suspended solids in the wastewater sample:

$$\begin{array}{rl} & 24.2048 \text{ g Dish and Suspended Solids} \\ - & \underline{24.1987} \text{ g Dish} \\ & 0.0061 \text{ g Suspended Solids} \end{array}$$

Now calculate mg/*L* suspended solids:

$$\frac{0.0061 \text{ g SS}}{25 \text{ m}L} \times \frac{1000 \text{ mg}}{1 \text{ g}} \times \frac{40}{40} = \boxed{244 \text{ mg}/L \text{ SS}}$$

b) Calculate the % volatile suspended solids:

$$\begin{array}{rl} & 24.2048 \text{ g Dish \& SS } \underline{\text{Before}} \text{ Burning} \\ - & \underline{24.2002} \text{ g Dish \& SS } \underline{\text{After}} \text{ Burning} \\ & 0.0046 \text{ g Solids Lost in Burning} \end{array}$$

$$\% \text{ VSS} = \frac{\text{Wt. of Vol. Solids}}{\text{Wt. of Susp. Sample}} \times 100$$

$$= \frac{0.0046 \text{ g VSS}}{0.0061 \text{ g SS}} \times 100$$

$$= \boxed{75\% \text{ VSS}}$$

<u>If a 25-m*L* sample is used,</u> mg/*L* suspended solids are calculated as follows: (25-m*L* samples are often used when samples filter slowly.)

$$\frac{\text{grams SS}}{25\text{-m}L \text{ Sample}} \times \overbrace{\frac{1000 \text{ mg}}{1 \text{ g}}}^{\text{Step 1}} \times \overbrace{\frac{40}{40}}^{\text{Step 2}} = \boxed{\text{mg}/L \text{ SS}}$$

In Step 1, grams suspended solids are re-expressed as milligrams suspended solids. In Step 2, 25 milliliters must be multiplied by 40 to equal 1000 m*L* (the desired 1 liter). Since the denominator of the fraction is to be multiplied by 40, the numerator must be multiplied as well.

<u>If a 50-m*L* sample is used,</u> the equation is similar. (The factor needed for Step 2 in this equation is 20 since 20 x 50 = 1000.)

$$\frac{\text{grams SS}}{50\text{-m}L \text{ Sample}} \times \frac{1000 \text{ mg}}{1 \text{ g}} \times \frac{20}{20} = \boxed{\text{mg}/L \text{ SS}}$$

To calculate % volatile suspended solids, the following equation is used:

$$\boxed{\% \text{ VSS} = \frac{\text{Vol. Solids Wt.}}{\text{Susp. Solids Wt.}} \times 100}$$

18.8 SLUDGE VOLUME INDEX (SVI) AND SLUDGE DENSITY INDEX (SDI)

Two variables are used to measure the settling characteristics of activated sludge. These are the **volume of the sludge** and the **density of the sludge**. Both variables indicate the settling characteristics of the sludge, one analysis is based on volume, and the other is based on density.

Because volume and density measurements are indirectly related, calculation of both SVI and SDI are not normally made. One of the two calculations is sufficient.

In the mathematical calculations of SVI and SDI, both volume and density factors will be part of the mathematical equation. The volume information is given by the milliliters (per liter) reported in the settleability test. The density information is given by the milligrams (per liter) reported in the Mixed Liquor Suspended Solids (MLSS) test. Both settleability and MLSS are tests taken from the activated sludge aeration tank.

Example 1: (SVI & SDI)

❑ The settleability test indicates that after 30 minutes, 210 m*L* of sludge settle in the 1-liter graduated cylinder. If the mixed liquor suspended solids (MLSS) concentration in the aeration tank is 2200 mg/*L*, what is the Sludge Volume Index?

$$SVI = \frac{\text{Volume}}{\text{Density}}$$

Volume: As given by the settleability test = 210 m*L* (per liter)

Density: As given by the MLSS concentration = 2200 mg (per liter)

$$SVI = \frac{210\ mL}{2200\ mg}$$

Because the basic definition of SVI requires **milliliters per gram**, milligrams must be converted to grams.*

$$SVI = \frac{210\ mL}{2200.\ mg} = \frac{210\ mL}{2.2\ g}$$

$$= \frac{95\ mL}{1\ g}$$

$$SVI = \boxed{95}$$

Example 2: (SVI & SDI)

❑ The activated sludge settleability test indicates 410 m*L* settling in the 2-liter graduated cylinder. If the MLSS concentration in the aeration tank is 2340 mg/*L*, what is the sludge volume index?

$$SVI = \frac{\text{Volume}}{\text{Density}}$$

Volume: As given by the settleability test = 410 m*L* (per 2 liters) or = 205 m*L* (per liter)

Density: As given by the MLSS concentration = 2340 m*L* (per liter)

$$SVI = \frac{205\ mL}{2340\ mg}$$

Converting mg to grams:

$$SVI = \frac{205\ mL}{2340.\ mg} = \frac{205\ mL}{2.34\ g}$$

$$= \frac{87.6\ mL}{1\ g}$$

$$SVI = \boxed{87.6}$$

* For a review of metric conversions, refer to Chapter 8.

Example 3: (SVI & SDI)

❑ The settleability test indicates that after 30 minutes, 390 m*L* of sludge settle in the 2-liter graduated cylinder. If the mixed liquor suspended solids (MLSS) concentration in the aeration tank is 2170 mg/*L*, what is the sludge volume index?

SVI = **Volume / Density**

Volume → As given by the settleability test = 195 m*L* (per liter)

Density → As given by the MLSS concentration = 2170 mg (per liter)

$$\text{SVI} = \frac{195 \text{ m}L}{2170 \text{ mg}}$$

Because the basic definition of SVI requires **milliliters per gram**, milligrams must be converted to grams.**

$$\text{SVI} = \frac{195 \text{ m}L}{2170. \text{ mg}} = \frac{195 \text{ m}L}{2.17 \text{ g}}$$

$$= \frac{90 \text{ m}L}{1 \text{ g}}$$

$$\text{SVI} = \boxed{90}$$

Example 4: (SVI & SDI)

❑ The activated sludge settleability test indicates 440 m*L* settling in the 2-liter graduated cylinder. If the MLSS concentration in the aeration tank is 2610 mg/*L*, what is the sludge volume index?

SVI = **Volume / Density**

Volume → As given by the settleability test = 440 m*L* (per 2 liters) or = 220 m*L* (per liter)

Density → As given by the MLSS concentration = 2610 m*L* (per liter)

$$\text{SVI} = \frac{220 \text{ m}L}{2610 \text{ mg}}$$

Converting mg to grams:

$$\text{SVI} = \frac{220 \text{ m}L}{2610. \text{ mg}} = \frac{220 \text{ m}L}{2.61 \text{ g}}$$

$$= \frac{84 \text{ m}L}{1 \text{ g}}$$

$$\text{SVI} = \boxed{84}$$

SLUDGE VOLUME INDEX (SVI)

In calculations of SVI, **volumes are being compared** and volume is in the numerator of the SVI equation:

$$\text{SVI} = \frac{\text{Volume, m}L}{\text{Density, g}}$$

SVI is defined as the milliliters volume occupied by 1 gram of activated sludge after a 30-minute settling period.*

$$\text{SVI} = \frac{\text{m}L}{1 \text{ gram}}$$

* Normally the SVI of a good quality activated sludge will range from 50-100. As the Index increases to 200 or more, the sludge is considered to be bulked sludge.

SLUDGE DENSITY INDEX

The calculation of sludge density index (SDI) is different from that of sludge volume index (SVI) in two respects:

- The numerators and denominators are "flipped". **For SDI, the density data is in the numerator**, whereas for SVI the volume data is in the numerator, and
- The **fraction must be multiplied by 100**, since SDI is essentially a percent calculation.* This is reflected by the definition of SDI.

SDI is defined as the concentration, in **percent solids**, which the activated sludge will assume after 30 minutes settling. Examples 5-8 illustrate the SDI calculation.

$$\text{SDI} = \frac{\text{Density, g}}{\text{Volume, mL}} \times 100$$

To remember whether density or volume goes in the numerator of SVI and SDI, simply remember that **volume** goes in the numerator of **sludge volume index (SVI)**, and **density** goes in the numerator for **sludge density index (SDI)**.

Example 5: (SVI & SDI)

❑ The MLSS concentration in the aeration tank is 2400 mg/*L*. If the activated sludge settleability test indicates 215 m*L* settled in the one-liter graduated cylinder, what is the sludge density index?

$$\text{SDI} = \frac{\text{Density}}{\text{Volume}}$$

Density: As given by the MLSS concentration = 2400 mg (per liter)

Volume: As given by the settleability test = 215 m*L* (per liter)

$$\text{SDI} = \frac{2400 \text{ mg}}{215 \text{ mL}} \times 100$$

Convert the mg to grams:

$$\text{SDI} = \frac{2400. \text{ mg}}{215 \text{ mL}} = \frac{2.4 \text{ g}}{215 \text{ mL}} \times 100$$

$$= \boxed{1.11}$$

Example 5: (SVI & SDI)

❑ The MLSS concentration in the aeration tank is 1980 mg/*L*. If the activated sludge settleability test indicates 175 m*L* settled in the one-liter graduated cylinder, what is the sludge density index?

$$\text{SDI} = \frac{1980 \text{ mg}}{175 \text{ mL}} \times 100$$

Convert the mg to grams:

$$\text{SDI} = \frac{1980. \text{ mg}}{175 \text{ mL}} = \frac{1.98 \text{ g}}{175 \text{ mL}} \times 100$$

$$= \boxed{1.13}$$

* For a review of percents, refer to Chapter 5 in *Basic Math Concepts*.

Example 7: (SVI & SDI)

❑ The sludge volume index for a sludge is known to be 85. What is the sludge density index for that sludge?

$$SDI = \frac{100}{SVI}$$

$$= \frac{100}{85}$$

$$= \boxed{1.18}$$

If the SVI value is known, the SDI can be calculated as follows:

$$\boxed{SDI = \frac{100}{SVI}}$$

Example 7 illustrates this type of calculation.

Example 8: (SVI & SDI)

❑ The SVI for a sludge is 72. If the settleability test indicates that 195 m*L* of sludge settles in the 1-liter graduated cylinder, what must be the MLSS concentration, in mg/*L*?

Use the same equation as usual, filling in the known information. First solve for grams, then convert to milligrams (mg/*L* MLSS):

$$SVI = \frac{\text{Volume, m}L}{\text{Density, g}}$$

$$72 = \frac{195 \text{ m}L}{x \text{ g}}$$

$$x = \frac{195}{72}$$

$$= 2.7 \text{ g}$$

$$\text{or} = \boxed{2700 \text{ mg}/L \text{ MLSS}}$$

CALCULATING OTHER UNKNOWN VALUES

SVI and SDI calculations involve three variables: SVI (or SDI), milliliters, and grams. In Examples 1-7, either SVI or SDI were the unknown variables. In fact, any one of the three variables can be the unknown factor. Example 8 illustrates calculations of this type.

18.9 TEMPERATURE

Since the temperature of the wastewater affects its general characteristics, this test is one of the most frequently performed tests. The formulas normally used to convert Fahrenheit (°F) to Celsius °C, are as follows:

Fahrenheit to Celsius:

$$°C = 5/9\,(°F - 32°)$$

Celsius to Fahrenheit:

$$°F = 9/5\,(°C) + 32°$$

Because these formulas have parentheses around different terms, and one requires the addition of 32° while the other requires the subtraction of 32°, they are very difficult conversions to remember unless used constantly.

Another method of converting these two terms is more easily remembered and can be used regardless of whether the conversion is from Fahrenheit to Celsius or vice versa. This method consists of three steps, as illustrated at the top of this page.

Step 2 involves the only variable in these conversions. The decision of whether to multiply by 5/9 or 9/5 is dependent whether a larger or smaller number is desired.

The Celsius temperature scale is a lower range scale than that of Fahrenheit. For example, the boiling point of water (at sea level) expressed in Fahrenheit is 212°; whereas the boiling point expressed in Celsius is 100°. Therefore, when converting 0° Celsius to Fahrenheit, the answer should be greater than 0°. When converting 80° Fahrenheit to Celsius, the answer should be less than 80°.

THE THREE-STEP METHOD OF TEMPERATURE CONVERSION

Step 1: Add 40°

Step 2: Multiply by 5/9 or 9/5
(Depending on direction of conversion.)

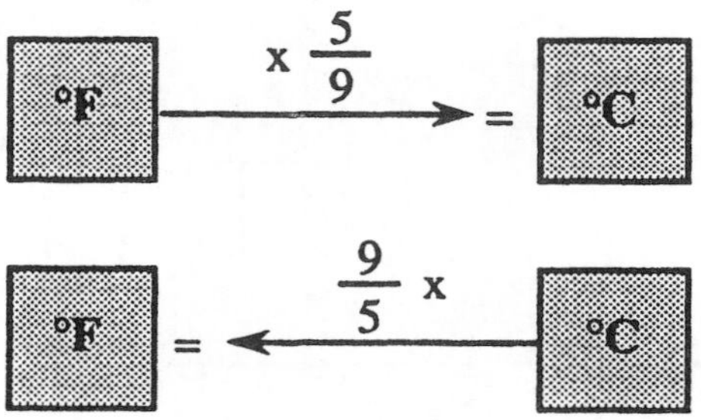

$$°F \xrightarrow{\times \frac{5}{9}} = °C$$

$$°F = \xleftarrow{\frac{9}{5} \times} °C$$

Step 3: Subtract 40°

THE EFFECT OF MULTIPLYING BY 5/9 OR 9/5

When **multiplying by 5/9** (approximately 1/2), the answer will be **less than the original number.** For example:

$$100 \times \frac{5}{9} = \frac{500}{9} = 56$$

Compare

When **multiplying by 9/5** (approximately 2), the answer will be **greater than the original number**. For example:

$$100 \times \frac{9}{5} = \frac{900}{5} = 180$$

Compare

Example 1: (Temperature)

❑ The influent to a treatment plant has a temperature of 75° F. What is this temperature expressed in degrees Celsius?

Step 1: (Add 40°)

$$\begin{array}{r} 75^\circ \\ +40^\circ \\ \hline 115^\circ \end{array}$$

Step 2: (Multiply by 5/9 or 9/5)

In this example the conversion is from Fahrenheit to Celsius. Since the answer should be a **smaller number**, multiply by 5/9:

$$\frac{(5)}{9}\ \frac{(115^\circ)}{1} = \frac{575^\circ}{9}$$

$$= 64^\circ$$

Step 3: (Subtract 40°)

$$\begin{array}{r} 64^\circ \\ -40^\circ \\ \hline 24^\circ \end{array}$$

$$\boxed{75^\circ\ F = 24^\circ\ C}$$

Example 2: (Temperature)

❑ The effluent of a treatment plant is 23° C. What is this expressed in degrees Fahrenheit?

Step 1: (Add 40°)

$$\begin{array}{r} 23^\circ \\ +40^\circ \\ \hline 63^\circ \end{array}$$

Step 2: (Multiply by 5/9 or 9/5)

In this example the conversion is from Celsius to Fahrenheit. Since the answer should be a **larger number**, multiply by 9/5:

$$\frac{(9)}{5}\ \frac{(63^\circ)}{1} = \frac{567^\circ}{5}$$

$$= 113^\circ$$

Step 3: (Subtract 40°)

$$\begin{array}{r} 113^\circ \\ -40^\circ \\ \hline 73^\circ \end{array}$$

$$\boxed{23^\circ\ C = 73^\circ\ F}$$

NOTES:

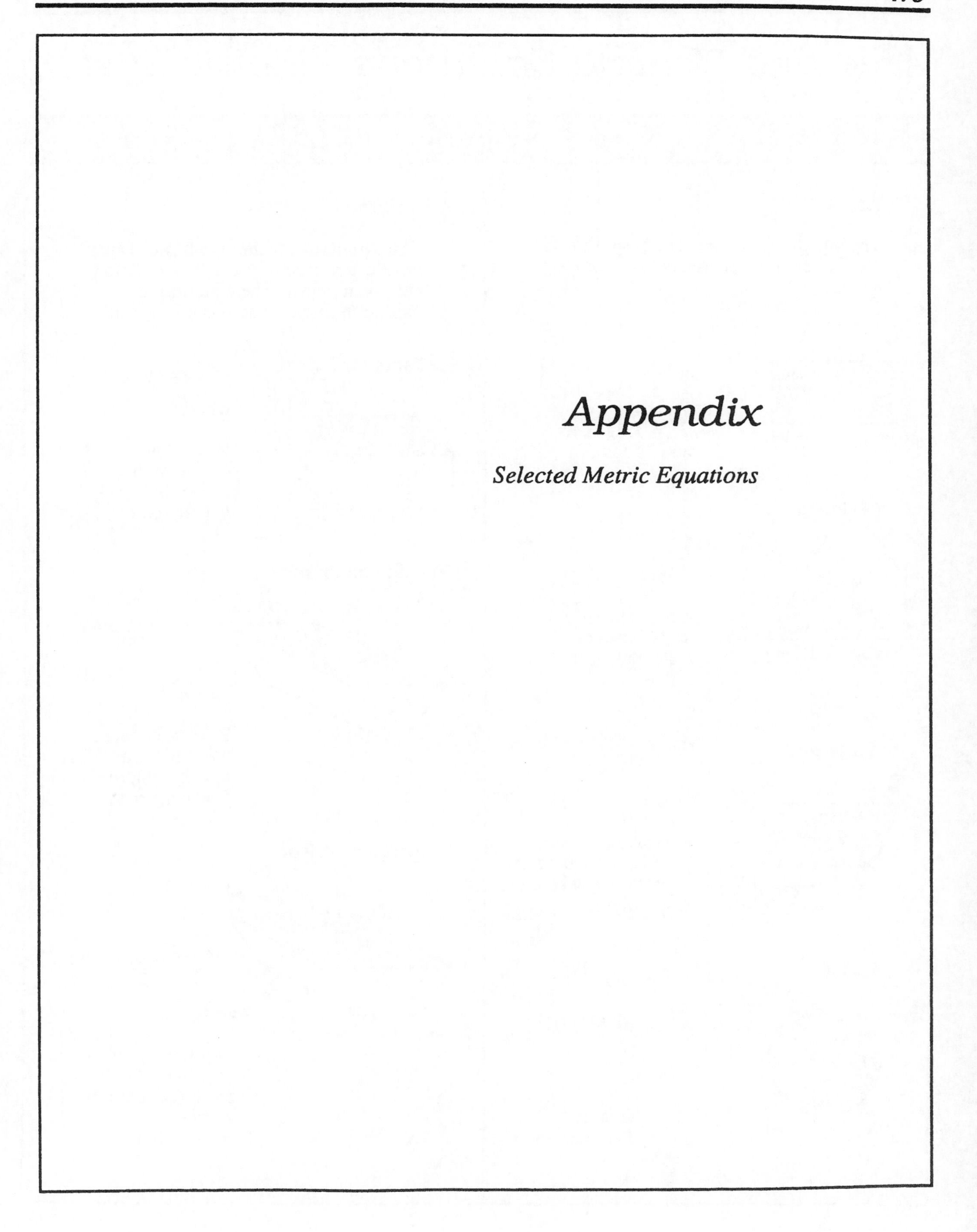

Appendix

Selected Metric Equations

Area and Volume Equations—Metric System

SUMMARY OF EQUATIONS

1. Areas

The equations for the four basic shapes most often used in area calculations are as follows:

Rectangle

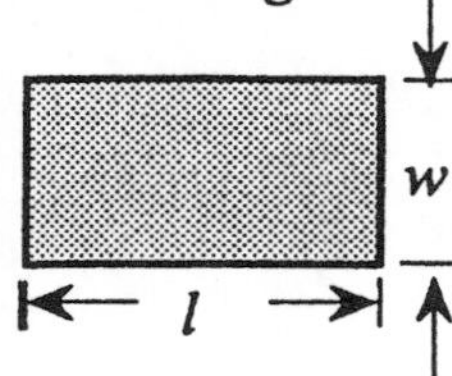

$$A = lw$$

where:

A = area, m^2
l = length, m
w = width, m

Triangle

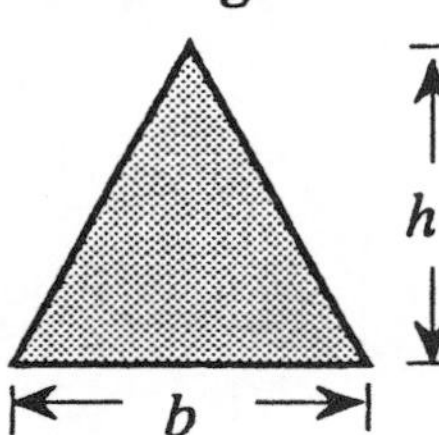

$$A = \frac{bh}{2}$$

where:

A = area, m^2
b = base, m
h = height, m

Trapezoid

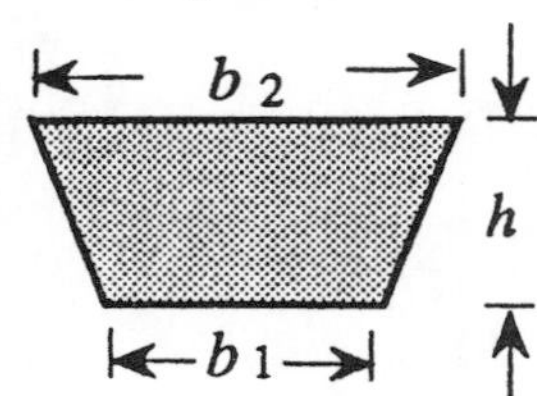

$$A = \frac{(b_1 + b_2)h}{2}$$

where:

A = area, m^2
b_1 = smaller base, m
b_2 = larger base, m
h = height, m

Circle

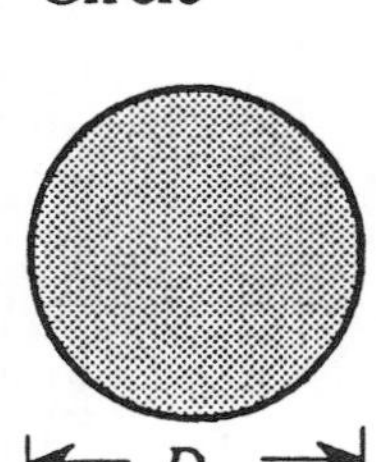

$$A = \frac{\pi D^2}{4}$$

or $A = (0.785)(D^2)$

where:

A = area, m^2
D = diameter, m

2. Volumes

The equations for the five basic shapes most often used in volume calculations are given below. The equation for oxidation ditch volume is also given.

Rectangular Prism

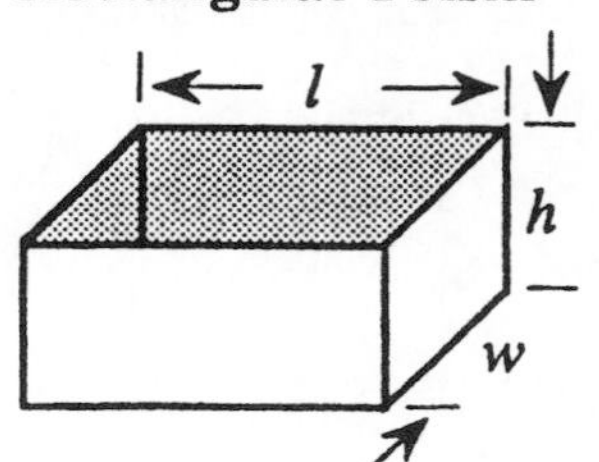

$$V = lwh$$

where:

V = volume, m^3
l = length, m
w = width, m
h = height, m

Triangular Prism

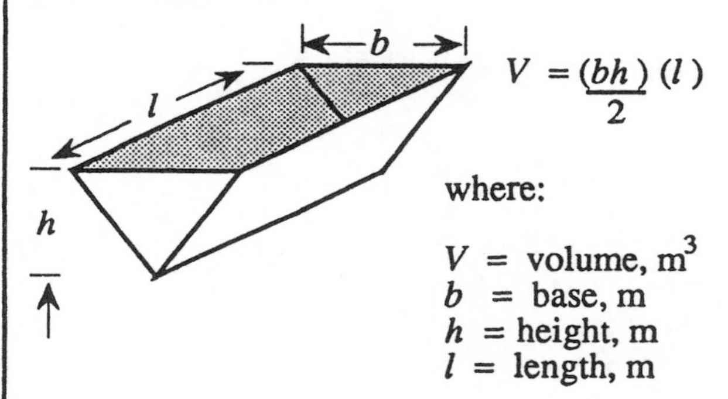

$$V = \frac{(bh)}{2}(l)$$

where:

V = volume, m^3
b = base, m
h = height, m
l = length, m

Trapezoidal Prism

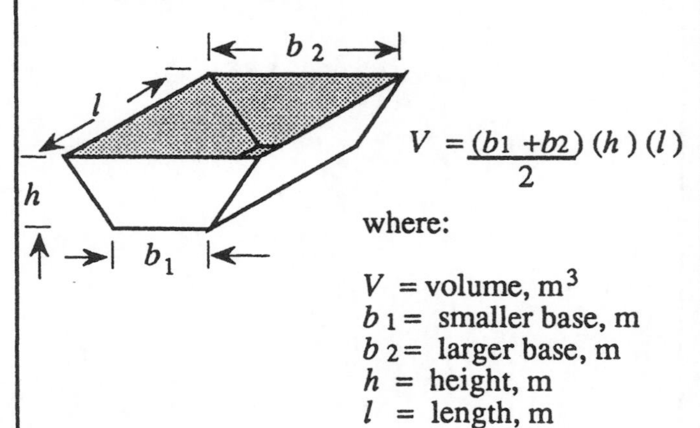

$$V = \frac{(b_1 + b_2)}{2}(h)(l)$$

where:

V = volume, m^3
b_1 = smaller base, m
b_2 = larger base, m
h = height, m
l = length, m

SUMMARY OF EQUATIONS

Cylinder

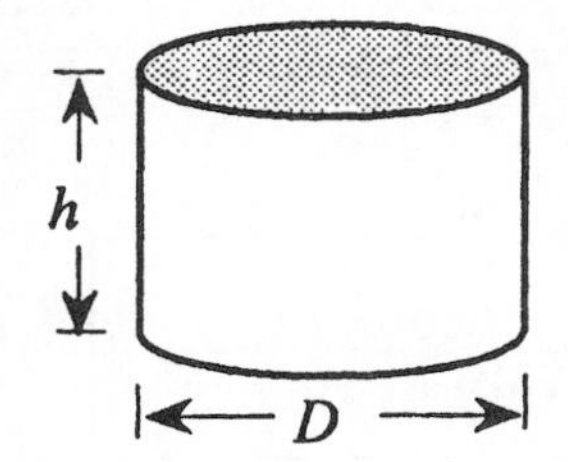

$V = (0.785)(D^2)(h.)$

where:

V = volume, m^3
D = diameter, m
h = height, m

Cone

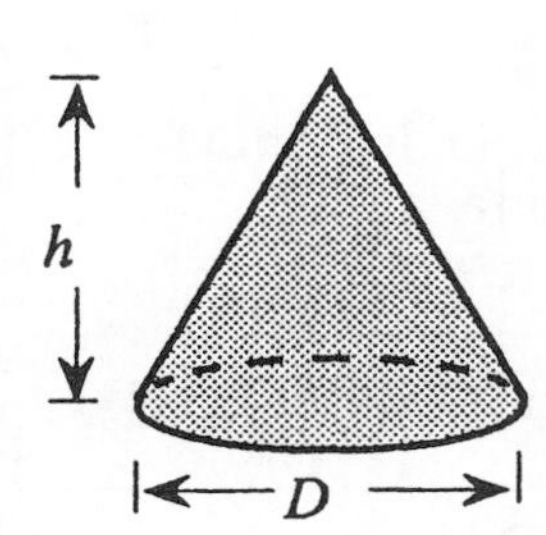

$V = \frac{1}{3}(0.785)(D^2)(h)$

where:

V = volume, m^3
D = diameter, m
h = height, m

Oxidation Ditch

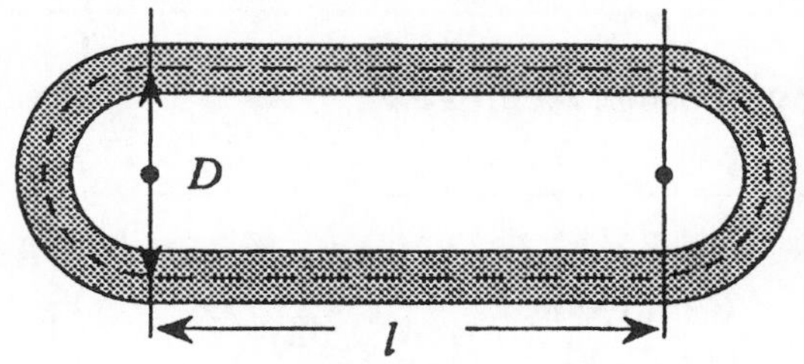

Top View

Dashed line represents total ditch length (L). This is equal to 2 half circumferences + 2 lengths

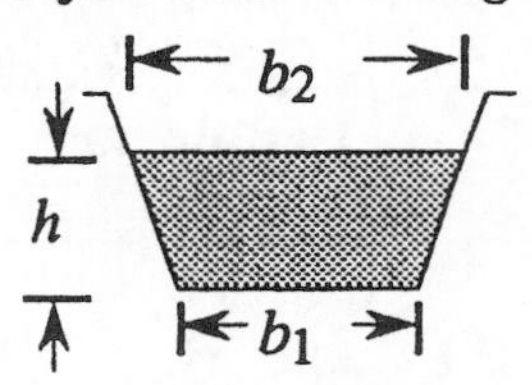

Cross-Section

Volume cu ft = (Trapezoidal Area) (Total Length)

$$V = \left[\frac{(b_1 + b_2)(h)}{2}\right]\left[\text{(Length of 2 Sides)} + \text{(Length Around 2 Half Circles)}\right]$$

$$V = \left[\frac{(b_1 + b_2)(h)}{2}\right]\left[(2l + \pi D)\right]$$

where:

V = volume, m^3
b_1 = smaller base, m
b_2 = larger base, m
h = height, m
l = length, m

Preliminary Treatment — Metric System

SUMMARY OF EQUATIONS

1. Screenings Removed

$$\text{Screenings Removed } (m^3/\text{day}) = \frac{\text{screenings, } m^3}{\text{day}}$$

OR

$$\text{Screenings Removed } (L/m^3) = \frac{\text{screenings, L}}{\text{flow, } m^3}$$

2. Screenings Pit Capacity

$$\text{Screening pit capacity, days} = \frac{\text{screening pit vol., } m^3}{\text{screening removed, } m^3/d}$$

3. Grit Channel Velocity

Two equations may be used to estimate the velocity of flow through the grit channel. The first equation is the $Q=AV$ equation.

$$Q_{m^3/s} = AV_{m/s}$$

$$\underset{m^3/s}{Q} = \underset{m}{(\text{width})}\ \underset{m}{(\text{depth})}\ \underset{m/s}{(\text{velocity})}$$

The second equation is required when using a float or dye to time the velocity of flow.

$$\text{Velocity} = \frac{\text{distance traveled, m}}{\text{test time, s}}$$

4. Particle Settling Rate

$$\text{Settling Rate, m/s} = \frac{\text{depth, m}}{\text{settling time, s}}$$

5. Water Depth, Velocity and Channel Length

The equation often used to determine required channel length to permit particle settling is:

$$\text{Length, m} = \frac{\underset{m}{(\text{channel depth})}\ \underset{m/sec}{(\text{flow velocity})}}{\text{Settling Rate, m/s}}$$

To make this equation easier to remember, it can be rearranged as shown below:

$$\frac{\text{channel length, m}}{\text{channel depth, m}} = \frac{\text{flow velocity, m/s}}{\text{settling rate, m/s}}$$

$$\frac{l_1}{l_2} = \frac{V_1}{V_2}$$

6. Grit Removal

$$\text{Grit Removal } L/m^3 = \frac{\text{grit, L}}{\text{flow, } m^3}$$

SUMMARY OF EQUATIONS

7. Flow measurement

Flow measurement for a particular moment can be determined using the *Q=AV* equation. Metering devices often have instrumentation to read and record flow rates. However, charts, graphs and nomographs can also be used to determine flow rates.

Flow measurement — Using *Q=AV*

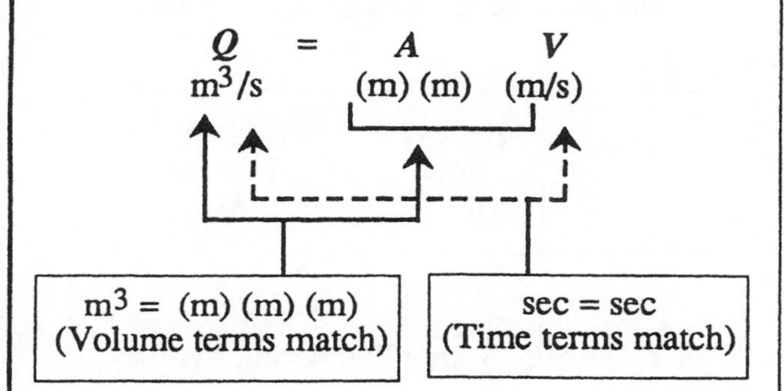

Sedimentation— Metric System

SUMMARY OF EQUATIONS

1. Detention Time

$$\text{Detention Time, hrs} = \frac{\text{tank volume, m}^3}{\text{flow, m}^3\text{/h}}$$

2. Weir Overflow Rate

$$\text{Weir Overflow Rate} = \frac{\text{flow, m}^3\text{/d}}{\text{weir, m}}$$

3. Surface Overflow Rate

$$\text{Surface Overflow Rate} = \frac{\text{flow, m}^3\text{/d}}{\text{area, m}^2}$$

(Surface overflow rate does not include recirculated flows.)

4. Solids Loading Rate on Clarifier

Simplified Equation:

$$\text{Solids Loading Rate} = \frac{\text{solids applied, kg/d}}{\text{surface area, m}^2}$$

Expanded Equation:

$$\text{Solids Loading Rate} = \frac{\dfrac{(\text{MLSS mg/L})(\text{Plant} + \text{RAS flow, m}^3\text{/d})}{1000}}{(0.785)(D^2)}$$

5. BOD and Suspended Solids Removed

First calculate BOD or SS removed, mg/L:

$$\text{Influent SS mg/L} - \text{Effluent SS mg/L} = \text{Removed SS mg/L}$$

Then calculate kg/d BOD or SS removed:

$$\frac{(\text{mg/L Removed})(\text{m}^3\text{/d flow})}{1000} = \text{BOD or SS Removed kg/d}$$

6. Dry Sludge Solids Produced, lbs/day

$$\frac{(\text{mg/L SS removed})(\text{m}^3\text{/d flow})}{1000} = \text{dry sludge solids kg/d}$$

7. Percent Solids and Clarifier Sludge Pumping

$$\%\ \text{Solids} = \frac{\text{solids, kg/d}}{\text{sludge, kg/d}} \times 100$$

This equation is sometimes rearranged for use in kg/d sludge calculations:

$$\text{Sludge, kg/d} = \frac{\text{solids, kg/d}}{\%\ \text{solids}} \times 100$$

SUMMARY OF EQUATIONS

8. Unit Process Efficiency

$$\text{\% SS Removed} = \frac{\text{SS removed, mg/L}}{\text{SS total, mg/L}} \times 100$$

$$\text{\% BOD Removed} = \frac{\text{BOD removed, mg/L}}{\text{BOD total, mg/L}} \times 100$$

Trickling Filters — Metric System

SUMMARY OF EQUATIONS

1. Hydraulic Loading Rate

$$\text{Hydraulic Loading Rate} = \frac{\text{Total Flow Applied, } m^3/d}{m^2 \text{ Area}}$$

(Hydraulic loading rate calculations include recirculated flows.)

2. Organic Loading Rate

Simplified Equation:

$$\text{Organic Loading Rate} = \frac{\text{kg/d BOD}}{m^3}$$

Expanded Equation:

$$\text{O. L. R.} = \frac{\dfrac{(\text{mg/L BOD})\,(m^3/\text{d flow})}{1000}}{m^3}$$

3. BOD and SS Removal, lbs/day

First calculate mg/L BOD or SS removed:

$$\begin{matrix}\text{BOD or SS} \\ \text{in Influent} \\ \text{(mg/L)}\end{matrix} - \begin{matrix}\text{BOD or SS} \\ \text{in Effluent} \\ \text{(mg/L)}\end{matrix} = \begin{matrix}\text{BOD or SS} \\ \text{Removed} \\ \text{(mg/L)}\end{matrix}$$

Then calculate kg/d BOD or SS removed:

$$\frac{\begin{matrix}\text{(mg/L BOD or SS)} \\ \text{Removed}\end{matrix}\ (m^3/\text{d flow})}{1000} = \begin{matrix}\text{kg/d} \\ \text{BOD or} \\ \text{SS Rem.}\end{matrix}$$

4. Unit Process or Overall Efficiency

Unit Process Efficiency:

$$\begin{matrix}\text{\% SS} \\ \text{Removed}\end{matrix} = \frac{\text{mg/L SS Removed}}{\text{mg/L SS Total}} \times 100$$

$$\begin{matrix}\text{\% BOD} \\ \text{Removed}\end{matrix} = \frac{\text{mg/L BOD Removed}}{\text{mg/L BOD Total}} \times 100$$

Overall Efficiency:

First calculate mg/L BOD or SS removed:

$$\begin{matrix}\text{BOD or SS} \\ \text{in Influent} \\ \text{(mg/L)}\end{matrix} - \begin{matrix}\text{BOD or SS} \\ \text{in Effluent} \\ \text{(mg/L)}\end{matrix} = \begin{matrix}\text{BOD or SS} \\ \text{Removed} \\ \text{(mg/L)}\end{matrix}$$

Then calculate the overall efficiency:

$$\begin{matrix}\text{\% Overall} \\ \text{Efficiency}\end{matrix} = \frac{\text{mg/L BOD or SS Removed}}{\text{mg/L BOD or SS in Influent}} \times 100$$

5. Recirculation Ratio

$$\text{Recirculation Ratio} = \frac{\text{Recirculated Flow, } m^3/d}{\text{Plant Influent Flow, } m^3/d}$$

Rotating Biological Contactors — Metric

SUMMARY OF EQUATIONS

1. Hydraulic Loading Rate

$$\text{Hydraulic Loading Rate} = \frac{\text{Total Flow Applied, m}^3\text{/d}}{\text{m}^2 \text{ Area}}$$

OR

$$\text{Hydraulic Loading Rate} = \frac{\text{Total Flow Applied, L/d}}{\text{m}^2 \text{ Area}}$$

(Hydraulic loading rate calculations include recirculated flows.)

2. Soluble BOD, mg/L

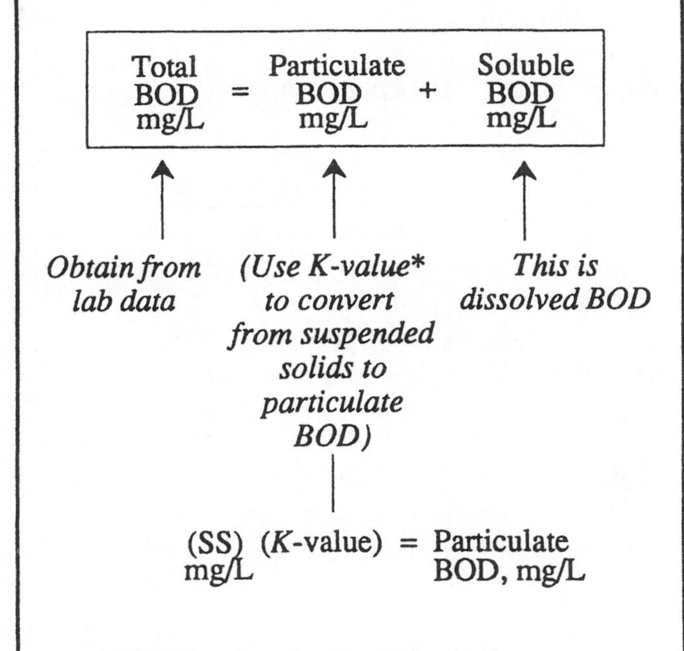

**(The K-value for most domestic wastewaters is 0.5-0.7)*

3. Organic Loading Rate

$$\text{System Organic Loading Rate, kg/d/1000-m}^2 = \frac{\text{kg/d Soluble BOD}}{\text{1000-m}^2 \text{ total stages}}$$

$$\text{First Stage Organic Loading Rate, kg/d/1000-m}^2 = \frac{\text{Soluble BOD, kg/d}}{\text{1000-m}^2 \text{ first stage}}$$

Activated Sludge — Metric System

SUMMARY EQUATIONS

1. BOD or COD Loading, kg/d

$$\frac{(\text{mg/L BOD})(\text{m}^3\text{/d flow})}{1000} = \text{BOD loading kg/d}$$

Or *

$$\frac{(\text{mg/L COD})(\text{m}^3\text{/d flow})}{1000} = \text{COD loading kg/d}$$

2. Food/Microorganism Ratio

Simplified Equation:

$$\text{F/M} = \frac{\text{BOD, kg/d}}{\text{MLVSS, kg}}$$

Expanded Equation:

$$\text{F/M} = \frac{\dfrac{(\text{mg/L BOD})(\text{m}^3\text{/d flow})}{1000}}{\dfrac{(\text{mg/L MLVSS})(\text{Aer Vol, m}^3)}{1000}}$$

3. Detention Time

$$\text{Detention Time, hrs} = \frac{\text{Volume, m}^3}{\text{Flow, m}^3\text{/h}}$$

4. Solids Inventory in Aeration Tank

$$\text{kg MLSS} = \frac{(\text{mg/L MLSS})(\text{Aer. Vol., m}^3)}{1000}$$

OR

$$\text{kg MLVSS} = \frac{\left(\text{mg/L MLSS}\right)\left(\text{Aer. Vol., m}^3\right)\left(\dfrac{\text{\% Vol. Sol.}}{100}\right)}{1000}$$

5. Return Sludge Rate —Using Settleability

The equation is written as a ratio:

$$\frac{R}{Q} = \frac{\text{Settled sludge volume, ml/L}}{\text{1000 ml/L} - \text{Settled sludge volume, ml/L}}$$

The RAS ratio can then be used to calculate RAS flow rate in m^3/d:

$$\frac{R}{Q} = \frac{\text{RAS flow rate, m}^3\text{/d}}{\text{Secondary influent flow rate, m}^3\text{/d}}$$

* COD can be used if there is generally a good correlation in BOD and COD characteristics of the wastewater.

SUMMARY EQUATIONS

6. Return Sludge Rate — Using Secondary Clarifier Mass Balance (Solids Balance)

Simplified Equation:

Suspended solids in = Suspended solids out

Expanded Equation:

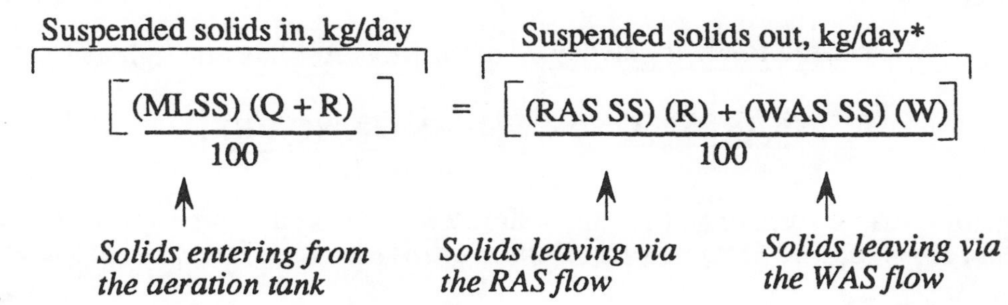

Where:

MLSS = mixed liquor suspended solids, mg/L
Q = secondary influent flow, m^3/d
R = return sludge flow, m^3/d
RAS SS = return activated sludge SS, mg/L
WAS SS = waste activated sludge SS, mg/L
W = waste sludge flow, m^3/d
R = return sludge flow, m^3/d

The equation may be rearranged so that both R terms are on the same side of the equation: *(Note that since the 100 factor is on both sides of the equation, it can be dropped.)*

$$(MLSS)(Q) + (MLSS)(R) = (RAS\ SS)(R) + (WAS\ SS)(W)$$

$$(MLSS)(Q) - (WAS\ SS)(W) = (RAS\ SS)(R) - (MLSS)(R)$$

$$(MLSS)(Q) - (WAS\ SS)(W) = \left[(RAS\ SS) - (MLSS)\right](R)$$

Then solve for R:

$$\frac{(MLSS)(Q) - (WAS\ SS)(W)}{(RAS\ SS) - (MLSS)} = R$$

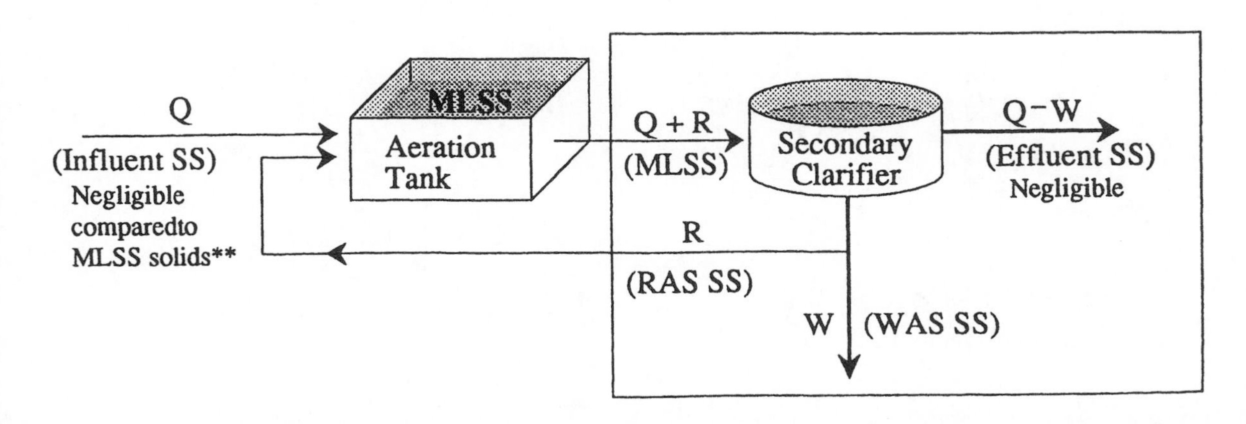

* This equation assumes negligible loss of solids in the effluent.
**Except for modified aeration processes which may have very low MLSS concentrations.

SUMMARY OF EQUATIONS

7. Return Sludge Rate — Using Aeration Tank Mass Balance (Solids Balance)

Simplified Equation:

$$\text{Suspended solids in} = \text{Suspended solids out *}$$

Expanded Equation:**

$$\underbrace{\left[\frac{(\text{RAS SS})(\text{R})}{100}\right]}_{\text{Suspended solids in, kg/day}} = \underbrace{\left[\frac{(\text{MLSS})(\text{Q}+\text{R})}{100}\right]}_{\text{Suspended solids out, kg/day}}$$

The equation may be rearranged so that both R terms are on the same side of the equation: (*Note that since the 100 factor is on both sides of the equation, it can be dropped.)*

$$(\text{RAS SS})(\text{R}) = (\text{MLSS})(\text{Q}) + (\text{MLSS})(\text{R})$$

$$(\text{RAS SS})(\text{R}) - (\text{MLSS})(\text{R}) = (\text{MLSS})(\text{Q})$$

$$\left[(\text{RAS SS}) - (\text{MLSS})\right](\text{R}) = (\text{MLSS})(\text{Q}) - (\text{WAS SS})(\text{W})$$

Then solve for R:

$$\text{R} = \frac{(\text{MLSS})(\text{Q})}{(\text{RAS SS}) - (\text{MLSS})}$$

Note that this is the **same equation** as for the secondary clarifier mass balance **except** that the aeration tank equation has no WAS term in the numerator (since there is no wasting from the aeration tank, as illustrated in the diagram to the left.

* For the aeration tank, this is true only when new cell growth in the tank is considered negligible.
** Abbreviation of terms is the same as that given for the secondary clarifier mass balance equation.

SUMMARY OF EQUATIONS

8. Solids Retention Time (SRT) (also called Mean Cell Residence Time, MCRT)

Simplified Equation:

$$\text{SRT, days} = \frac{\text{SS in System, kg}}{\text{SS Leaving System, kg/d}}$$

OR

$$\text{SRT, days} = \frac{\text{SS in System, kg}}{\text{WAS SS, kg/d} + \text{S.E. SS, kg/d}}$$

Expanded Equation:

$$\text{SRT, days} = \frac{\dfrac{(\text{MLSS mg/L})(\text{Aer. Vol., m}^3)}{\cancel{1000}} + \dfrac{(\text{CCSS mg/L})(\text{Fin. Clar.Vol., m}^3)}{\cancel{1000}}}{\dfrac{(\text{WAS SS mg/L})(\text{WAS Flow, m}^3\text{/d})}{\cancel{1000}} + \dfrac{(\text{S.E. SS, mg/L})(\text{Plant Flow, m}^3\text{/d})}{\cancel{1000}}}$$

(Since 1000 is found in each denominator, it can be divided out, thus simplifying the equation.)

Note: There are four ways to account for system solids in the SRT calculation. The preferred and most accurate calculation of system solids is given in the SRT equation above. The other three methods are shown below. Use the method which works best for your plant and stay with it.

2. To measure aeration tank solids and estimate clarifier solids:

$$\frac{(\text{MLSS, mg/L})(\text{Aer. Vol., m}^3)}{1000} + \frac{\left(\dfrac{\text{MLSS mg/L} + \text{RAS SS mg/L}}{2}\right)(\text{Sludge Blanket Vol., m}^3)}{1000} = \text{kg MLSS}$$

3. To measure aeration tank solids and estimate clarifier solids:

$$\frac{(\text{MLSS mg/L})(\text{Aer. Vol., m}^3 + \text{Fin. Clar., m}^3)}{1000} = \text{kg MLSS}$$

4. To measure aeration tank solids only:

$$\frac{(\text{MLSS mg/L})(\text{Aer. Vol., m}^3)}{1000} = \text{kg MLSS}$$

* CCSS is the average clarifier SS concentration of the entire water column sampled by a core sampler.

SUMMARY OF EQUATIONS

9. Wasting Rate (kg/d)– Using Solids Retention Time, SRT (also called Mean Cell Residence Time)

(Use the SRT equation to determine the WAS SS kg/d are the kg/d SS to be wasted.)

Simplified Equation:

$$\text{SRT, days} = \frac{\text{Suspended Solids in System, kg}}{\text{WAS SS, kg/d} + \text{S.E. SS, kg/d}}$$

Expanded Equation:

$$\text{SRT, days} = \frac{\dfrac{(\text{MLSS mg/L})(\text{Aer. Vol., m}^3)}{1000} + \dfrac{(\text{CCSS mg/L})(\text{Clar. Vol., m}^3)}{1000}}{\text{WAS SS kg/d} + \dfrac{(\text{S.E. SS, mg/L})(\text{m}^3\text{/d plant flow})}{1000}}$$

10. Wasting Rate (kg/day)– Using Constant F/M Ratio

Use the ***desired F/M*** ***ratio*** *and BOD or COD applied (food) to calculate the* ***desired kg MLVSS.*** *:*

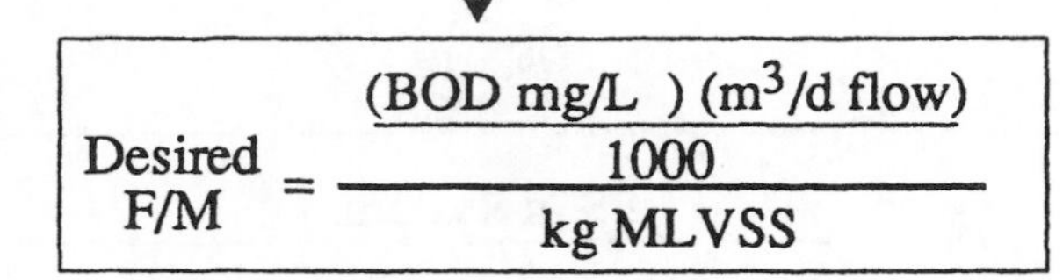

Then determine the ***desired kg MLSS*** *using % volatile solids:*

$$\text{Desired MLSS, kg} = \frac{\text{Desired MLVSS, kg}}{\dfrac{\%\ \text{VS}}{100}}$$

Compare the desired and actual MLSS to determine kg SS to be wasted:

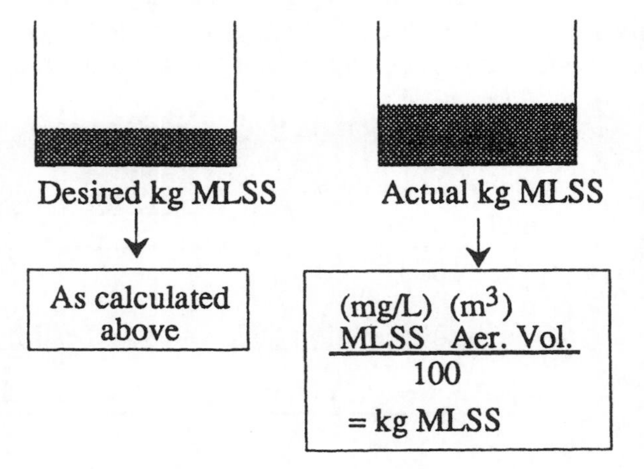

The kg SS to be wasted are therefore:

$$\text{Actual kg MLSS} - \text{Desired kg MLSS} = \text{kg SS to be Wasted}$$

SUMMARY OF EQUATIONS

11. Wasting Rate (kg/day)– Using Constant MLSS

Compare the desired and actual kg MLSS to determine kg SS to be wasted:

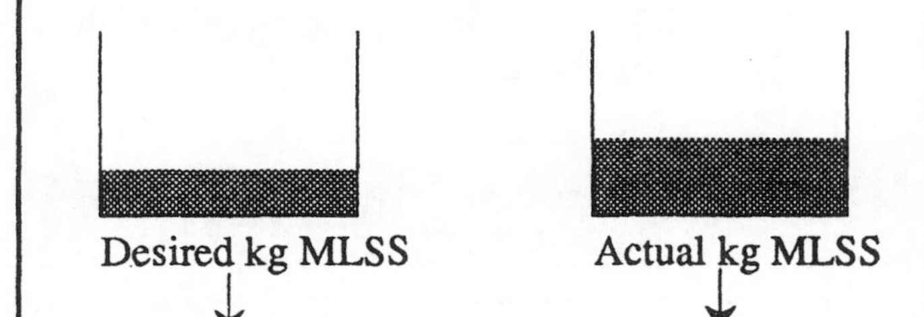

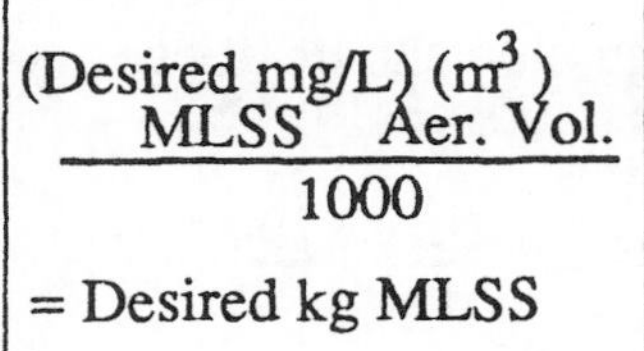

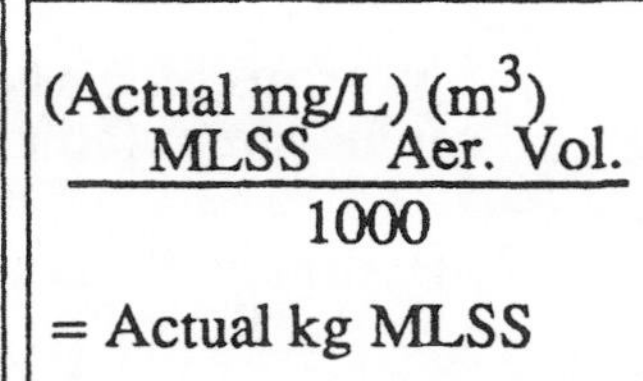

After calculating the desired and actual kg MLSS, **subtract** to determine kg SS to be wasted:

$$\text{Actual kg MLSS} - \text{Desired kg MLSS} = \text{kg SS to be Wasted}$$

12. WAS Pumping Rate — Using the mg/L to kg/d equation

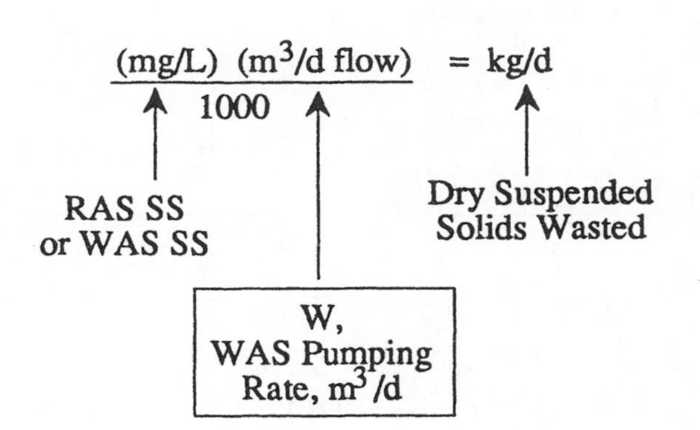

13. WAS Pumping Rate — Using the SRT Equation

$$\text{SRT, days} = \frac{\dfrac{(\text{MLSS mg/L})(\text{Aer. Vol., m}^3)}{\not{1000}} + \dfrac{(\text{CCSS mg/L})(\text{Fin. Clar. Vol., m}^3)}{\not{1000}}}{\dfrac{(\text{WAS SS mg/L})(\text{WAS Flow, m}^3/\text{d})}{\not{1000}} + \dfrac{(\text{S.E. SS, mg/L})(\text{Plant Flow, m}^3/\text{d})}{\not{1000}}}$$

WAS Pumping Rate, W

(Since 1000 is found in each denominator, it can be divided out, thus simplifying the equation.)

Waste Treatment Ponds — Metric System

SUMMARY OF EQUATIONS

1. BOD Loading

$$\frac{(\text{mg/L BOD})(\text{m}^3\text{/d flow})}{1000} = \text{kg/d BOD}$$

2. Organic Loading Rate

Simplified Equation:

$$\text{Organic Loading} = \frac{\text{kg/d BOD}}{\text{m}^2}$$

Expanded Equation:

$$\text{Organic Loading} = \frac{\frac{(\text{mg/L BOD})(\text{m}^3\text{/d flow})}{1000}}{\text{m}^2}$$

3. BOD Removal Efficiency

$$\%\text{ BOD Removed} = \frac{\text{BOD Removed, mg/L}}{\text{BOD Total, mg/L}} \times 100$$

4. Hydraulic Loading Rate

Hydraulic loading rate can be calculated three different ways depending on which units are desired.

$$\text{Hydraulic Loading Rate} = \frac{\text{m}^3\text{/d flow}}{\text{m}^2\text{ Area}}$$

The terms of the equation shown above can be simplifed as cm/day:

$$\text{Hydraulic Loading Rate} = \frac{(\text{m})}{\text{day}} \frac{(100\text{ cm})}{\text{m}}$$

$$\text{Hydraulic Loading Rate} = \frac{\text{cm}}{\text{day}}$$

$$\text{Hydraulic Loading Rate} = \frac{\text{Depth of Pond, cm}}{\text{Detention Time, days}}$$

5. Population Loading and Population Equivalent

$$\text{Population Loading} = \frac{\text{persons}}{\text{m}^2}$$

$$\text{Population Equivalent} = \frac{\text{kg/d BOD}}{\text{kg/d BOD/person}}$$

SUMMARY OF EQUATIONS

6. Detention Time

$$\text{Detention Time, days} = \frac{\text{Volume of Pond, m}^3}{\text{Flow Rate, m}^3\text{/d}}$$

Chemical Dosage — Metric System

SUMMARY OF EQUATIONS

1. Chemical Feed Rate, lbs/day

$$\frac{(\text{mg/L Chem.})(\text{m}^3\text{/d Flow})}{1000} = \text{kg/d Chemical}$$

2. Chlorine Dose, Demand and Residual

$$Cl_2 \text{ Dose} = Cl_2 \text{ Demand} + Cl_2 \text{ Residual}$$

3. Chemical Feed Rate, kg/day
(Dosing hypochlorites and other chemicals less than full strength)

$$\frac{\dfrac{(\text{mg/L Chem.})(\text{m}^3\text{/d Flow})}{1000}}{\dfrac{\text{\% Strength of Chemical}}{100}} = \text{kg/d Chemical}$$

4. Percent Strength of Solutions

Percent strength using dry chemicals:

$$\text{\% Strength} = \frac{\text{Chemical, kg}}{\text{Solution, kg}} \times 100$$

Percent strength using liquid chemicals:

$$(\text{Liquid Polymer, kg})\frac{(\text{\% Strength of Liq. Polym.})}{100} = (\text{Polym. Sol'n kg})\frac{(\text{\% Strength of Poly. Sol'n.})}{100}$$

5. Mixing Solutions of Different Strength

$$\text{\% Strength of Mixture} = \frac{\text{Chemical in Mixture, kg}}{\text{Solution Mixture, kg}} \times 100$$

OR, if target strength is desired:

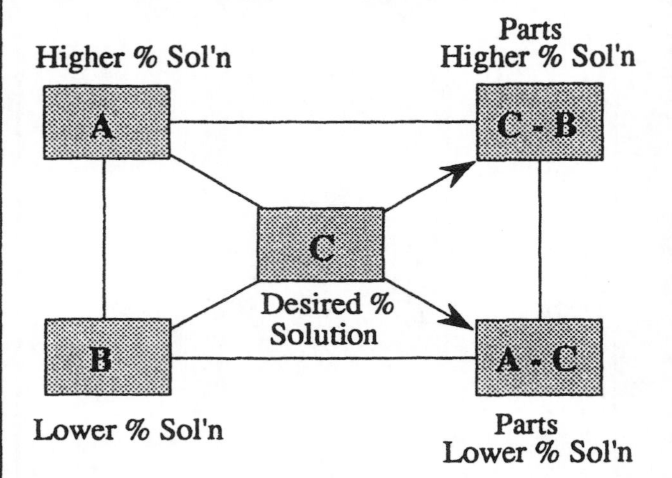

$$(\text{Fraction of Higher \% Sol'n})(\text{Total Desired lbs of Sol'n}) = \text{Wt of Higher \% Sol'n}$$

$$(\text{Fraction of Lower \% Sol'n})(\text{Total Desired lbs of Sol'n}) = \text{Wt of Lower \% Sol'n}$$

6. Solution Chemical Feeder Setting, kg/d

Simplified Equation:

$$\text{Desired Dose, kg/d} = \text{Actual Dose, kg/d}$$

Expanded Equation:

$$\frac{(\text{mg/L Dose})(\text{m}^3\text{/d Flow Treated})}{1000} = \frac{(\text{mg/L Sol'n})(\text{m}^3\text{/d Sol'n})}{1000}$$

SUMMARY OF EQUATIONS

7. Chemical Feed Pump—% Stroke Setting

$$\text{\% Setting} = \frac{\text{Required Feed Pump, ml/min}}{\text{Maximum Feed Pump, ml/min}} \times 100$$

8. Solution Chemical Feeder Setting, ml/min

First, calculate the m^3/d flow rate using the mg/L to kg/d equation:

$$\frac{(\text{mg/L Chem.})\ (m^3/\text{d Flow})}{1000} = \text{kg/d Chemical}$$

Then convert m^3/d flow rate to ml/min flow rate:

$$\frac{(m^3)}{d} \frac{(1000\ L)}{1\ m^3} \frac{(1000\ ml)}{1\ L} \frac{(1\ day)}{1440\ min} = \frac{ml}{min}$$

9. Dry Chemical Feeder Calibration

$$\text{Chemical Feed Rate, kg/d} = \frac{\text{Chemical Used, kg}}{\text{Application Time, days}}$$

10. Solution Chemical Feeder Calibration (Given ml/min Flow)

First convert ml/min flow to m^3/d flow:

$$\frac{(ml)}{min} \frac{(1440\ min)}{day} \frac{(1\ L)}{1000\ ml} \frac{(1\ m^3)}{1000\ L} = \frac{m^3}{d}$$

Then calculate chemical dosage, kg/day:

$$\frac{(\text{mg/L Chem.})\ (m^3/\text{d Flow})}{1000} = \text{kg/d Chemical}$$

11. Solution Chemical Feeder Calibration (Given Drop in Solution Tank Level)

First, calculate m^3/min pumped:

$$\text{Flow } m^3/\text{min} = \frac{\text{Volume Pumped, } m^3}{\text{Duration of Test, min}}$$

Then convert m^3/min to ml/min pumping rate:

$$\frac{(m^3)}{min} \frac{(1000\ L)}{1\ m^3} \frac{(1000\ ml)}{1\ L} = \text{Flow ml/min}$$

12. Average Use Calculations

First determine the average chemical use:

$$\text{Average Use kg/d} = \frac{\text{Total Chem. Used, kg}}{\text{Number of Days}}$$

OR

$$\text{Average Use ml/d} = \frac{\text{Total Chem. Used, ml}}{\text{Number of Days}}$$

Then calculate day's supply in inventory

$$\text{Day's Supply in Inventory} = \frac{\text{Total kg Chem. in Inventory}}{\text{Average Use, kg/d}}$$

OR

$$\text{Day's Supply in Inventory} = \frac{\text{Total ml Chem. in Inventory}}{\text{Average Use, ml/d}}$$

Sludge Digestion & Solids Handling—Metric

SUMMARY OF EQUATIONS

1. Sludge Thickening

Simplified Equation:

$$\text{kg/d Solids} = \text{kg/d Solids}$$

Expanded Equation:

$$\left(\begin{matrix}\text{Prim. or Sec}\\ \text{Sludge,}\\ \text{kg/d}\end{matrix}\right)\frac{(\%\text{ Solids})}{\cancel{100}} = \left(\begin{matrix}\text{Thickened}\\ \text{Sludge,}\\ \text{kg/d}\end{matrix}\right)\frac{(\%\text{Solids})}{\cancel{100}}$$

(Since the 100 factor is on **both sides** of the equation, it can be divided out, if desired.)

2. Mixing Different % Solids Sludges

Simplified Equation:

$$\begin{matrix}\%\text{ Solids of}\\ \text{Sludge Mixture}\end{matrix} = \frac{\text{kg/d Solids in Mixture}}{\text{kg/d Sludge Mixture}} \times 100$$

Expanded Equation:

$$\begin{matrix}\%\text{ Solids}\\ \text{of Sludge}\\ \text{Mixture}\end{matrix} = \frac{\begin{matrix}\text{kg/d}\\ \text{Prim. Sl. Sol.}\end{matrix} + \begin{matrix}\text{kg/d}\\ \text{Sec. Sl. Sol.}\end{matrix}}{\begin{matrix}\text{kg/d}\\ \text{Prim. Sludge}\end{matrix} + \begin{matrix}\text{kg/d}\\ \text{Sec. Sludge}\end{matrix}} \times 100$$

Or

$$\begin{matrix}\%\text{ Solids}\\ \text{of Sludge}\\ \text{Mixture}\end{matrix} = \frac{\left(\begin{matrix}\text{Prim.}\\ \text{Sludge}\\ \text{kg/d}\end{matrix}\right)\frac{(\%\text{ Sol.})}{100} + \left(\begin{matrix}\text{Sec.}\\ \text{Sludge}\\ \text{kg/d}\end{matrix}\right)\frac{(\%\text{ Sol.})}{100}}{\begin{matrix}\text{Prim. Sludge}\\ \text{kg/d}\end{matrix} + \begin{matrix}\text{Sec. Sludge}\\ \text{kg/d}\end{matrix}} \times 100$$

3. Sludge Volume Pumped

$$\begin{matrix}\text{Sludge}\\ \text{Pumped}\\ (m^3/\text{min})\end{matrix} = \left(\begin{matrix}m^3\text{ pumped}\\ \text{each Stroke}\end{matrix}\right)\left(\begin{matrix}\text{No. of Strokes}\\ \text{each Minute}\end{matrix}\right)$$

The pumping rate may be expressed as L/min or ml/min, if desired:

$$\begin{matrix}\text{Sludge}\\ \text{Pumped}\\ (\text{L/min})\end{matrix} = \frac{(m^3)}{\text{min}}\ \frac{(1000\text{ L})}{1\ m^3}$$

OR

$$\begin{matrix}\text{Sludge}\\ \text{Pumped}\\ (\text{ml/min})\end{matrix} = \frac{(m^3)}{\text{min}}\ \frac{(1000\text{ L})}{1\ m^3}\ \frac{(1000\text{ ml})}{1\text{ L}}$$

4. Volatile Solids to the Digester, kg/d

If kg/d solids have already been calculated, either of the following two equations may be used to calculate kg/d volatile solids:

$$\%\text{ Vol. Solids} = \frac{\text{kg/d Vol. Solids}}{\text{kg/d Tot. Solids}} \times 100$$

Or

$$\left(\begin{matrix}\text{Tot. Sol.}\\ \text{kg/d}\end{matrix}\right)\frac{\left(\begin{matrix}\%\text{ Vol.}\\ \text{Solids}\end{matrix}\right)}{100} = \begin{matrix}\text{Vol. Sol.}\\ \text{kg/d}\end{matrix}$$

If kg/d solids have not been determined yet, the following equation can be used to calculate kg/d volatile solids:

$$\left(\begin{matrix}\text{Sludge}\\ \text{kg/d}\end{matrix}\right)\frac{\left(\begin{matrix}\%\\ \text{Solids}\end{matrix}\right)}{(100)}\ \frac{\left(\begin{matrix}\%\text{ Vol.}\\ \text{Solids}\end{matrix}\right)}{(100)} = \begin{matrix}\text{Vol. Sol.}\\ \text{kg/d}\end{matrix}$$

SUMMARY OF EQUATIONS

5. Digester Loading Rate

Simplified Equation:

$$\text{Digester Loading} = \frac{\text{kg/d VS Added}}{\text{m}^3\ \text{Volume}}$$

Expanded Equation:

$$\text{Digester Loading} = \frac{(\text{kg/d Sludge})\ \frac{(\%\ \text{Solids})}{100}\ \frac{(\%\ \text{VS})}{100}}{(0.785)\ (\text{D}^2)\ (\text{Water Depth, m}^3)}$$

6. Digester Volatile Solids Loading Ratio

Simplified Equation:

$$\text{VS Loading Ratio} = \frac{\text{kg/d VS Added}}{\text{kg VS in Digester}}$$

Expanded Equation:

$$\text{VS Loading Ratio} = \frac{(\text{kg/d Sludge})\ \frac{(\%\ \text{Sol})}{\cancel{100}}\ \frac{(\%\ \text{VS})}{\cancel{100}}}{(\text{kg Sludge in Dig.})\ \frac{(\%\ \text{Sol})}{\cancel{100}}\ \frac{(\%\ \text{VS})}{\cancel{100}}}$$

*(When 100's are shown in **both** the numerator and the denominator of the equation, they may be divided out, if desired.)*

7. Seed Sludge Based on Digester Capacity

$$\%\ \text{Seed Sludge} = \frac{\text{Seed Sludge, m}^3}{\text{Total Digester Capacity, m}^3} \times 100$$

8. Volatile Acids/Alkalinity Ratio

$$\text{Vol. Acids/Alkalinity Ratio} = \frac{\text{Vol. Acids, mg/L}}{\text{Alkalinity, mg/L}}$$

9. Lime Required for Neutralization

Volatile Acids, kg = Lime Required, kg

$$\frac{(\text{mg/L VA})\ (\text{Dig. Vol., m}^3)}{1000} = \text{kg Volatile Acids} = \text{kg Lime Required}$$

10. Percent Volatile Solids Reduction

There are two equations that may be used to calculate percent volatile solids reduction.

$$\%\ \text{VS Reduction} = \frac{(\%\ \text{VS In} - \%\ \text{VS Out})}{\%\ \text{VS}_{\text{In}} - \frac{(\%\ \text{VS}_{\text{In}})\ (\%\ \text{VS}_{\text{Out}})}{100}} \times 100$$

In the second equation, the "In" and "Out" data are written as **decimal fractions.** For example, 70% volatile solids entering the digester would be written as 0.70, and 52% leaving the digester would be written as 0.52

$$\%\ \text{VS Reduction} = \frac{\text{In} - \text{Out}}{\text{In} - (\text{In} \times \text{Out})} \times 100$$

SUMMARY OF EQUATIONS

11. Digester Gas Production

$$\frac{\text{Gas Produced}}{\text{m}^3\text{/kg VS Destroyed}} = \frac{\text{m}^3 \text{ gas}}{\text{kg VS Destroyed}}$$

12. Sludge Withdrawal To Drying Beds

$$\underset{\text{m}^3}{\text{Sludge Withdrawn}} = \underset{\text{m}^3}{\text{Sludge to Drying Beds}}$$

OR

$$(0.785)\,(D^2)\,(\text{Draw-down, m}) = (\text{length, m})\,(\text{width, m})\,(\text{depth, m})$$

NOTES

Index

Index

Index—Cont'd

Index—Cont'd

Index—Cont'd